T0313362

Engineering Tolerance in Crop Plants Against Abiotic Stress

Footprints of Climate Variability on Plant Diversity

Series Editor: Shah Fahad

Climate Change and Plants: Biodiversity, Growth, and Interactions
Shah Fahad, Osman Sönmez, Shah Saud, Depeng Wang, Chao Wu, Muhammad Adnan, and Veysel Turan

Developing Climate Resilient Crops: Improving Global Food Security and Safety
Shah Fahad, Osman Sönmez, Shah Saud, Depeng Wang, Chao Wu, Muhammad Adnan, and Veysel Turan

Sustainable Soil and Land Management and Climate Change
Shah Fahad, Osman Sönmez, Veysel Turan, Muhammad Adnan, Shah Saud, Chao Wu, and Depeng Wang

Plant Growth Regulators for Climate-Smart Agriculture
Shah Fahad, Osman Sönmez, Veysel Turan, Muhammad Adnan, Shah Saud, Chao Wu, and Depeng Wang

Engineering Tolerance in Crop Plants Against Abiotic Stress
Shah Fahad, Osman Sönmez, Shah Saud, Depeng Wang, Chao Wu, Muhammad Adnan, Muhammad Arif, and Amanullah

Engineering Tolerance in Crop Plants Against Abiotic Stress

Edited by
Shah Fahad
Osman Sönmez
Shah Saud
Depeng Wang
Chao Wu
Muhammad Adnan
Muhammad Arif
and Amanullah

CRC Press
Taylor & Francis Group
Boca Raton London New York

CRC Press is an imprint of the
Taylor & Francis Group, an **informa** business

First edition published 2022

by CRC Press
6000 Broken Sound Parkway NW, Suite 300, Boca Raton, FL 33487-2742

and by CRC Press
2 Park Square, Milton Park, Abingdon, Oxon, OX14 4RN

Library of Congress Cataloging-in-Publication Data
Names: Fahad, Shah (Assistant professor of agriculture), editor. |
Sönmez, Osman, editor. | Saud, Shah, editor. | Wang, Depeng (Professor of agriculture), editor. |
Wu, Chao (Associate research fellow in agriculture), editor. |
Adnan, Muhammad (Lecturer in agriculture), editor. | Arif, Muhammad (Agronomist), editor. | Amanullah, 1973- editor.
Title: Engineering tolerance in crop plants against abiotic stress / Shah Fahad, Osman Sönmez, Shah Saud, Depeng Wang, Chao Wu, Muhammad Adnan, Muhammad Arif, and Amanullah.
Other titles: Footprints of climate variability on plant diversity.
Description: First edition. | Boca Raton, FL : CRC Press, 2022. |
Series: Footprints of climate variability on plant diversity | Includes bibliographical references and index.
Identifiers: LCCN 2021019412 (print) | LCCN 2021019413 (ebook) | ISBN 9780367750091 (hardback) |
ISBN 9780367750114 (paperback) | ISBN 9781003160717 (ebook)
Subjects: LCSH: Crops--Effect of stress on. | Plants--Effect of stress on. | Plants, protection of.
Classification: LCC SB112.5 ,E54 2022 (print) | LCC SB112.5 (ebook) | DDC 631.4/52--dc23
LC record available at https://lccn.loc.gov/2021019412
LC ebook record available at https://lccn.loc.gov/2021019413

ISBN: 978-0-367-75009-1 (hbk)
ISBN: 978-0-367-75011-4 (pbk)
ISBN: 978-1-003-16071-7 (ebk)

DOI: 10.1201/9781003160717

Typeset in Times
by MPS Limited, Dehradun

Contents

Acknowledgement

Words are bound and knowledge is limited to praise ALLAH, the Instant and Sustaining Source of all Mercy and Kindness, and the Sustainer of the Worlds. My greatest and ultimate gratitude is due to ALLAH (Subhanahu wa Taqadus). I thank ALLAH with all my humility, for everything that I can think of. His generous blessing and exaltation succeeded my thoughts and thrived my ambition to have the cherished fruit of my modest efforts in the form of this piece of literature from the blooming spring of blossoming knowledge. May ALLAH forgive my failings and weaknesses, strengthen and enliven my faith in HIM and endow me with knowledge and wisdom. All praises and respects are for **Holy Prophet Muhammad** *Salle Allah Alleh Wassalam,* the greatest educator, the everlasting source of guidance and knowledge for humanity. He taught the principles of morality and eternal values and enabled us to recognize our Creator. I have a deep sense of obligation to my parents, my brothers, sisters and son. Their unconditional love, care, and confidence in my abilities helped me achieve this milestone in my life. For this and much more, I am forever in their debt. It is to them that I dedicate this book. In this arduous time, I also appreciate the patience and serenity of my wife, who brought joy to my life in so many different ways. It is indeed on account of her affections and prayers that I was able to achieve something in my life.

Shah Fahad

Editors

Dr Shah Fahad is an Assistant Professor in the Department of Agronomy, University of Haripur, Khyber Pakhtunkhwa, Pakistan. He obtained his PhD in Agronomy from Huazhong Agriculture University, China, in 2015. After doing his postdoctoral research in Agronomy at the Huazhong Agriculture University (2015–17), he accepted the position of Assistant Professor at the University of Haripur. He has published over 270 peer-reviewed papers (Impact factor 810.45) with more than 240 research and 30 review articles, on important aspects of climate change, plant physiology and breeding, plant nutrition, plant stress responses and tolerance mechanisms, and exogenous chemical priming-induced abiotic stress tolerance. Having contributed 50 book chapters to various book editions published by Springer, Wiley-Blackwell, and Elsevier, Dr. Fahad has also edited fifteen book volumes, including this one, published by CRC press, Springer, and Intech Open. He won the Young Rice International Scientist award and distinguished scholar award in 2014 and 2015, respectively, and has won projects, comprising a portfolio of 15 projects, from international and national donor agencies. The name of Dr. Shah Fahad figured among the top two percent scientists in a global list compiled by the Stanford University,USA. He has worked on and is presently engaged in studying a wide range of topics, including climate change, greenhouse gas emission, abiotic stresses tolerance, roles of phytohormones and their interactions in abiotic stress responses, heavy metals contamination, and regulation of nutrient transport processes.

Prof. Dr. Osman Sönmez is a Professor in the Department of Soil Science, Faculty of Agriculture, Erciyes University, Kayseri, Turkey. He obtained his MS and Ph.D in Agronomy from Kansas State University, Manhattan-KS, USA in 1996-2004. In 2014, he accepted the position of Associate Professor at the University of Erciyes. Since 2014, he has worked in Department of Soil Science, Faculty of Agriculture at Erciyes University. He has published over 90 peer-reviewed papers, research, and review articles on soil pollution, plant physiology, and plant nutrition.

Dr. Shah Saud received his Ph.D. in Turf grasses (Horticulture) from Northeast Agricultural University, Harbin, China, and is currently working as a Post Doctorate researcher in the Department of Horticulture, Northeast Agricultural University, Harbin, China. Having published 125 research papers in peer-reviewed journals, Dr. Shah Saud has also edited 3 books and written 25 book chapters on important aspects of plant physiology, plant stress responses, and environmental problems in relation to agricultural plants. According to Scopus®, Dr. Shah Saud's publications have received roughly 2500 citations with an h-index of 24.

Dr. Depeng Wang has completed Ph.D. in 2016, in the field of Agronomy and Crop Physiology from Huazhong Agriculture University, Wuhan, China. Presently, he is serving as a professor in the College of Life Sciences, Linyi University, Linyi, China, and the principal investigator of Crop Genetic Improvement, Physiology & Ecology Center in Linyi University. His current research broadly encompasses crop ecology and physiology, and agronomy, with a focus on the key characteristics associated with high yielding crop, the effect of temperature on crop grain yield and solar radiation utilization, morphological plasticity to agronomic manipulation in leaf dispersion and orientation, and optimal integrated crop management practices for maximizing crop grain yield. Dr. Depeng Wang has published over 36 papers in reputed journals.

Dr. Chao Wu is engaged in studying field crop cultivation and physiology, and plant phenomics. He received his Ph.D for work done during 2013–2016 from Huazhong Agricultural University, Wuhan, China, and completed his post-doctorate during 2017–2019 from Nanjing Agricultural University, Nanjing, China. Presently, he is associate research fellow in Guangxi Institute of Botany, Guangxi Zhuang Autonomous Region and the Chinese Academy of Sciences, Guilin, China. He chairs a Natural Science Foundation of Jiangsu Province, and leads two Postdoctoral Science Foundation researches, with a focus on physiological mechanisms of abiotic-stress tolerance (heat, drought) in crops and medicinal plants.

Dr. Muhammad Adnan is a lecturer in the Department of Agriculture at the University of Swabi (UOS), Pakistan. He has completed his PhD (soil fertility and microbiology) from the Department of Soil and Environmental Sciences (SES), the University of Agriculture, Peshawar, Pakistan and the Department of Plant, Soil and Microbial Sciences, Michigan State University, USA. He has received his MSc and BSc (Hons) in Soil and Environmental Sciences from Department of SES, the University of Agriculture, Peshawar-Pakistan.

Dr. Muhammad Arif is Professor in the department of Agronomy, the University of Agriculture, Peshawar, Pakistan. He started his career in the same department as farm manager/lecturer after MS in 2000, and received his Ph.D. degree in 2005 from the same department and a post-doctorate from the UK in 2008. His main area of interest is seed priming, crop nutrition, and dual purpose technology development and using biochar and compost as soil amendments for improving soil fertility and productivity. He got a Silver Medal for studies in MS in 1999 and a Research Productivity Award for the year 2011 and 2012. He is also the member of many national and international societies, and the reviewer for more than a dozen national and international journals. With more than 140 research articles in internatial reputed jorunals, Dr. Muhammad Arif has been published in 8 international book chapters of edited books. He has been actively involved in conducting farmers', students', and researchers' trainings, and has sucessfully completed nine national and international research projects as PI mainly sponsored by USAID, PSF, DoST, and HEC. Officiating during different times in the capacity of Administrative officer, Director Administration, and Manager ORICs, he is currently working as Director Farms and Additional Director ORICs in addition to fulfilling his own professorial duties. Professor (Dr.) Arif has vast experience of project proposal writing, project execution, report writing, and as consultant to agriculture-related projects.

Dr. Amanullah is currently working as Associate Professor in the Department of Agronomy, Faculty of Crop Production Sciences, The University of Agriculture Peshawar, Pakistan. With a Ph.D in Agronomy from The University of Agriculture Peshawar in 2004 and a post-doctorate from Dryland Agriculture Institute, WTAMU, Canyon Texas, USA in 2010, Dr. Amanullah has published more than 20 books and more than 200 research papers in peer-reviewed journals, including 100 papers in impact factor journals. He is the co-author of three recent books published by FAO (1): Soil and Pulses: Symbiosis for Life (2016) (2): Unlocking the Potential of Soil Organic Carbon (2017) and (3): Soil Pollution: a hidden reality (2018). Dr. Amanullah edited three books with Intech: (1) Rice - Technology and Production (2017), (2) Nitrogen in Agriculture-Updates (2018) and (3) Corn: Production and Human Health in Changing Climate (2018). Awarded three Research Productivity Awards by the Pakistan Council for Science and Technology (PCST), Islamabad, in 2011–12, 2012–13, and 2015–16, Dr. Amanullah represented Pakistan in the FAO Intergovernmental technical panel on soil of Global Soil Partnership (2015–2018). He also won the first prize in the innovative research proposal competition arranged by DICE at the University of Gujarat in 2013–14. With a wide array of interests, Dr. Amanullah's focus of study is directed to Agronomy, Field Crops Production, Crop Physiology & Growth Analysis, Inter-Cropping & Plants Competition, Biodiversity, Carbon Estimation in Field Crops, Crop Nutrition, Fertilizer and Water Use Efficiency, Dryland Agriculture & Drought, Organic Farming, Crops Management under Stressful Environments, Sustainable Crop, Sustainable Soil Management and Water Management, and Farmers training, among others.

Contributors

Saghir Abbas
Department of Botany, Faculty of Life Sciences, Government College University, Faisalabad
Pakistan

Muhammmad Adnan
Department of Agriculture, The University of Swabi, Swabi
Pakistan

Zahoor Ahmad
University of Central Punjab, Bahawalpur Campus,
Pakistan

Mukhtar Ahmed
Department of Agronomy, PMAS-Arid Agriculture University Murree Road Rawalpindi

Sharif Ahmed
International Rice Research Institute, Bangladesh Office
Bangladesh

Adnan Akbar
National Key Laboratory of Crop Genetic Improvement, Huazhong Agricultural University

Kashif Akhtar
Institute of Nuclear Agricultural Sciences, Key Laboratory of Nuclear Agricultural Sciences of Ministry of Agriculture and Zhejiang Province, Zhejiang University, Hangzhou
China

Ahmad Ali
National Engineering Research Center for Sugarcane, Fujian Agriculture and Forestry University, Fuzhou
China

Md Panna Ali
Entomology Division, Bangladesh Rice Research Institute, Gazipur
Bangladesh

Qasim Ali
Department of Botany, Faculty of Life Sciences, Government College University, Faisalabad
Pakistan

Shafaqat Ali
Department of Environmental Sciences and Engineering, Government College University Faisalabad
Pakistan

Sikandar Ali
Department of Irrigation and Drainage, University of Agriculture Faisalabad
Pakistan

Amna
Department of Bioinformatics and Biotechnology, Government College University, Faisalabad
Pakistan

Muhammad Arif
Department of Agronomy, the University of Agriculture, Peshawar
Pakistan

Adnan Arshad
College of Resources & Environment Sciences, China Agricultural University, Beijing,
PR China

Doğan Arslan
Department of Field Crops, Faculty of Agriculture, Siirt University
Turkey

Hüseyin Arslan
Department of Agronomy, Faculty of Agriculture, Kafrelsheikh University, Kafr El-Shaikh
Egypt

Mehtab Muhammad Aslam
Center for Plant Water-Use and Nutrition Regulation, College of Life Sciences, Joint International Research Laboratory of Water and Nutrient in Crops, Fujian Agriculture and Forestry University, Fuzhou, Fujian
China

Muhammad Azeem
Department of Botany, Faculty of Life Sciences, Government College University, Faisalabad
Pakistan

Amna Bari
National Key Laboratory of Crop Genetic Improvement, College of Informatics, Huazhong Agricultural University, Wuhan
P. R. China

Marian Brestic
Department of Plant Physiology, Slovak University of Agriculture, Nitra
Slovakia

Panchali Chakraborty
Agricultural Genomics Institute at Shenzhen, Chinese Academy of Agricultural Sciences, Shenzhen
China

Swapan Chakrabarty
Agricultural Genomics Institute at Shenzhen, Chinese Academy of Agricultural Sciences, Shenzhen
China

Arzu Çığ
Department of Horticulture, Siirt University, Siirt, Turkey

Fatih Çığ
Department of Field Crops, Faculty of Agriculture, Siirt University, Siirt
Turkey

M Kaium Chowdhury
Agricultural Training Institute, Department of Agricultural Extension, Gaibanda, Bangladesh

Subhan Danish
Department of Soil Science, Faculty of Agricultural Sciences and Technology, Bahauddin Zakariya University, Multan, Punjab
Pakistan

Juel Datta
Entomology Division, Bangladesh Rice Research Institute, Gazipur
Bangladesh

Rahul Datta
Department of Geology and Pedology, Faculty of Forestry and Wood Technology, Mendel University in Brno, Zemedelska1, Brno Czech Republic

Anamika Dubey
Metagenomics and Secretomics Research Laboratory, Department of Botany, Dr. Harisingh

Gour Central University, Sagar
India

Murat Erman
Department of Field Crops, Faculty of Agriculture, Siirt University
Turkey

Shah Fahad
Department of Agronomy, The University of Haripur
Pakistan

Sajid Fiaz
Department of Plant Breeding and Genetics, University of Haripur, Khyber Pakhtunkhwa
Pakistan

Aamir Hamid
National Key Laboratory of Crop Genetic Improvement, Huazhong Agricultural University

Afsana Hossain
State Key Laboratory of Rice Biology, Institute of Biotechnology, Zhejiang University, Hangzhou
China

Akbar Hossain
Bangladesh Wheat and Maize Research Institute, Dinajpur
Bangladesh

Mohammad Anwar Hossain
Department of Genetics and Plant Breeding, Bangladesh Agricultural University, Mymensingh
Bangladesh

Sajjad Hussain
Department of Environmental Sciences, COMSATS University, Islamabad, Vehari Campus, Vehari
Pakistan

Paul Ola Igboji
Department of Soil Science and Environmental Management, Faculty of Agriculture and Natural Resources Management, Ebonyi State University, PMB Abakaliki
Nigeria

Muhammad Aamir Iqbal
Department of Agronomy, Faculty of Agriculture, University of Poonch Rawalakot (AJK)
Pakistan

Mohammad Sohidul Islam
Department of Agronomy, Hajee Mohammad
Danesh Science and Technology University,
Dinajpur
Bangladesh

Tasmiya Jabeen
Department of Environmental Sciences, COMSATS
University Islamabad
Vehari Campus-Pakistan

Muhammad Tariq Javed
Department of Botany, Faculty of Life Sciences,
Government College University, Faisalabad
Pakistan

Bello H. Jakada
Key Laboratory of Genetics, Breeding and Multiple
Utilization of Crops, Ministry of Education, Fujian
Provincial Key Laboratory of Haixia Applied Plant
Systems Biology, College of Crop Science, Fujian
Agriculture and Forestry University, Fuzhou
China

Muhammad Kamran
Key Laboratory of Crop Physio-Ecology and Tillage
Science in North-Western LoessPlateau, Ministry of
Agriculture/College of Agronomy, Northwest A&F
University, Yangling
P.R. China

Cetin Karademir
Department of Agronomy, Faculty of Agriculture,
Kafrelsheikh University, Kafr El-Shaikh
Egypt

Emine Karademir
Department of Agronomy, Faculty of Agriculture,
Kafrelsheikh University, Kafr El-Shaikh
Egypt

Joseph K. Karanja
Center for Plant Water-Use and Nutrition
Regulation, College of Life Sciences, Joint
International Research Laboratory of Water and
Nutrient in Crops, Fujian Agriculture and Forestry
University, Fuzhou, Fujian
China

Abdul Saboor Khan
National Key Laboratory of Crop Genetic
Improvement, Huazhong Agricultural University

Nazish Huma Khan
Department of Environmental Sciences,
University of Swabi, Pakistan

Ashwani Kumar
Metagenomics and Secretomics Research
Laboratory, Department of Botany, Dr. Harisingh
Gour Central University, Sagar
India

Arpna Kumari
Department of Botanical and Environmental
Sciences, Guru Nanak Dev University, Amritsar
Punjab, India

Aziz Khan
College of Agriculture, Guangxi University,
Nanning
China

Wajid Mahboob
Plant Physiology Division, Nuclear Institute of
Agriculture, Tando Jam
Pakistan

Sonia Mbarki
Laboratory of Valorisation of Unconventional
Waters, National Institute of Research in Rural
Engineering, Water and Forests (INRGREF), BP
Ariana, Tunisia

Iqra Mehmood
Department of Bioinformatics and Biotechnology,
Government College University Faisalabad,
Faisalabad
Pakistan

Skalický Milan
Department of Botany and Plant Physiology, Faculty
of Agrobiology, Food, and Natural Resources, Czech
University of Life Sciences Prague, Kamycka 129,
Prague, Czech Republic

Md Fuad Mondal
Department of Entomology, Sylhet Agricultural
University, Sylhet
Bangladesh

Mahmuda Binte Monsur
State Key Laboratory of Rice Biology, China
National Rice Research Institute, Hangzhou
China

Muhammad Mubeen
Department of Environmental Sciences, COMSATS
University Islamabad, Vehari Campus-Pakistan

Mohammad Nafees
Department of Environmental Sciences, University
of Peshawar, Pakistan

Misbah Naz
State Key Laboratory of Crop Genetics and Germplasm Enhancement, MOA Key Laboratory of Plant Nutrition and Fertilization in Low-Middle Reaches of the Yangtze River, Nanjing Agricultural University, Nanjing
China

Misbah Naz
State Key Laboratory of Wheat and Maize Crop Science, College of Life Sciences, Henan Agriculture University, Zhengzhou,
China

Ghulam Shah Nizamani
Plant breeding and Genetics Division Nuclear Institute of Agriculture, Tandojam, Sindh
Pakistan

Muhammad Noor
Department of Agriculture, Hazara University, Mansehra
Pakistan

Noor-ul-Ain
FAFU and UIUC-SIB Joint Center for Genomics and Biotechnology, College of Agriculture, Fujian Agriculture and Forestry University, Fuzhou
China

Witness J. Nyimbo
Juncao Research Center, College of Life Sciences, Fujian Agriculture and Forestry University, Fuzhou, Fujian
China

Eyalira Jacob Okal
Juncao Research Center, College of Life Sciences, Fujian Agriculture and Forestry University, Fuzhou, Fujian,
China

Sytar Oksana
Department of Plant Biology, Institute of Biology, Kiev National University of Taras Shevchenko, Volodymyrska, Kyiv
Ukraine

Muhammad Tahir ul Qamar
State Key Laboratory for Conservation and Utilization of Subtropical Agro-bioresources, College of Life Science and Technology, Guangxi University, Nanning
P. R. China

Muhammad Habib Ur Rahman
Department of Agronomy, MNS- University of Agriculture
Multan Pakistan

Karthika Rajendran
School of Agricultural Innovations and Advanced Learning, Vellore Institute of Technology (VIT), Vellore, Tamil Nadu
India

Haroon Rasheed
Department of Civil Engineering, Khawaja Fareed University of Engineering and Information Technology, Rahim Yar Khan

Madiha Rashid
Department of Botany, University of Education, Township Campus, College Road, Lahore
Pakistan

Disna Ratnasekera
Department of Agricultural Biology, Faculty of Agriculture, University of Ruhuna
Sri Lanka

Muhammad Riaz
Department of Environmental Sciences & Engineering, Government College University, Faisalabad, Pakistan

Ali Raza
Key Lab of Biology and Genetic Improvement of Oil Crops, Oil Crops Research Institute, Chinese Academy of Agricultural Sciences (CAAS), Wuhan, China

Muhammad Ali Raza
College of Agronomy, Sichuan Agricultural University, Chengdu
China

Ayman EL Sabagh
Department of Agronomy, Kafrelsheikh University
Kafrelsheikh, Egypt

Tooba Saeed
National Centre of Excellence in Physical Chemistry, University of Peshawar
Pakistan

Shah Saud
College of Horticulture, Northeast Agricultural University, Harbin
China

Kulvir Singh
Punjab Agricultural University, Regional Research
Station, Faridkot
Punjab, India

Rajesh Kumar Singhal
ICAR-Indian Grassland and Fodder Research
Institute, Jhansi
India

Junaite Bin Gias Uddin
Center for Plant Water-Use and Nutrition
Regulation, College of Life Sciences, Joint
International Research Laboratory of Water and
Nutrient in Crops, Fujian Agriculture and Forestry
University, Fuzhou, Fujian
China

Fazli Wahid
Department of Agriculture, The University of
Swabi, Swabi
Pakistan

Mirza Waleed
Department of Environmental Sciences, COMSATS
University Islamabad
Vehari Campus-Pakistan

Muhammad Mohsin Waqas
Center for Climate Change and Hydrological
Modeling Studies Water Management and
Agricultural Mechanization Research Center

Ejaz Waraich
Department of Agronomy, University of Agriculture,
Faisalabad,
Pakistan

Allah Wasaya
College of Agriculture, BZU, Bahadur
Sub-Campus Layyah
Pakistan

Muhammad Waseem
Horticulture, South China Agricultural University,
Guangzhou,
China

Afifa Younas
Department of Botany, Lahore College for
Women University, Jail Road, Lahore
Pakistan

Qian Zhang
Center for Plant Water-Use and Nutrition Regulation,
College of Life Sciences, Joint International Research
Laboratory of Water and Nutrient in Cops, Fujian
Agriculture and Forestry University, Fuzhou, Fujian,
China

Fazli Zuljalal
National Centre of Excellence in Physical Chemistry,
University of Peshawar
Pakistan

1

Biochar: An Adsorbent to Remediate Environmental Pollutants

Iqra Mehmood, Amna Bari, Mehtab Muhammad Aslam, Eyalira Jacob Okal, Muhammad Riaz, Muhammad Tahir ul Qamar, Muhammad Adnan, Mukhtar Ahmed, Shah Saud, Fazli Wahid, Muhammad Noor, and Shah Fahad

CONTENTS

1.1 Introduction

Water is one of the most important natural resources that are very essential for survival of all living organisms. As WHO (world health organization) and some other studies revealed that of the 100% of water is randomly distributed on earth, 97.5% is sea water concentrated with salt (Millero et al. 2008; Pawlowicz 2015). Therefore, the larger percentage of water present on the earth's surface cannot be used for drinking purposes without treatment. In general, 2.5% of water on earth is fresh, out of which 70% is frozen in the form of either glaciers or ice or occurs as underground water. It is worth noting that less than 1% of available water is suitable for human consumption and other household uses (Gupta et al. 2009). The small percentage of available clean water is often contaminated by pollutants that are generated from human activities such as mining, industrialization, sewage leakage, and use of agrochemicals (Ayuso and Foley 2016; Christophoridis et al. 2019; Khan et al. 2013; Vareda et al. 2019; Yang et al. 2020). The majority of anthropogenic activities strongly depend on water availability and are known to have a negative impact on water resources. Furthermore, agricultural production heavily relies on water availability (Schwarzenbach et al. 2010). The use of chemicals and dyes in industries, and pesticides in agricultural practices continuously contribute to water pollution due to toxic organic and inorganic compounds released into the ecosystem (Fatta-Kassinos et al. 2011; Li et al. 2011; O'Connor 1996). According to United Nations world water development, two million tons of waste are disposed of daily into natural water resources. These waste materials mainly originate from industries, households and agricultural activities, which contribute to water pollution due to emitted pesticides, herbicides, insecticides, fertilizers, and human waste (Programme 2003). In addition, heavy metals significantly contribute to the pollution of natural water resources and also pose serious harm to human health (Akinci et al. 2013; Al-Musharafi et al. 2013; Fujita et al. 2014; Naser 2013). Due to their subtle nature, heavy metals are not easily degraded by microbes, and therefore, remain persistent in the environment for many years (Xu et al. 2012a). Heavy metals in the environment leach down into natural water resources from where they enter into the food chain through plants and later affect animal and human health (Mashhadizadeh and Karami 2011; Nassar 2010; Zhong et al. 2007). Different methods utilized to remove heavy metals from water include membrane separation (Doke and Yadav 2014), constructed wetlands (Sultana et al. 2014), ion exchange (Cavaco et al. 2007), chemical precipitation (Kurniawan et al. 2006), ultra-filtration (Chakraborty et al. 2014), reverse osmosis, synthetic coagulants, and photocatalytic oxidation (Dimitrov 2006; Friedrich et al. 1998). These methods are, however, costly and time-consuming. Pros and cons of these strategies have already been discussed in various studies (Clifford et al. 1986; Kurniawan et al. 2006; Mohan and Pittman Jr 2006; Owlad et al. 2009). In contrast, adsorption is cheaper, environment-friendly and adaptive approach that is often preferred in removing organic and inorganic pollutants from waste water. Adsorbents are porous materials that adhere or adsorb waste materials to remove them from the water. In this chapter, we propose to discuss various types of absorbents used to remove different types of aquatic pollutants. Interest has been developed to manipulate low cost and more effective adsorbents with minute size to act upon a large surface area and a huge amount of pollutants at the same time (Hao et al. 2010).

1.2 Types of Water Pollutants Treated by Adsorbents

Water pollutants are categorized into various types that include **biological, chemical, radiations and heat** (Gupta et al. 2009). Chemical pollutants can be of organic and inorganic nature (Walker et al. 2005), and biological pollutants involve any type of disease-causing microbes.

1.2.1 Biological Pollutants

Sewage water contains diverse disease-causing microbes such as *Salmonella typhosa, Shigella dysenteriae, Yersinia enterocolitica, Vibrio comma, Escherichia coli* (Han et al. 2018; Liu et al. 2018). Waterborne diseases include cholera, typhoid, jaundice, gastroenteritis, dysentery (Gupta et al. 2009) and hepatitis A. Waste water from industries and sewerages should first be treated before being discharged into natural water resources to avoid life-threatening diseases. Poor sewage and sanitary system are also major contributors of biological pollutants into water resources.

1.2.2 Soluble and Non-soluble Contaminants

Water discharged from industries and agriculture and domestic effluents contain a huge amount of organic and inorganic waste. This waste may be in a soluble or non-soluble form. Non-soluble wastes persist in water for a longer period and remain scattered in water as colloidal particles. They form a suspension in water and disturb the overall quality of water. Non-soluble contaminants may also settle down in water as sediment and give water a muddy form (Gupta et al. 2009). They block waterways, fill dams, and are very harmful to aquatic life (Kupchella and Hyland 1993). The most common pollutants found in water are pesticides, herbicides, dyes, microbes, PCBs, Crude oil, heavy metals like copper, lead, mercury, arsenic, etc.

1.2.3 Heavy Metals

Heavy metal contamination is a major source of water pollution. Heavy metals may enter into water naturally or through human activities (Adaikpoh et al. 2005; Akoto et al. 2008; Bem et al. 2003; Wong et al. 2003). Industrialization is one of the major causes of heavy metal contamination, especially in developed countries due to urbanization. For meeting the needs of the growing population, industrialization increased day by day resulting in severe environmental damage. Due to poor sanitation systems, contaminated water enters into natural water resources and pollutes them. These systems are not properly developing as they should, especially in urbanized countries (Ahmad et al. 2010; Akoto et al. 2008; Karbassi et al. 2007; Sundaray et al. 2006). Human activities that mainly release heavy metals in the environment are due to excessive use of fertilizers and pesticides-containing heavy metals for improving agriculture, waste discharge that is in-completely treated or not treated at all containing heavy metal contamination, and released chelating agents from industries (Ammann 2002; Hatje et al. 1998; Nouri et al. 2008; Nouri et al. 2006) that can attach with more than one metallic compound. Polluted water contains heavy metals like copper, mercury, arsenic, lead, cadmium, cobalt, molybdenum, vanadium, and bismuth. These metals are released from leather, electro-plating, dyeing, glassware, and mining industries. Waste water contain heavy metals beyond the threshold level that why it becomes necessary to treat waste water (Gupta et al. 2009).

1.2.4 Dyes

Synthetic dyes had been discovered in 1856, and till now, more than 8000 dyes have been formulated (Venkataraman 2012). They are used industrially and domestically for particular purposes. Many industries use synthetic dyes daily for different purposes. For instance, paper pulp, dyeing, textile, tannery, and paint industries discharge considerable amounts of dyes and chemical wastes that pollute water. Plant manufactured dyes and those released from industries have toxic effects on human life and bestow colour to water. These dyes can be water-soluble and insoluble (Anthony 1977; Gupta et al. 2009; Walsh et al. 1980).

1.2.5 Phenols

Besides dyes and heavy metals, phenols are also of primary concern because of their unpleasant taste to water even their low concentration is highly toxic (Caturla et al. 2000; Weiner 2012); therefore, their removal from water becomes necessary. Phenols added to the water from various industrial effluents, for example, resin, gas, paint, pesticides, and chemical industries (Gupta et al. 2009), continuously release phenolic compounds into water and damage its quality.

1.2.6 Other Substances

Apart from inorganic compounds, dyes, and phenolic substances, and many other organic compounds like pesticides, PCBs and detergents are also released by industries leading to pollution of potable water. **PCBs** (Polychlorinated biphenyls) are persistent and highly toxic organic pollutants. Due to their hydrophobic nature, they are insoluble in water and also in soil (Beyer and Biziuk 2009; Campanella et al. 2002). Pesticides applied to agricultural fields to control pests also contribute to water pollution. Organochlorine is one of a class of pesticides included DDT, endosulfan, and lindane (Hong et al. 1995; Lee et al. 1997; Loganathan and Kannan 1994). When pollutants exceed a certain threshold limit, they became highly toxic (Gupta et al. 2009). In the past, many conventional approaches were used to control water pollution, although they have limitations. The use of adsorbents is a major advancement in the era of environmental remediation, and various adsorbents are currently being used for environmental remediation. In this article, we aim to discuss the production and use of biochar adsorbent for environmental remediation.

1.3 Adsorbents

Adsorbents can be synthetic (polystyrene, polypropylene, polyethylene, nylon fibres, and polyurethane), natural organic (vegetable fibres, sawdust, peat, hay, feathers, straw, biochar) and natural inorganic (clay, wool, sand, glass, volcanic ash, and nanoparticles) (Ndimele et al. 2018). Various studies (Buss et al. 2012; Fazlzadeh et al. 2017; Hagner et al. 2016; Rajput et al. 2016; Tan et al. 2015) revealed that nanoparticles and biochar are used extensively as adsorbents

1.3.1 Biochar

Biochar is a carbon rich product which is obtained when waste, like leaves, animals manure or wood, are carbonized in a closed container with no or little aeration. Biochar used in environmental management is used in four different ways:

1. Soil's quality enhancement
2. Waste management and remediation
3. Energy production
4. Climate change alleviation.

Biochar improves soil in different ways. For instance, it decreases nutrient leaching and enhances the water holding capacity of the soil. Water holding capacity increase up to 18% with the addition of biochar has been reported (Glaser et al. 2002; Sohi et al. 2009). It also improves structure, physico-chemical, and biological properties of soil. Seed germination, plant growth, crop yield also enhanced due to the implementation of biochar on soil (Glaser et al. 2002). Biochar is also used effectively in waste management produced from animals and plants. Plants and animal waste are used to produce biochar which is valuable, fuel-efficient, and inexpensive. Benefits include energy production and climate change alleviation (Barrow 2012). Various types of wastes are used for biochar production, e.g., paper mill's waste, sewage slurry, food processing waste, municipal solid waste, crops remnant, forestry waste, and animal manure (Brick and Lyutse 2010; Cantrell et al. 2012; Chen et al. 2011a; Enders et al. 2012).

Climate change alleviation by biochar is possible due to the seclusion of carbon present in soil (Lehmann et al. 2008). The sustainability of biochar in soil is a crucial factor in reducing carbon dioxide discharge in the atmosphere (Cheng et al. 2008; Kuzyakov et al. 2009; Singh et al. 2012). Biochar is also used in the production of energy that replaces the use of fossil fuels with low CO_2 emissions (Bolan et al. 2013b). Biochar is also used in environmental management since it reduces the bioavailability of waste, thereby protecting plants and animals (Basir et al. 2015; Sohi 2012).

1.4 Production of Biochar

1.4.1 Pyrolysis Process

Biomass resources for production of biochar are usually limited because it is obtained from agricultural (plants and crops). Excessive use of biomass for biochar production would lead to less fertile and nutrient-deficient soil. Cutting trees for biomass would cause soil erosion and deforestation. Brick and Lyutse (2010) categorized biomass into two groups, the first category consists of fundamentally produced biomass for production of biochar and use as an energy source, while the second one is made up of the remaining byproducts as waste biomass. The pyrolysis process is defined as the roasting of wastes at different temperature ranges of (200–900 °C) with a small amount of absence of oxygen (Demirbas and Arin 2002).

On the basis of different occupancy times and temperatures, the pyrolysis process can be fast, medium, or slow (Mohan et al. 2006). **Fast pyrolysis** usually occurs at 400–500°C temperature

(Lima et al. 2010). It requires dry biomass and usually a short time (less than 2 seconds). In this type of pyrolysis, bio-oil usually produced with production rate of up to 75% (Mohan et al. 2006). **Intermediate or slow pyrolysis** duration is from few hours to days. It has been used for a long period to produce charcoal and biochar (Major et al. 2010). It requires dry biomass and usually operates at 500°C temperature with 25–35% biochar production.

Gasification is a different process from pyrolysis and is used to produce fuel gas. Biomass treated at high temperature (above 700°C) under oxygen controlled environment, converted into gas rich in hydrogen and carbon monoxide (Mohan et al. 2006).

1.5 Factors Affecting Biochar Properties

There are various factors that affect the yield of biochar, such as feedstock type, pyrolysis temperature, as well as heating rate. Mostly, animal manure produces a larger amount of biochar as compared to wood and crop biomass (Enders et al. 2012). Similarly, biomass with high inorganic contents would give high biochar yield due to more ash production. Feedstock with high lignin contents produce a higher amount of biochar (Ibrahim et al. 2016; Sohi et al. 2010). Heating duration is the least consideration factor for biochar's yield. Minute change occurs in biomass yield of biochar with changing of heating rate 5–15°C per minute (Karaosmanoğlu et al. 2000). Pyrolysis temperature is a major factor influencing biochar's yield. Various studies have indicated how pyrolysis temperature affects biochar's yield; for example, Uchimiya et al. (2011a) carried out the production of biochar from cottonseed at different temperatures ranging from 200 to 800°C. Biochar yield quickly decreased when the temperature dropped to 400°C or <400°C. At temperatures above 400°C, biochar yield enhanced due to low lignin concentration in cottonseed. A comparative study between distinct biochars derived from wood and grass was done by Keiluweit et al. (2010). At temperatures lower than 300°C, a decrease in biochar production occurred due to initial dehydration reactions. Untimely degradation occurred at low pyrolysis temperatures (200–400°C) due to low lignin contents in grass. Carbon contents increased when pyrolysis temperature got raised. On the other hand, hydrogen and oxygen decreased at the same time. At temperature above or equal to 700°C, 90% of carbon is produced in biochar from different varieties of feedstock. Increased carbon and decreased oxygen and hydrogen contents result in decreased molar ratios (H/C and O/C); hence, deoxygenation and dehydration of biomass is demonstrated (Chen et al. 2008; Keiluweit et al. 2010; Lian et al. 2011; Uchimiya et al. 2011a; Zeeshan et al. 2016). Effect of pyrolysis temperature on nitrogen was not observed. But different feedstocks have different nitrogen contents. Feedstocks obtained from sewage slurry generally contain higher nitrogen contents. Similarly, on the P and S contents of biochar, no remarkable consequences of pyrolysis temperature were found.

Pyrolysis temperature also affects the structure and morphology of biochar (Liu et al. 2010; Uchimiya et al. 2011a). It was observed that with an increase in pyrolysis temperature, the surface area of biochar got increased. But some studies revealed that with an increase in pyrolysis temperature (700°C/>700°C), the surface area got decreased (Uchimiya et al. 2011a). Some of the following reasons are suggested for the increase in surface area.

1. Exposure of aromatic lignin core and destruction of ester and aliphatic alkyl groups at high temperature (Chen and Chen 2009).
2. Pore size distribution (suggested by association between surface area and microscopic volume) (Joseph and Lehmann 2009).
3. Biochar obtained from agricultural biomass (wood, plants, crops) has a high surface area as compared to biochar obtained from animal manure and sewage slurry even at high temperatures. Reason behind this may be the minute amount of carbon contents and high H/O, H/C ratio in agricultural biomass, which leads to immense cross-linkages (Bourke et al. 2007).
4. Biochar produced at a low temperature or more than 700°C possesses a low surface area. The reason behind this can be deformation, blockage, and splitting of small pores in the biochar structure (Lian et al. 2011; Liu et al. 2010).

1.6 Remediation of Organic Contaminants from Water and Soil using Biochar

Various biochar types used for adsorption of distinct types of waste are shown in Table 1.1. Organic contaminant's removal from waste water depends on their high surface area and the small porous structure of biochar (Lou et al. 2011; Yang et al. 2010; Yu et al. 2009). The efficiency of biochar relies on its pyrolysis temperature. For instance, biochar produced at a temperature above 400°C effectively adsorbs organic contaminants because of its microporous structure and high surface area (Ahmad et al. 2012; Uchimiya et al. 2010; Yang et al. 2010). Adsorption of organic contaminants into carbonized, non-carbonized biochar's snippet was effective at high (400–700°C) and low (100–300°C) pyrolysis temperatures, respectively. Surface polarity and aromaticity are considerable features of biochar that affect their adsorption capacity (Chen et al. 2008). Limited work done for remediation of organic matter from contaminated soil using biochar. One study revealed an everlasting effect of biochar on simazine contaminated soil. Biochar suppresses simazine's degradation as well as reduces its leaching into groundwater (Jones et al. 2011). Yang et al. (2010) and Yu et al. (2009) reported that biochar obtained from cotton straw and woodchips at a high pyrolysis temperature (850°C), when applied on contaminated soil, decreases the degradation of fipronil, chlorpyrifos, and carbofuran, and hence, reduced their bioavailability. They also reported pesticide uptake reduction by plants grown in that area (Zhang et al. 2010). Hence, biochar with small pores and more surface area is effective in remediating soil contaminated with organic matter. Pyrolysis temperature is an important consideration for the production of effective biochar. Low pyrolysis temperature is applied to produce biochar from agricultural waste (Das et al. 2013; Keiluweit et al. 2010). Pyrolysis temperature reaching up to 700°C caused an increase in the surface area of biochar (Ahmad et al. 2014). Properties of biochar should be carefully examined before being applied to a larger contaminated area.

1.7 Remediation of Inorganic Contaminants from Water and Soil using Biochar

Heavy metals contaminate soil and water when added to the environment through different sources (Adriano 2001; Lim et al. 2013; Ok et al. 2011; Usman et al. 2012). These heavy metals added to soil and water and impose serious harm to living organisms (Adriano 2001; Zhang et al. 2013). Biochar is an active ingredient to remove heavy metals from the environment. This carbonaceous material has the capacity to actively adsorb and remove heavy metals. Table 1.2 shows various biochar originated from different sources for adsorption of inorganic pollutants. (Lima et al. 2010) disclosed comparison of eight different biochar adsorbents derived from alfalfa stems, corn stover, corn cobs, switchgrass, broiler litter, soybean straw, guayule bagasse, and guayule shrub with their activated analogues to remove copper, zinc, nickel, and cadmium from water. Biochar has a greater affinity to remove heavy metals due to their larger surface area and also easy approach to their functional groups. Cu (copper) binds with biochar more effectively than other metals that made surface complexes with biochar (Tong et al. 2011). Adsorptive interaction of copper with biochar is PH dependent process, which is more effective at PH 6 and 7 (Ippolito et al. 2012). Another parameter that is considerable for metal adsorption is their atomic/ ionic size. Metals with small ionic radius generally more effectively adsorb into the small porous structure of biochar (Ko et al. 2004; Ngah and Hanafiah 2008). Biochar has different mechanisms for the mobility of metals in soil, as compared to water. One research was conducted on polluted soil with multiple metals, e.g., arsenic, copper, cadmium, and zinc. Copper and arsenic were mobilized in soil treated with biochar, whereas cadmium and zinc were immobilized in untreated soil. Copper leaching is PH dependent process, i.e., with an increase in PH and addition of biochar, copper leaching is enhanced. In literature, an increase in copper mobility with the inclusion of chicken manure-derived biochar and organic carbon has been reported. However, solubility of zinc and cadmium decreased with the inclusion of biochar due to increased PH (Beesley et al. 2010). Arsenic and antimony show a higher mobility with the addition of biochar and broiler litter-derived biochar respectively (Hartley et al. 2009; Uchimiya et al. 2012). Pyrolysis temperature is an important factor for lead removal. Biochar produced at low temperature (350–650°C) effectively removed Pb (Uchimiya et al. 2012). Phosphorus, potassium,

TABLE 1.1

Biochar Types Used in the Remediation of Organic Contaminants From Water and Soil (Ahmad et al. 2014)

Contaminant	Biochar Type	Matrix	Effect	References
Agro Chemicals				
Atrazine	Dairy manure (450 °C)	Soil	Sorption	(Cao et al. 2011)
Atrazine	Dairy manure (200 °C)	Water	Partitioning into organic C/sorption	(Cao and Harris 2010)
Atrazine and simazine	Green waste (450 °C)	Water	Partitioning into organic C/sorption	(Zheng et al. 2010)
Chloropyrifos and carbofuran	Woodchips (450 and 850 °C)	Soil	Adsorption due to high surface area and nanoporosity	(Yu et al. 2009)
Chlorpyrifos and fipronil	Cotton straw (450 and 850 °C)	Water	Adsorption due to high surface area and microporosity	(Yang et al. 2010)
Deisopropylatrazine	Broiler litter (350 and 700 °C)	Water	Sorption due to high surface area and aromaticity; sorption on non-carbonized fraction	(Uchimiya et al. 2010)
Pentachlorophenol	Bamboo (600 °C)	Soil	Reduced leaching due to diffusion and partition	(Xu et al. 2012b)
Pentachlorophenol	Rice straw	Soil	Adsorption due to high surface area and microporosity	(Lou et al. 2011)
Pyrimethanil	Red gum woodchips (450 and 850 °C)	Water	Adsorption due to high surface area and microporosity	(Yu et al. 2010)
Simazine	Hardwood (450 and 600 °C)	Soil	Sorption due to abundance of micropores	(Jones et al. 2011)
Norflurazon and fluridone	Grass and wood (200–600 °C)	Water	Sorption on amorphous C phase	(Sun et al. 2011)
Antibiotics				
Sulfamethazine	Hardwood (600 °C)	Water	Adsorption due to π–π electron donor–acceptor interaction; negative charge assisted H-bonding	(Teixidó et al. 2011)
Sulphamethoxazole	Bamboo (450 and 600 °C) Pepperwood (450 and 600 °C) Sugarcane bagasse (450 and 600 °C) Hickory wood (450 and 600 °C)	Water	Sorption	(Yao et al. 2012)
Tylosin	Pulpgrade hardwood and softwood chips (850 and 900 °C)	Soil	Sorption	(Jeong et al. 2012)
Tetracycline	Rice husk (450–500 °C)	Water	Formation of π–π interactions between ring structure of tetracycline molecule and graphite-like sheets of biochars	(Liu et al. 2012)

(Continued)

TABLE 1.1 (Continued)

Biochar Types Used in the Remediation of Organic Contaminants From Water and Soil (Ahmad et al. 2014)

Contaminant	Biochar Type	Matrix	Effect	References
Other hydrocarbons				
Brilliant blue and rhodanine dyes	Rice and wheat straw	Water	Electrostatic attraction/repulsion and intermolecular hydrogen bonding	(Qiu et al. 2009)
Catechol and humic acid	Hard wood, softwood, and grass (250, 400 and 650 °C)	Water	Adsorption due to presence of nanopores	(Kasozi et al. 2010)
m-Dinitrobenzene	Pine needles (100–700 °C)	Water	Transitional adsorption and partition	(Chen et al. 2008)
Methyl violet	Crop residue (350 °C)	Water	Electrostatic attraction; interaction between dye and carboxylate and phenolic hydroxyl groups; surface precipitation	(Xu et al. 2011)
Naphthalene	Pine needles (100–700 °C)	Water	Transitional adsorption and partition	(Chen et al. 2008)
Naphthalene	Orange peel (250, 400 and 700 °C)	Water	Adsorption and partition	(Chen et al. 2011a)
Naphthalene and 1-naphthol	Orange peel (150–700 °C)	Water	Adsorption and partition	(Chen and Chen 2009)
Nitrobenzene	Pine needles (100–700 °C)	Water	Transitional adsorption and partition	(Chen et al. 2008)
Phenanthrene	Pine wood (350 and 700 °C)	Soil	Entrapment in micro- or meso-pores	(Zhang et al. 2010)
Phenanthrene	Soybean stalk (300–700 °C)	Water	Partitioning	(Kong et al. 2011)
p-Nitrotoluene	Orange peel (250, 400 and 700 °C)	Water	Adsorption and partition	(Chen et al. 2011a)
Polycyclic aromatic hydrocarbons (PAHs)	Hardwood	Soil	Sorption and biodegradation	(Beesley et al. 2010)
Polycyclic aromatic hydrocarbons (PAHs)	Sewage sludge (500 °C)	Soil	Partitioning	(Khan et al. 2013)
Pyrene	Corn stover (600 °C)	Water	Adsorption due to nanoporosity	(Hale et al. 2011)
Pyrene	Saw dust (400 and 700 °C)	Water	Sorption	(Zhang et al. 2011)
Trichloroethylene	Soybean stover (300 and 700 °C) Peanut shell (300 and 700 °C)	Water	Sorption	(Ahmad et al. 2012)

TABLE 1.2

Biochar Types and Their Adsorption Capacities Against Different Heavy Metals

Biochar's Type	Metals Adsorbed	Adsorption Capacity (mg/g)	References
Pine wood biochar	As3+	4.13	(Mohan and Pittman Jr 2007)
	Pb2+	1.20	
Oak wood biochar	Pb2+	2.62	
	Cd2+	0.37	
	As3+	5.85	
Pine bark biochar	Pb2+	3.00	
	Cd2+	0.34	
	As3+	12.15	
Oak bark biochar	Pb2+	13.10	
	Cd2+	5.40	
	Ar3+	7.40	
Hardwood biochar	Zn2+	4.54	(Chen et al. 2011b)
	Cu2+	6.79	
Corn straw biochar	Cu2+	12.52	
	Zn2+	11.00	
Peanut straw biochar	Cu2+	0.09	(Tong et al. 2011)
	Cu2+	0.05	
Soybean straw biochar	Cu2+	0.05	
		0.03	
Canola straw biochar	Cu2+	0.04	
		0.03	
Rice husks biochar	Cu2+	6.26	(Pellera et al. 2012)
		3.49	
		4.57	
		0.27	
Dried olive pomace biochar	Cu2+	7.07	
	Cu2+	1.44	
	Cu2+	5.12	
Orange waste biochar	Cu2+	0.66	
	Cu2+	10.26	
	Cu2+	5.81	
	Cu2+	4.92	
Compost biochar	Cu2+	0.42	
	Cu2+	10.14	
	Cu2+	7.72	
	Cu2+	7.94	
	Cu2+	3.38	
Dry manure biochar	Pb2+	109.40	(Cao et al. 2009)
	Pb2+	132.81	
	Pb2+	93.65	
Pinewood biochar	Pb2+	3.89	(Liu and Zhang 2009)
	Pb2+	4.03	
	Pb2+	4.25	
Rice husk biochar	Pb2+	1.84	
	Pb2+	2.25	
	Pb2+	2.40	
Digested sugarcane bagasse biochar	Pb2+	135.40	(Inyang et al. 2011)
Raw sugarcane bagasse biochar	Pb2+	81.90	

(Continued)

TABLE 1.2 (Continued)

Biochar Types and Their Adsorption Capacities Against Different Heavy Metals

Biochar's Type	Metals Adsorbed	Adsorption Capacity (mg/g)	References
Buffalo weed biochar	Pb2+	333.33	(Yakkala et al. 2013)
	Cd2+	11.63	
Sugar beet tailing biochar	Cr6+	123	(Dong et al. 2011)
Soybean Stalk-based Biochar	Hg2+	0.67	(Kong et al. 2011)
Empty fruit branch magnetic biochar	Zn2+	1.18	(Mubarak et al. 2013)
Oak Wood biochar	Cr6+	3.03	(Mohan et al. 2011)
		4.08	
		4.93	
Oak Bark biochar	Cr6+	4.62	
		7.43	
		7.52	
Rice straw biochar	Cd2+	34.13	(Han et al. 2013)
Pinewood biochar	Cu2+	4.46	(Liu et al. 2010)
		2.75	
Dairy manure biochar	Cu2+	48.41	(Xu et al. 2013)
	Zn2+	32.95	
	Cd2+	32.03	
	Cu2+	51.46	
	Zn2+	31.84	
	Cd2+	54.63	
Pig manure biochar	Cu2+	78.36	(Kołodyńska et al. 2012)
	Zn2+	62.13	
	Cd2+	107.08	
	Cu2+	88.23	
	Zn2+	79.62	
	Cd2+	117.01	
	Pb2+	175.44	
Cow manure biochar	Pb2+	230.70	
		212.77	
		219.34	
		151.52	
		154.60	
	Cu2+	76.12	
		88.50	
		81.50	
		83.97	
	Zn2+	58.11	
		61.91	
		49.06	
		51.28	
	Cd2+	114.75	
		118.40	
		78.19	
		82.30	
Digested dairy waste biochar	Pb2+	51.38	(Inyang et al. 2012)
Digested sugar beet biochar	Pb2+	51.38	

and calcium released from biochar played a role in the high stabilization of lead. Oxygen-containing functional groups in biochar have the capacity to bind with metals. Cottonseed hull-derived biochar with high oxygen contents produced at 350°C has the potential to uptake copper, zinc, lead, and cadmium (Uchimiya et al. 2011b).

Different types of biochar separately used for organic and inorganic contaminants. Biochar is utilized for organic pollutants produced at high temperatures. It has a highly porous structure with a huge surface area. In contrast, biochar for inorganic contaminants is baked at low temperatures for higher oxygen contents and more release of cations (Ahmad et al. 2014). Different biochars have different sorption capacities towards different contaminants. Not all contaminants can be effectively removed with one biochar type. Similarly, no single type of biochar can be effectively uses against various contaminants. Biochar properties have to be carefully observed before being applied to a larger soil area. The following table represents various biochar types used to remove heavy metals from the environment.

1.8 Interaction Mechanisms of Biochar with Contaminants

1.8.1 Organic Contaminants

Adsorption and electrostatic interactions are considerable mechanisms by which biochar and organic contaminants interact with each other. These interactions are interpretative for use of biochar in environmental remediation.

Polar organic compounds, e.g., norflurazon and fluridone, interact with biochar by forming a **hydrogen bonding**. This hydrogen bonding occurs by the forming of a hydrogen bond between the hydrogen-containing contaminant and oxygen-containing portions of biochar (Sun et al. 2011). While non-polar organic compounds e.g., trichloroethylene, interact with **hydrophobic** parts of biochar without any hydrogen bonding between H and O containing groups of contaminant and biochar respectively (Ahmad et al. 2012), **electrostatic interaction** between organic pollutants and biochar is another process by which adsorption can occur. Normally, the biochar surface has negative charges that attract positively charged organic compounds. One study showed adsorption of positively charged dyes e.g., rhodanine and methyl violet with biochar from a contaminated aqueous environment. Different studies revealed electrostatic interactions for eradication of aqueous cationic dyes, e.g., rhodanine, as well as methyl violet (Qiu et al. 2009; Xu et al. 2011). In highly polar biochar (produced at 400C), aromatic π systems are usually electron extracted (Keiluweit et al. 2010). Hence electron shortages occur between them, and they interact with electron contributors. Electron deficient and rich functional groups are present in biochar produced at high temperatures. Such a type of biochar can interact with both electron donor and acceptor compounds (Sun et al. 2012). Electrostatic interaction enhanced between electron contributor graphene surface of biochar and electron receivers cationic organic compound (Qiu et al. 2009; Sun et al. 2012; Teixidó et al. 2011). There are also chances of repulsive forces between negatively charged organic compounds and biochar. Adsorption of such organic compounds then occurs by strong hydrogen bonding (Teixidó et al. 2011). As one study revealed sulfamethazine adsorption through hardwood-derived biochar at 600°C. Sulfamethazine loses protons under high PH released OH- resulting in strong hydrogen bonding between sulfamethazine and phenolate/carboxylate groups present on biochar (Teixidó et al. 2011).

1.8.2 Inorganic Contaminants

This figure illustrates the overall mechanism of the adsorption of heavy metals into biochar. Lead adsorption into sewage slurry-derived biochar was reported (Lu et al. 2012). Four possible mechanisms were proposed for this purpose.

1. Outer surface electrostatic convolution due to heavy metals interaction with Na+ and K+ ions present in biochar.

2. Heavy metals precipitation and inner surface convolution due to interaction with mineral oxides present in biochar.

3. Precipitation as lead phosphate silicate

4. Formation of the tangled surface with biochar's hydroxyl and carboxyl functional groups (Lu et al. 2012).

Biochar derived from dairy manure to precipitate lead from phosphate contaminated water has also been reported (Cao and Harris 2010). Mercury also precipitated on alkaline surfaced biochar with high chlorine contents as $HgCl_2$ and $Hg(OH)_2$ (Kong et al. 2011). Hexavalent chromium, when converting to Cr(III), bind to negative parts of biochar caused by presence of oxygen-containing functional groups (Bolan et al. 2013a; Choppala et al. 2012; Dong et al. 2011). Oak bark and oak wood-derived biochar actively adhere and remove hexavalent chromium due to their swelled structure produced by the fast pyrolysis process (Mohan et al. 2011). An aqueous medium swelled structure opened the pores that are usually closed in dry biochar, and hence, actively enhanced Cr adsorption (Mohan et al. 2011).

There are possibly two mechanisms for Cr adsorption

1. Directly adsorption of hexavalent chromium on biochar's surface.

2. Reduction of chromium within biochar from hexavalent to Cr(III) (Hsu et al. 2009).

Chromium, reduced by hydroxyl, allows carboxyl functional group formation on biochar surface (Wang et al. 2010). Reduced chromium then adsorb or precipitate again into the biochar surface (Hsu et al. 2009). Hence biochar reduced toxic chromium Cr(VI) to less toxic Cr(III) in an aqueous environment. Some microbes which use carbon as an energy source also contribute toward chromium reduction. These microbes utilize carbon from biochar and aid in chromium reduction; moreover less solubility of Cr(VI) and Cr(III) results in low mobility and transport of these heavy metals (Choppala et al. 2012). Similarly, biochar affects the mobility of other metals in soil, e.g., broiler litter-derived biochar, when applied to soil, increases antimony (Sb) mobility (Uchimiya et al. 2012). Copper mobility, also enhanced with the addition of biochar, which depends on the organic contents of biochar (Beesley et al. 2010; Park et al. 2011). Higher organic contents in biochar can block copper adsorption; however, low carbon contents enhanced both mobility and adsorption of Cu (Ali et al., 2018; Bolan et al. 2011; Cao et al. 2011; Uchimiya et al. 2011c).

Chemical and physical characteristics of biochar affect its ability to adsorb pollutants. Biochar differs on the basis of their adsorption capacity toward different contaminants. Mostly, biochars produced at high temperatures are efficient for organic compounds adsorption, and biochars produced at low temperatures are effective for inorganic contaminants. Hence, the type of biochar should be carefully selected and information about its properties is also necessary before its interaction with a vast contaminated area.

1.9 Modified Types of Biochar

1.9.1 Magnetic Biochar

New revolutions are made in this era of remediation by induced nanoparticles into biochar for better soil, water treatment, and carbon seclusion. This revolution overcomes problems related to filtration of non-magnetized biochar. Magnetic biochar can easily be removed from the water after its treatment by applying low-power magnetic fields. Many studies have been made on magnetic biochar's which revealed their properties characterization and relevant applications to soil or water. Following Table 1.3 represents studies carried out on magnetic biochar types and their various properties.

All these studies highlighted the use of magnetic biochar in environmental remediation. Magnetic biochar had an excellent capacity to adsorb and reduce organic, as well as inorganic contaminants, from

TABLE 1.3

Production Process and Properties of Magnetic Biochar

Name of Magnetic Biochar	Production Process	Properties	Contaminant Type Removed by Magnetic Biochar	Reference
MOP700, MOP250 and MOP400	Produce by chemically precipitation of iron on orange peels at different pyrolysis temperatures (250°C, 400°C and 700°C)	Higher organic contents, high iron oxide contents, efficient pollutants removal,	Naphthalene, p-nitrotoluene and phosphate	(Chen et al. 2011a)
Corncob magnetic biochar	Produce by using corncob waste and magnetized it under precise, controlled conditions.	They have a high surface area and microporous structure for efficient adsorbance. They have an adsorption capacity of 163.93 mg/g for both dyes.	Anionic and cationic dyes	(Ma et al. 2015)
Magnetic biochar	Produce from herbs.	Reduced surface area from 111.48 to 59.34m2/g, maximum chromium adsorption (23.85 ± 0.23 mg/g) at PH 2.	Hexavalent chromium	(Shang et al. 2016)
Magnetic biochar	Produce by precipitation of strontium hexaferrite (SrFe$_{12}$O$_{19}$) onto sewage biochar surface under high temperature and anaerobic conditions.	This biochar was efficient with appropriate adsorption of 388.65mg/g with biochar's concentration 500mg/L for 40min of contact time at PH 7	Malachite green dye	(Zhang et al. 2016)
Magnetic biochar	Manufactured from sawdust using iron oxide (Fe$_x$O$_y$). Oxidative hydrolysis of iron chloride (FeCl$_2$) done to produce magnetized biochar.	Highly magnetized 47.8A.m2 /kg, which allow magnetic biochar to collect easily from outside using magnets, stable at 4-9PH.	Sulfamethoxazole (SMX)	(Reguyal et al. 2017)
Ferric oxide magnetic biochar derived from mangosteen peel.	Produced by pyrolysis process which was carried out at 800°C for 20 min with modified muffle furnace under strictly anaerobic conditions.	Highly efficient for methylene blue, cadmium (II) as well as for organic pollutants.	Methylene blue and cadmium (II)	(Ruthiraan et al. 2017)

(Continued)

TABLE 1.3 (Continued)

Production Process and Properties of Magnetic Biochar

Name of Magnetic Biochar	Production Process	Properties	Contaminant Type Removed by Magnetic Biochar	Reference
Tea waste-derived magnetic porous carbonaceous (MPC) material	Magnetic biochar produced by applying Fe_2O_3 on carbonaceous material or biochar	Exhibited maximum adsorption capacity for chromium and arsenic (21.23 mg/g and 38.03 mg/g, respectively).	Fe (iron) particles, chromium, arsenic, anionic, and cationic dyes.	(Wen et al. 2017)
3 magnetic biochars	Produced by pyrolysis reaction at 400, 600 and 800°C temperature.	Large surface area (14.5 greater as compared to native biochar particles), high adsorption capacity (50.24, 41.71, and 34.06 mg/g for 3 types), and reusable.	17β-estradiol from aqueous environment	(Dong et al. 2018)
Haematite-modified magnetic biochar	Magnetic biochar was produced by pyrolysis process haematite mineral and pinewood biomass.	Strong magnetic properties, high Fe contents (118 times more than unmodified biochar due to haematite) and high arsenic sorption (429 mg kg^{-1}).	Arsenic	(Wang et al. 2015)
CeO_2–MoS_2 hybrid magnetic biochar (CMMB)	CMMB produced by pyrolysis process of CeO_2–MoS_2 hybrid at 240 °C.	High removal efficiency (>99% humate and Pb (II) removed), easily recoverable by using magnets after use.	Humate and Pb (II)	(Li et al. 2019)
Humic acid-coated magnetic biochar (HAB)	Produced by shaking humic acid and magnetic biochar at 25 °C.	Have more oxygen-containing functional groups and high adsorption efficiency	Fluoroquinolone antibiotics from water	(Zhao et al. 2019)

the environment. This procedure was economical due to the reusability of magnetic biochars, as they can be recovered after the remediation procedure by applying external magnetic fields. With magnetic coating, the adsorption capacity of biochar is amazingly enhanced, and they adsorbed huge concentrations of heavy carcinogenic metals, as well as other organic pollutants. Hence it was a magnificent approach for environmental remediation.

1.9.2 Chemically Modified Biochar

For enhanced use of biochar in the future, it requires modifications either in structural or chemical properties. Various chemically modified biochars have been developed for better environmental remediation purposes. Table 1.4 represents such types of biochar which are chemically modified and used against various pollutants.

From all these studies, we concluded that modified biochar's either modified chemically or in some other way had the ability to adsorb large concentrations of pollutants. They were cost-effective, reusable, and saved time.

1.9.3 Biochar Coated Nanoparticles

To enhance the adsorption efficacy of biochar, they are used in combination with nanoparticles. Nanoparticles have a large surface area and very small sizes up to nanometres. They have very small porous structure, and therefore, high adsorption capacity towards contaminants. Heavy metals, dyes, and other organic contaminants effectively removed from water using biochar coated nanoparticles. In following Table 1.5 some studies are coded that are carried out on biochar coated nanoparticles.

No doubt that magnetic and chemically modified biochar enhanced adsorption of inorganic also organic pollutants. But, the innovation of nanoparticles in this era is magnificent. Nanoparticles in combination with biochar proved effective adsorbents for pollutants removal. Nanoparticles or nanomaterial having very small size and large surface area are cost-efficient with excellent adsorption ability. A larger surface area of nanoparticles allows them to adsorb a huge amount of pollutants.

1.10 Conclusion and Future Perspectives

Biochar is an economical adsorbent with excellent pollutants remediation efficiency. Due to its complex structure, it is widely used for waste water treatment. Biochar is produced from waste materials like sewage sludge, agricultural wastes, manures, and forestry waste through pyrolysis and thermal conversion process. Its properties are strongly influenced by feedstock type, pre and post treatment of feedstock and biochar, respectively. Biochar interact with pollutants through hydrophobic and hydrophilic interactions, which depend on properties and types of pollutants. Apart from water, it can also be used to treat and remediate the contaminated soil. Some modifications in biochar either chemically or use in combination with nanoparticles considerably enhanced its adsorption ability against various pollutants. However, it also helps to reduce waste concentration from the environment. Different types of biochar require different conditions or parameters to act against inorganic and organic contaminants separately. Even same kind of biochar needs different conditions to act against different pollutants. Production of such kind of biochar which need same conditions against various pollutants will save time and resources. Although much research has been done on the production and application of biochar in wastewater treatment, there are still knowledge gaps that need to be filled. No doubt, with advent of modified types of biochar, wide range of pollutants can be treated efficiently. But new revolutions and understanding about its molecular structure physical properties will lead to the success of using biochar at a larger scale (especially for the treatment of municipal and industrial wastewater). On the other hand, in the future, it could be used as a promising approach to reduce contaminants concentration from environment by saving time, money, and resources.

TABLE 1.4

Various Studies that were Carried Out at Chemically Modified Biochar

Chemically Modified Biochar	Production Process	Properties	Contaminant Type	References
9 distinct coconut fibre-derived biochars	Produced at different pyrolysis temperature and chemically modified with nitric acid, hydrogen peroxide and ammonia.	Adsorption capacity depends on temperature (more temperature, less adsorption).	Pb-montmorillonite, $PbSO_4$, $Pb_3(PO_4)_2$, $Pb-Al_2O_3$ and $Pb(C_2H_3O_2)_2$	(Wu et al. 2017)
Methanol-modified rice husk-derived biochar.	–	Enhanced adsorption by 45.6% and more oxygen-containing functional groups in modified biochar.	Tetracycline	(Jing et al. 2014)
Amino-modified biochar	–	Excellent adsorption capacity (70% more than unmodified biochar).	Aquatic copper	(Yang and Jiang 2014)
Magnesium-modified biochar	–	Increase in adsorption achieved from (2.1%–3.6%) to (66.4%–70.3%).	Phosphate	(Takaya et al. 2016)
Organic acid-modified biochar	Natural adsorbents (Cinnamomum camphora sawdust) converted to modified form with inclusion of different organic acids (tartaric acid, citric acid, and oxalic acid)	Adsorption ability was 15.75, 222.8, and 280.3 mg/g with tartaric acid, citric acid and oxalic acid-modified biochars, respectively.	Hazardous malachite green	(Wang et al. 2014)
Biochar supported zerovalent iron	Chitosan helped to adhere zerovalent iron on carbonaceous porous bamboo biochar surface and the resultant composite was known as BBCF.	High adsorption capacity, BBCF can be recovered by using magnets after use	Hexavalent chromium, Pb (II), As(V), methylene blue and phosphates.	(Zhou et al. 2014)
Titanium dioxide-coated biochar composites (TBCs)	Composites were produced by operating TiO_2 coated biochar through pyrolysis process and their performance for safranine removal evaluated under UV/ without UV environment.	Highly efficient, highly stable, and effective at a wide range of PH, large surface area and enhanced photocatalytical activity.	Safranine T	(Cai et al. 2018)
Modified palm-derived biochar	Produced by using nZVI (nanoscale zerovalent iron)	Maximum adsorption (80mg/g), huge surface area, complex ligands and oxygenated functional groups	Glyphosate	(Jiang et al. 2018)
Reed-derived biochar modified by hydroxyapatite (HAP)		High porous structure, surface area, and suitable adsorption capacity and cost-effective.	Methylene Blue	(Li et al. 2018)
Modified municipal sewage sludge-derived biochar	Produced by acid base modification of biochar under pyrolysis process.	Highly efficient (optimum removal was 286.9129 mg/g), cost-effective, easily manageable and reusable.	Organic contaminants	(Tang et al. 2018)

TABLE 1.5

Biochar Coated Nanoparticles (BCNs) and their Properties

Biochar Coated Nanoparticles (BCNs)	Production Process	Properties	Contaminant Type Removed by (BCNs)	References
Activated carbon loaded Ag (silver) nanoparticles	–	Highly efficient (remove 98% dye), cost-effective and promising towards dye removal.	Direct yellow 12 dye	(Ghaedi et al. 2012)
Biochar/AlOOH nanocomposites	In nitrogen presence, under slow pyrolysis process (600C temperature) nanocomposites with AlOOH manufactured from biomass already treated with aluminium chloride.	These nanocomposites have 30nm thickness and length and width of about 100 nm with excellent adsorbance efficiency.	Aqueous contaminants e.g., phosphate, arsenic, and methylene blue.	(Zhang and Gao 2013)
Activated carbon-loaded gold nanoparticles.	Procedure involved ultrasonic mediated dye adsorption onto absorbent surface. Different parameters like PH, sonication time, initial dye concentration, absorbent dosage were under consideration	Had excellent potential to remove dye (removed 99% dye), cost-effective produce and simple to apply.	Malachite green dye	(Roosta et al. 2014)
ZVI-MBC (zero-valent iron magnetic biochar composites).	These nanocomposites were produced by biochar derived from paper mill sewage. Sodium borohydride (NaBH4) was used as a reducing agent, and (CTAB) cetyl trimethyl ammonium bromide as a surfactant during ZVI-MBC production.	Highly efficient procedure and complete removal of contaminants were achieved.	Natural as well as synthetic sewage sludge-containing pentachlorophenol (PCP)	(Devi and Saroha 2014)
Biochar loaded with nanoscale zerovalent iron (B- nZVI)	These nanoparticles were produced by the liquid-phase reduction method. Nanoscale zerovalent iron than well dispersed on biochar surface having particular surface area of 52.21 m2 /g	98.3% AO7 (20mg/L) removal achieved by B- nZVI (2g/L) within initial 10 min, specific surface area and favourable material to remove dyes.	Acid orange 7 (AO7)	(Quan et al. 2014)
Carbon nanotube (CNT)-modified biochar	These modified nanoparticles were produced at slow pyrolysis process (600C) in the presence of nitrogen. Bagasse and Hickory biomass dipped into CNT suspension with or without presence of sodium dodecylbenzenesulfonate (SDBS).	Enhanced Pb and SPY removal (adsorption was 86% and 71% for SPY and lead respectively).	Lead (Pb) and SPY (sulfayridine).	(Inyang et al. 2015)
Biochar loaded with nanoscale zerovalent iron (B- nZVI)	–	Increased surface area and highly efficient (with nZVI/BC 98.51% methyl orange removal has achieved).	Industrial sewage (methyl orange)	(Han et al. 2015)

(Continued)

TABLE 1.5 (Continued)

Biochar Coated Nanoparticles (BCNs) and their Properties

Biochar Coated Nanoparticles (BCNs)	Production Process	Properties	Contaminant Type Removed by (BCNs)	References
Copper oxide nanoparticles impregnated on activated carbon (CuO-NPs–AC)	–	Excellent absorption capacity and very promising for dye removal.	Various dyes e.g., eosin yellow (EY), auramine O (AO), brilliant green (BG) also methylene blue (MB).	(Dashamiri et al. 2016)
Ferric oxide biochar nanocomposite (Fe2O3- BC)	This composite produced from the sludge of paper and pulp industry by pyrolysis process at temperature 750°C	Small surface area, porous structure, adsorption efficacy was 52.79% higher than native biochar and highly efficient (remove >97% dye).	Methyl orange	(Chaukura et al. 2017)
Composite nanoparticles	Procedure involves implementation of (Fe3O4) ferric oxide on bamboo-derived biochar to produce composite (Fe3O4-BB).	Highly efficient and cost-effective method with removal efficacy of 86%.	Remediate PAH (Polyaromatic hydrocarbons)	(Dong et al. 2017)
Iron oxide nanoparticles, biochar composites in addition with photosynthetic microbes (PSB)	Procedure involved the preparation of nanocomposites by biochar and iron oxide nanoparticles. Then nanocomposites are loaded by PSB (plant photosynthetic bacteria). Nanocomposites exhibit a huge amount of PSB in a concentration of (5.45 × 109 cells/g).	Highly stable and re-useable as they remained stable up to continuous 5 cycle's also excellent adsorption efficiency (removed phosphate and ammonia 92.1 and 87.5%, respectively).	Phosphate and ammonia	(He et al. 2017)
Nanocomposites	Nanocomposites made by combine CeO2 (cerium dioxide) and biochar	Highly efficient and remove 98% dye from waste water.	Textile dye Reactive Red 84 (RR84)	(Khataee et al. 2018)
Activated carbon modified with magnetic nanoparticles (Ce and Fe)	–	Can separate from media using external magnets and can be reused up to 10 cycles successfully.	Rhodamine B (RhB)	(Tuzen et al. 2018)
Nanocomposites	Nanocomposites produced by impregnated TiO2 (titanium dioxide) on biochar derived from *Salvinia molesta*	High adsorption capacity (46.5% at 60 min) and had photocatalytical efficacy (57.6% at 180 min), highly active and can be reused up to 6 cycles.	AO7 (acid orange 7) dye	(Silvestri et al. 2019)

REFERENCES

Adaikpoh E, Nwajei G, Ogala J (2005) Heavy metals concentrations in coal and sediments from River Ekulu in Enugu, Coal City of Nigeria. *J AEMS* 9:5–8

Adriano DC (2001) Arsenic. In: *Trace elements in terrestrial environments*. New York, Berlin, Heidelberg: Springer, pp 219–261

Ahmad M et al. (2014) Biochar as a sorbent for contaminant management in soil and water: a review. *Chemosphere* 99:19–33

Ahmad M, Islam S, Rahman M, Haque M, Islam M (2010) Heavy metals in water, sediment and some fishes of Buriganga River, Bangladesh. *Int J EnvRes* 4:321–332

Ahmad M, Lee SS, Dou X, Mohan D, Sung J-K, Yang JE, Ok YS (2012) Effects of pyrolysis temperature on soybean stover-and peanut shell-derived biochar properties and TCE adsorption in water. *Bioresour Technol* 118:536–544

Akinci G, Guven DE, Ugurlu SK (2013) Assessing pollution in Izmir Bay from rivers in western Turkey: heavy metals. *Environ Sci Process* 15:2252–2262

Akoto O, Bruce T, Darko D (2008) Heavy metals pollution profiles in streams serving the Owabi reservoir. *Afr J Environ Sci Technol* 2:354–359

Ali S, Xua S, Ahmad I, Jia Q, Ma X, Ullah H, Alam M, Adnan M, Daur I, Ren X, Cai T, Zhang J, Jia Z (2018) Tillage and deficit irrigation strategies to improve winter wheat production through regulating root development under simulated rainfall conditions. *Agric Water Manag* 209:44–54

Al-Musharafi S, Mahmoud I, Al-Bahry S (2013) Heavy metal pollution from treated sewage effluent. *APCBEE Procedia* 5:344–348

Ammann AA (2002) Speciation of heavy metals in environmental water by ion chromatography coupled to ICP–MS. *Anal Bioanal Chem* 372:448–452

Anthony AJ (1977) Characterization of the impact of coloured waste waters on free-flowing streams. In: 37 Proceedings of 32 nd Industrial Waste Conference. Purdue University, pp 288–293

Ayuso, RA, Foley, NK (2016) Pb-Sr isotopic and geochemical constraints on sources and processes of lead contamination in well waters and soil from former fruit orchards, Pennsylvania, USA: a legacy of anthropogenic activities. *J Geochem Explor* 170:125–147

Barrow C (2012) Biochar: potential for countering land degradation and for improving agriculture. *Appl Geogr* 34:21–28

Basir A, Ali R, Alam M, Shah AS, Khilwat A, Adnan M, Ibrahim M, Rehman I, Adnan T (2015) Potential of wheat (Tritium aestivum L.) advanced lines for yield and yield attributes under different planting dates in Peshawar valley. *Am Eurasian J Agric Environ Sci* 15(12):2484–2488.

Beesley L, Moreno-Jiménez E, Gomez-Eyles JL (2010) Effects of biochar and greenwaste compost amendments on mobility, bioavailability and toxicity of inorganic and organic contaminants in a multi-element polluted soil. *Environ Pollut* 158:2282–2287

Bem H, Gallorini M, Rizzio E, Krzemińska M (2003) Comparative studies on the concentrations of some elements in the urban air particulate matter in Lodz City of Poland and in Milan, Italy. *Environ Int* 29:423–428

Beyer A, Biziuk M (2009) Environmental fate and global distribution of polychlorinated biphenyls. *Environ Contam Tox* 201: 137–158. Springer

Bolan NS, Choppala G, Kunhikrishnan A, Park J, Naidu R (2013a) Microbial transformation of trace elements in soils in relation to bioavailability and remediation. *Environ Contam Tox* 1–56. Springer

Bolan NS, Adriano DC, Kunhikrishnan A, James T, McDowell R, Senesi N (2011) Dissolved organic matter: biogeochemistry, dynamics, and environmental significance in soils. *Adv Agron* 110: 1–75. Elsevier

Bolan NS, Thangarajan R, Seshadri B, Jena U, Das K, Wang H, Naidu R (2013b) Landfills as a biorefinery to produce biomass and capture biogas. *Biores Technol* 135:578–587

Bourke J, Manley-Harris M, Fushimi C, Dowaki K, Nunoura T, Antal MJ (2007) Do all carbonized charcoals have the same chemical structure? 2. A model of the chemical structure of carbonized charcoal. *Ind Eng Chem Res* 46:5954–5967

Brick S, Lyutse S (2010) Biochar: Assessing the promise and risks to guide US policy Natural Resources Defense Council: NRDC Issue Paper

Buss W, Kammann C, Koyro H-W (2012) Biochar reduces copper toxicity in Chenopodium quinoa Willd. In a sandy soil. *J Environ Qual* 41:1157–1165

Cai X et al. (2018) Titanium dioxide-coated biochar composites as adsorptive and photocatalytic degradation materials for the removal of aqueous organic pollutants. *J Chem Technol Biotechnol* 93:783–791

Campanella BF, Bock C, Schröder P (2002) Phytoremediation to increase the degradation of PCBs and PCDD/Fs. *Environ Sci Pollut Res* 9:73–85

Cantrell KB, Hunt PG, Uchimiya M, Novak JM, Ro KS (2012) Impact of pyrolysis temperature and manure source on physicochemical characteristics of biochar. *Biores Technol* 107:419–428

Cao X, Harris W (2010) Properties of dairy-manure-derived biochar pertinent to its potential use in remediation. *Biores Technol* 101:5222–5228

Cao X, Ma L, Gao B, Harris W (2009) Dairy-manure derived biochar effectively sorbs lead and atrazine. *Environ Sci Technol* 43:3285–3291

Cao X, Ma L, Liang Y, Gao B, Harris W (2011) Simultaneous immobilization of lead and atrazine in contaminated soils using dairy-manure biochar. *Environ Sci Technol* 45:4884–4889

Caturla F, Martin-Martinez J, Molina-Sabio M, Rodriguez-Reinoso F, Torregrosa R (2000) Adsorption of substituted phenols on activated carbon. *J Colloid Interface Sci* 124:528–534

Cavaco SA, Fernandes S, Quina MM, Ferreira LM (2007) Removal of chromium from electroplating industry effluents by ion exchange resins. *J Hazard Mater* 144:634–638

Chakraborty S, Dasgupta J, Farooq U, Sikder J, Drioli E, Curcio S (2014) Experimental analysis, modeling and optimization of chromium (VI) removal from aqueous solutions by polymer-enhanced ultrafiltration. *J Mem Sci* 456:139–154

Chaukura N, Murimba EC, Gwenzi W (2017) Synthesis, characterisation and methyl orange adsorption capacity of ferric oxide–biochar nano-composites derived from pulp and paper sludge. *Appl Water Sci* 7:2175–2186

Chen B, Chen Z (2009) Sorption of naphthalene and 1-naphthol by biochars of orange peels with different pyrolytic temperatures. *Chemosphere* 76:127–133

Chen B, Zhou D, Zhu L (2008) Transitional adsorption and partition of nonpolar and polar aromatic contaminants by biochars of pine needles with different pyrolytic temperatures. *Environ Sci Technol* 42:5137–5143

Chen B, Chen Z, Lv S (2011a) A novel magnetic biochar efficiently sorbs organic pollutants and phosphate. *Biores Technol* 102:716–723

Chen X, Chen G, Chen L, Chen Y, Lehmann J, McBride MB, Hay AG (2011b) Adsorption of copper and zinc by biochars produced from pyrolysis of hardwood and corn straw in aqueous solution. *Biores Technol* 102:8877–8884

Cheng CH, Lehmann J, Thies JE, Burton SD (2008) Stability of black carbon in soils across a climatic gradient. *J. Geophys. Res.* 113

Choppala G, Bolan N, Megharaj M, Chen Z, Naidu R (2012) The influence of biochar and black carbon on reduction and bioavailability of chromate in soils. *J Environ Qual* 41:1175–1184

Christophoridis, C, Bourliva, A, Evgenakis, E, Papadopoulou, L, Fytianos, K (2019) Effects of anthropogenic activities on the levels of heavy metals in marine surface sediments of the Thessaloniki Bay, northern Greece: spatial distribution, sources and contamination assessment. *Microchem J* 149:104001.

Clifford D, Subramonian S, Sorg TJ (1986) Water treatment processes. III. Removing dissolved inorganic contaminants from water. *Environ Sci Technol* 20:1072–1080

Das L, Kolar P, Classen JJ, Osborne JA (2013) Adsorbents from pine wood via K2CO3-assisted low temperature carbonization for adsorption of p-cresol. *Ind Crops Prod* 45:215–222

Dashamiri S, Ghaedi M, Dashtian K, Rahimi MR, Goudarzi A, Jannesar R (2016) Ultrasonic enhancement of the simultaneous removal of quaternary toxic organic dyes by CuO nanoparticles loaded on activated carbon: central composite design, kinetic and isotherm study. *Ultrason Sonochem* 31:546–557

Demirbas A, Arin G (2002) An overview of biomass pyrolysis *Energy Sources* 24:471–482

Devi P, Saroha AK (2014) Synthesis of the magnetic biochar composites for use as an adsorbent for the removal of pentachlorophenol from the effluent. *Biores Technol* 169:525–531

Dimitrov DS (2006) Interactions of antibody-conjugated nanoparticles with biological surfaces. *Colloids Surf A Physicochem Eng Asp* 282–283:8–10

Doke SM, Yadav GD (2014) Process efficacy and novelty of titania membrane prepared by polymeric sol–gel method in removal of chromium (VI) by surfactant enhanced microfiltration. *J Chem Eng* 255:483–491

Dong X, Ma LQ, Li Y (2011) Characteristics and mechanisms of hexavalent chromium removal by biochar from sugar beet tailing. *J Hazard Mater* 190:909–915

Dong C-D, Chen C-W, Hung C-M (2017) Synthesis of magnetic biochar from bamboo biomass to activate persulfate for the removal of polycyclic aromatic hydrocarbons in marine sediments. *Biores Technol* 245:188–195

Dong X, He L, Hu H, Liu N, Gao S, Piao Y (2018) Removal of 17β-estradiol by using highly adsorptive magnetic biochar nanoparticles from aqueous solution. *J Chem Eng* 352:371–379

Enders A, Hanley K, Whitman T, Joseph S, Lehmann J (2012) Characterization of biochars to evaluate recalcitrance and agronomic performance. *Biores Technol* 114:644–653

Fatta-Kassinos D, Kalavrouziotis I, Koukoulakis P, Vasquez M (2011) The risks associated with wastewater reuse and xenobiotics in the agroecological environment. *Sci Total Environ* 409:3555–3563

Fazlzadeh M, Khosravi R, Zarei A (2017) Green synthesis of zinc oxide nanoparticles using Peganum harmala seed extract, and loaded on Peganum harmala seed powdered activated carbon as new adsorbent for removal of Cr (VI) from aqueous solution. *Ecol Eng* 103:180–190

Friedrich K, Henglein F, Stimming U, Unkauf W (1998) Investigation of Pt particles on gold substrates by IR spectroscopy particle structure and catalytic activity. *Colloids Surf A Physicochem Eng Asp* 134:193–206

Fujita M, Ide Y, Sato D, Kench PS, Kuwahara Y, Yokoki H, Kayanne H (2014) Heavy metal contamination of coastal lagoon sediments. *Chemosphere* 95:628–634

Ghaedi M, Sadeghian B, Pebdani AA, Sahraei R, Daneshfar A, Duran C (2012) Kinetics, thermodynamics and equilibrium evaluation of direct yellow 12 removal by adsorption onto silver nanoparticles loaded activated carbon. *J Chem Eng* 187:133–141

Glaser B, Lehmann J, Zech W (2002) Ameliorating physical and chemical properties of highly weathered soils in the tropics with charcoal–a review. *Biol Fertil Soils* 35:219–230

Gupta V, Carrott P, Ribeiro Carrott M, Suhas (2009) Low-cost adsorbents: growing approach to wastewater treatment—a review. *Crit Rev Environ Sci Technol* 39:783–842.

Hagner M, Kemppainen R, Jauhiainen L, Tiilikkala K, Setälä H (2016) The effects of birch (Betula spp.) biochar and pyrolysis temperature on soil properties and plant growth. *Soil Till Res* 163:224–234

Hale S, Hanley K, Lehmann J, Zimmerman A, Cornelissen G (2011) Effects of chemical, biological, and physical aging as well as soil addition on the sorption of pyrene to activated carbon and biochar. *Environ Sci Technol* 45:10445–10453

Han X, Liang C-f, Li T-q, Wang K, Huang H-g, Yang X-e (2013) Simultaneous removal of cadmium and sulfamethoxazole from aqueous solution by rice straw biochar. *J Zhejiang Univ Sci* 14:640–649

Han L, Xue S, Zhao S, Yan J, Qian L, Chen M (2015) Biochar supported nanoscale iron particles for the efficient removal of methyl orange dye in aqueous solutions. *PloS One* 10:e0132067

Han D, Hung YC, Bratcher CL, Monu EA, Wang Y, Wang L (2018) Formation of sublethally injured Yersinia enterocolitica, Escherichia coli O157: H7, and Salmonella enterica serovar enteritidis cells after neutral electrolyzed oxidizing water treatments. *Appl Environ Microbiol* 84: 6–18.

Hao Y-M, Man C, Hu Z-B (2010) Effective removal of Cu (II) ions from aqueous solution by amino-functionalized magnetic nanoparticles. *J Hazard Mater* 184:392–399

Hartley W, Dickinson NM, Riby P, Lepp NW (2009) Arsenic mobility in brownfield soils amended with green waste compost or biochar and planted with Miscanthus. *Environ Pollut* 157:2654–2662

Hatje V, Bidone E, Maddock J (1998) Estimation of the natural and anthropogenic components of heavy metal fluxes in fresh water Sinos River, Rio Grande do Sul State. *South Braz Environ Technol* 19:483–487

He S, Zhong L, Duan J, Feng Y, Yang B, Yang L (2017) Bioremediation of wastewater by iron oxide-biochar nanocomposites loaded with photosynthetic bacteria. *Front Microbiol* 8:823

Hong H, Xu L, Zhang L, Chen J, Wong Y, Wan T (1995) Special guest paper: environmental fate and chemistry of organic pollutants in the sediment of Xiamen and Victoria Harbours. *Mar Pollut Bull* 31:229–236

Hsu N-H, Wang S-L, Lin Y-C, Sheng GD, Lee J-F (2009) Reduction of Cr (VI) by crop-residue-derived black carbon. *Environ Sci Technol* 43:8801–8806

Ibrahim M, Jamal Y, Basir A, Adnan M, I.-u-Rahman, Khan IA, Attaullah (2016) Response of Sesame (*Sesamumindicum*l.) to various levels of Nitrogen and Phosphorus in agro-climatic condition of Peshawar. *Pure Appl Biol* 5(1):121–126

Inyang M, Gao B, Zimmerman A, Zhou Y, Cao X (2015) Sorption and cosorption of lead and sulfapyridine on carbon nanotube-modified biochars. *Environ Sci Pollut Res* 22:1868–1876

Inyang M, Gao B, Ding W, Pullammanappallil P, Zimmerman AR, Cao X (2011) Enhanced lead sorption by biochar derived from anaerobically digested sugarcane bagasse Separation. *Sci Technol* 46: 1950–1956

Inyang M, Gao B, Yao Y, Xue Y, Zimmerman AR, Pullammanappallil P, Cao X (2012) Removal of heavy metals from aqueous solution by biochars derived from anaerobically digested biomass. *Biores Technol* 110:50–56

Ippolito J, Strawn D, Scheckel K, Novak J, Ahmedna M, Niandou M (2012) Macroscopic and molecular investigations of copper sorption by a steam-activated biochar. *J Environ Qual* 41:1150–1156

Jeong CY, Wang JJ, Dodla SK, Eberhardt TL, Groom L (2012) Effect of biochar amendment on tylosin adsorption–desorption and transport in two different soils. *J Environ Qual* 41:1185–1192

Jiang X, Ouyang Z, Zhang Z, Yang C, Li X, Dang Z, Wu P (2018) Mechanism of glyphosate removal by biochar supported nano-zero-valent iron in aqueous solutions. *Colloids Surf A Physicochem Eng Asp* 547:64–72

Jing X-R, Wang Y-Y, Liu W-J, Wang Y-K, Jiang H (2014) Enhanced adsorption performance of tetracycline in aqueous solutions by methanol-modified biochar. *Chem Eng J* 248:168–174

Jones D, Edwards-Jones G, Murphy D (2011) Biochar mediated alterations in herbicide breakdown and leaching in soil. *Soil Biol Biochem* 43:804–813

Joseph S, Lehmann J (2009) Biochar for environmental management. *Sci Technol* Earthscan, London, GB. 631.422 B615bi.

Karaosmanoğlu F, Işığıgür-Ergüdenler A, Sever A (2000) Biochar from the straw-stalk of rapeseed plant. *Ener Fuels* 14:336–339

Karbassi A, Nouri J, Ayaz G (2007) Flocculation of trace metals during mixing of Talar river water with Caspian Seawater. *Int J Environ Res* 1:66–73

Kasozi GN, Zimmerman AR, Nkedi-Kizza P, Gao B (2010) Catechol and humic acid sorption onto a range of laboratory-produced black carbons (biochars). *Environ Sci Technol* 44:6189–6195

Keiluweit M, Nico PS, Johnson MG, Kleber M (2010) Dynamic molecular structure of plant biomass-derived black carbon (biochar). *Environ Sci Technol* 44:1247–1253

Khan S, Wang N, Reid BJ, Freddo A, Cai C (2013b) Reduced bioaccumulation of PAHs by Lactuca satuva L. grown in contaminated soil amended with sewage sludge and sewage sludge derived biochar. *Environ Pollut* 175:64–68

Khan A, Sharif M, Ali AS, Shah NM, Mian IA, Wahid F, Jan B, Adnan M, Nawaz S, Ali N (2013a) Potential of AM fungi in phytoremediation of heavy metals and effect on yield of wheat crop. *Am J Plant Sci* 5:1578–1586.

Khataee A, Gholami P, Kalderis D, Pachatouridou E, Konsolakis M (2018) Preparation of novel CeO2-biochar nanocomposite for sonocatalytic degradation of a textile dye. *Ultrason Sonochem* 41:503–513

Ko DC, Cheung CW, Choy KK, Porter JF, McKay G (2004) Sorption equilibria of metal ions on bone char. *Chemosphere* 54:273–281

Kołodyńska D, Wnętrzak R, Leahy J, Hayes M, Kwapiński W, Hubicki Z (2012) Kinetic and adsorptive characterization of biochar in metal ions removal. *Chem Eng J* 197:295–305

Kong H, He J, Gao Y, Wu H, Zhu X (2011) Cosorption of phenanthrene and mercury (II) from aqueous solution by soybean stalk-based biochar. *J Agric Food Chem* 59:12116–12123

Kupchella CE, Hyland MC (1993) *Environmental Science: Living within the system of nature.* Prentice Hall International

Kurniawan TA, Chan GY, Lo W-H, Babel S (2006) Physico–chemical treatment techniques for wastewater laden with heavy metals. *Chem Eng J* 118:83–98

Kuzyakov Y, Subbotina I, Chen H, Bogomolova I, Xu X (2009) Black carbon decomposition and in-corporation into soil microbial biomass estimated by 14C labeling. *Soil Biol Biochem* 41:210–219

Lee JS, Tanabe S, Takemoto N, Kubodera T (1997) Organochlorine residues in deep-sea organisms from Suruga Bay, Japan. *Mar Pollut Bull* 34:250–258

Lehmann J et al. (2008) Australian climate–carbon cycle feedback reduced by soil black carbon. *Nat Geosci* 1:832

Li R et al. (2019) High-efficiency removal of Pb (II) and humate by a CeO2–MoS2 hybrid magnetic biochar. *Biores Technol* 273:335–340

Li Y, Zhang Y, Wang G, Li S, Han R, Wei W (2018) Reed biochar supported hydroxyapatite nanocomposite: characterization and reactivity for methylene blue removal from aqueous media. *J Mol Liq* 263:53–63

Li X, Zeng G-M, Huang J-H, Zhang D-M, Shi L-J, He S-B, Ruan M (2011) Simultaneous removal of cadmium ions and phenol with MEUF using SDS and mixed surfactants. *Desalination* 276:136–141

Lian F, Huang F, Chen W, Xing B, Zhu L (2011) Sorption of apolar and polar organic contaminants by waste tire rubber and its chars in single-and bi-solute systems. *Environ Pollut* 159:850–857

Lim JE et al. (2013) Effects of natural and calcined poultry waste on Cd, Pb and As mobility in contaminated soil. *Environ Earth Sci* 69:11–20

Lima IM, Boateng AA, Klasson KT (2010) Physicochemical and adsorptive properties of fast-pyrolysis bio-chars and their steam activated counterparts. *J Chem Technol Biotechnol* 85:1515–1521

Liu Z, Zhang F-S (2009) Removal of lead from water using biochars prepared from hydrothermal liquefaction of biomass. *J Hazard Mater* 167:933–939

Liu Z, Zhang F-S, Wu J (2010) Characterization and application of chars produced from pinewood pyrolysis and hydrothermal treatment. *Fuel* 89:510–514

Liu H, Whitehouse CA, Li B (2018) Presence and persistence of Salmonella in water: the impact on microbial quality of water and food safety. *Public Health Front* 6, 159

Liu P, Liu W-J, Jiang H, Chen J-J, Li W-W, Yu H-Q (2012) Modification of bio-char derived from fast pyrolysis of biomass and its application in removal of tetracycline from aqueous solution. *Biores Technol* 121:235–240

Loganathan BG, Kannan K (1994) Global organochlorine contamination trends: an overview *Ambio*:187–191

Lou L et al. (2011) Sorption and ecotoxicity of pentachlorophenol polluted sediment amended with rice-straw derived biochar. *Biores Technol* 102:4036–4041

Lu H, Zhang W, Yang Y, Huang X, Wang S, Qiu R (2012) Relative distribution of Pb2+ sorption mechanisms by sludge-derived biochar. *Water Res* 46:854–862

Ma H, Li J-B, Liu W-W, Miao M, Cheng B-J, Zhu S-W (2015) Novel synthesis of a versatile magnetic adsorbent derived from corncob for dye removal. *Biores Technol* 190:13–20

Major J, Rondon M, Molina D, Riha SJ, Lehmann J (2010) Maize yield and nutrition during 4 years after biochar application to a Colombian savanna oxisol. *Plant Soil* 333:117–128

Mashhadizadeh MH, Karami Z (2011) Solid phase extraction of trace amounts of Ag, Cd, Cu, and Zn in environmental samples using magnetic nanoparticles coated by 3-(trimethoxysilyl)-1-propantiol and modified with 2-amino-5-mercapto-1, 3, 4-thiadiazole and their determination by ICP-OES *J Hazard Mater* 190:1023–1029

Millero FJ, Feistel R, Wright DG, McDougall TJ (2008) The composition of Standard Seawater and the definition of the Reference-Composition Salinity Scale. *Deep Sea Res Part I Oceanogr Res Pap* 55:50–72

Mohan D, Pittman Jr CU (2006) Activated carbons and low cost adsorbents for remediation of tri-and hexavalent chromium from water *J Hazard Mater* 137:762–811

Mohan D, Pittman Jr CU (2007) Arsenic removal from water/wastewater using adsorbents—a critical review. *J Hazard Mater* 142:1–53

Mohan D, Pittman CU, Steele PH (2006) Pyrolysis of wood/biomass for bio-oil: a critical review. *Ener Fuels* 20:848–889

Mohan D, Rajput S, Singh VK, Steele PH, Pittman Jr CU (2011) Modeling and evaluation of chromium remediation from water using low cost bio-char, a green adsorbent. *J Hazard Mater* 188:319–333

Mubarak N, Alicia R, Abdullah E, Sahu J, Haslija AA, Tan J (2013) Statistical optimization and kinetic studies on removal of Zn2+ using functionalized carbon nanotubes and magnetic biochar. *J Environ Chem Eng* 1:486–495

Naser HA (2013) Assessment and management of heavy metal pollution in the marine environment of the Arabian Gulf. *Mar Pollut Bull* 72:6–13

Nassar NN (2010) Rapid removal and recovery of Pb (II) from wastewater by magnetic nanoadsorbents *J Hazard Mater* 184:538–546

Ndimele PE, Saba AO, Ojo DO, Ndimele CC, Anetekhai MA, Erondu ES (2018) Remediation of Crude Oil Spillage. In: *The Political Ecology of Oil and Gas Activities in the Nigerian Aquatic Ecosystem*. Elsevier, pp 369–384

Ngah WW, Hanafiah Makm (2008) Removal of heavy metal ions from wastewater by chemically modified plant wastes as adsorbents. *Biores Technol* 99:3935–3948

Nouri J, Mahvi AH, Babaei A, Ahmadpour E (2006) Regional pattern distribution of groundwater fluoride in the Shush aquifer of Khuzestan County, *Iran Fluoride* 39:321

Nouri J, Mahvi A, Jahed G, Babaei A (2008) Regional distribution pattern of groundwater heavy metals resulting from agricultural activities. *Environ Geo* 55:1337–1343

O'Connor GA (1996) Organic compounds in sludge-amended soils and their potential for uptake by crop plants. *Sci Total Environ* 185:71–81

Ok YS, Usman AR, Lee SS, El-Azeem SAA, Choi B, Hashimoto Y, Yang JE (2011) Effects of rapeseed residue on lead and cadmium availability and uptake by rice plants in heavy metal contaminated paddy soil. *Chemosphere* 85:677–682

Owlad M, Aroua MK, Daud WAW, Baroutian S (2009) Removal of hexavalent chromium-contaminated water and wastewater. *Water Air Soil Pollut* 200:59–77

Park JH, Choppala GK, Bolan NS, Chung JW, Chuasavathi T (2011) Biochar reduces the bioavailability and phytotoxicity of heavy metals. *Plant Soil* 348:439

Pawlowicz R (2015) The absolute salinity of seawater diluted by riverwater. *Deep Sea Res Part I Oceanogr Res Pap* 101: 71–79

Pellera F-M, Giannis A, Kalderis D, Anastasiadou K, Stegmann R, Wang J-Y, Gidarakos E (2012) Adsorption of Cu (II) ions from aqueous solutions on biochars prepared from agricultural by-products. *J Environ Manag* 96:35–42

Programme WWA (2003) *Water for people, water for life: the United Nations World Water Development Report: executive summary.* Unesco Pub.

Qiu Y, Zheng Z, Zhou Z, Sheng GD (2009) Effectiveness and mechanisms of dye adsorption on a straw-based biochar. *Biores Technol* 100:5348–5351

Quan G, Sun W, Yan J, Lan Y (2014) Nanoscale zero-valent iron supported on biochar: characterization and reactivity for degradation of acid orange 7 from aqueous solution. *Water Air Soil Pollut* 225:2195

Rajput S, Pittman Jr CU, Mohan D (2016) Magnetic magnetite (Fe3O4) nanoparticle synthesis and applications for lead (Pb2+) and chromium (Cr6+) removal from water. *J Colloid Interface Sci* 468:334–346

Reguyal F, Sarmah AK, Gao W (2017) Synthesis of magnetic biochar from pine sawdust via oxidative hydrolysis of FeCl2 for the removal sulfamethoxazole from aqueous solution. *J Hazard Mater* 321:868–878

Roosta M, Ghaedi M, Shokri N, Daneshfar A, Sahraei R, Asghari A (2014) Optimization of the combined ultrasonic assisted/adsorption method for the removal of malachite green by gold nanoparticles loaded on activated carbon: experimental design. *Spectrochim. Acta A* 118:55–65

Ruthiraan M, Abdullah E, Mubarak N, Noraini M (2017) A promising route of magnetic based materials for removal of cadmium and methylene blue from waste water. *J Environ Chem Eng* 5:1447–1455

Schwarzenbach RP, Egli T, Hofstetter TB, Von Gunten U, Wehrli B (2010) Global water pollution and human health. *Annu Rev Environ Resour* 35:109–136

Shang J, Pi J, Zong M, Wang Y, Li W, Liao Q (2016) Chromium removal using magnetic biochar derived from herb-residue. *J Taiwan Inst Chem Eng* 68:289–294

Silvestri S, Gonçalves MG, da Silva Veiga PA, da Silva Matos TT, Peralta-Zamora P, Mangrich AS (2019) TiO2 supported on Salvinia molesta biochar for heterogeneous photocatalytic degradation of Acid Orange 7 dye. *J Environ Chem Eng* 7:102879

Singh BP, Cowie AL, Smernik RJ (2012) Biochar carbon stability in a clayey soil as a function of feedstock and pyrolysis temperature. *Environ Sci Technol* 46:11770–11778

Sohi SP (2012) Carbon storage with benefits *Science* 338:1034–1035

Sohi S, Lopez-Capel E, Krull E, Bol R (2009) Biochar, climate change and soil: A review to guide future research CSIRO. *Land Water Sci Rep* 5:17–31

Sohi SP, Krull E, Lopez-Capel E, Bol R (2010) A review of biochar and its use and function in soil. *Adv Agron* 105: 47–82. Elsevier

Sultana M-Y, Akratos CS, Pavlou S, Vayenas DV (2014) Chromium removal in constructed wetlands: a review. *Int Biodeter Biodeg* 96:181–190

Sun K, Keiluweit M, Kleber M, Pan Z, Xing B (2011) Sorption of fluorinated herbicides to plant biomass-derived biochars as a function of molecular structure. *Biores Technol* 102:9897–9903

Sun K, Jin J, Keiluweit M, Kleber M, Wang Z, Pan Z, Xing B (2012) Polar and aliphatic domains regulate sorption of phthalic acid esters (PAEs) to biochars. *Biores Technol* 118:120–127

Sundaray SK, Panda UC, Nayak BB, Bhatta D (2006) Multivariate statistical techniques for the evaluation of spatial and temporal variations in water quality of the Mahanadi river–estuarine system (India)–a case study. *Environ Geochem Health* 28:317–330

Takaya C, Fletcher L, Singh S, Okwuosa U, Ross A (2016) Recovery of phosphate with chemically modified biochars. *J EnviroN Chem Eng* 4:1156–1165

Tan M, Qiu G, Ting Y-P (2015) Effects of ZnO nanoparticles on wastewater treatment and their removal behavior in a membrane bioreactor. *Biores Technol* 185:125–133

Tang L et al. (2018) Sustainable efficient adsorbent: alkali-acid modified magnetic biochar derived from sewage sludge for aqueous organic contaminant removal. *Chem Eng J* 336:160–169

Teixidó M, Pignatello JJ, Beltrán JL, Granados M, Peccia J (2011) Speciation of the ionizable antibiotic sulfamethazine on black carbon (biochar). *Environ Sci Technol* 45:10020–10027

Tong X-j, Li J-y, Yuan J-h, Xu R-k (2011) Adsorption of Cu (II) by biochars generated from three crop straws. *Chem Eng J* 172:828–834

Tuzen M, Sarı A, Saleh TA (2018) Response surface optimization, kinetic and thermodynamic studies for effective removal of rhodamine B by magnetic AC/CeO2 nanocomposite. *J Enviorn Manag* 206:170–177

Uchimiya M, Chang S, Klasson KT (2011a) Screening biochars for heavy metal retention in soil: role of oxygen functional groups. *J Hazard Mater* 190:432–441

Uchimiya M, Wartelle LH, Lima IM, Klasson KT (2010) Sorption of deisopropylatrazine on broiler litter biochars. *J Agri Food Chem* 58:12350–12356

Uchimiya M, Klasson KT, Wartelle LH, Lima IM (2011b) Influence of soil properties on heavy metal sequestration by biochar amendment: 1. Copper sorption isotherms and the release of cations. *Chemosphere* 82:1431–1437

Uchimiya M, Wartelle LH, Klasson KT, Fortier CA, Lima IM (2011c) Influence of pyrolysis temperature on biochar property and function as a heavy metal sorbent in soil. *J Agri Food Chem* 59:2501–2510

Uchimiya M, Bannon DI, Wartelle LH, Lima IM, Klasson KT (2012) Lead retention by broiler litter biochars in small arms range soil: impact of pyrolysis temperature. *J Agri Food Chem* 60:5035–5044

Usman AR, Lee SS, Awad YM, Lim KJ, Yang JE, Ok YS (2012) Soil pollution assessment and identification of hyperaccumulating plants in chromated copper arsenate (CCA) contaminated sites, Korea. *Chemosphere* 87:872–878

Vareda JP, Valente AJ, Durães L (2019) Assessment of heavy metal pollution from anthropogenic activities and remediation strategies: a review. *J Environ Manage* 246:101–118.

Venkataraman K (2012) *The chemistry of synthetic dyes* 4. Elsevier

Walker CH, Sibly R, Hopkin SP, Peakall DB (2005) *Principles of ecotoxicology*. CRC Press

Walsh GE, Bahner LH, Horning WB (1980) Toxicity of textile mill effluents to freshwater and estuarine algae, crustaceans and fishes mental. *Environ Pollut Ser Ecol Biol* 21:169–179

Wang H et al. (2014) Removal of malachite green dye from wastewater by different organic acid-modified natural adsorbent: kinetics, equilibriums, mechanisms, practical application, and disposal of dye-loaded adsorbent. *Environ Sci Pollut Res* 21:11552–11564

Wang XS, Chen LF, Li FY, Chen KL, Wan WY, Tang YJ (2010) Removal of Cr (VI) with wheat-residue derived black carbon: reaction mechanism and adsorption performance. *J Hazard Mater* 175:816–822

Wang S, Gao B, Zimmerman AR, Li Y, Ma L, Harris WG, Migliaccio KW (2015) Removal of arsenic by magnetic biochar prepared from pinewood and natural hematite. *Biores Technol* 175:391–395

Weiner ER (2012) *Applications of environmental aquatic chemistry: a practical guide*. CRC Press

Wen T, Wang J, Yu S, Chen Z, Hayat T, Wang X (2017) Magnetic porous carbonaceous material produced from tea waste for efficient removal of As (V), Cr (VI), humic acid, and dyes. *ACS Sustainable Chem Eng* 5:4371–4380

Wong C, Li X, Zhang G, Qi S, Peng X (2003) Atmospheric deposition of heavy metals in the Pearl River Delta, China. *Atmo Environ* 37:767–776

Wu W et al. (2017) Unraveling sorption of lead in aqueous solutions by chemically modified biochar derived from coconut fiber: a microscopic and spectroscopic investigation. *Sci Total Environ* 576:766–774

Xu P et al. (2012a) Use of iron oxide nanomaterials in wastewater treatment. *Sci Total Environ* 424:1–10

Xu R-k, Xiao S-c, Yuan J-h, Zhao A-z (2011) Adsorption of methyl violet from aqueous solutions by the biochars derived from crop residues. *Biores Technol* 102:10293–10298

Xu T, Lou L, Luo L, Cao R, Duan D, Chen Y (2012b) Effect of bamboo biochar on pentachlorophenol leachability and bioavailability in agricultural soil. *Sci Total Environ* 414:727–731

Xu X, Cao X, Zhao L, Wang H, Yu H, Gao B (2013) Removal of Cu, Zn, and Cd from aqueous solutions by the dairy manure-derived biochar. *Environ Sci Pollut Res* 20:358–368

Yakkala K, Yu M-R, Roh H, Yang J-K, Chang Y-Y (2013) Buffalo weed (Ambrosia trifida L. var. trifida) biochar for cadmium (II) and lead (II) adsorption in single and mixed system. *Desalin Water Treat* 51:7732–7745

Yang X-B et al. (2010) Influence of biochars on plant uptake and dissipation of two pesticides in an agricultural soil. *J Agri Food Chem* 58:7915–7921

Yang G-X, Jiang H (2014) Amino modification of biochar for enhanced adsorption of copper ions from synthetic wastewater. *Water Res* 48:396–405

Yang L, Zhou Y, Shi B, Meng J, He B, Yang H, Wang T (2020) Anthropogenic impacts on the contamination of pharmaceuticals and personal care products (PPCPs) in the coastal environments of the Yellow and Bohai seas. *Environ Int* 135: 105306.

Yao Y et al. (2012) Adsorption of sulfamethoxazole on biochar and its impact on reclaimed water irrigation. *J Hazard Mater* 209:408–413

Yu X-Y, Ying G-G, Kookana RS (2009) Reduced plant uptake of pesticides with biochar additions to soil. *Chemosphere* 76:665–671

Yu X, Pan L, Ying G, Kookana RS (2010) Enhanced and irreversible sorption of pesticide pyrimethanil by soil amended with biochars. *J Environ Sci* 22:615–620

Zeeshan M, Ahmad, Khan I, Shah B, Naeem A, Khan N, Ullah W, Adnan M, Shah SRA, Junaid K, Iqbal M (2016) Study on the management of *Ralstoniasolanacearum* (Smith) with spent mushroom compost. *J Ento Zool Studies* 4(3):114–121

Zhang X et al. (2013) Using biochar for remediation of soils contaminated with heavy metals and organic pollutants. *Environ Sci Pollut Res* 20:8472–8483

Zhang M, Gao B (2013) Removal of arsenic, methylene blue, and phosphate by biochar/AlOOH nanocomposite. *Chem Eng J* 226:286–292

Zhang W, Wang L, Sun H (2011) Modifications of black carbons and their influence on pyrene sorption. *Chemosphere* 85:1306–1311

Zhang H, Lin K, Wang H, Gan J (2010) Effect of Pinus radiata derived biochars on soil sorption and desorption of phenanthrene. *Environ Pollut* 158:2821–2825

Zhang J, Liu M, Yang T, Yang K, Wang H (2016) A novel magnetic biochar from sewage sludge: synthesis and its application for the removal of malachite green from wastewater. *Water Sci Technol* 74:1971–1979

Zhao J, Liang G, Zhang X, Cai X, Li R, Xie X, Wang Z (2019) Coating magnetic biochar with humic acid for high efficient removal of fluoroquinolone antibiotics in water. *Sci Total Environ* 688:1205–1215

Zheng W, Guo M, Chow T, Bennett DN, Rajagopalan N (2010) Sorption properties of greenwaste biochar for two triazine pesticides. *J Hazard Mater* 181:121–126

Zhong L-S, Hu J-S, Cao A-M, Liu Q, Song W-G, Wan L-J (2007) 3D flowerlike ceria micro/nanocomposite structure and its application for water treatment and CO removal. *Chem Mater* 19:1648–1655

Zhou Y, Gao B, Zimmerman AR, Chen H, Zhang M, Cao X (2014) Biochar-supported zerovalent iron for removal of various contaminants from aqueous solutions. *Biores Technol* 152:538–542

2

Understanding the Physiological and Genetic Responses of Plant Root to Phosphorus-deficient Condition

Mehtab Muhammad Aslam, Joseph K. Karanja, Noor ul Ain, Kashif Akhtar, Muhammad Arif, Junaite Bin Gias Uddin, Ahmad Ali, and Qian Zhang

CONTENTS

2.1 Introduction

Phosphorous is an essential nutrient that plays a critical role in maintaining several plant metabolic processes e.g., respiration and photosynthesis (Sánchez-Calderón et al. 2010). Agricultural production has become completely dependent on the use of chemical fertilizers (Akhtar et al. 2018b; Mehra et al. 2019), and the overuse or less use of fertilizer affects agricultural production (Akhtar et al. 2019b; Akhtar et al. 2020; Mo et al. 2019). However, due to the increase in population and intensive farming, the substantial application of phosphate fertilizer is inevitable because of the limited availability of infinite phosphate ore (Vance et al. 2003). Inorganic phosphate (P) is naturally present at a very low concentration of approximately 10 µM in soil solution (Hinsinger 2001). Phosphorus availability is limited because it forms insoluble complexes with cations (Al, Mn, Fe), making it a key limiting factor for plant productivity (Jalal et al. 2020; Bargaz et al. 2012; Cordell et al. 2009).

Being sessile, plants adopt various mechanisms (Aslam et al. 2020b) to accomplish P assimilation, transport, and recycling under P stress conditions (Sanders and Tinker 1971; Ryan et al. 2014). Plants exposed to low P conditions exhibit various biochemical and physiological changes in the root system architecture (RSA) e.g., root length, structure, increased number of lateral roots, the formation of dense cluster roots, ethylene synthesis, secretion of organic acids, auxin transport, activation PSI (phosphate starvation-induced) genes (Li et al. 2016b; Vengavasi et al. 2017), and established symbiotic associations with microbes (Aslam et al. 2019). These root architectural modifications enable plants to explore topsoil where phosphorus tends to be more abundant (Maharajan et al. 2018; Aslam et al. 2020a). Diverse plant species have shown RSA modifications (Aslam et al. 2020a), which are linked with substitute adaptive mechanisms to cope with P stress (Zeng et al. 2017). Symbiotic association of the plant with beneficial fungi (ectomycorrhizal), for example, is a universal and coherent response to enhance P uptake in plants (Behie and Bidochka 2014; Aslam et al. 2019). In the absence of symbiotic associations, few plant species of Fabaceae, Proteaceae, and Casuarinaceae families can form dense cluster roots (Lambers et al. 2018; Shane and Lambers 2005; Neumann and Martinoia 2002). For example, white lupin (*Lupinus albus*) is a key model plant for understanding plant adaptation to low-phosphorus availability that is capable of surviving under extreme P deficiency (Aslam et al. 2020b), particularly through the development of cluster roots (Xu et al. 2020). Additionally, plants secrete organic acid, acid phosphatases and nucleases into the rhizosphere to solubilize P from other organic complexes such as Fe, Al, Ca and Mn (Uhde-Stone 2017; Aslam et al. 2020a).

All of these adaptive responses are mainly mediated by complex genetic mechanisms underlying P starvation, involved in the upregulated expression of certain genes and transcription factors to preserve internal P (Plaxton and Tran 2011; Cheng et al. 2011a). The remobilization of internal P becomes important under P starvation conditions. It is demonstrated that overexpression of the *GmPT5* gene is involved in mobilization and transportation of P towards nodules in soybean (Xu et al. 2013; Khan et al. 2016; Wu et al. 2018). P starvation-induced genes are upregulated in the *pho2* mutant of *Arabidopsis thaliana* (Quaghebeur and Rengel 2004). The elevated level of acid phosphatases responsible for P uptake in several crops (Tran et al. 2010; Sun et al. 2020). It is reported that P-efficient genotype of *Phaseolus vulgaris* developed shallow root to forage more P content from the topsoil. Contrastingly, *Lupinus albus* developed dense cluster and lateral roots with more number of root hairs to acquire P from soil (Shen et al. 2011). It suggests that RSA play a vital role in improving P uptake efficiency, larger root surface provides more chance to explore the soil. Another study has been conducted on soybean showed that the purple acid phosphatase gene significantly contributes to improving P efficiency and photosynthetic activity of plants (Li et al. 2016a). Therefore, a better understanding of plant root adaptation to low P would appear as a helpful strategy to improve the more efficient use of P in plant growth and further preserve limited phosphorus ore stock.

This review highlights the major physiological and genetic traits associated with mechanisms underlying P sensing, acquisition, and mitigation strategies to P stress conditions. We summarized some effective key genes and transcription factors involved in improving P uptake and remobilization in the plant and elucidated the role of phytohormones, LncRNAs, and micro-RNAs in response to P-deficient condition. The main aim of this review is to anticipate general root adaptability mechanisms of plant P deficiency responses. It will open a new insight to screen outcrops that are better adapted to low P availability environment, especially for agriculture researchers and breeders whose work is not related to this direction. Future studies on determining signalling pathways that regulate P sensing/response and identifying the proteins that are involved are of great interest (Figure 2.1).

2.2 P Sensing at the Root Tip

Naturally, plants grow under conditions with heterogeneous nutrient availability (Akhtar et al. 2019a; Akhtar et al. 2018a; Fatima et al. 2016; Zhou et al. 2012). Thus, to maintain cellular homeostasis, growth and reproduction, plants have to evolve various biochemical and physiological adaptations (Yang et al. 2004). The root tip is the first core site in the whole plant where P sensed and transport signals to shoots via xylem (Svistoonoff et al. 2007). The shoots perceive nutrient signals from roots and send them to shoots apices and back to root via the phloem to adjust P use efficiency by the whole plant (Lucas et al. 2013). Root tip encounters a region of low P, a primary signal or stimulus (P in the apoplast) is perceived by plasma membrane-localized sensors as below in Figure 2.2. Alternatively, in the root tip cells, an internal P-deficiency signal (P in the cytoplast) is perceived by internal sensors (Lucas et al. 2013). Limited studies have been carried out on the molecular mechanisms of internal/external P sensor (Chiou and Lin 2011). In tomato cell culture

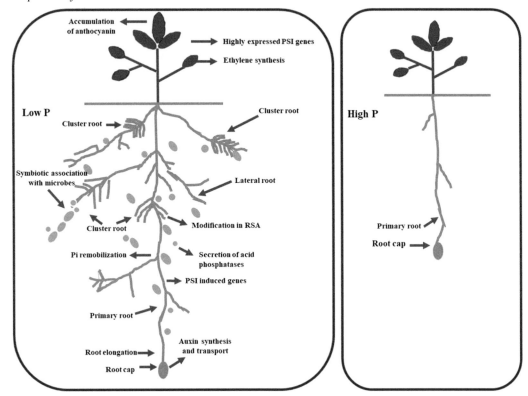

FIGURE 2.1 Modification of root system architecture (RSA) to high and low phosphorus availability. In low P conditions, root surface area increases through the formation of dense cluster roots, increased number of lateral roots, secretion of organic acids in the rhizosphere, and expression of PSI (phosphate starvation-induced) genes, subsequently improving P transportation efficiency. RSA modifications significantly improve P uptake and remobilization, thus minimizing P fertilizer overuse, which reduces cost, as well as disastrous environmental effects. High P conditions strongly reduce primary root length and inhibit lateral and cluster roots formation.

assay, the external P concentration in the medium was constant, while the internal P concentration was reduced by incubating the cells with D-mannose and other nutrients, which are known to sequester P, intracellularly, into organic compounds. In this assay, several RNase transcripts were induced, which normally do not express in optimum P conditions. These results support the existence of an intracellular P sensing mechanism (Köck et al. 1998). When plants are exposed to P starvation conditions, the hormone production, transportations, and signalling rate may change, which leads to the upregulation of phosphate starvation responsive genes (PSR) and modification in RSA (Chiou and Lin 2011). Conventional breeding strategies made some progress in improving P efficiency (Shenoy and Kalagudi 2005). P efficient varieties of soybean have been produced and performed well compared to that of local varieties and in field conditions in RSA, which are now certified for commercial use in South China (Wang et al. 2010).

The downstream P adaptive signalling mechanism involved P sensing and perception by plant root cells (Svistoonoff et al. 2007). However, great efforts have been made in understanding phosphate starvation responsive (PSR) genes, while the molecular bases of local P sensing and signalling cascade needed further exploration. P sensor recognizes particular components involved in mediating P sensing and induction of other downstream mechanisms (Zhang et al. 2014). The response of a plant to low and high P levels is shown in Figure 2.3, Plasma membrane-localized transporters sense and activates PSR gene under low P level, resulting in increased synthesis of PHT1 in the plasma membrane, while high P level leads to the degradation of PHT1transporter in the vacuole, resulting in inactivation/repression of PSR in the nucleus.

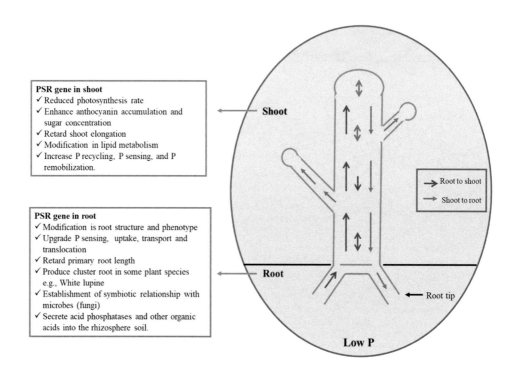

PSR gene in shoot
✓ Reduced photosynthesis rate
✓ Enhance anthocyanin accumulation and sugar concentration
✓ Retard shoot elongation
✓ Modification in lipid metabolism
✓ Increase P recycling, P sensing, and P remobilization.

PSR gene in root
✓ Modification is root structure and phenotype
✓ Upgrade P sensing, uptake, transport and translocation
✓ Retard primary root length
✓ Produce cluster root in some plant species e.g., White lupine
✓ Establishment of symbiotic relationship with microbes (fungi)
✓ Secrete acid phosphatases and other organic acids into the rhizosphere soil.

FIGURE 2.2 Responses of PSR (phosphate starvation-responsive) genes in shoot and root under Low P soil. To maintain cellular P homeostasis plants, deliver and exchange their nutrients through the xylem and phloem transpiration stream. The root tip is the most important part of plant roots to sense P deficiency, and transportation of nutrients from the xylem to phloem and phloem to xylem are indicated with colour arrows, and two-headed arrows represent nutrients movement on both sides.

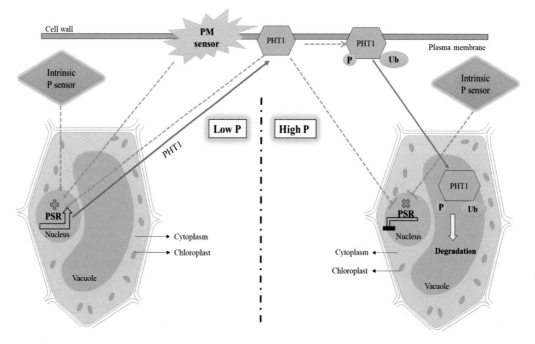

FIGURE 2.3 Cellular based P sensing under low and high P levels in plants. When plants sense low P, the cellular sensor of plants is thought to highly upregulate expression of PSR genes, resulting in enhanced expression of PHT1 on the plasma membrane. When plants sense high P, plasma membrane-localized PHT1 transporters are considered to sense high external P levels, leading to induced post-transcriptional and translation modifications in plant cells.

2.3 Changes in Plant Physiology under P Deficit Condition

Comprehending the mechanism underlying plant adaptability to a P stress condition has become an interesting talk for scientists and agronomists to acquire deep insight into the underground part of a plant (Sulieman and Tran 2015). Tremendous advancement has been made in understanding this phenomenon in the *Arabidopsis* model plant. Interestingly, higher plants share common developmental responses, which led to the formation of the narrow and proliferated root system and improved the efficiency of P uptake from soil (Wang et al. 2018). It has been demonstrated that *Phaseolus vulgaris, Arabidopsis thaliana, Oryza sativa,* and *Zea mays* have very different root architecture system, but they have many conserved regulators among these species. *Oryza sativa* showed improved primary root growth, regulated by *OsPHR1* transcription factor orthologue of *AtPHR1* gene under low P soil (Panigrahy et al. 2009; Yang and Finnegan 2010).

The nutrient-sensing and transportation mechanism of the plant is a key step towards improving P uptake efficiency (Giovannetti et al. 2017). Plasma membrane transporters (G- protein-coupled receptors) play an important role in sensing phosphate and other nutrient and translocate into the different parts of a plant (Ludwig et al. 2003). It is reported that *Arabidopsis* has a high-affinity nitrate reductase transporter (*NTR-1.1*) to detect and transport nutrients, while many other P transporters (e.g., *PHTI, PHO1*) are also involved (Gojon et al. 2011). The Pho84 (phosphate transporter) in yeast (*Saccharomyces cerevisiae*) is involved in P perception and P sensing machinery (Zhang et al. 2014). These findings have paved the system for identifying molecular and functional characterization of P perception machinery. Considerably, these key findings are integrated into the higher plants to understand the plant root adaptability phenomenon.

Plants adapted two types of responses under P starvation condition, i.e., plant developmental response is controlled at the local level, and P homeostasis is controlled at a systematic level (Figure 2.4).

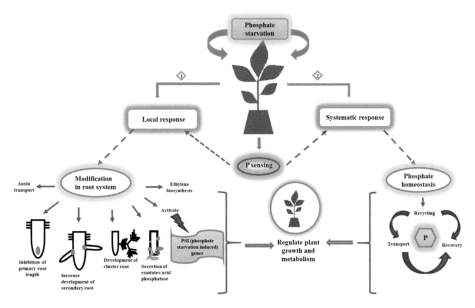

FIGURE 2.4 Plant adaptive physiological responses to P deficiency. When plants are exposed to P starvation condition, they adopt two different types of responses; one is a local response in which root structural architecture is modified, with root structure and morphology altered. Physiological responses are the functions performed due to alteration in RSA such as auxin transport, inhibition of primary root length, increased development of secondary root, cluster root development, exudate secretion, activation of PSI genes, and ethylene biosynthesis. The systematic response involves molecular mechanisms and signalling molecules, expression of a high-affinity phosphate transporter (e.g., *PHT-1*) involved in the transportation of P, secretion of acid phosphatase/exudates involved in P recovery, and catabolism of phospholipids involved in the P recycling mechanism.

Although these two types of pathways are integrated by crosstalk at some levels, P homeostasis mechanisms have been described extensively in several studies (Rouached et al. 2010; Luan et al. 2019; Sakuraba et al. 2018). All of these studies tend to exhibit that P provokes a switch in developmental and physiological status to improve plant fitness. As it is illustrated that *Arabidopsis* is the best organism to know the mechanism of P observation and plant response in root development, this model plant is unable to undergo symbiosis/cluster root formation, which limits the developmental adaptation to P limited environment (Lavenus et al. 2013). However, researchers are gaining more interest in those plants that have cluster root forming abilities and show extreme adaptability to low P. For instance, *Lupinus albus* (white lupin) used as a model plant for cluster root formation and nitrogen fixation in a low P condition (Xu et al. 2020). Additionally, numerous studies had shown that in monocot species, rice was used as a model to study responses and P perception machinery (Takehisa et al. 2013; Ma et al. 2012). Future studies on identifying signalling molecules, P sensing, and transport mechanisms would help to understand the plant root adaptive responses under phosphate starvation conditions.

2.4 Genetic Responses to P Deficiency

To cope with P-deficient condition, changes in the expression of thousands of genes occurred in a spatiotemporal manner that ensured P uptake, transport, recycling, and stress protection (Wang et al. 2019b). The first step of surviving in P-deficient conditions is to identify the scarcity and pass the signal throughout the plant (Hammond and White 2011). The plant roots are the first body part to identify the shortage of phosphate resulting in regulation of root-to-shoot and shoot-to-root signalling, which certifies the crucial structural and functional changes of plant tissues (Svistoonoff et al. 2007). From the identification of the P-deficient condition to signalling and changes of a plant (structural and functional), tissues are carried out by many different genes (Ajmera et al. 2019; Jung and McCouch 2013).

Considerable progress has been achieved in understanding the genetic responses to Low P condition on Arabidopsis (Lin et al. 2011; Bari et al. 2006; Hammond et al. 2004), rice (Heuer et al. 2009), white lupin (Uhde-Stone et al. 2003; O'Rourke et al. 2013), tomato (Wang et al. 2002), maize (Calderón-Vázquez et al. 2009; Calderón-Vázquez et al. 2011), mustard (Hammond et al. 2005), and bean (Hernández et al. 2007). Several regulatory genes (RG) and phosphorus starvation responsive (PSR) genes have been recognized from transcriptomic based studies involved in mediating several processes such as P uptake, remobilization, signalling, recycling, transcriptional regulation, and plant growth and development (Liang et al. 2010). Some promotor components enriched in PSR genes also investigated (Kerwin and Wykoff 2012). We address some important genes associated with plant developmental adaptation to P deficiency.

P is well known to play a critical role in regulating all biological processes. P excretion not only causes P deficiency signalling events, but it may also induce several alterations in metabolism and differentiate between early and late responses to P deficiency. However, the most significant step is to examine PSR gene expression during transitioning of P-replete to P-static condition. PSR genes are characterized into early (during P excretion) and late (after 1-day of P excretion) responsive groups (Hammond et al. 2003; Misson et al. 2005), and transcriptomic analysis has been investigated at organ-specific levels, e.g., shoot, root, or leaves (Amtmann et al. 2005).

Early PSR genes, non-specifically and highly upregulated, are assumed to respond under P stress condition. Generally, the PSR genes are related to cytochrome P450, peroxidase, glutathione S-transferase, ethylene, salicylic acid, JA mediated biotic and abiotic and rapidly induced during early exposure to low P (Misson et al. 2005; Lin et al. 2011). Genes underlying signal transduction are 14-3-3 protein, MAP kinase, MAP kinase kinases, calmodulin, CDPKs (Calmodulin dependent protein kinase) (Lin et al. 2011; Hammond et al. 2003). Several transcription factors involving WRKY, AP2/ERF, CCAAT, MYB, NAC, and bHLH, are upregulated upon exposure to P deficiency (Misson et al. 2005). Lin et al. (2011) revealed that WRKY18 and WRKY40 transcription factors are activated within 1 hour under low P condition. Although PSR genes encode acid phosphatases, phosphate transporter, SPX domain-related proteins, ribonucleases are also induced during early exposure. Surprisingly, it was observed that after 6-hour initiation, several root-related genes were downregulated under low P stress (Hammond et al. 2003).

Late PSR genes are involved in mediating downstream related genes accessing P uptake, transportation, recycling, synthesis of primary and secondary metabolism, and protein biosynthesis. Generally, these developmental adaptations improve P uptake and P use efficiency. Root development-related genes coding enzymes involved in the synthesis and exudation of organic acid, genes related to glycolysis, ensures that carbon supply was highly expressed under low P soil (Cordell et al. 2009). Consistently, P deficiency signalling and modification in RSA, gene expression related to hormone synthesis, response, and metabolism are altered by low P. Moreover, auxin-responsive genes and transcription factors were observed to be influenced by P stress, and inhibited primary and lateral root formation (O'Rourke et al. 2013). Interestingly, white lupin and Arabidopsis exposed to P stress exhibited repression of gibberellin related genes (Yang and Finnegan 2010). Several transcription factors in rice are considered as a promising candidate to regulate P uptake and signalling mechanism such as *OsPHR2, OsPTF1, OsARF12/16, OsMYB2P-1, OsMYB4P, OsPHO2* (Liu et al. 2010; Yi et al. 2005; Dai et al. 2012; Yang et al. 2014), while in Arabidopsis, *AtAVP1, AtPAP10, and AtPAP15* genes control P acquisition as well as improve plant biomass and yield (Kuang et al. 2009; Li et al. 2005; Wang et al. 2011). Similarly, a wide range of genes involved in regulating P uptake, utilization, and increasing crop yield has been widely characterized in several crop species. Some novel key genes related to low P signalling and adaptive mechanisms in plants are listed below in Table 2.1.

2.5 Phytohormonal Responses to P Deficiency

Plant hormones serve as a chemical messenger in regulating cell division, elongation, and signal transduction, and they play an important role in mediating plant adapted responses to nutrient stress conditions (Lucena et al. 2006). P deficiency can modulate hormones synthesis, sensitivity, and transport (Chiou and Lin 2011); modification in root elongation, root tip, and root hair density emerging is an important P foraging strategy (Lynch 2011). To maximize P uptake and perception plant displayed great plasticity to low P to modulate RSA. It has been revealed that Arabidopsis confronted with P deprivation showed reduced primary root length and increased the number of lateral roots (López-Bucio et al. 2005; Müller and Schmidt 2004).

Plant phenotypic responses to P deficiency are distinctly varied among different plant species, e.g., soybean, maize, and common bean developed shallow root structure under low P availability (Lynch 2011). Contrastingly, barley and wheat develop longer and highly dense root structure and improve P acquisition under P stressed condition (Haling et al. 2013). White lupin has shown extreme tolerance to low P due to its ability to develop highly dense cluster root with thick root hairs (Cheng et al. 2011b). It indicates that plant species differ in their tolerance or susceptibility to P deficiency.

Phytohormonal synthesis and signalling mechanism have potentially correlated with low P conditions such as modification in root development. Auxin, cytokinin, ethylene, gibberellic acid, abscisic acid, and strigolactone all have been implicated in the induction of PSR genes and RSA modification (Nie et al. 2017). Auxin played a critical role in altering RSA under P-deficient condition. In Arabidopsis, exogenous application auxin shows the increase in the number of lateral root and root hairs and reduced primary root elongation under low P condition (Casimiro et al. 2001). Upon exposure to P deficiency, expression of auxin receptor transport inhibitor response-1 (TIR1) upregulated, and facilitates auxin repressor AUX/IAA degradation, activates ARF19 auxin response factor 19 that controls the expression of other auxin response-related genes, and thus significantly stimulate lateral root formation. Consistent with this finding, OsARF12 and OsARF16 correlated with auxin signalling and PSR gene in rice (Qi et al. 2012; Shen et al. 2013; Sana et al. 2015).

Cytokinin has been well documented in Arabidopsis and negatively regulates PSR genes, e.g., AtIPS1, At4, and AtPT1 under P limiting condition. Strigolactone is newly developed class of plant

TABLE 2.1

Genes and Transcription Factor (TF) Involved in P-Deficient Signalling and Required Adaptive Changes

Gene and TF	Plant Name	Role under P Deficiency	References
OsPHR2 *OsPTF1* *OsARF12/16* *OsMYB2P-1* *OsMYB4P*	*Oryza sativa*	Improves P uptake	(Liu et al. 2010; Yi et al. 2005; Dai et al. 2012; Yang et al. 2014)
OsPHO2	*Oryza sativa*	Improves P uptake, signalling and transportation	
OsSPX1 *OsSPX3/5* *OsARF12*	*Oryza sativa*	Maintains P homeostasis, and stress signalling	(Liu et al. 2010; Qi et al. 2012)
OsPT1/6/8/9/10	*Oryza sativa*	Increased P uptake and Upregulate H^+ transporter	
OsPHF1	*Oryza sativa*	Enhanced P uptake efficiency	
OsWRKY74	*Oryza sativa*	Improves P uptake and increase root length	(Dai et al. 2012)
AtPAP10	*Arabidopsis thaliana*	Controls P utilization	(Wang et al. 2011; Li et al. 2005; Kuang et al. 2009)
AVP1		Improves biomass and yield	
AtPAP15	*Arabidopsis thaliana*	Controls P content and crop yield	(Kuang et al. 2009)
AtPHO1	*Arabidopsis thaliana*	Maintains P homeostasis, improves P uptake and remobilization	
ZmPHT1;6 *ZmPTF*	*Zea mays*	Regulates P uptake, transfer P root to shoot, and maintain P homeostasis	(Liu et al., 2010)
NtPT1	*Nicotiana tabacum*	Improved P uptake and grain yield	(Park et al. 2007)
TaPHR1	*Triticum aestivum*	Improves P uptake	
TaPht1;4	*Triticum aestivum*	Increased P uptake	(Liu et al., 2010)
GmEXPB2	*Glycine max*	Modifies root architecture system	(Zhou et al., 2012)
GmPT5	*Glycine max*	Maintenance growth, nodulation, and phosphorous equilibrium.	
LaGPX-PDE1/2	*Lupinus albus*	Alteration in root architectural system, Improves P availability	
PvPAP1 *PvPAP3*	*Phaseolus vulgaris*	Controls P secretion from dNTPs.	
PvPHR1	*Phaseolus vulgaris*	Improves P uptake	
StPT3 *StPT4*	*Solanum tuberosum*	Regulates P uptake, transfer P root to shoot, and maintain P homeostasis	
BnPHR1	*Brassica napus*	Controls P homeostasis, and acquisition	
MtPT4	*Medicago truncatula*	Controls uptake, transfer P root to shoot, and maintain P homeostasis	
LePT3LePT4	*Lycopersicon esculentum*	Controls P uptake, transfer root to shoot, and maintain homeostasis	

hormones involved in mediating several plant processes such as establishing symbiotic associations with fungi, control shoot branching, and promote root growth (Czarnecki et al. 2013). It is concluded that plant hormones significantly contribute to induce PSR expression and promoting root development under low P conditions.

2.6 Role of lncRNAs as P Stabilizers

When plants are exposed to an environmental stress condition, the expression of a large number of genes is regulated at transcriptional and post-transcriptional levels in a spatiotemporal manner. The altered genes expression of structural genes resulted in certain developmental modifications named phenotypic plasticity. In phosphorous starved conditions, certain genetic and molecular processes modulate hormones concentration and promote root branching for the formation of lateral roots. With the introduction of improved molecular and sequencing techniques, it was revealed that more than 90% of the eukaryotic genome undergoes transcription, generating protein-coding and non-coding RNAs, but a small proportion (1–2%) can translate into functional proteins (Kapranov et al. 2007). These non-coding RNAs (ncRNAs), based upon their length, are categorized into small and long non-coding RNAs (lncRNAs). LncRNAs are the largest groups of ncRNAs, having more than 200 nucleotides and a low capacity for translation (Chekanova 2015).

Pachnis et al. (1988), for the first time, discovered lncRNA H19 and regarded it as 'transcriptional noise' like other ncRNAs. These lncRNAs are transcribed by RNA polymerase II, and their expression is specific in organs and tissues. Based on their associated locations with protein-coding genes in the genome are sub-grouped into sense, antisense, bidirectional, intronic, and intergenic lncRNAs (Ponting et al. 2009; Hammad et al. 2015). Research experiments conducted on lncRNAs revealed that they act as scaffolds for gene regulation by epigenetic or post-transcriptional modifications.

lncRNAs play a vital part in the certain biological process at different growth stages (growth, flowering, fruit ripening) in plants, e.g., binding chromatin-modifying complexes (Zhao et al. 2010; Khalil et al. 2009). At the cellular level, lncRNAs interact with CIS regulatory elements in the promoter region of co-expressing genes, and in return, modulate the genes expression at the transcription or post-transcriptional level (Ponting et al. 2009). This interaction may overlap with TF binding sites and modify gene expression patterns (de la Fuente 2010). Moreover, lncRNAs may assist the TFs to enhance their binding activity after attaching in the vicinity of their binding sites.

2.7 lncRNAs: Regulators of P Metabolism to P Deficiency

lncRNAs perform roles in metabolic reprogramming during sensing and combating P deficiency in plants. In Arabidopsis, P deficiency strongly induced the lncRNA named At4 and IPS1 and loss of function mutant showed lack of redistribution of root and shoot P, leading to unbalanced homeostasis ISP1 is a well-studied and characterized lncRNA, which binds to miRNA399 and ceases miR399 dependent cleavage of target genes. During phosphate deficiency, the interaction between ISP1 and miRNA399 increases, and the resultant expression of mRNA PHO2 (phosphate genes) also increases, which is otherwise the target of miRNA399 (Wang et al. 2019a). IPS1 contains similar sequences like that of PHO2 and mimics in binding with miRNA399 (P starvation-induced). Thus, IPS1 is stable and does not undergo cleavage and finally causes sequestration of miRNA399. This mimicry region in ISP1 is conserved in many species hence proving its role in P metabolism in diverse plants (Hou et al. 2005).

Experimental evidence concluded that lncRNAs were shown to be involved in signal transduction, energy metabolism and phosphate metabolism (Figure 2.5) (Wang et al. 2017). P starvation in *M. truncatula* induced expression of 3 LncRNAs in their corresponding Tnt1 mutants. PDL1 played a role in decreased degradation of MtPHO2, which encodes E2 (ubiquitin-conjugating enzyme) regulated by miRNA399. Additionally, PDIL2 and PDIL3 were involved in P movement within the plant at the transcriptional level (Wang et al. 2017). In an experiment conducted on maize, GMOs overexpressing MIR399b showed highly accumulated levels of P in shoots. In P in-efficient lines, ZmPHO2 was repressed due to interaction with miR399 at the post-transcriptional level. PILNCR1 is induced in P starvation and transient expression led to PILNCR1 interaction with ZmmiR399, and preventing the cleavage of ZmPHO2. The profound expression of PILNCR1 in P in-efficient lines is an indication of their regulatory roles in P tolerance in maize (Du et al. 2018).

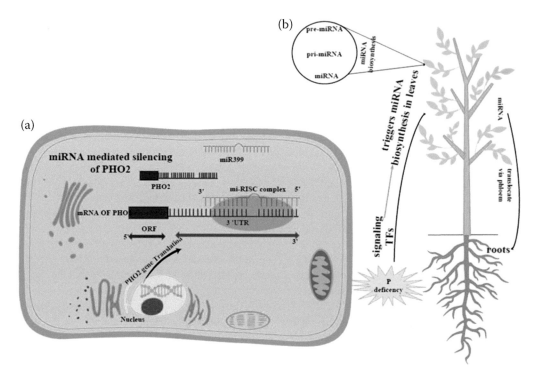

FIGURE 2.5 Role of non-coding RNA in regulation P homeostasis under P-deficient condition. (a) describes the post-transcriptional silencing of the *PHO2* gene mediated by miRNA at the cellular level. (b) P deficiency is accessed by plant roots afterwards that signalling transcription factors (TFs) trigger the formation of MicroRNAs in aerial parts of a plant. miRNA biosynthesis involves three steps, i.e., pri-miRNA, pre-miRNA and finally mature miRNA and finally translocated to plant roots by phloem.

2.8 MicroRNAs serve as Sensors and Regulator to Low P Levels

MicroRNAs are newly discovered players regarding their function in phosphorous metabolism. These are endogenous, comprising of 20–24 nt and usually produced by ssRNA precursor. Usually, miRNAs are reported to pair with their respective target genes mRNAs after transcription causing cleavage and down-regulation of the respective gene in plants (Reinhart et al. 2002). Several microarray analyses showed significant differential expression of miRNAs in response to varying the amounts of supplied P (Pant et al. 2009). NLA was reported to be involved in N homeostasis upon exposure to N_2 starved condition (Peng et al. 2007). Interestingly experiments involved map-based cloning revealed that the expression of two P transport genes PHT1;1 and PHF1 is regulated by NLA. The results show the role of NLA in maintaining P by restricting over acquisition, as NLA mutant showed early senescence due to P hyperaccumulation (Kant et al. 2011). In Arabidopsis, P starvation leads to induction of miR827, thereby suppressing RING (Really Interesting New Gene) type E3 ubiquitin ligase, which encodes NLA. It exhibits an inverse correlation between miR827 and NLA-mRNA regarding P availability and regulation. Overexpression analysis of miR827, as well as the loss of function NLA mutant, both showed P accumulation up to toxic levels (Lin et al. 2013). These results were further confirmed by the plasma membrane sub-cellular localization of NLA, causing ubiquitination of membrane-localized PHT1s, leading to clathrin directed endocytosis and negligible P intake at the cellular level. While in rice, instead of suppressing homologs of NLA, miR827 withholds the SPX-MFS family of solute transporters (Wang et al. 2012). Likewise, miR827-overexpressing plants and spx-mfs1 mutants showed increased accumulation in rice.

In another experiment, 12 miRNAs were identified using high throughput sequencing in P stressed seedlings. It was concluded that both miRNA399b and Zma-miR3 were induced under low phosphorus stress and regulated their target genes (Zmpt1 and Zmpt2) (Zhang et al. 2012). Moreover, Zma-miR3 is a newly identified miRNAs in maize, and miRNA399b in Arabidopsis and rice may likely be directly involved in maintaining P homeostasis. These results indicated that both conserved and maize-specific miRNAs play an important role in stress and other physiological processes and correlated with P starvation, regulated by their target genes. Identification of these differentially expressed miRNAs will facilitate us to uncover the molecular mechanisms underlying the progression of maize seedling roots development under low-level phosphorus stress (Zhang et al. 2012). Therefore, we concluded that both conserved and crop-specific RNAs may act as regulators of P related genes. Moreover, we can exploit microRNA as a tool for improving P efficiency in P-deficient soils using advanced molecular and biotechnological approaches.

2.9 Conclusions

The excess use of phosphate fertilizer affected qualitative and quantitative attributes of crops and also not friendly to the environment. Therefore, the limited availability and non-renewable nature of P reserves, agronomist researchers are gaining more interest to improve P efficiency and generation of P efficient crops. Sustainable agriculture systems can improve P uptake and P use efficiency by plant roots. The improvement of P uptake can be uplifted either by enriching P uptake from soil (P uptake efficiency) or by increasing crop production/unit P (P use efficiency). Under nutrient-limited environment, plants have evolved different physiological, biochemical, and genetic mechanisms such as modification in RSA, development of cluster roots (*Lupinus albus*), exudate secretion, root colonization with fungi, and activation of PSR genes. A deep molecular signalling mechanism is needed to comprehend the plant sensing, signalling and adaptability mechanism to nutrient deficiency. Future studies could help to identify the phosphate transporter and signalling molecules that are involved in plant adaptation response.

Acknowledgments

This work was supported by grants from a Newton Advanced Fellowship (NSFC-RS: NA160430) and the Education Department of Fujian Province grant (KLA19016A).

Author Contribution

MA designed the manuscript and wrote the first draft. MA, JK, NA, MA contributed to writing. KA, ZY, JB, AA, and ZQ critically reviewed and improved the manuscript. MA and NA designed the diagrams. All other authors read and approved the final manuscript.

Conflicts of Interest

The authors declare no conflict of interest.

REFERENCES

Ajmera I, Hodgman TC, Lu C (2019) An integrative systems perspective on plant phosphate research. *Genes* 10 (2):139

Akhtar K, Wang W, Khan A, Ren G, Afridi MZ, Feng Y, Yang G (2018a) Wheat straw mulching with fertilizer nitrogen: an approach for improving soil water storage and maize crop productivity. *Plant Soil Environ* 64: 330–337

Akhtar K, Wang W, Ren G, Khan A, Feng Y, Yang G (2018b) Changes in soil enzymes, soil properties, and maize crop productivity under wheat straw mulching in Guanzhong, China. *Soil Tillage Res* 182: 94–102

Akhtar K, Wang W, Khan A, Ren G, Afridi MZ, Feng Y, Yang G (2019a) Wheat straw mulching offset soil moisture deficient for improving physiological and growth performance of summer-sown soybean. *Agric Water Manag* 211: 16–25

Akhtar K, Wang W, Ren G, Khan A, Feng Y, Yang G, Wang H (2019b) Integrated use of straw mulch with nitrogen fertilizer improves soil functionality and soybean production. *Environ Int* 132: 94–102

Akhtar K, Wang W, Ren G, Khan A, Nie E, Khan A, Feng Y, Yang G, Wang H (2020) Straw mulching with inorganic nitrogen fertilizer reduces soil CO_2 and N_2O emissions and improves wheat yield. *Sci Total Environ* 741:140488

Amtmann A, Hammond JP, Armengaud P, White PJ (2005) Nutrient sensing and signalling in plants: potassium and phosphorus. *Adv BoT Res* 43:209–257

Aslam MM, Karanja J, Bello SK (2019) Piriformospora indica colonization reprograms plants to improved P-uptake, enhanced crop performance, and biotic/abiotic stress tolerance. *PMPP* 106:232–237

Aslam MM, Akhtar K, Karanja JK, Haider FU (2020a) Understanding the adaptive mechanisms of plant in low phosphorous soil. In: *Plant stress physiology*. IntechOpen

Aslam MM, Karanja JK, Zhang Q, Lin H, Xia T, Akhtar K, Liu J, Miao R, Xu F, Xu W (2020b) In vitro regeneration potential of White Lupin (*Lupinus albus*) from cotyledonary nodes. *Plants* 9 (3):318

Bargaz A, Faghire M, Abdi N, Farissi M, Sifi B, Drevon J-J, Cherkaoui Ikbal M, Ghoulam C (2012) Low soil phosphorus availability increases acid phosphatases activities and affects P partitioning in nodules, seeds and rhizosphere of Phaseolus vulgaris. *Agriculture* 2 (2):139–153

Bari R, Pant BD, Stitt M, Scheible W-R (2006) PHO2, microRNA399, and PHR1 define a phosphate-signaling pathway in plants. *Plant Physiol* 141 (3):988–999

Behie SW, Bidochka MJ (2014) Nutrient transfer in plant–fungal symbioses. *Trends Plant Sci* 19 (11): 734–740

Calderón-Vázquez C, Alatorre-Cobos F, Simpson-Williamson J, Herrera-Estrella L (2009) Maize under phosphate limitation. In: *Handbook of maize: Its biology*. Springer, pp 381–404

Calderón-Vázquez C, Sawers RJ, Herrera-Estrella L (2011) Phosphate deprivation in maize: genetics and genomics. *Plant Physiol* 156 (3):1067–1077

Casimiro I, Marchant A, Bhalerao RP, Beeckman T, Dhooge S, Swarup R, Graham N, Inzé D, Sandberg G, Casero PJ (2001) Auxin transport promotes Arabidopsis lateral root initiation. *Plant Cell* 13 (4): 843–852

Chekanova JA (2015) Long non-coding RNAs and their functions in plants. *Curr Opin Plant Biol* 27:207–216

Cheng L, Bucciarelli B, Liu J, Zinn K, Miller S, Patton-Vogt J, Allan D, Shen J, Vance CP (2011a) White lupin cluster root acclimation to phosphorus deficiency and root hair development involve unique glycerophosphodiester phosphodiesterases. *Plant Physiol* 156 (3):1131–1148

Cheng L, Bucciarelli B, Shen J, Allan D, Vance CP (2011b) Update on white lupin cluster root acclimation to phosphorus deficiency update on lupin cluster roots. *Plant Physiol* 156 (3):1025–1032

Chiou T-J, Lin S-I (2011) Signaling network in sensing phosphate availability in plants. *Annu Rev Plant Biol* 62:185–206

Cordell D, Drangert J-O, White S (2009) The story of phosphorus: global food security and food for thought. *Global Environ Change* 19 (2):292–305

Czarnecki O, Yang J, Weston DJ, Tuskan GA, Chen J-G (2013) A dual role of strigolactones in phosphate acquisition and utilization in plants. *Int J Mol Sci* 14 (4):7681–7701

Dai X, Wang Y, Yang A, Zhang W-H (2012) OsMYB2P-1, an R2R3 MYB transcription factor, is involved in the regulation of phosphate-starvation responses and root architecture in rice. *Plant Physiol* 159 (1):169–183

Du Q, Wang K, Zou C, Xu C, Li W-X (2018) The PILNCR1-miR399 regulatory module is important for low phosphate tolerance in maize. *Plant Physiol* 177 (4):1743–1753

Fatima, Shah M, Usman A, Sohail K, Afzaal M, Shah B, Adnan M, Ahmed N, Junaid K, Shah SRA, Rahman LU (2016) Rearing and identification of *Callosobruchusmaculatus* (Bruchidae: Coleoptera) in Chickpea. *J Ento Zoo Studies* 4(2):264–266

de la Fuente A (2010) From 'differential expression' to 'differential networking'–identification of dysfunctional regulatory networks in diseases. *Trends Genet* 26 (7):326–333

Giovannetti M, Volpe V, Salvioli A, Bonfante P (2017) Fungal and plant tools for the uptake of nutrients in arbuscular mycorrhizas: a molecular view. In: *Mycorrhizal mediation of soil*. Elsevier, pp 107–128

Gojon A, Krouk G, Perrine-Walker F, Laugier E (2011) Nitrate transceptor (s) in plants. *J Exp Bot* 62 (7):2299–2308

Haling RE, Brown LK, Bengough AG, Young IM, Hallett PD, White PJ, George TS (2013) Root hairs improve root penetration, root–soil contact, and phosphorus acquisition in soils of different strength. *J Exp Bot* 64 (12):3711–3721

Hammad M, Ahmed N, Khan IA, Shah B, Khan A, Ali M, Rasheed MT, Junaid K, Adnan M, Zaki AB, Ahmed S (2015) Studies on the efficacy of selected insecticides against anopheles mosquitoes of village Goth. Bhoorji (Sindh) Pakistan. *J Ento Zool Studies* 3(6):169–173

Hammond JP, Bennett MJ, Bowen HC, Broadley MR, Eastwood DC, May ST, Rahn C, Swarup R, Woolaway KE, White PJ (2003) Changes in gene expression in Arabidopsis shoots during phosphate starvation and the potential for developing smart plants. *Plant Physiol* 132 (2):578–596

Hammond JP, Broadley MR, White PJ (2004) Genetic responses to phosphorus deficiency. *Ann Bot* 94 (3):323–332

Hammond JP, Broadley MR, Craigon DJ, Higgins J, Emmerson ZF, Townsend HJ, White PJ, May ST (2005) Using genomic DNA-based probe-selection to improve the sensitivity of high-density oligonucleotide arrays when applied to heterologous species. *Plant Methods* 1 (1):10

Hammond JP, White PJ (2011) Sugar signaling in root responses to low phosphorus availability. *Plant Physiol* 156 (3):1033–1040

Hernández G, Ramírez M, Valdés-López O, Tesfaye M, Graham MA, Czechowski T, Schlereth A, Wandrey M, Erban A, Cheung F (2007) Phosphorus stress in common bean: root transcript and metabolic responses. *Plant Physiol* 144 (2):752–767

Heuer S, Lu X, Chin JH, Tanaka JP, Kanamori H, Matsumoto T, De Leon T, Ulat VJ, Ismail AM, Yano M (2009) Comparative sequence analyses of the major quantitative trait locus phosphorus uptake 1 (Pup1) reveal a complex genetic structure. *Plant Biotechnol J* 7 (5):456–471

Hinsinger P (2001) Bioavailability of soil inorganic P in the rhizosphere as affected by root-induced chemical changes: a review. *Plant Soil* 237 (2):173–195

Hou X, Wu P, Jiao F, Jia Q, Chen H, Yu J, Song X, Yi K (2005) Regulation of the expression of OsIPS1 and OsIPS2 in rice via systemic and local Pi signalling and hormones. *Plant Cell Environ* 28 (3): 353–364

Jalal F, Arif M, Akhtar K, Khan A, Naz M, Said F, Zaheer S, Hussain S, Imtiaz M, Khan MA, Ali M, Fan W (2020) Biochar integration with legume crops in summer gape synergizes nitrogen use efficiency and enhance maize yield. *Agronomy* 10: 58

Jung Jkhm, McCouch SRM (2013) Getting to the roots of it: genetic and hormonal control of root architecture. *Front Plant Sci* 4:186

Kant S, Peng M, Rothstein SJ (2011) Genetic regulation by NLA and microRNA827 for maintaining nitrate-dependent phosphate homeostasis in Arabidopsis. *PLoS Genet* 7 (3)

Kapranov P, Cheng J, Dike S, Nix DA, Duttagupta R, Willingham AT, Stadler PF, Hertel J, Hackermüller J, Hofacker IL (2007) RNA maps reveal new RNA classes and a possible function for pervasive transcription. *Science* 316 (5830):1484–1488

Kerwin CL, Wykoff DD (2012) De novo generation of a phosphate starvation-regulated promoter in Candida glabrata. *FEMS Yeast Res* 12 (8):980–989

Khalil AM, Guttman M, Huarte M, Garber M, Raj A, Morales DR, Thomas K, Presser A, Bernstein BE, van Oudenaarden A (2009) Many human large intergenic noncoding RNAs associate with chromatin-modifying complexes and affect gene expression. *PNAS* 106 (28):11667–11672

Khan IA, Shah B, Khan A, Zaman M, MDin MMU, Junaid K, Shah SRA, Adnan M, Ahmad N, Akbar R, Fayaz W, Rahman IU (2016) Study on the population densities of grass hopper and armyworm on different maize cultivars at Peshawar. *J Ento Zool Studies* 4(1):28–331

Köck M, Theierl K, Stenzel I, Glund K (1998) Extracellular administration of phosphate-sequestering metabolites induces ribonucleases in cultured tomato cells. *Planta* 204 (3):404–407

Kuang R, Chan K-H, Yeung E, Lim BL (2009) Molecular and biochemical characterization of AtPAP15, a purple acid phosphatase with phytase activity, in Arabidopsis. *Plant Physiol* 151 (1):199–209

Lambers H, Albornoz F, Kotula L, Laliberté E, Ranathunge K, Teste FP, Zemunik G (2018) How belowground interactions contribute to the coexistence of mycorrhizal and non-mycorrhizal species in severely phosphorus-impoverished hyperdiverse ecosystems. *Plant Soil* 424 (1-2):11–33

Lavenus J, Goh T, Roberts I, Guyomarc'h S, Lucas M, De Smet I, Fukaki H, Beeckman T, Bennett M, Laplaze L (2013) Lateral root development in Arabidopsis: fifty shades of auxin. *TreNDS Plant Sci* 18 (8):450–458

Li H, Yang Y, Zhang H, Chu S, Zhang X, Yin D, Yu D, Zhang D (2016a) A genetic relationship between phosphorus efficiency and photosynthetic traits in soybean as revealed by QTL analysis using a high-density genetic map. *Fron Plant Sci* 7:924

Li L, Yang H, Ren W, Liu B, Cheng D, Wu X, Gong J, Peng L, Huang F (2016b) Physiological and biochemical characterization of sheepgrass (Leymus chinensis) reveals insights into photosynthetic apparatus coping with low-phosphate stress conditions. *J Plant Biol* 59 (4):336–346

Li J, Yang H, Peer WA, Richter G, Blakeslee J, Bandyopadhyay A, Titapiwantakun B, Undurraga S, Khodakovskaya M, Richards EL (2005) Arabidopsis H+-PPase AVP1 regulates auxin-mediated organ development. *Science* 310 (5745):121–125

Liang C, Tian J, Lam H-M, Lim BL, Yan X, Liao H (2010) Biochemical and molecular characterization of PvPAP3, a novel purple acid phosphatase isolated from common bean enhancing extracellular ATP utilization. *Plant Physiol* 152 (2):854–865

Lin W-Y, Huang T-K, Chiou T-J (2013) Nitrogen limitation adaption, a target of microRNA827, mediates degradation of plasma membrane–localized phosphate transporters to maintain phosphate homeostasis in Arabidopsis. *Plant Cell* 25 (10):4061–4074

Lin W-D, Liao Y-Y, Yang TJ, Pan C-Y, Buckhout TJ, Schmidt W (2011) Coexpression-based clustering of Arabidopsis root genes predicts functional modules in early phosphate deficiency signaling. *Plant Physiol* 155 (3):1383–1402

Liu F, Wang Z, Ren H, Shen C, Li Y, Ling HQ, Wu C, Lian X, Wu P (2010) OsSPX1 suppresses the function of OsPHR2 in the regulation of expression of OsPT2 and phosphate homeostasis in shoots of rice. *Plant J* 62 (3):508–517

López-Bucio J, Hernández-Abreu E, Sánchez-Calderón L, Pérez-Torres A, Rampey RA, Bartel B, Herrera-Estrella L (2005) An auxin transport independent pathway is involved in phosphate stress-induced root architectural alterations in Arabidopsis. Identification of BIG as a mediator of auxin in pericycle cell activation. *Plant Physiol* 137 (2):681–691

Luan M, Zhao F, Han X, Sun G, Yang Y, Liu J, Shi J, Fu A, Lan W, Luan S (2019) Vacuolar phosphate transporters contribute to systemic phosphate homeostasis vital for reproductive development in Arabidopsis. *Plant Physiol* 179 (2):640–655

Lucas WJ, Groover A, Lichtenberger R, Furuta K, Yadav SR, Helariutta Y, He XQ, Fukuda H, Kang J, Brady SM (2013) The plant vascular system: evolution, development and functions. *J Integr Plant Biol* 55 (4):294–388

Lucena C, Waters BM, Romera FJ, García MJ, Morales M, Alcántara E, Pérez-Vicente R (2006) Ethylene could influence ferric reductase, iron transporter, and H+-ATPase gene expression by affecting FER (or FER-like) gene activity. *J Exp Bot* 57 (15):4145–4154

Ludwig M-G, Vanek M, Guerini D, Gasser JA, Jones CE, Junker U, Hofstetter H, Wolf RM, Seuwen K (2003) Proton-sensing G-protein-coupled receptors. *Nature* 425 (6953):93–98

Lynch JP (2011) Root phenes for enhanced soil exploration and phosphorus acquisition: tools for future crops. *Plant Physiol* 156 (3):1041–1049

Ma T-L, Wu W-H, Wang Y (2012) Transcriptome analysis of rice root responses to potassium deficiency. *BMC Plant Biol* 12 (1):161

Maharajan T, Ceasar SA, Ajeesh krishna TP, Ramakrishnan M, Duraipandiyan V, Naif Abdulla AD, Ignacimuthu S (2018) Utilization of molecular markers for improving the phosphorus efficiency in crop plants. *Plant Breed* 137 (1):10–26

Mehra P, Pandey BK, Verma L, Giri J (2019) A novel glycerophosphodiester phosphodiesterase improves phosphate deficiency tolerance in rice. *Plant Cell Environ* 42 (4):1167–1179

Misson J, Raghothama KG, Jain A, Jouhet J, Block MA, Bligny R, Ortet P, Creff A, Somerville S, Rolland N (2005) A genome-wide transcriptional analysis using Arabidopsis thaliana Affymetrix gene chips determined plant responses to phosphate deprivation. *PNAs* 102 (33):11934–11939

Mo X, Zhang M, Liang C, Cai L, Tian J (2019) Integration of metabolome and transcriptome analyses highlights soybean roots responding to phosphorus deficiency by modulating phosphorylated metabolite processes. *Plant Physiol Biochem* 139:697–706

Müller M, Schmidt W (2004) Environmentally induced plasticity of root hair development in Arabidopsis. *Plant Physiol* 134 (1):409–419

Neumann G, Martinoia E (2002) Cluster roots–an underground adaptation for survival in extreme environments. *Trends Plant Sci* 7 (4):162–167

Nie P, Li X, Wang S, Guo J, Zhao H, Niu D (2017) Induced systemic resistance against Botrytis cinerea by Bacillus cereus AR156 through a JA/ET-and NPR1-dependent signaling pathway and activates PAMP-triggered immunity in Arabidopsis. *Front Plant Sci* 8:238

O'Rourke JA, Yang SS, Miller SS, Bucciarelli B, Liu J, Rydeen A, Bozsoki Z, Uhde-Stone C, Tu ZJ, Allan D (2013) An RNA-Seq transcriptome analysis of orthophosphate-deficient white lupin reveals novel insights into phosphorus acclimation in plants. *Plant Physiol* 161 (2):705–724

Pachnis V, Brannan C, Tilghman SM (1988) The structure and expression of a novel gene activated in early mouse embryogenesis. *EMBO J* 7 (3):673–681

Panigrahy M, Rao DN, Sarla N (2009) Molecular mechanisms in response to phosphate starvation in rice. *Biotechnol Adv* 27 (4):389–397

Pant BD, Musialak-Lange M, Nuc P, May P, Buhtz A, Kehr J, Walther D, Scheible WR (2009) Identification of nutrient-responsive Arabidopsis and rapeseed microRNAs by comprehensive real-time polymerase chain reaction profiling and small RNA sequencing. *Plant Physiol* 150 (3):1541–1555

Peng M, Hannam C, Gu H, Bi YM, Rothstein SJ (2007) A mutation in NLA, which encodes a RING-type ubiquitin ligase, disrupts the adaptability of Arabidopsis to nitrogen limitation. *Plant J* 50 (2):320–337

Plaxton WC, Tran HT (2011) Metabolic adaptations of phosphate-starved plants. *Plant Physiol* 156 (3):1006–1015

Ponting CP, Oliver PL, Reik W (2009) Evolution and functions of long noncoding RNAs. *Cell* 136 (4):629–641

Qi Y, Wang S, Shen C, Zhang S, Chen Y, Xu Y, Liu Y, Wu Y, Jiang D (2012) OsARF12, a transcription activator on auxin response gene, regulates root elongation and affects iron accumulation in rice (Oryza sativa). *New Phytol* 193 (1):109–120

Quaghebeur M, Rengel Z (2004) Arsenic uptake, translocation and speciation in pho1 and pho2 mutants of Arabidopsis thaliana. *Physiol Plant* 120 (2):280–286

Reinhart BJ, Weinstein EG, Rhoades MW, Bartel B, Bartel DP (2002) MicroRNAs in plants. *Genes Dev* 16 (13):1616–1626

Rouached H, Arpat AB, Poirier Y (2010) Regulation of phosphate starvation responses in plants: signaling players and cross-talks. *Mol Plant* 3 (2):288–299

Ryan PR, James RA, Weligama C, Delhaize E, Rattey A, Lewis DC, Bovill WD, McDonald G, Rathjen TM, Wang E (2014) Can citrate efflux from roots improve phosphorus uptake by plants? Testing the hypothesis with near-isogenic lines of wheat. *Physiol Plant* 151 (3):230–242

Sakuraba Y, Kanno S, Mabuchi A, Monda K, Iba K, Yanagisawa S (2018) A phytochrome-B-mediated regulatory mechanism of phosphorus acquisition. *Nat Plants* 4 (12):1089–1101

Sana A, Ahmed N, Khan IA, Shah B, Khan A, Adnan M, Junaid K, Rasheed MT, Zaki AB, Huma Z, Ahmed S (2015) Evaluating larvicidal action of *Coriandrumsativum* (Dhania) and Mentha (mint) plant extracts against *Aedesaegypti*and *Aedesalbopictus* larvae under laboratory conditions. *J Ento Zool Studies* 3(6):156–159

Sánchez-Calderón L, Chacon-López A, Pérez-Torres C-A, Herrera-Estrella L (2010) Phosphorus: plant strategies to cope with its scarcity. In: *Cell biology of metals and nutrients*. Springer, pp 173–198

Sanders F, Tinker P (1971) Mechanism of absorption of phosphate from soil by Endogone mycorrhizas. *Nature* 233 (5317):278–279

Shane MW, Lambers H (2005) Cluster roots: a curiosity in context. In: *Root physiology: from gene to function*. Springer, pp 101–125

Shen C, Wang S, Zhang S, Xu Y, Qian Q, Qi Y, Jiang DA (2013) OsARF16, a transcription factor, is required for auxin and phosphate starvation response in rice (Oryza sativa L.). *Plant Cell Environ* 36 (3):607–620

Shen J, Yuan L, Zhang J, Li H, Bai Z, Chen X, Zhang W, Zhang F (2011) Phosphorus dynamics: from soil to plant. *Plant Physiol* 156 (3):997–1005

Shenoy V, Kalagudi G (2005) Enhancing plant phosphorus use efficiency for sustainable cropping. *Biotechnol Adv* 23 (7-8):501–513

Sulieman S, Tran LSP (2015) Phosphorus homeostasis in legume nodules as an adaptive strategy to phosphorus deficiency. *Plant Sci* 239:36–43

Sun B, Gao Y, Wu X, Ma H, Zheng C, Wang X, Zhang H, Li Z, Yang H (2020) The relative contributions of pH, organic anions, and phosphatase to rhizosphere soil phosphorus mobilization and crop phosphorus uptake in maize/alfalfa polyculture. *Plant Soil*:1–17

Svistoonoff S, Creff A, Reymond M, Sigoillot-Claude C, Ricaud L, Blanchet A, Nussaume L, Desnos T (2007) Root tip contact with low-phosphate media reprograms plant root architecture. *Nature Gene* 39 (6):792–796

Takehisa H, Sato Y, Antonio BA, Nagamura Y (2013) Global transcriptome profile of rice root in response to essential macronutrient deficiency. *Plant Signal Behav* 8 (6):e24409

Tran HT, Hurley BA, Plaxton WC (2010) Feeding hungry plants: the role of purple acid phosphatases in phosphate nutrition. *Plant Sci* 179 (1–2):14–27

Uhde-Stone C (2017) White lupin: a model system for understanding plant adaptation to low phosphorus availability. In: *Legume nitrogen fixation in soils with low phosphorus availability*. Springer, pp 243–280

Uhde-Stone C, Zinn KE, Ramirez-Yáñez M, Li A, Vance CP, Allan DL (2003) Nylon filter arrays reveal differential gene expression in proteoid roots of white lupin in response to phosphorus deficiency. *Plant Physiol* 131 (3):1064–1079

Vance CP, Uhde-Stone C, Allan DL (2003) Phosphorus acquisition and use: critical adaptations by plants for securing a nonrenewable resource. *New Phytol* 157 (3):423–447

Vengavasi K, Pandey R, Abraham G, Yadav RK (2017) Comparative analysis of soybean root proteome reveals molecular basis of differential carboxylate efflux under low phosphorus stress. *Genes* 8 (12):341

Wang C, Huang W, Ying Y, Li S, Secco D, Tyerman S, Whelan J, Shou H (2012) Functional characterization of the rice SPX-MFS family reveals a key role of OsSPX-MFS1 in controlling phosphate homeostasis in leaves. *New Phytol* 196 (1):139–148

Wang F, Deng M, Xu J, Zhu X, Mao C (2018) Molecular mechanisms of phosphate transport and signaling in higher plants. In: *Seminars in cell & developmental biology*. Elsevier, pp 114–122

Wang L, Li Z, Qian W, Guo W, Gao X, Huang L, Wang H, Zhu H, Wu J-W, Wang D (2011) The Arabidopsis purple acid phosphatase AtPAP10 is predominantly associated with the root surface and plays an important role in plant tolerance to phosphate limitation. *Plant Physiol* 157 (3):1283–1299

Wang X, Wang Z, Zheng Z, Dong J, Song L, Sui L, Nussaume L, Desnos T, Liu D (2019b) Genetic dissection of Fe-dependent signaling in root developmental responses to phosphate deficiency. *Plant Physiol* 179 (1):300–316

Wang X, Yan X, Liao H (2010) Genetic improvement for phosphorus efficiency in soybean: a radical approach. *Ann Bot* 106 (1):215–222

Wang Y-H, Garvin DF, Kochian LV (2002) Rapid induction of regulatory and transporter genes in response to phosphorus, potassium, and iron deficiencies in tomato roots. Evidence for cross talk and root/rhizosphere-mediated signals. *Plant Physiol* 130 (3):1361–1370

Wang T, Zhao M, Zhang X, Liu M, Yang C, Chen Y, Chen R, Wen J, Mysore KS, Zhang W-H (2017) Novel phosphate deficiency-responsive long non-coding RNAs in the legume model plant Medicago truncatula. *J Exp Bot* 68 (21-22):5937–5948

Wang T, Xing J, Liu Z, Zheng M, Yao Y, Hu Z, Peng H, Xin M, Zhou D, Ni Z (2019a) Histone acetyltransferase GCN5-mediated regulation of long non-coding RNA At4 contributes to phosphate starvation response in Arabidopsis. *J Exp Bot* 70 (21):6337–6348

Wu W, Lin Y, Liu P, Chen Q, Tian J, Liang C (2018) Association of extracellular dNTP utilization with a GmPAP1-like protein identified in cell wall proteomic analysis of soybean roots. *J Exp Bot* 69 (3):603–617

Xu F, Liu Q, Chen L, Kuang J, Walk T, Wang J, Liao H (2013) Genome-wide identification of soybean microRNAs and their targets reveals their organ-specificity and responses to phosphate starvation. *BMC Genomics* 14 (1):66

Xu W, Zhang Q, Yuan W, Xu F, Aslam MM, Miao R, Li Y, Wang Q, Li X, Zhang X (2020) The genome evolution and low-phosphorus adaptation in white lupin. *Nat Commun* 11 (1):1–13

Yang XJ, Finnegan PM (2010) Regulation of phosphate starvation responses in higher plants. *Ann Bot* 105 (4):513–526

Yang X, Wang W, Ye Z, He Z, Baligar VC (2004) Physiological and genetic aspects of crop plant adaptation to elemental stresses in acid soils. In: *The red soils of China*. Springer, pp 171–218

Yang WT, Baek D, Yun D-J, Hwang WH, Park DS, Nam MH, Chung ES, Chung YS, Yi YB, Kim DH (2014) Overexpression of OsMYB4P, an R2R3-type MYB transcriptional activator, increases phosphate acquisition in rice. *Plant Physiol Biochem* 80:259–267

Yi K, Wu Z, Zhou J, Du L, Guo L, Wu Y, Wu P (2005) OsPTF1, a novel transcription factor involved in tolerance to phosphate starvation in rice. *Plant Physiol* 138 (4):2087–2096

Zeng G, Wan J, Huang D, Hu L, Huang C, Cheng M, Xue W, Gong X, Wang R, Jiang D (2017) Precipitation, adsorption and rhizosphere effect: the mechanisms for phosphate-induced Pb immobilization in soils—a review. *J Hazard Mater* 339:354–367

Zhang Z, Liao H, Lucas WJ (2014) Molecular mechanisms underlying phosphate sensing, signaling, and adaptation in plants. *J Integr Plant Biol* 56 (3):192–220

Zhang Z, Lin H, Shen Y, Gao J, Xiang K, Liu L, Ding H, Yuan G, Lan H, Zhou S (2012) Cloning and characterization of miRNAs from maize seedling roots under low phosphorus stress. *Mol Biol Rep* 39 (8):8137–8146

Zhao J, Ohsumi TK, Kung JT, Ogawa Y, Grau DJ, Sarma K, Song JJ, Kingston RE, Borowsky M, Lee JT (2010) Genome-wide identification of polycomb-associated RNAs by RIP-seq. *Mol Cell* 40 (6):939–953

Zhou J, Dong B-C, Alpert P, Li H-L, Zhang M-X, Lei G-C, Yu F-H (2012) Effects of soil nutrient heterogeneity on intraspecific competition in the invasive, clonal plant Alternanthera philoxeroides. *Ann Bot* 109 (4):813–818

3

Environmental Upheaval: Consequences and Management Strategies

Misbah Naz, Muhammad Aamir Iqbal, Mohammad Sohidul Islam, Akbar Hossain,
Subhan Danish, Rahul Datta, Shah Fahad, Disna Ratnasekera, Mohammad Anwar Hosssain,
Muhammad Habib Ur Rahman, Shah Saud, Muhammad Kamran, Wajid Mahboob,
Rajesh Kumar Singhal, Sharif Ahmed, Sonia Mbarki, Doğan Arslan, Murat Erman, and
Ayman EL Sabagh

CONTENTS

3.1 Introduction

Environmental changes and disruptions especially water diversion affect the structure and function of delta communities (Ibanez and Prat 2003). Overall, it is not possible to understand environmental upheavals by looking into one factor as only comprehending all aspects can lead to the precise estimation of factors responsible. Recently, the external environment is changing rapidly and imparting adverse influences on all business enterprises and anthropogenic activities. There are efforts to understand environmental changes and our understanding is increasing over time due to the evolution of technology, particularly digital information processing power (Malhotra 2000; Ibanez and Prat 2003). Technology itself does not change anything, but how it empowers people is changing the landscape one by one adding to our environmental dilemmas (Jallat 2004). Our current industries can potentially destroy our environment, but if we use them wisely and develop them sustainably, they can also help us to create solutions to solve the ecological problems. However, current major environmental problems, include climate change, characterized by global warming, shifting of seasons and rainfall patterns, pollution caused by industrial functioning, and farm input wastage from agricultural fields leading to environmental degradation and severe depletion of nonrenewable resources. Conservation campaigns lobby for the conservation of species and the protection of any ecologically valuable natural areas, genetically modified foods, and global warming (Heath and Gifford 2006). Environmental changes in the form of changing climate, drought, and increased soil depletion may lead to alterations in settlement patterns. At the end of the relative drought in the late 16th and early 17th centuries, agricultural shortages and insecurity of large sedentary populations may have explained the decline in new settlements (Khan et al. 2016; Degroot 2018).

However, new developments in biological technology and nanotechnology, and information and communication technologies have shaped the research environment over the past decade. Highly educated experts in the research and development sector are increasingly working with scientists and researchers at universities and research institutions to develop new technologies (Roco et al. 2011). Multinational companies operating in different countries need to manage various research developments and marketing strategies in diversified cultures. The new knowledge economy permeates companies, universities, research institutions, and countries, creating an interdisciplinary global environment (Buckley and Ghauri 2004). It becomes evident that the management of evolving technology presents significant challenges to technical management (TM) and trainers in this new environment. The discipline of TM addresses potential challenges pertaining to innovation, operations, and strategy development (Horwitch and Stohr 2012). Nowadays, organizations find themselves competing to keep up with the competition by adopting new technologies. However, there are two questions that need to be answered: Are these organizations ready for the advancement of each new technology, and are the technical challenges and management of the organizations focused on the new developments and challenges that organizations face today in the field of human resources and business, on account of the highly complex changes continuing in technical resources (Bishop 2006). Organizations need more efficient and flexible management to meet their human and business needs. Researchers, academics, and the company's understanding and contribution to the company's research interests are vary. The focus should be on the exchange of practical management experience, as well as the current technical challenges and management research status (Andersen and Strandskov 2008). Because of these important facts, the benefits of technological advances, such as the operation of technology can bring products faster to the market, and expand employee skills, allowing organizations to move to areas such as telecommunicating. In addition, technology can increase the involvement and participation of remote employees. This chapter provides an overview of the consequences of hostile environmental conditions and management strategies for food and environmental security.

3.2 Challenges and Management Technologies

The choice of alliance partners highlights two major issues in the area of information technology (IT). First, identifying IT requirements that require a technology partner. Second, the alliance often requires the implementation of integrated information systems (Rothaermel and Deeds 2006). It has been argued that the impact of agriculture on air and water should be greater than previous estimates, but technological developments have assisted in reducing these drastic impacts through efficient utilization of farm-applied resources.

3.2.1 Information Technology Management Challenges

Technology may eventually provide a sustainable solution to environmental dilemmas over time, provided targeted research is done and concerted efforts are made for the dissemination of results to stakeholders. The prevalent practices adopted by commercial business enterprises, including agriculture, have contributed to the destruction of the environment, which necessitate developing sustainable approaches that offer economically feasible solutions to control ecological deteriorations (Clayton and Radcliffe 1996). Conservation campaigns are an effective option to protect ecologically valuable but vulnerable species in their habitats (Sanne 2000).

Opening a campaign will implement an important IT infrastructure because of the need to access the organization's information. Sub-information is easily a piece of work required to implement a successful start-up strategy (Ortiz 2013). As current demand grows, it is difficult to predict future system demand, making this problem even worse. If it coordinates all developments and blocks automation silos, it can at least manage this problem, even if it immediately necessitates that such interconnects are not required. In the early days of the company's existence, assigning this responsibility to the top management of the organization will put the company in a strong position in any future opening plan (Ryu et al. 2009). Technology can expand the engagement in project work and a pool of risk by sharing a network, internal and Internet networks, as well as challenges such as information security, conduct, Internet, Internet issues, risk management and governance, information technology integration, and upgrading (Palensky and Dietrich 2011). Resource management, behavioural monitoring, business continuity/disaster recovery are discussed, and possible solutions for the future are presented.

In addition to financial issues, senior management should also be concerned because IT may limit a company's strategic positioning or lock it into a particular business process. Some decisions, such as introducing enterprise-wide applications or maintaining too, can be damaging. Senior managers must continue to be involved in information technology-related decisions, as investments in specific information systems, if not considered strategic, can have high opportunity costs and may limit the management's ability to deliver critical information.

3.2.2 Alliance Agreements and Implementation of Integrated Information Systems

Alliances have emerged as an effective tool to cope with environmental upheavals sustainably. However, most of those have proved inherently risky leading to partial or complete failure of the alliances (Cano-Kollmann et al. 2018); the impact of Integrated Information Systems (IIS) and Federation Formalization (i.e., formal alliance control) on information exchange risk and overall alliance risk. We developed a model that predicts that when companies are involved in alliances with larger bodies (i.e., alliances with multiple goals), they are more likely to use IIS and a broad portfolio of formal controls to manage alliance risk (Kehrel et al. 2016; Ali et al. 2013). Information systems are negatively correlated with the risk of information exchange (Christ and Nicolaou 2016), but this impact is regulated through the formalization of alliances, indicating that formal controls implemented within the alliance and within integrated information systems are an important factor in reducing the risk of information exchange, followed by the risk of the entire alliance (Abbeele 2016; Christ and Nicolaou 2016). This section attempts to add tools and strategies for the effective management of strategic alliances that are discussed in the following sub-headings:

3.3 Strategies for the Effective Management of Strategic Alliances

3.3.1 Pre-alliance Planning

Business models are going through a redefinition process as they strive to develop new business structures to support competition. The result is the development of virtual organizations based on alliance structure (Nielsen 2007). This article uses case studies to describe how telecommunications organizations and other retail power organizations initiate strategic alliances to improve their market position, finances, and customer base (Langfield-Smith 2008). The case study describes how telecommunication organizations explore to build up strategic alliances with retail power organizations, pre-union activities and their impact on alliance formation, such as the design of an information system framework. The case studies identify learning opportunities for organizations that seek strategic alliances to seize new business opportunities (Scaringella and Burtschell 2017).

3.3.2 Alliance Agreements and Agricultural Inconsistencies

Agricultural cooperatives consider their options for using strategic alliances to identify differences in cooperative practices with suppliers, customers and competitors, in particular, differences in the interests and characteristics of cooperatives (Disdier et al. 2008). For achieving this goal, four agricultural cooperatives were subjected to exploratory qualitative studies on how strategic alliances are almost unknown and have little basis for the theory. There is evidence that the choice of strategic alliances is an important decision for cooperatives, regardless of the alliance (Meier 2011). Besides, the use of strategic alliances often has a positive impact not only on the outcomes of cooperatives but also on the most expected outcomes in the years owned by cooperatives. As the competition among organizations unfolds, strategic alliances have gained a foothold and have proved extremely effective in strengthening the organizations, thus, affecting their survival. This strategy will be a result not only in the enterprise but also in the years and will want to compete with the organization.

3.4 Environmental Challenges Faced by Science

3.4.1 Water Challenges in the Delta, River Basin, and Low-riparian Area

The current planning of work beyond the delta area and its basic research are merely assertions that the objectives are equal. Efforts are needed to achieve varying degrees of goals for balancing the resources devoted to each goal. The current planning work in the Gulf-Delta region and its basic research is only equal assertions of equality in objectives. A fundamental problem is in the distribution of scarce water (Jalongo 2007). The legislature provides an opportunity for planners to create the necessary balance without specifying the common equality objectives. For now, however, it has been one of the planning objectives. It seems to assume that additional water must be found to serve the goal of common equality. When water is scarce, it is not possible to distribute water to support the water requirement of one side without reducing the distribution of water to the other side of the water resource. Of course, additional water can always be found by redistributing the other uses of water unrelated to the uses envisaged in the common equality objective (Aleem 1993). In the face of all these ecological and environmental constraints, effective planning and management systems must consider a wider range of water management options than in the past. Perhaps more importantly, it is a reality that all deltas and exporters need to spread the country's water scarcity more broadly.

3.4.2 Water Planning and Management

For centuries, water systems have benefited people and their economies. The services provided by these systems are multiple (Pedro-Monzonis et al. 2015). Water systems also fail to support and sustain resilient raw materials. Typical causes include inadequate and/or degraded river flows, excessive reflux of river flows, pollution from industrial and agricultural activities, eutrophication of nutrient loads, irrigation reflux salinization, infestation of exotic plants and animals, excessive low-conductive fish tape, flood plains, and changes in the amount of water and sediment flow systems from development activities (Liu 2009). Recently, there are several technologies for the treating of toxic pollutants, and the major techniques are costly to conduct full scale to control the performance optimization (Saleem et al. 2020).

3.5 Environmental Problems

3.5.1 Ecosystem Degradation

Aquatic and riverbank markets may face some threats. The most important include habitat loss due to river training and land reclamation in urban and industrial development, poor water quality due to pesticide, fertilizer and wastewater discharges, and harmful aquatic species (Forslund et al. 2009). Environmental degradation is the dreadful conditions of the environment caused by the depletion of resources such as air, water and soil.

3.5.2 Pollution

Air, water, land or noise are harmful to the environment, regardless of any form of pollution (Vesilind et al. 2013). The presence of pollution in the air we breathe leads to health problems. Water pollution will reduce the quality of the water we drink. Land pollution causes human activities to degrade the earth's surface (Schweitzer and Noblet 2018). Noise pollution can cause irreparable damage to the ears due to constant loud noises, such as vehicles screaming on busy roads or machines making loud noises in factories (Hammond 2000).

3.5.3 Impact on Human Health

Due to environmental degradation, human health may be at risk with the emergence of deadly diseases. Areas exposed to toxic pollutants can cause respiratory problems such as pneumonia and asthma (Laumbach and Kipen 2012). It is understood that millions of people will be at risk owing to the indirect effects of environmental pollution (D'Amato 2011). It is important to maintain the balance of its lines in the form of pollution control, nutrient restoration, water conservation and climate stability; and deforestation, global warming, overpopulation and pollution are the main causes of loss (Al-Awadhi et al. 2003).

3.5.4 Environmental Impact of Economic Growth

A country can have a huge impact on environmental degradation and a huge economic impact when it comes to greening, cleaning landfills and protecting species (Watson et al. 2014). Economic impacts may also be linked to economic losses. Many factors can have an impact on the environment (Beniston 2007). If the people are not careful, they may lead to environmental degradation that is taking place around the world. However, man can take action to stop it, take care of the world we live in, provide people with environmental education, help them choose familiar environments around them, so that they can deal with the issues and become more familiar with the environment (Holmes et al. 2009). For the benefit and protection of our children and other future generations. Economic growth means real (and large) growth in U.S. output. Therefore, as consumption increases, it is likely to see costs imposed on the environment (Peters and Hertwich 2008). The environmental impacts of economic growth include increased consumption of non-renewable resources, higher levels of pollution, global warming and potential loss of environmental habitats. However, not all forms of economic growth are environmentally damaging. As real incomes increase, individuals have a greater capacity to devote their resources to protecting the environment and mitigating the harmful effects of pollution (Ambec et al. 2013). Also, economic growth resulting from technological improvements can increase output and reduce pollution.

3.5.5 Climate Change and Temperature

Global temperature is increasing as a consequence of climate change (Huang et al. 2020). It is predicted that the average global temperature will rise in the next few years due to some of the forces affecting the climate (Immerzeel et al. 2010; EL Sabagh 2020b, 2020c, 2020d). The amount of carbon dioxide (CO_2) in the atmosphere also will rise, both of which will affect water resources; evaporation depends largely on temperature and availability of water, which ultimately affects the amount of water supplied to groundwater (Wallace 2000). Plant steaming may be affected by an increase in CO_2 in the atmosphere, which can reduce the use of water but also increase the area of the leaves (Tague and Peng 2013). Higher temperatures can reduce the winter snow season, increase the melting fields that lead to runoff peaks, and affect soil moisture, flood and drought risk, and the storage capacity of the area-based.

Increased winter temperatures can lead to condensed snow cover, ultimately lead to reduced water resources in summer (Chang et al. 2002). This is particularly important in mid-latitudes and in mountainous regions, which rely on glacier runoff to replenish their river systems and groundwater supplies, making these areas increasingly vulnerable to water shortages over time.

Water's heat and ocean glaciers are melting as temperatures rise, giving way to rising sea levels. This also affects the supply of fresh water in coastal areas. As the more salinating estuaries and deltas move further inland, the intrusion of seawater leads to increased salinity in reservoirs and aquifers. As a result, sea-level rise may also be caused by groundwater depletion (Chang et al. 2002), as climate change can affect the hydrological cycle in many ways. Global temperature rise and uneven distribution of precipitation lead to overwater oversupply and deficits, but the decline in global groundwater indicates that sea levels rise even after the increase in melting water and thermal expansion, which can provide positive feedback on the problems of fresh-water supply caused by sea-level rise (Aleksandrova et al. 2014). Warmer temperatures lead to higher water temperatures, which are also important in water degradation because water is more susceptible to bacterial growth (Koss 1994). Increased water temperatures can also affect a species's sensitivity to temperature, and the decrease in dissolved oxygen in the water due to higher temperatures causes the water's self-system to grow.

3.5.6 Agriculture under Changing Climate

The atmospheric greenhouse gases assist in providing the requisite warmth to carry out agriculture-related activities, especially commercial farming. However, climate change has given rise to global warming of the earth that is equivalent to the greenhouse effect (de Richter et al. 2016). The absence of the greenhouse effect is bound to lead to a terrible cooling of the globe, as occurred about 1,500 years ago. During its incidence, two volcanic eruptions threw so much black dust into the upper atmosphere that temperatures fell sharply, leading to complete crop failure. The event was followed by numerous causalities, and that is why it is known as the 'black death' (de Richter et al. 2016). It has been revealed that the dust slowly fell on the earth, and subsequently, the sun was once again able to warm the earth leading to the revival of life. However, human development has put us in different sort of problem (overheating of earth called 'global warming') caused by trapping of radiations having longer wavelength emitted from the earth on cooling down (Milfont 2012). The heat in the greenhouse is kept in the greenhouse so that the earth's temperature is rising faster than at any time. Anthropogenic activities have also resulted in the rising of CO_2 concentration in the atmosphere, and as a result, crop productivity has been projected to be greatly influenced by the combined effect of increased temperature and CO_2 concentration (Iqbal et al. 2016). Researchers are of the view that productivity of C_4 crops (sorghum, millets, sugar cane etc.) is expected to be multiplied under the changing climate while C_3 crops (wheat, rice etc.) might witness a sharp decline in their growth and production globally (Iqbal 2019a, 2019b, 2019c). The rising temperature has been projected to boost temperate crops productivity in Polar Regions owing to temperature increment. The melting of permafrost is expected to bring new fertile areas under cultivation. However, shifting rainfall patterns have disturbed the sowing and harvesting times of crops that have made it necessary to revise prevalent cropping systems (Iqbal 2017, 2018a, 2018b; EL Sabagh et al. 2020). In addition to shifting of rainfall patterns, erratic rainfall will result in either flooding or drought occurrence leading to compromising the food security of the rapidly increasing human population across the globe. The melting of glaciers has been projected to raise sea level, and thus, vast coastal areas would submerge in seawater, leading to deprivation of wetlands and shifting of human settlements.

3.6 Do Environmental Policies Affect Global Value Chains?

3.6.1 What causes Global Warming?

Global warming occurs when carbon dioxide (CO_2) and other air pollutants and greenhouse gases collect in the atmosphere and absorb sunlight and solar radiation that have bounced off the earth's surface (Ward 2015). Normally, this radiation would escape into space – but these pollutants, which can last for years to centuries in the atmosphere, trap the heat and cause the planet to get hotter (Milfont 2012).

3.6.2 Climate Change and Precipitation

The rise in global temperatures is also expected to be associated with increased global precipitation, but water quality is highly likely to decline due to increased runoff, flooding, increased soil erosion, and mass movement of land, as the water will carry more purchases, but it will also carry more pollutants (Jeppesen et al. 2009). While much of the focus on climate change is on global warming and the greenhouse effect, some of the most likely sources of climate change come from the temperature of precipitation, evapotranspiration, and runoff, and soil moisture (Garcia-Ruiz 2010). Global precipitation is expected to increase in tropical and high latitudes, while a sharp decline is expected in subtropical regions (Polade et al. 2014). This will eventually result in a change in the latitude of the water distribution. Areas with more precipitation are also expected to increase in winter and will become drier in summer, resulting in more changes in precipitation distribution. The information of precipitation affects the time and extent of floods and droughts, changes the runoff process, and changes the groundwater recharge rate. Vegetation patterns and growth rates will be directly affected by changes in precipitation and distribution, which, in turn, will affect agriculture and natural hospitals.

The temperature rise may lead to increased transpiration and evaporation leading to depletion of groundwater reserves along with multiplication of salt-affected soils (Basharat et al. 2014). Natural science itself is not well recognized for sustainable development, as is economics. Economists alone do not have the necessary knowledge base of natural science to understand the complex ecosystems in which economic systems operate and the impact of economic activities on them (Viviroli et al. 2011). Progress in sustainable development requires cooperation between social scientists, including economists and natural scientists. Of course, achieving sustainable development requires institutional and political coordination, which goes far beyond the scientific knowledge generated by the pooling of integrated scientific knowledge (Sahlberg and Oldroyd 2010). The production of special crops in controlled agro-environmental conditions such as hydroponics may provide food that is beneficial for human health (Hirel et al. 2011). Functional foods can be produced by professionals by simply modifying the nutritional composition and environmental controls. Besides, the development and replenishment of special dietary ingredients provide several human health benefits in addition to basic nutrition (Navarro et al. 2006). New targeted research is needed to be conducted about the sustainable production of crops under changing climate along with nutritional quality maintenance, including secondary metabolites and antioxidants concentrations in a controlled environment. The changes in the nutritional quality and antioxidant properties of crops are also introduced (Alvarez-Jubete et al. 2010). The sensory perception of fruits and vegetables grown in hydroponics compared to soil cultivation. Further, a new approach to the sustainable production of functional foods using professional cultural techniques and demand-oriented distribution (Godfrey et al. 2010). This chapter brings together interesting research on controlled environmental agriculture from around the world, providing valuable resources for teachers, researchers, commercial growers, and senior students in the science of plant biology.

The challenge of global warming that is a slow rise in the average temperature of the Earth's atmosphere, as the increase in energy (heat) from the sun, hitting the Earth, is captured in the atmosphere instead of being radiated back into space, must be addressed on a priority basis

(Stefanon et al. 2013). The Earth's atmosphere always captures the sun's heat like a greenhouse, ensuring the temperature the earth enjoys and allowing the appearance of life forms because human beings know them.

3.6.3 Dairy Farming: A Fundamental Source of Environmental Pollution

Agriculture relies on soil moisture directly affected by climate dynamics, and precipitation is an input to the system, with processes such as evaporation, surface runoff, drainage, and infiltration into groundwater (Bruijnzeel 2004). Climate change, especially the precipitation and evasive changes predicted by climate models, will have a direct impact on soil moisture, surface runoff, and groundwater recharge.

Soil moisture may be greatly reduced in areas where climate models predict a decrease in precipitation. With this in mind, agriculture in most areas already requires irrigation, as water is degraded by actual water use and agriculture (Cooper et al. 2013), which depletes freshwater supplies. Irrigation increases the salt and nutrient content of areas that are not normally affected and destroys damming and clearing rivers. Fertilizer sits into the waste stream of humans and livestock and eventually into groundwater, while nitrogen, phosphorus, and other chemicals in fertilizer can acidify soil and water (Nardone et al. 2010). The demand for staple food is bound to be multiplied consistently owing to the increasing global population, while global freshwater supplies are projected to decrease substantially. Cows need water to drink if the temperature is high, the humidity is low, and more if the cow's production system is extensive because finding food requires more effort. Water is needed for both meat processing and livestock feed production (Mok et al. 2014). Feces contaminates the body of freshwater, and abattoirs, depending on management, cause waste such as blood, fat, hair, and other body components to the supply of freshwater.

The shift from agriculture to urban and suburban water has raised concerns about agricultural sustainability, the rural socioeconomic recession, food security, an increased carbon footprint of imported food, and a decline in the balance of foreign trade (Maldonado 2012). The depletion of freshwater is applied to more specific and densely populated areas, exacerbating freshwater shortages and making populations vulnerable to economic, social, and political conflicts in many ways; Climate change is an important cause of involuntary migration and forced displacement (Aeschbach-Hertig and Gleeson 2012; Rahman et al. 2016). According to the Food and Agriculture Organization (FAO) of the United Nations, global animal agriculture emits more greenhouse gases than transport.

3.6.4 Water Management

By increasing water management, the problem of freshwater depletion can be solved. While water management systems are often flexible, adapting to new hydrological systems can be very expensive, and technological limitations can make it more difficult to be adopted by less developing nations (Ma et al. 2010). A preventive approach is needed to avoid efficient costs and the need to restore the water supply. Innovations in overall demand reduction may be important for planning water sustainability water systems, as they exist now, based on assumptions about the current climate and adapted for existing river flows and flood frequencies (Wu and Chen 2013; Naseer et al. 2016). Reservoirs operate according to past hydrological records and irrigation systems based on historical temperature, water availability, and crop water needs; A re-examination of engineering operations, optimization, and planning, as well as a reassessment of legal, technical, and economic approaches to managing water resources, are critical for future water management to address water degradation under changing climate (Black et al. 2011; Ali et al. 2013). Another approach is the privatization of water, which, despite its economic and cultural effects, can be easier to control and distribute in the quality of services and the overall quality of water. Rationality and sustainability are appropriate and require limits on overexploitation and pollution, as well as conservation efforts.

3.7 Environmental Degradation: Driver of Disasters

Environmental degradation is the driver and major reason behind emerging disasters that render the environment unable to meet the social and ecological needs of humans and other life on Earth (Ciccarese et al. 2012). Excessive consumption of natural resources can lead to environmental degradation and reduce the effectiveness of basic ecosystem services, such as flood reduction and landslides (Ciccarese et al. 2012). This leads to an increased risk of disasters that further reduces (Schipper and Pelling 2006). Environmental degradation is the deprivation of the environment by the exhaustion of resources such as air, water, and soil. It is defined as any change or interference with the environment that is considered harmful or undesirable (York and Venkataraman 2010). Environmental degradation is one of the greatest threats to the world today that we need to focus and try to work on it to save living organism from environmental degradation effects (Bracke 2004).

3.8 Conclusion

The environment gets affected by anthropogenic activities, and in return, drastically influences human health and business enterprises through global warming, degradation of ecology, pollution, and contamination of non-renewable resources. All these disasters adversely impact human life through forced alterations in political stability, need for new settlements, food insecurity resulting from decreased agricultural productivity, uncontrolled poverty, etc. Among the biggest challenges confronted with humanity right now is the management of environmental upheaval. The evolving technology through targeted research and appropriate dissemination of findings to stakeholders can pave the way towards sustainable utilization of renewable and non-renewable resources on the earth.

Conflicts of Interest

The authors declare that they have no conflicts of interest to report regarding the present study.

REFERENCES

Abbeele AVD (2016) Discussion of integrated information systems, alliance formation, and the risk of information exchange between partners. *J Manag Account Res* 28(3): 19–23

Aeschbach-Hertig W, Gleeson T (2012) Regional strategies for the accelerating global problem of groundwater depletion. *Nature Geosci* 5(12): 853–861 doi: 10.1038/ngeo1617

Al-Awadhi JM, Misak RF, Omar SS (2003) Causes and consequences of desertification in Kuwait: a case study of land degradation. *Bull Eng Geol Environ* 62(2):107–115

Aleem D (1993) *Review of Research on Ways To Attain Goal Six: Creating Safe, Disciplined, and Drug-Free Schools.* OERI, Washington, DC. (ERIC Document Reproduction Service No. ED357446)

Aleksandrova M, Lamers JPA, Martius C, Tischbein B (2014) Rural vulnerability to environmental change in the irrigated lowlands of Central Asia and options for policy-makers: a review. *Environ Sci Policy* 41:77–88 doi: 10.1016/j.envsci.2014.03.001

Ali K, Arif M, Khan Z, Tariq M, Waqas M, Gul B, Bibi S, Zia-ud-Din, Ali M, Shafi B, Adnan M (2013) Effect of cutting on productivity and associated weeds of canola. *Pak J Weed Sci Res* 19(4):393–401

Alvarez-Jubete, L, Wijngaard, H, Arendt EK, Gallagher E (2010) Polyphenol composition and in vitro antioxidant activity of amaranth, quinoa buckwheat and wheat as affected by sprouting and baking. *Food Chem* 119(2): 770–778

Ambec S, Cohen MA, Elgie S, Lanoie P (2013) The porter hypothesis at 20: can environmental regulation enhance innovation and competitiveness? *Rev Environ Econ Policy* 7(1): 2–22 doi: 10.1093/reep/res016

Andersen PH, Strandskov J (2008) The innovator's dilemma: when new technologies cause great firms to fail. *Acad Manag Rev* 33(3): 790–794

Basharat M, Hassan D, Bajkani AA, Sultan SJ (2014) Surface wateGroundwater management options for indus basin irrigation system. *International Waterlogging and Salinity Research Institute (IWASRI) Lahore Pakistan Water and Power Development Authority Publication* 299:155

Beniston M (2007) Linking extreme climate events and economic impacts: examples from the Swiss Alps. *Energy Policy* 35(11):5384–5392

Bishop PC (2006) Tech mining: exploiting new technologies for competitive advantage. *Technol Forecast Social Change* 73(1):91–93 doi: 10.1016/j.techfore.2005.08.001

Black R, Adger WN, Arnell NW, Dercon S, Geddes A, Thomas D (2011) The effect of environmental change on human migration. *Global Environ Change* 21(supp-S1)

Bracke MS (2004) World agriculture and the environment: a commodity-by-commodity guide to impacts and practices. *Library J* 129(2):116-116

Bruijnzeel LA (2004) Hydrological functions of tropical forests: not seeing the soil for the trees? *Agric Ecosyst Environ* 104(1):185–228

Buckley PJ, Ghauri PN (2004) Globalisation, economic geography and the strategy of multinational enterprises 35: 81. *J Int Bus Stud* 35(3):255-255 doi: 10.1057/palgrave.jibs.8400088

Cano-Kollmann M, Awate S, Hannigan TJ, Mudambi R (2018) Burying the hatchet for catch-up: open innovation among industry laggards in the automotive industry. *Calif Manag Rev* 60(2):17–42 doi: 10.11 77/0008125617742146

Chang H, Knight CG, Staneva MP, Kostov D (2002) Climate change in Central and Eastern Europe Water resource impacts of climate change in southwestern Bulgaria. *Geojournal* 57(3): 159–168

Christ MH, Nicolaou AI (2016) Integrated information systems, alliance formation, and the risk of information exchange between partners. *J Manag Account Res* 28(3):1–18

Ciccarese L, Mattsson A, Pettenella D (2012) Ecosystem services from forest restoration: thinking ahead. *New Forests* 43(5-6):543–560 doi: 10.1007/s11056-012-9350-8

Clayton AMH, Radcliffe NJ (1996) *Sustainability: a systems approach.* 1st edn. CRC Press, USA

Cooper SD, Lake PS, Sabater S, Melack JM, Sabo JL (2013) The effects of land use changes on streams and rivers in Mediterranean climates. *Hydrobiologia* 719(1):383–425

D'Amato G (2011) Effects of climatic changes and urban air pollution on the rising trends of respiratory allergy and asthma. *Multidisc Respir Med* 6(1):28–37 doi: 10.1186/2049-6958-6-1-28

Degroot D (2018) Climate change and society in the 15th to 18th centuries. *Wiley Interdisc Rev Climate Change* 9(3): e518. 10.1002/wcc.518

EL Sabagh A et al (2020a) Consequences and mitigation strategies of heat stress for sustainability of soybean (*Glycine max L. Merr.*) production under the changing climate. *Plant stress physiology*. IntechOpen, UK

Disdier AC, Fontagne L, Mimouni M (2008) The impact of regulations on agricultural trade: evidence from the SPS and TBT agreements. *Amer J Agric Econ* 90(2):336–350 doi: 10.1111/j.1467-8276.2 007.01127.x

EL Sabagh A et al (2020d) Drought and heat stress in cotton (*Gossypium hirsutum L.*): consequences and their possible mitigation strategies. In: *Agronomic crops*. Springer, Singapore, pp 613–634

EL Sabagh A, Hossain A, Islam MS, Iqbal MA, Raza A, Karademir Ç, Karademir E, Rehman A, Rahman MA, Singhal RK, Llanes A, Raza MA, Mubeen M, Nasim W, Barutçular C, Meena RS, Saneoka H (2020b) Elevated CO2 concentration improves heat-tolerant ability in crops. In: *Abiotic stress in plants*. IntechOpen. 10.5772/intechopen.94128

EL Sabagh A, Hossain A, Aamir Iqbal M, Barutçular C, Islam MS, Çig F, Erman M, Sytar O, Brestic M, Wasaya A, Jabeen T, Asif Bukhari M, Mubeen M, Athar HR, Azeem F, Akdeniz H, Konuskan O, Kizilgeci F, Ikram M, Sorour S, Nasim W, Elsabagh M, Rizwan M, Swaroop Meena R, Fahad S, Ueda U, Liu L, Saneoka H (2020c) Maize adaptability to heat stress under changing climate. In: *Plant stress physiology*. IntechOpen, 2020, 10.5772/intechopen.92396

Forslund A, Renöfält BM, Barchiesi S, Cross K, Smith M (2009) Securing water for ecosystems and human well-being: the importance of environmental flows. https://www.iucn.org/downloads/securing_water_ for_ecosystems_and_human_well_being.pdf

Garcia-Ruiz JM (2010) The effects of land uses on soil erosion in Spain: a review. *Catena* 81(1):1–11 doi: 10.1016/j.catena.2010.01.001

Godfrey D, Hawkesford MJ, Powers, SJ, Millar S, Shewry PR (2010) Effects of crop nutrition on wheat grain composition and end use quality. *J Agric Food Chem* 58(5):3012–3021

Hammond GP (2000) Energy, environment and sustainable development: a UK perspective. *Proc Saf Environ Prot* 78(4): 304–323

Heath Y, Gifford R (2006) Free-market ideology and environmental degradation - The case of belief in global climate change. *Environ Behav* 38(1):48–71 doi: 10.1177/0013916505277998

Hirel B, Tétu T, Lea PJ Dubois F (2011) Improving nitrogen use efficiency in crops for sustainable agriculture. *Sustainability* 3(9):1452–1485

Holmes TP, Aukema JE, Von Holle B, Liebhold A, Sills E (2009) Economic impacts of invasive species in forests past, present, and future. In: Ostfeld RS, Schlesinger WH (eds). *Year in ecology and conservation biology*. Wiley , vol. 1162, pp 18–38

Horwitch M, Stohr EA (2012) Transforming technology management education: value creation-learning in the early twenty-first century. *J Eng Technol Manag* 29(4):489–507 doi: 10.1016/j.jengtecman. 2012.07.003

Huang B, Menne MJ, Boyer T, Freeman E, Gleason BE, Lawrimore JH, Liu C, Rennie JJ, Schreck III CJ, Sun F, Vose R (2020) Uncertainty estimates for sea surface temperature and land surface air temperature in NOAAGlobalTemp version 5. *J Climate* 33(4): 1351–1379. doi: 10.1175/JCLI-D-19-0395.1

Ibanez C, Prat N (2003) The environmental impact of the Spanish National Hydrological Plan on the lower Ebro River and delta. *Int J Water Resour Dev* 19(3):485–500 doi: 10.1080/0790062032 000122934

Immerzeel WW, van Beek LPH, Bierkens MFP (2010) Climate change will affect the Asian Water Towers. *Science* 328(5984):1382–1385 doi: 10.1126/science.1183188

Iqbal MA, Iqbal A, Abbas RN (2018a) Spatio-temporal reconciliation to lessen losses in yield and quality of forage soybean (*Glycine max* L.) in soybean-sorghum intercropping systems. *Bragantia* 77(2): 283–291

Iqbal MA, Iqbal A, Ayub M, Akhtar J (2016) Comparative study on temporal and spatial complementarity and profitability of forage sorghum-soybean intercropping systems. *Custos e Agronegocio* 12(4):2–18

Iqbal MA, Siddiqui MH, Afzal, S, Ahmad Z, Maqsood Q, Khan RD (2018b) Forage productivity of cowpea [*Vigna unguiculata* (L.) Walp] cultivars improves by optimization of spatial arrangements. *Revista Mexicana De Ciencias Pecuarias* 9(2):203–219

Iqbal MA, Bethune, BJ, Iqbal A, Abbas RN, Aslam Z, Khan HZ, Ahmad B (2017) Agro-botanical response of forage sorghum-soybean intercropping systems under atypical spatio-temporal patterns. *Pakistan J Botany* 49(3):987–994

Iqbal MA, Hamid A, Ahmad T, Hussain I, Ali S, Ali A, Ahmad Z (2019a) Forage sorghum-legumes intercropping: effect on growth, yields, nutritional quality and economic returns. *Bragantia* 78(1):82–95

Iqbal MA, Hamid A, Hussain I, Siddiqui MH, Ahmad T, Khaliq A, Ahmad Z (2019b) Competitive indices in cereal and legume mixtures in a South Asian environment. *Agron J* 111(1):242–249

Iqbal MA, Hamid A, Siddiqui MH, Hussain I, Ahmad T, Ishaq S, Ali A (2019c) A meta-analysis of the impact of foliar feeding of micronutrients on productivity and revenue generation of forage crops. *Planta Daninha* 37:e019189237

Jallat F (2004) Reframing business: when the map changes the landscape. *Int J Service Ind Manag* 15(1):122–125 doi: 10.1108/09564230410523367

Jalongo MR (2007) Beyond benchmarks and scores: reasserting the role of motivation and interest in children's academic achievement an ACEI position paper. *Childhood Educ* 83(6):395–407

Jeppesen E, Kronvang B, Meerhoff M, Sondergaard M, Hansen KM, Andersen HE, Olesen JE (2009) Climate change effects on runoff, catchment phosphorus loading and lake ecological state, and potential adaptations. *J Environ Qual* 38(5):1930–1941 doi: 10.2134/jeq2008.0113

Kehrel U, Kai K, Sick N (2016) Why research partnerships fail in the biotechnology sector - an empirical analysis of strategic partnerships. *Int J Innov Technol Manag* 13(1):1–23 1650003.1650001-1650003.1650023

Khan A, Yousafzai AM, Shah N, Muhammad, Ahmad MS, Farooq M, Aziz F, Adnan M, Rizwan M, Jawad SM (2016) Enzymatic profile activity of grass carp (CtenopharyngodonIdella) after exposure to the pollutant named Atrazine (Herbicide). *Pol J Environ Stud* 25(5):2001–2006

Koss DR (1994) The effects of temperature and time of first feeding on egg and fry development in Atlantic salmon, Salmo salar L. PhD Thesis, University of Stirling

Langfield-Smith K (2008) The relations between transactional characteristics, trust and risk in the start-up phase of a collaborative alliance. *Manag Account Res* 19(4):344–364

Laumbach RJ, Kipen HM (2012) Respiratory health effects of air pollution: update on biomass smoke and traffic pollution. *J Allergy Clin Immunol* 129(1):3–11

Liu SX (2009) *Food and agricultural wastewater utilization and treatment.* Wiley Online Library

Ma H, Yang D, Tan SK, Gao B, Hu Q (2010) Impact of climate variability and human activity on streamflow decrease in the Miyun Reservoir catchment. *J Hydrol* 389(3-4): 317–324 doi: 10.1016/j.jhydrol.2010.06.010

Maldonado JK (2012) A new path forward: researching and reflecting on forced displacement and resettlement: report on the International Resettlement Conference: economics, social justice, and ethics in development-caused involuntary migration, the Hague, 4-8 October 2010. *J Refugee Stud* 25(2):193–220

Malhotra Y (2000) Knowledge management for e-business performance: advancing information strategy to 'internet time'. *Inf Strategy Exec J* 16(4):5–16

Meier M (2011) Knowledge management in strategic alliances: a review of empirical evidence. *Int J Manag Rev* 13(1):1–23 doi: 10.1111/j.1468-2370.2010.00287.x

Milfont TL (2012) The interplay between knowledge, perceived efficacy, and concern about global warming and climate change: a one-year longitudinal study. *Risk Anal* 32(6):1003–1020

Mok HF, Williamson VG, Grove, JR, Burry K, Barker SF, Hamilton AJ (2014) Strawberry fields forever? *Urban Agric Dev Countries A Rev* 34:21–43

Nardone A, Ronchi B, Lacetera N, Ranieri MS, Bernabucci U (2010) Effects of climate changes on animal production and sustainability of livestock systems. *Livestock Sci* 130(1-3):0–69

Naseer F, Baig J, Din SU, Nafees MA, Alam A, Hameed N, Adnan M, Arshad M, Khan MA, Romman M, Mian IA, Shah SRA (2016) Impact of water quality on distribution of macro-invertebrate in Jutialnallah, Gilgit-Baltistan, Pakistan. *Int J Biosci* 9(6):451–459.

Navarro JM, Flores P, Garrido C, Martinez V (2006) Changes in the contents of antioxidant compounds in pepper fruits at different ripening stages, as affected by salinity. *Food Chem* 96(1):66–73 doi: 10.1016/j.foodchem.2005.01.057

Nielsen BB (2007) Determining international strategic alliance performance: a multidimensional approach. *Int Bus Rev* 16(3):337–361

Palensky P, Dietrich D(2011) Demand side management: demand response, intelligent energy systems, and smart loads. *IEEE Trans Ind Inform* 7(3):381–388

Pedro-Monzonis M, Solera A, Ferrer J, Estrela T, Paredes-Arquiola J (2015) A review of water scarcity and drought indexes in water resources planning and management. *J Hydrol* 527:482–493 doi: 10.1016/j.jhydrol.2015.05.003

Peters GP, Hertwich EG (2008) CO_2 embodied in international trade with implications for global climate policy. *Environ Sci Technol* 42(5):1401–1407 doi: 10.1021/es072023k

Polade SD, Pierce DW, Cayan DR, Gershunov A, Dettinger MD (2014) The key role of dry days in changing regional climate and precipitation regimes. *Sci Rep* 4:4364. doi: 10.1038/srep04364

Rahman I, Ali S, Rahman I, Adnan M, Ullah H, Basir A, Malik FA, Shah AS, Ibrahim M, Arshad M (2016) Effect of pre-storage seed priming on biochemical changes in okra seed. *Pure Appl Biol* 5(1):165–171

de Richter RK, Ming T, Caillol S, Liu W (2016) Fighting global warming by GHG removal: destroying CFCs and HCFCs in solar-wind power plant hybrids producing renewable energy with no-intermittency. *Int J Greenhouse Gas Control* 49:449–472 doi: 10.1016/j.ijggc.2016.02.027

Roco MC, Mirkin CA, Hersam MC (2011) Nanotechnology research directions for societal needs in 2020: summary of international study. *J Nanopart Res* 13(3): 897–919 doi: 10.1007/s11051-011-0275-5

Rothaermel FT, Deeds DL (2006) Alliance type, alliance experience and alliance management capability in high-technology ventures. *J Bus Ventur* 21(4):429–460 doi: 10.1016/j.jbusvent.2005.02.006

Ryu SJ, Tsukishima T, Onari H (2009) A study on evaluation of demand information-sharing methods in supply chain. *Int J Prod Econ* 120(1):162–175 doi: 10.1016/j.ijpe.2008.07.030

Sahlberg P, Oldroyd D (2010) Pedagogy for economic competitiveness and sustainable development. *Eur J Educ* 45(2):280–299 doi: 10.1111/j.1465-3435.2010.01429.x

Saleem MH, Fahad S, Khan SU, Din M, Ullah A, EL Sabagh A, Hossain A, Llanes A, Liu L (2020) Copper-induced oxidative stress, initiation of antioxidants and phytoremediation potential of flax (Linum usitatissimum L.) seedlings grown under the mixing of two different soils of China. *Environ Sci Pollut Res* 27:5211–5221

Sanne C (2000) Dealing with environmental savings in a dynamical economy - how to stop chasing your tail in the pursuit of sustainability. *Energy Policy* 28(6-7):487–495

Scaringella L, Burtschell F (2017) The challenges of radical innovation in Iran: knowledge transfer and absorptive capacity highlights - Evidence from a joint venture in the construction sector. *Technol Forecast Social Change* 122:151–169 doi: 10.1016/j.techfore.2015.09.013

Schipper L, Pelling M (2006) Disaster risk, climate change and international development: scope for, and challenges to, integration. *Disasters* 30(1):19–38. doi: 10.1111/j.1467-9523.2006.00304.x

Schweitzer L, Noblet J (2018) Water contamination and pollution. In: *Green chemistry*. Elsevier, pp 261–290

Stefanon M, Drobinski P, D'Andrea F, Lebeaupin-Brossier C, Bastin S (2013) Soil moisture-temperature feedbacks at meso-scale during summer heat waves over Western Europe. *Climate Dyn* 42(5-6): 1309–1324

Tague C, Peng H (2013) The sensitivity of forest water use to the timing of precipitation and snowmelt recharge in the California Sierra: implications for a warming climate. *J Geophys Res Biogeosci* 118(2): 875–887 doi: 10.1002/jgrg.20073

Vesilind PA, Peirce JJ, Weiner RF (2013) *Environmental pollution and control.* 3rd edn. Elsevier

Viviroli D, Archer DR, Buytaert W, Fowler HJ, Greenwood GB, Hamlet AF,... Woods R (2011) Climate change and mountain water resources: overview and recommendations for research, management and policy. *Hydrol Earth Syst Sci* 15(2):471–504 doi: 10.5194/hess-15-471-201

Wallace JS (2000) Increasing agricultural water use efficiency to meet future food production. *Agric Ecosyst Environ* 82(1-3):105–119 doi: 10.1016/S0167-8809(00)00220-6

Ward PL (2015) *What really causes global warming?: greenhouse gases or ozone depletion?* Morgan James Publishing, 5 Penn Plaza, 23rd Floor, New York, NY 10001, USA, p 268

Watson JEM, Dudley N, Segan DB, Hockings M (2014) The performance and potential of protected areas. *Nature* 515(7525):67–73

Wu Y, Chen J (2013) Estimating irrigation water demand using an improved method and optimizing reservoir operation for water supply and hydropower generation: a case study of the Xinfengjiang reservoir in southern China. *Agric Water Manag* 116:110–121. doi: 10.1016/j.agwat.2012.10.016

York JG, Venkataraman S (2010) The entrepreneur-environment nexus: uncertainty, innovation, and allocation. *J Bus Ventur* 25(5):449–463. doi: 10.1016/j.jbusvent.2009.07.007

4

Salinity Stress in Cotton: Adverse Effects, Survival Mechanisms and Management Strategies

Ayman EL Sabagh, Mohammad Sohidul Islam, Muhammad Aamir Iqbal, Akbar Hossain,
Muhammad Mubeen, Tasmiya Jabeen, Mirza Waleed, Allah Wasaya,
Muhammad Habib Ur Rahman, Disna Ratnasekera, Muhammad Arif, Ali Raza,
Subhan Danish, Arpna Kumari, Murat Erman, Cetin Karademir, Emine Karademir,
Hüseyin Arslan, Muhammad Ali Raza, and Shah Fahad

CONTENTS

DOI: 10.1201/9781003160717-4

4.1 Introduction

Globally, abiotic stresses such as drought, salinity, heat, soil erosion, waterlogging, heavy metals toxicity, etc. are posing serious challenges to modern farming systems under the changing climate (Iqbal 2020; Iqbal et al. 2015; Iqbal and Iqbal 2015; Raza et al. 2019; EL Sabagh et al. 2020). Among abiotic stresses, salinity is a global problem and is mostly found in arid and semi-arid regions due to the accumulation of free salt (Hussain et al. 2019a; Bange et al. 2004; EL Sabagh et al. 2020b; Hafeez et al. 2021; Hossain et al. 2020a, 2020b; Tariq et al. 2013). Around 6% of the total land area of Asia, the Pacific, and Australia is severely affected by salinity. Soil salinity is the single most pertinent factor, restraining about 20% of agricultural productivity of the cultivated area and one-half of the irrigated area throughout the world (Machado and Serralheiro 2017). Saline soils are defined in terms of electrical conductivity (EC) of saturated paste extract with soils whose EC is greater than 4 dS m^{-1} at 25°C (Simões et al. 2016; Abrol et al. 1988). Various crops respond differently to salinity threshold levels (Table 4.1).

Cotton (*Gossypium hirsutum* L.) is tolerant to soil salinity with a threshold level of 7.7 dS m^{-1} (Zhang et al. 2013; Hou et al. 2009; Gillham et al. 1995; Lu and Zeiger 1994) and thus can give reasonable productivity if grown on saline soil and will eventually result in economic development in high salinity regions. Cotton is a natural fibre crop used globally as fuel, edible oil, and manufacturing garments (Gormus et al. 2017a, 2017b; Omar et al. 2018; Ahmad et al. 2018; Han and Kang 2001). During its life cycle, it faces numerous biotic as well as abiotic stresses, and sustainable cotton production is severely affected due to salt stress. Stages of cotton crop vary in their capability to stand different salt stress levels (Santhosh and Yohan 2019). Due to saline conditions, salt stress at the cellular level interrupts the osmotic, as well as ionic homoeostasis, constrains photosynthesis, and decreases the cellular energy, which ultimately leads to redox imbalance (Gupta et al. 2020). Stress due to soil salinity constrains plant growth, primarily by inhibiting leaf expansion and reducing net photosynthetic rate (Pn). This reduced Pn is primarily due to stomatal closure or feedback from

TABLE 4.1

An overview of Important Crops Regarding Their Salt Tolerance Ability

Crops	The Threshold Level of EC (dSm^{-1})	Status of Crop	Sensitive Stages	Yield Loss due to Salt Stress (%)	References
Cotton	7.7	Tolerant	Germination, emergence of first reproductive branches	50–90	Zhang et al. 2013
Wheat	6	Moderately tolerant	Tillering, plume formation, head, and grain formation	13.4	Mojid and Wyseure (2013)
Sorghum	2.8	Moderately sensitive	Emergence, flowering, and soft dough stages throughout the growing season	50	Banerjee and Roychoudhury (2018)
Rice	2	Moderately Sensitive	Tillering, plume formation, head, and grain formation	30–35	
Sugarcane	1.7	Moderately sensitive	Vegetative growth and sucrose formation	10–100	Simões et al. (2016)
Maize	1.3	Moderately sensitive	Earlier vegetative growth stages	16–22	Katerji et al. (2000)
Onion	1.21	Sensitive	Seedling, vegetative growth, bulb size, and bulb diameter	50	Sta-Baba et al. (2010)

enhanced sucrose levels in source leaves. Salt stress adversely influences cotton physiology from germination to boll development, emergence rate, plant height, and leaf number. Subsequently, reduced photosynthesis affects the metabolism activity of the cell and eventually results in irregular growth of the plant (Zhang et al. 2016).

The current study intends to synthesize numerous adverse effects of salinity on the growth, lint production, and fibre quality parameters of cotton. In addition, the tolerance mechanism to cope with the saline environment has been objectively highlighted. Moreover, various management approaches for boosting lint yield and fibre quality by ameliorating the adverse effects of salinity stress have also been critically assessed.

4.2 Morpho-physiological Mechanisms of Cotton in Response to Abiotic Stresses

4.2.1 Effect of Abiotic Stresses on Cotton

Cotton crop is sensitive to abiotic stresses like drought, temperature, and salinity, especially the incidence of abiotic stresses at the reproductive stage leading to senescence of flowers and ultimately lowering lint yield. Among various effects of abiotic stresses on reproductive parts, spike denaturing is the most common effect caused by ethylene toxicity. It is important to prevent senescence of reproductive parts due to this ethylene toxicity. Using ethylene inhibitor is the best technique to prevent this loss of reproductive parts (Costa and Azevedo 2010).

4.2.2 Effect of Drought Stress on Cotton Yield

Moisture deficit condition is one of the imperative abiotic stresses, which causes a severe decline in crop production potential (EL Sabagh et al. 2020b). Limited soil moisture contents reduce plant capability towards yield and inhibit plant intercellular activities (Muller et al. 2011; Carmo-Silva et al. 2012; Raza et al. 2020). Therefore, drought stress is considered the main stress among abiotic stresses that hampers plant yield and quality (Basal et al. 2009; Schittenhelm 2010). Cotton is an imperative crop susceptible to water shortage (Loka 2012). In many cotton developing areas, moisture deficiency is frequently implemented at the boll formation stage. Cotton is more susceptible to drought stress at the boll formation phase than other phases (Loka and Oosterhuis 2012). Duration and intensity of moisture stress depend on the weather forecast (Giorgi and Lionello 2008). For making sure the long-time spam of cotton production, it is necessary to know better about the consequences of moisture stress on reproductive growth, yield, and fibre quality. Many researchers have explored adverse effects of moisture stress on yield and fibre quality of cotton, whereby reproductive growth of cotton was especially prone to salinity (Rontain et al. 2002; Basal et al. 2009; Lokhande and Reddy 2014). Water stress can lower the boll formation and boll weight with the aid of using various levels of field capacity (Basal et al. 2009; Lokhande and Reddy 2014). The increase and reduction in various yield-related traits are associated with moisture deficiency (Snowden et al. 2013; Rahman et al. 2018).

It is pertinent to mention that all yield-related traits of crops, including cotton, are not equally affected by suboptimal moisture conditions. Lint yield is the main plant character that is reduced under moisture deficit stress, and it usually depends upon the condition of distinctive fruit branches. Boll biomass distribution and seed number per boll on upper fruiting branches are also altered by moisture deficit stress. Fibre length and quality are declined with the decrease in water availability at better fruiting branches. Under identical moisture conditions, the micronaire efficiency at better fruiting branches is better, and drought has no lasting impact on the micronaire efficiency (Wang et al. 2016; Rahman et al. 2019; Rahman et al. 2016a).

4.2.3 Effect of Salt Stress on Cotton Yield

Salt toxicity negatively impacts the physiology from germination to boll development, which properly explains the resistance mechanism (Manikandan et al. 2019; Hafeez et al. 2021). It has been verified that germination rate, plant canopy, and leaves density are the high-quality morphological signs for pointing out salt tolerance. Cotton is highly vulnerable to yield reduction up to 30% in salt-affected soils (Ouda et al. 2014). After the threshold salinity of 7.7 dS m^{-1} in *Gossypium hirsutum*, every 1 dS m^{-1} increase in EC decreased the yield up to 5.2% (Maas 1990). Cotton yield decline under saline conditions was related to fluctuations in dissolved salts in soil solution (Ahmad et al. 2002) and plant leaves (Zhang et al. 2014). A comparable yield reduction is found in fields irrigated with saline water. Cotton yields have reduced by using 9% gypsum because of alternate weather in saline areas. It has been found that cotton yield decreases by 10–20% at EC value of 5 dS m^{-1} and by 27% at 8 dS m$^{-1.}$ (Hebbar et al. 2005). Although the plant is classed as salt-tolerant, this tolerance is not limited; however, it varies with the advancement in growth stages of the plant (Qidar and Shams 1997). Several types of experiments had been carried out to assess the impact of salinity on germination, vegetative growth cycle, and yield of cotton (Guo et al. 2012).

4.2.4 Effect of Heat Stress on Cotton Yield

The temperature necessities of cotton species range from the phenological duration of the genus. This distinction can extrude now no longer simplest with the phenological duration, however additionally with the time spam of temperature, physiological morphology and genetic shape of the plant. Many studies describe the temperature effects on cotton yield (Burke and Wanjura 2009). Cotton is a plant that can tolerate a hot climate; however, an increase in temperature above optimum hampers cotton productivity (Oosterhuis 2002; Rahman et al. 2018). Heat stress reduces the growth period of cotton, which badly impacts many agronomic and morphological characteristics of cotton such as sympodial branches, monopodial branches, seed numbers per boll, lint yield, and fibre quality (Khan et al. 2008; Rahman et al. 2016b), especially for economically important genotypes (Lu and Zeiger 1994). The improvement in growth characteristics of cotton plants, inclusive of shoot development, flowering, and fibre quality, is significantly declined by excessive temperature (Noshair Khan et al. 2014; Saifullah et al. 2015; Farooq et al. 2015; Wajid et al. 2014). Plant height is significantly increased by heat stress (Pace et al. 1999).

Plant size, sympodial branches can maintain the first position in the field due to high-temperature stress decreases under controlled conditions (Zeeshan et al. 2010; Akhtar et al. 2013). Heat stress reduces pollination (Burke et al. 2004) and subsequent fertilization, resulting in fewer seeds per ball (Snider et al. 2010). Oosterhuis (1999) has described an adverse relation between yield and heat stress during boll formation. Heat stress being associated with low yield, the seeds per boll, boll retention rate (Maggio et al. 2000), and high temperature decreased the fruiting rate and number of nodes (Akhtar et al. 2013). Earlier research suggested that warm temperature declined the initiation of flowers (Saifullah et al. 2015; Farooq et al. 2015). Noshair Khan et al. (2014) found that heat stress also reduced the shoot density and flower initiation in cotton crops.

4.2.5 Effect of Heavy Metals on Cotton Yield

Heavy metals such as cadmium (Cd), copper (Cu), lead (Pb), and zinc (Zn) have numerous negative effects on plant growth, cottonseed yield, and fibre quality (Wang et al. 2007). Very few studies have been conducted to assess the impact of heavy metals toxicity on cotton. In contrast, Cd, which is a non-essential element, is highly toxic to cotton plants. It is absorbed by roots, transported across plant tissues, and finally gets accumulated in roots, buds, fruits, and grains (Qian et al. 2009). The adverse results of Cd on plant life can be associated with the interference of Cd with numerous metabolic processes (Li et al. 2012). A massive variety of studies suggested that immoderate cadmium in plant life can cause plant growth retardation, chlorosis, leaf shrinkage, and necrosis (Khan et al. 2016; Xue et al. 2013).

4.2.6 Effect of Light/Radiation Intensity on Cotton Growth and Yield

Light is the primary restricting factor in cotton production. Although the impact of shading on cotton yield has been verified in several studies, light intensity in field-grown cotton causes disturbance in physiological mechanisms, which ultimately causes boll shading. Due to the uncertainty of cotton plant growing patterns, the results of mild low pressure at exceptional levels increase cotton yield (Zhao and Oosterhuis 1998). Of the various electromagnetic radiation spectra (ERS) that reach the Earth's atmosphere, the human eye can best stumble on a small part of this spectrum. This component intently corresponds to the photosynthetic radiation, and with its range of 400–700, photosynthetically active radiation is reasonable for the photosynthetic efficiency and electron transport response in plants and the long-run increase in plant yields. However, radiation other than this range additionally influences the body structure of vegetation. Infrared and purple radiation may be detected with plant pigments (a blue pigment). Photomorphogenesis refers back to the plant pigments, crypto-chromes, and phototrophic proteins in vegetation that increase with mild changes. Ultraviolet rays have a terrible effect on plant growth and yield.

Insufficient light will decrease the intensity of electrons released during photosynthetic activity, as well as a hydrolytic activity, during the photosynthetic process and finally decrease cotton yield. More light intensity may be the reason for photo-inhibition, a photo-damaging responsive mechanism that will also decrease plant yield. Changes in light intensity affect the growth and transpiration rate of *Gossypium hirsutum* plants, and chlorophyll contents are highly susceptible to light intensity. Low light intensity has adverse effects on the growth and yield of cotton, and the lack of light leads to stunt growth (Santosh and Yohan 2019).

Studies have indicated that early growth stages like germination, emergence, and seedling are more affected by salinity compared to later growth stages of cotton. Salinity causes late flowering, fewer fruits set, loss of fruit, and a decreased bulb weight, which eventually affects the yield of seed cotton. It suppressed metabolic enzyme activities, namely: Acidic invertase, alkaline invertase, and sucrose phosphate synthase leading to the deterioration of the salt content of the fibre quality. The detail is shown in the following lines.

4.2.6.1 Germination

Seedlings are greater prone to salinity than the juvenile phase. In the cotton crop, a severe drop in germination percentage was found above 10 dSm^{-1}. Due to salt stress, cotton sprouting, as well as emergence phases, is severely delayed. At 15–20 dSm^{-1}, the emergence of the plant was observed to delay by 4–5 days as compared to a plant grown under normal conditions; because of poor germination, the plant population decreases, resulting in an overall reduction in cotton yield (Sharif et al. 2019). At the germination stage, salt tolerance can be estimated by analyzing germination potential, germination rate, fresh mass, and vigour index. High salinity affects the cell wall and membrane permeability and inhibits the imbibition of water and other ions. Seed germination is a good indicator of salt tolerance in cotton; however, using germination percentage as a sole indicator of salinity tolerance could be misleading.

4.2.6.2 Growth

Salinity significantly affects root morphology and root growth. However, soil salinity was reported to bring change in root characteristics, such as orientation (anisotropy), development of root cell pattern, cell elongation rate, root length, root density, and shape. Salinity stress inhibits the cotton tissue development pattern at the root tip and stimulates the root endodermis and exodermis with delayed primary root growth and poor lateral roots. According to Soares et al. (2018), irrigation with saline water at the vegetative and reproductive phases can be utilized in cotton cultivation with minimum losses in growth, biomass buildup, and fibre quality.

4.2.6.3 Root Growth

Soil salinity normally results in decreased root length, and the secondary roots are also affected, contributing to reduced growth of roots. As the magnitude of salinity in soil increases, there is a gradual decrease in primary roots' growth while the length of secondary roots starts to decrease at mild salt stress. The ionic influx in the roots and its movement toward the shoot is mainly responsible for developing plants. Comparatively decreased sodium ion retention in the roots results in lower root growth than shoot.

4.2.6.4 Shoot Growth

Research on cotton and salt stress confirms that cotton can withstand salt stress, but too much salt will negatively affect yield. However, some researchers have also observed the improvement of plants due to a reduction in salt pressure. This may be due to the sparing effect of nutrients or due to the presence of micronutrients as impurities in the saline growth medium. Salinity reduces the shoot/root ratio, as observed by Babu et al. (1987). Khan et al. (1998b) studied the behaviour of cotton types D-9, MNH-93, NIAB-78, and Ravi under different salt solutions. They found that salt stress-induced by NaCl slowed down bud growth, and the effect of salt stress can be ameliorated by adding Ca^{2+} to the rooting medium. It was found that NIAB-78 was more salt-tolerant among all the strains. This information shows that cotton cultivation is highly dependent on salt stress conditions, but different genetic lines respond differently to salt stress.

4.2.7 Cotton Boll Development and Yield

Cotton bolls and boll size are premier yield contributing parameters. The diminished number of boll and boll size due to the expansion of soil saltiness basically reduces cotton crop yield. The overdue flowering is primarily due to the effect of salt stress in the vegetative growth stage. In cotton, 60–87% synthesized sucrose is transported from subtending leaf of cotton boll (SLCB) to developing bolls, and it plays a prominent role in cotton yield. Under saline conditions, sucrose accumulation in SLCB is not affected, but its efficient transportation towards developing bolls is retarded, resulting in reduced boll weight (Peng et al. 2016a; Rahman et al. 2016b). Cotton crop outcome was decreased to 50% at 17 dS m^{-1}; however, lower levels of salinity had no negative consequence on crop development. On the other hand, early leaf agedness and shedding occurred with the upsurge in the soil salinity. It was observed that irrigation containing high salt concentration in the cotton crop at the budding stage reduced the overall cotton yield up to 90% and declined the cotton fibres' quality during bulb development (Soares et al. 2018).

4.2.8 Fibre Quality

Fibre quality is a hereditary characteristic that is also influenced by an adverse environment. Salinity reduces fibre length, quality, and development while enhances fibre fineness. Fibre length and quality are diminished with the increasing levels of sodium ions. Processes like cellulose deposition, photosynthesis process, and sugar transport are affected by higher EC values, ultimately affecting fibre maturity. Decreased mature fibre is produced due to decreased cross-sectional area, which is a consequence of lower cellulose deposition. Cellulose deposition is responsible for fibre quality, while the metabolism of sucrose is responsible for cellulose synthesis. More than 85% of the cellulose is present in the mature fibre. The synthesis and buildup of mature fibre occur during the thickening of fibre, and next to fibre elongation, its deposition begins due to an upsurge in cellulose synthesis. Cellulose content and sucrose transformation rate are diminished as salt stress increases. In saline conditions, sucrose is available, but cellulose conversion is inhibited due to suppressed processes of metabolic enzymes such as acidic invertase, phosphate synthase and alkaline invertase. Saline water irrigation during boll development results in poor quality cotton. Salt sensitive cultivars show high Cl^- in leaves and get poor quality fibre. Salt tolerant cultivars accumulate high K^+ and Ca^{2+} in leaves. High soil salinity increases the ginning out-turn and micronaire, decreasing the staple length, fibre maturity, and fibre strength. It confirms that salt tolerance is associated with nutrient accumulation in plant tissues. Under salinity, poor fibre quality is obtained, and it may be due to the decreased maturity of individual fibres. Higher salinity decreased fibre length, fibre strength, and fineness. Internationally the main purpose of cotton breeding and its various genetic programs is to improve fibre quality. A summary of the effects of salinity on various growing phases of cotton is shown in Table 4.2.

TABLE 4.2

Effect of Salinity on Various Cotton Growth Stages

Growth Stages	Causes	Effect on Growth	References
Germination	Less and delayed germination due to reduced germination potential, vigour index and fresh mass.	Negative	(Guo-Wei et al. 2011)
Emergence	Delayed emergence results in the non-availability of nutrients. Less plant vigour results in poor crop establishment.	Negative	(Ahmad et al. 2002)
Seedling stage	Reduction in root vigour, root dry weight, shoot dry weight, plant height, leaf expansion, and net photosynthetic rate.	Negative	(Guo-Wei et al. 2011)
Root growth	Reduction in root length and number of secondary roots reduced fresh and dry weight of the roots.	Negative	(Shaheen et al. 2012)
Flowering and Boll development stage	Delayed start of flowering, less fruit-bearing position, fruit shedding, reduced photosynthetic rate, sucrose transformation rate, boll weight and boll size.	Negative	(Peng et al. 2016a)
Fibre quality	Reduced sucrose and cellulose contents, sucrose transformation rate and less activity of metabolic enzymes (alkaline invertase and SPS).	Negative	(Peng et al. 2016b)

4.3 Physiological Mechanisms of Cotton in Response to Salinity Stress

Salinity is secondary stress-causing directly on plant growth by reducing water uptake due to high osmotic potential and absorption of excessive ions creating salt specific ion toxicity (Greenway and Munns 1980; Hafeez et al. 2021). Salinity affects almost all characteristics of plants, including morphological, physiological, and biochemical aspects. It reduces water potential, which causes the closing of stomata, thereby limiting CO_2 fixation (Munns et al. 2006). The most common adverse physiological events are membrane damage, nutrient imbalance, enzymatic inhibition, metabolic dysfunction, photosynthesis inhibition, and hinder other major physiological and biochemical processes leading to retarded growth or death of the plant (Munns and Tester 2008; Alamri et al. 2020). Salinity impaired photosynthesis leads to the destruction of cellular metabolism and ultimately results in abnormal plant growth in cotton (Zhang et al. 2016). The common physiological phenomena in cotton plants are reported as reduction or inhibition of growth and development, reducing major primary metabolic processors such as photosynthesis, respiration, and protein synthesis, etc. (Meloni et al. 2003; Ali et al. 2016). Salinity significantly impacts germination, vegetative growth, and cotton yield with great spatial and temporal variation (Guo et al. 2012). The flowering and boll-forming stages are the key yield-determinant period of upland cotton, which are highly sensitive to salinity (Han and Kang 2001) and thus affect fibre yield directly. The cellular level direct responses to salinity are the imbalance of cellular ions causing ion toxicity, osmotic stress, and reactive oxygen species (ROS) (Khan et al. 2000; Hasanuzzaman et al. 2020). Reports showed that salinity causes the excessive generation of reactive oxygen species (ROS) such as superoxide anion (O^-), hydrogen peroxide (H_2O_2), and the hydroxyl radicals ($OH^•$) in plants (Mittler 2002; Masood et al. 2006), which are harmful or cause deadly damages in cellular environments. Thus, protection against damaging effects from ROS is regulated by the antioxidant enzyme systems of the plants. Superoxide dismutase (SOD), ascorbate peroxidase (APX), and glutathione reductase (GR) are commonly available antioxidants that act against ROS damage (Prochazkova and Wilhelmova 2007; Hasanuzzaman et al. 2020). Proline and glycine betaine are certain osmolytes that accumulate in cellular environments protecting plants from salinity stress (Mitysik et al. 2002; Rontain et al. 2002). The impact of salinity on photosynthesis is a secondary effect that is facilitated by low CO_2 concentrations in the leaves leading to stomatal closure (Meloni et al. 2003; DeRidder and Salvucci 2007).

4.3.1 Growth Responses of Cotton to Salinity Stress

There are two main plant groups (halophyte and glycophyte) based on their level of tolerance to salinity (Cheeseman 2015; Hafeez et al. 2021). Halophytes have a relatively higher tolerance to high salts concentrations (400 mM NaCl), while glycophytes show severely retarded growth under high salinity conditions. Plants evolved certain salt tolerance mechanisms, such as specific salt on the leaf surface for salt secretion. The capacity for secreting salts (by salt glands) is highly correlated with salt tolerance in halophytes (Tester and Davenport 2003; Hafeez et al. 2021). Cotton is reported as a moderately salt-tolerant crop (Ashraf et al. 2010). The varied genotypic differences of response at different growth stages are reported by Ashraf et al. (2008). According to Maas and Hoffman (1977), cotton is classified as the second most salt-tolerant crop, though the degree of tolerance varies among genotypes. The salt threshold level of cotton has been reported as 7.7 dS m^{-1} (Maas and Hoffman 1977; Zhang et al. 2013). Salinity tolerance mechanisms are complex, depending on a wide range of traits controlled by different genetic expressions. Different saline responses were detected in different saline regimes. The variation of growth in cotton genotypes is closely associated with physiological characteristics such as photosynthesis, water use efficiency, and respiration, indicating differential genetic behaviour in biochemical mechanisms (Ziaf et al. 2009).

Cotton responses to salinity are greatly varied with genotypes and growth stages. Seed germination, seedling emergence, and young seedling stage of cotton are more sensitive to salinity (Ahmad et al. 2002), showing delayed germination and seedling emergence (Khorsandi and Anagholi 2009). Reduced plant growth observed in cotton cultivars might be directly associated with a salt-induced reduction in photosynthetic rate and imbalanced water status (Shaheen et al. 2012). Drastic reduction in cotton yield under saline conditions is due to a decline in the number of mature balls as a result of delayed flowering, a high percentage of shedding flowers, and immature balls (Anagholi et al. 2013). Sucrose accumulation in leaf cotton balls is critical for cotton yield, and retarded ball development is noticed under salinity stress due to declined sucrose transportation from leaf cotton balls to developing balls (Peng et al. 2016a). Fibre quality and quantity are economically important parameters of cotton. Quality parameters such as thickness and length of fibre mainly depend on the deposition of cellulose, and hence sucrose metabolism directly involves in cellulose synthesis (Fernandes et al. 2004). Salinity impaired sucrose translocation and thereby cellulose deposition, causing deterioration in fibre quality. A recent report showed that available sucrose is not efficiently converted to cellulose because of altered enzymatic activities under salt stress (Peng et al. 2016b).

4.3.2 Physiological Responses of Cotton to Salinity Stress

Photosynthesis is the key process directly associated with yield. Salinity causes a reduction in chlorophyll contents due to retarded production of enzymes responsible for chlorophyll (Chl) synthesis (Lee et al. 2013), hence directly affecting photosynthesis (Zhang et al. 2014). Further decrease in carotenoids biosynthesis was reported to hinder photosynthetic rate under salt stress (Zhang et al. 2014; Shah et al. 2017). However, carotenoids' degradation rate is lower compared to Chl under salinity stress (Rafique et al. 2003). Cotton genotypes showed a differential response in chlorophyll fluorescence in terms of quantum yield of PSII (Fv/Fm) under high salt concentration (Shaheen et al. 2012). Varietal differences in mitigating oxidative stress by keeping higher antioxidant activities, variations in the Chl a, Chl b, and Chl (a+b) contents, net photosynthetic rate and stomatal conductance, the differential magnitude of activities of SOD, catalase (CAT), APX, and GR were observed in cotton cultivars (Zhang et al. 2014).

Accumulation of salts in different tissues is said to be a salt-resistant mechanism in halophytes. Salt accumulation in tissues varies with genotypes. Certain genotypes showed high concentrations of Na$^+$ in shoot tissues (Gouia et al. 1994), but in some cases, in roots and leaves (Sun and Liu 2001). Salt secretion through salt glands and leaf glandular trichomes in the leaves to lower salt concentration is another mechanism for salinity tolerance in cotton plants (Peng et al. 2016a). Ion compartmentalization is another strategy for salt tolerance in plants. Species-specific Na$^+$ and Cl$^-$ compartmentalization is well understood. In particular, high ion concentration was confined to inactive tissues while showing lower ion distributions in active and more sensitive tissues designating ion compartmentalization in tolerant cotton genotypes (Peng et al. 2016a). The higher peak of Na^+ than Cl^- was previously reported in cotton under salt stress (Gouia et al. 1994). Peng et al. (2016c) reported that Na^+ accumulation is significantly high in root vascular cylinder cells of cotton temporarily and transported to shoot later after 48 hours. Thus, cotton shoots have a considerable role in Na^+ compartmentalization capacity for maintaining an optimal cytosolic K^+/Na^+ ratio. The high selective absorption of K^+ over Na^+ is regulated by *GhSOS1*, *AKT1* and *HAK-5* genes (downregulation), and *GhHKT1* and *GhNXH1* genes (upregulation) in salt-tolerant cotton genotypes through compartmentalization of Na^+ ion into vacuole facilitating K^+ uptake (Wang et al. 2017).

Also, mesophyll cells contributed to extrude Na^+ and H^+ at high salt environments contributing to Na^+ homoeostasis at the cellular level through Na^+/H^+ transporter and H^+-ATPase in the plasma membrane (Peng et al. 2016b). The net Na^+ efflux of the cotton root at salinity was positively correlated with the net influx of H^+ due to an active Na^+/H^+ antiporter across plasma membrane H^+-ATPase (Kong et al. 2012). The osmotic adjustment by the Na^+ compartment in the vacuole, decreasing Na^+ in the cytosol significantly contributed to salt tolerance in plants (Maathuis 2014). Accumulation of osmo-protectants such as amino acids, sugars, glycine betaine, polyols, and polyamines regulate metabolic processors under salt stress stabilizing the plasma membrane and preventing protein denaturation, maintaining cell turgor facilitating water uptake (Naidoo and Naidoo 2001; Rontain et al. 2002). High expressions of the choline monooxygenase (CAM) gene, which is related to the catalytic pathway for conversion of choline into betaine was detected in cotton transgenics in response to salinity tolerance (Zhang et al. 2009). The choline monooxygenase (CMO) gene (*AhCMO*) overexpressed cotton plants showed enhanced photosynthetic capability (Fv/Fm value) with slighter leaf damages upon exposure to salinity (Zhang et al. 2009).

Under saline environments, many genes have been reported as salinity tolerant viz. ZFP (Guo et al. 2009), MKK (Lu et al. 2013), ERF (Johnson et al. 2003), NAC (Meng et al. 2009), DREB (Gao et al. 2009), MPK (Zhang et al. 2011), *GhMT3a* (Xue et al. 2008), tonoplast Na^+/H^+ antiporter (Wu et al. 2004) and 109 WRKY genes (*GarWRKYs*) (Fan et al. 2015). Further, increased expressions of *GhSOS2*, *GhSOS1*, *GhSOS3*, *GhPMA1*, and *GhPMA2* genes have been detected in cotton plants under salinity stress (Peng et al. 2016a). The upregulation of five ABC transporters (Gh_A12G1090, Gh_A10G0583, Gh_A05G1089, and Gh_Sca006272G01), which may have a possible role in salt tolerance, and ABC2 (Gh_A09G1286) in response to Na_2CO_3, NaCl, and NaOH stresses in cotton has been reported (Zhang et al. 2018). Certain aquaporin proteins such as tonoplast intrinsic proteins (TIPs) and plasma membrane intrinsic proteins (PIPs) in cotton were downregulated to conserve water under salt stress (Li et al. 2015). The overexpression of *GhMT3a, GhSOD1,* and *GhCAT1* genes downregulated the ROS and reported high tolerance to salt in transgenic cotton (Lu et al. 2013). Cheng et al. (2018) reported that co-expression of *AtNHX1-TsVP* genes showed greater seed yields in cotton under salinity, contributing Na, K, and Ca ion accumulations in salt-affected cotton leaves adjusting osmotic potential to maintain turgor and carbon fixation.

4.3.3 Salinity Tolerance Mechanism in Cotton

Salinity adversely affects the seed germination process (rate and delay of germination), leading to sub-optimal crop establishment and finally reduces crop yield. It has been claimed that salt stress delays and reduces the germination percentage (GP) in cotton due to osmotic stress and Na+ and Cl's negative effect (Sattar et al. 2010). Seed priming is a pre-sowing treatment used to advance germi-nation through controlled hydration but to a level not sufficient for complete germination. It is used to increase the germination rate and reduce the time required for germination and emergence by alle-viating the adverse impacts of saline stress. Several priming techniques, viz., hydro-priming, osmo-priming, halo-priming, matric priming, thermo-priming, priming with plant growth hormones, bio-priming, and drum priming are used. Priming contributes to metabolic repair in seeds and increases the germination rate, induces osmotic adjustment, and reduces germination time under salt stress (Ashraf and Foolad 2005; Bilqees et al. 2019). Seed priming practice overcomes the adverse effects of salinity stress during germination (Bradford 1986). Hydro-primed cottonseeds increased germination and emergence percentage, plumule and radicle lengths, dry seedling weight, and reduced the total germination period, as well as enhanced stand establishment, showing superior plant height, dry weight, and leaf area (Ahmad et al. 2012). Priming of cottonseed with H_2O_2 improved germination by enhancing the concentration of abscisic acid (ABA), and gibberellic acid (GA) through down-regulation of *NCED5* and *NCED9*, and *GA2ox1* genes, respectively (Kong et al. 2017). In addition, priming with potassium nitrate also remained effective in increasing the germination, emergence vigour, length of plumule and radicle, dry weight of seedling and plant, plant length, and leaf area plant^{-1} (Bozcuk 1981). Moreover, priming with kinetin boosted germination rate (Bozcuk 1981), seed priming with calcium led to higher shoot length (Kent and Läuchli 1985).

   Higher concentrations of salts in soil solution inhibit cotton growth, reproductive development, and lint yield, especially in arid and semi-arid regions of the world. Irrigation water having higher salt concentrations results in their accumulation of Na^+ and Cl^- ions leading to the severe decline in the yield of seed cotton. Boosting cotton tolerance to salt stress can impart yield stability and limit the salinization process through drastic reductions in inputs. To date, targeted breeding for improving salinity tolerance in cotton has remained limited in scope, especially regarding attaining reliable botanical traits and physiological mechanisms against salinity. Among salinity tolerance mechanisms of cotton, reduced accumulation of salt ions and organic solutes synthesis are the leading mechanisms that help the stressed plants in offsetting the effects of salinity (Naidoo and Naidoo 2001). For instance, limited uptake of Na^+ from soil solution is one of the most pertinent traits which imparts salinity tolerance to cotton plants.

4.3.3.1 Production of Organic Solutes

To cope with the adverse effects of salinity, cotton plants accumulate numerous osmoprotectants like glycine betaine, polyamines, amino acids, sugars, proline, and polyols. These osmoprotectants assist plants in metabolic adjustment and ultimately impart salt tolerance. Inside the vacuole, balancing the osmotic potential is done through the compartmentalization of organic ions. Besides, organic solutes offer a shield against protein degradation by imparting stability to membranes and maintaining cell turgor and the output of gradient force, which triggers uptake of water (Rontain et al. 2002). Certain inorganic ions like K^+ and Na^+ and organic osmolytes like proline play key roles in salt tolerance. Plants can protect themselves from salt toxicity by maintaining higher K^+ content and K^+/Na^+ ratio and/or organic solutes.

4.3.3.1.1 Proline

Among organic solutes, proline assists in RUBISCO stabilization and its functionality maintenance under severe salt stress (Ahmad et al. 2002). The concentration of proline in Egyptian cotton, as well as tree cotton, gets multiplied (36 and 121% in roots and leaves, respectively) in the presence of NaCl (Meloni et al. 2001). Contrarily, no significant alterations in proline content under salt stress have also been reported in cotton (Ullah et al. 2016; Golan-Goldhirsh et al. 1990). Although proline performs a vital role in offsetting salinity effects if it remains contained in the cytosol, it cannot adjust osmotic pressure on its own.

4.3.3.1.2 Glycine Betaine

Glycine betaine is an osmoprotectant that plays an essential function and fast accumulates in various plants throughout environment stress (EL Sabagh et al. 2019a, 2019b; Ali et al. 2020). In response to heat and salt stresses, glycine betaine is another prime osmoprotectant synthesized by cotton plants (Quan et al. 2004). Osmotic adjustment is the major role of glycine betaine under salinity stress (Khan et al. 1998b). In transgenic cotton, the CAM gene has been reported to be responsible for more elevated glycine betaine production, which imparts salinity tolerance (Zhang et al. 2009).

4.3.3.2 Membrane and Transport

Cotton plants tend to regulate the flux of ions, which leads to considerably lower retention of salt ions. Contrary to salt-stressed conditions, a high K:Na ratio is usually maintained in the cytosol of plant cells having lower Na concentration and higher K level (Higinbotham 1973). However, disturbance in K:Na ratio, especially in favour of sodium ions that get accumulated in plants roots causing hyperosmotic stress and ion toxicity. Under the saline condition, the ionic imbalance gets disrupted as hydrated forms of Na and K become similar, rendering K influx pathway unable to differentiate between Na and K ions. Thus, Na ion influx passes through the K influx pathway leading to Na toxicity in the cell cytoplasm. To maintain K:Na ratio, cotton plants reduce Na influx in roots, compartmentalize Na in cell cytosol to vacuole, and make Na efflux from root cells. The primary active transport facility is utilized by plant cells to trigger a salt over-sensitive (SOS) pathway to initiate Na efflux, leading to a high K:Na ratio in the cytosol (Zhao et al. 2013). Along with K:Na ratio, the maintenance of a higher ratio of Ca:Na is of utmost importance, enabling plants to cope with salt stress. Salt tolerant genotypes of cotton have higher K absorption compared to Na, which is maintained and downregulated by *GhSOS1, HAK-5,* and *AKT1* along with *GhHKT1* and *GhNXH1* upregulation. Thus, it becomes evident that cotton genotypes salt tolerance may be regarded as directly linked with K:Na compartmentalization and higher uptake of K (Wang et al. 2017).

4.3.3.3 Antioxidants

Salt tolerant genotypes of cotton exhibit higher antioxidants activity, which is directly associated with salt tolerance levels (Noreen and Ashraf 2009). Salinity causes the initiation of ROS, including hydrogen peroxide, hydroxyl radical, and superoxide synthesis (Monsur et al. 2020). The accumulation of ROS under saline conditions causes oxidative stress, nucleic acid mutation, protein denaturation, and destruction, along with severely disturbing the metabolism process (Yassin et al. 2019). Contrary to salt stress conditions, these ROS get neutralized by intercellular antioxidants under normal conditions (Czegeny et al. 2014). To alleviate the drastic effects of salinity-induced oxidative stress, cotton plants utilize enzymatic and non-enzymatic antioxidant systems. The enzymatic system of antioxidants carries CAT, peroxidases (POD), SOD, and glutathione peroxidase, and numerous enzymes of APX, ascorbate–glutathione peroxidase, and GR. Among antioxidant enzymes, SOD is considered vital as it regulates the concentration of O_2^- and H_2O_2. Also, the scavenging ability for ROS has been demonstrated by APX and CAT. However, their scavenging ability multiplies in the presence of SOD (Zhang et al. 2016). While among non-enzymatic antioxidants, carotenoids ascorbic acid, glutathione, and tocopherols are most prominent (Ashraf et al. 2008; Foyer and Noctor 2000). Under saline conditions, cotton exhibits intensified activity of SOD and POD activity, which leads to a higher photosynthetic rate and ultimately enables plants to offset the drastic effects of salinity (Zhang et al. 2014). Ascorbic acid is a vital non-enzymatic antioxidant that tends to multiply its concentration in chloroplast and cytosol under saline conditions. It also plays a role in protecting photosynthetic machinery. Furthermore, it has been revealed that enzymatic antioxidant activity increased during salt stress at the fibre development stage of cotton (Rajguru et al. 1999). Thus, both enzymatic and non-enzymatic antioxidants systems tend to impart tolerance in cotton plants against a saline environment.

4.3.4 Improving the Root Zone Environment

Cotton plants are susceptible to salt stress at the emergence and young seedling stages than any other growth stage. Good stand establishment is a prerequisite for higher yield of cotton, and thus proper field management should be taken for good emergence and stand establishment under saline conditions (Dong et al. 2008). Poor stand establishment and seedling growth are often encountered in saline soils. Salinity stress is originated from the root zone soil environment. Unequal salt distribution in the root zone alleviates salt damage to cotton plants (Dong et al. 2010a). Therefore, any practice that improves at the least part of the root zone environment can alleviate salt damage. Ways to improve root zone environment include reduction of soil salinity by increasing soil moisture and temperature. The entire root system of cotton plants exposed to NaCl significantly reduced the shoot dry weight, leaf area, plant biomass, leaf chlorophyll, photosynthesis, and transpiration, economic and biological yields compared to the NaCl-free control. In contrast, the inhibition effect of salinity on growth and yield was significantly reduced when only half of the root system was exposed to low-salinity (Dong 2012a). Dong et al. (2008) found that furrow-bed seeding induced unequal salt distribution in saline fields and significantly improved the plant growth, yield, and earliness than flat beds seeding in cotton. The improvement of yield and earliness was mainly due to the unequal distribution of salts in the root zone (Dong 2012a). Unequal salt distribution increased water use efficiency, K^+ and K^+/Na^+ ratio, and decreased Na^+ accumulation in leaves (Kong et al. 2012). Recirculation of Na^+ from the shoot to the low-salinity side of roots through the phloem is an important mechanism for reducing Na^+ accumulation in leaves. Enhanced Na^+ efflux from the low-salinity root side induced by the high-salinity root side might also play an important role in decreasing foliar Na^+ accumulation. The Na^+ extrusion in salt-stressed cotton roots is mainly attributed to an active Na^+/H^+ antiport across the plasma membrane (Dong 2012a).

4.4 Agronomic Management

4.4.1 Mulching and Furrow Seeding

Mulching can alleviate salt damage of cotton plants in saline fields. Plastic mulching (covering row with polyethylene film) is common in many countries to conserve soil moisture. It enhanced plant growth and cotton lint yield by increasing soil temperature, water conservation (Dong et al. 2007), controlling weeds, and saline toxicity in the root zone (Dong 2012b). Plastic mulching developed the root system of cotton plants in relatively low-saline soil, reducing the damage to plant growth under salt conditions (Bezborodov et al. 2010; Dong et al. 2010a). It is also reported that the integration of plastic mulching with furrow seeding more effectively enhanced the stand establishment, earliness, yield components, and yield of cotton than mulching or furrow seeding alone (Dong 2012b). A combination of plastic mulching with furrow seeding augmented the unequal salt distribution. It elevated soil temperature and moisture in the root zone soon after seeding, resulting in reduced uptake of Na^+ in roots and leaves, peroxidation of lipids in cotton tissues, and increased Pn. It has been argued that the integration of plastic mulching with furrow bed seeding is a promising cotton production technique in saline areas (Dong et al. 2008; Dong et al. 2010). Generally, plastic mulching is used after sowing, but pre-sowing evaporation in spring enhances the accumulation of salts and moisture loss from the surface layer of saline-affected soils. Early mulching is also a promising cotton production technique in the saline areas, which reduces moisture loss, elevates soil temperature, and controls root zone soil salinity (Dong et al. 2009). Hence, both conventional and early mulching could effectively improve stand establishment, plant growth, earliness, and lint yield of cotton. Early mulching was more beneficial to stand establishment, plant growth, and yield (Dong 2012b).

4.4.2 Late Planting of Short-Season Cotton

Late planting of short-season cotton is a promising system for growing cotton in saline-affected areas. Normal planting of full-season cotton in saline fields in temperate areas is faced with poor stand establishment, late maturity, and increased inputs cost. Dong et al. (2010b) showed that late planting of short-season cotton significantly improved seed emergence and seedling growth due to increased temperature and reduced Na^+ concentration in cotton tissues relative to normal planting in a saline field. The yield from late-planted short-season cotton performed better in earliness and required less input than normal-planted full-season cotton. Therefore, the net returns from late-planting of short-season cotton are greater than those from normal-planting of full-season cotton.

4.4.3 Plant Density Management

Many earlier studies (Francois 1982; Feinerman 1983) have indicated that increased plant density under salinity stress considerably increased the cotton yield. Usually, growth and plant size is reduced in excessive saline soils (Khan et al. 2004), and the smaller plant size left a significant space between plant canopies, which could support additional plants to grow (Francois 1982). It has been reported that increasing plant population enhanced the earliness of cotton (Fowler and Ray 1977). Dong (2012a) concluded that the seed cotton yield is greatly improved by increasing plant density under strong salinity conditions. It is suggested that increased plant density would be necessary for enhancing the yield and earliness of cotton in highly saline fields.

4.4.4 Fertilizer Management

Plant growth, nutrient absorption and metabolism, protein synthesis, and water absorption are greatly altered under salt stress and ultimately reduced the uptake and full utilization of plant nutrients (Ella and Shalaby 1993; Pessarakli 2001). Proper fertilizer management in saline-affected soil increased the yield of cotton (Xin et al. 2010). Application of over-dose fertilizers as in soil, foliar, or both successfully alleviated the salt stress effects and increased cotton yield. Inhibition of cotton growth due to salinity is alleviated by the application of nitrogenous fertilizer (Chen et al. 2009). At the beginning of an irrigation cycle, the application of N enhanced yield and fertilizer use efficiency (Hou et al. 2009). Both soil and foliar application of N in cotton plants under saline condition (12.5 dS m^{-1}) improved plant growth and salinity tolerance by increasing uptake and balanced distribution of N across tissues, K$^+$ and ratio of K$^+$/Na$^+$ (Dong 2012b). These findings may be of significance for nitrogen management for cotton in highly saline soils. The uptake of K is decreased in saline soil. Keshavarz et al. (2004) indicated that the application of K in soil improved the growth and yield of cotton under saline conditions. Foliar application of KCl (@500 mg/L) and NH$_4$NO$_3$ (@500 mg/L) alone or in mixture alleviated the detrimental effects of salt stress by improving vegetative and reproductive parameters in cotton (Jabeen and Ahmad 2009). Cotton plants treated with NPK fertilizer as soil application under saline conditions significantly increased nutrient uptake, decreased Na$^+$ uptake, and produced the highest biomass and lint yield (Xin et al. 2010).

4.4.5 Increasing Soil Moisture and Temperature

It has been declared that seed emergence and seedling growth is improved due to increasing soil moisture and temperature under salinity stress condition (Dong 2012b). Cottonseeds were sown in pots containing different levels of saline soils collected from saline fields in the Yellow River Delta with different moisture content (12, 16, and 18%) by Dong (2012c), and reported that the emergence and seedling growth is increased with increasing soil moisture levels. The higher level of moisture diluted the saline toxicity effect and decreased the osmotic stress and Na$^+$ accumulation in leaves, resulting in improved seedlings' emergence and growth (Dong 2012c). In another study, Dong (2012c) sown cottonseeds in potted saline soils at varying dates to determine soil temperature effects and depicted that soil temperature ranging from 20 to 30°C is beneficial for seedling emergence and growth under salinity stress.

4.4.6 Water Management Strategies

Plant growth of cotton is reduced due to low-quality irrigation water with high EC, sodium adsorption ratio (SAR), residual sodium carbonate (RSC), and pH value, which are the reasons for salinity stress (Murtaza et al. 2006). Therefore, a judicious water management strategy is crucial for successful cotton cultivation against salinity stress. The use of good quality water is crucial for better soil management, plant growth, and productivity under a saline environment because it drains out or leaches down soluble salts from the root zone (Ezeaku et al. 2015). Surface water should be used instead of salty sub-surface water for irrigation in crop fields. In salty sub-surface water, it is advised to use gypsum, which increased grain yield in rice and wheat under salinity (Zaka et al. 2009). In addition to this, the use of gypsum with surface irrigation water helps to reduce ECe, pH value, and SAR of soil at 0–30 cm depths (Mehdi et al. 2013).

4.5 Application of Growth Hormones

Hormonal imbalance is one of the important impacts that salinity has on plants. There are many plant growth regulators (PGRs) being used to induce plant growth and bolls development in cottons, such as aminoethoxy vinyl glycine (AVG), ethephon, and 1-methyl cyclopropane (1-MCP) for ethylene inhibitor under salt stress (Hussain et al. 2019b). Likewise, exogenous applications of ABA, brassinosteroids (BRs), or their analogs (D-31, D-100, etc.) are good options to improve plant performance under salinity stress (Singh et al. 1987; Ashraf et al. 2010).

4.6 Conclusion

Among abiotic stresses, soil salinity has emerged as one of the most serious threats that significantly reduce the yield and quality traits of cotton by altering the physiological and biochemical processes at different growth stages. Cotton plants tend to cope with soil salinity, generally activate different physiological and biochemical mechanisms by changing their morphology, anatomy, water relations, photosynthesis, protein synthesis, primary and secondary metabolism, and biochemical adaptations such as the antioxidative metabolism response. Globally, there is a dire need to increase our understanding of soil salinity's adverse impacts on cotton growth and lint yield along with fibre quality, particularly in arid and semi-arid regions, in order to ensure sustainable production of the fibre under changing climate. The underlying mechanisms of salinity tolerance like biosynthesis of antioxidants and other chemical compounds Different agronomic management practices such as late planting of short-season varieties, seed priming with growth regulators, using mulches, planting density management as per pedo-climatic conditions, and fertilization management have the potential to boost the cotton ability to cope with drastic effects of the saline environment. Therefore, breeders and producers need to understand the influence of salinity on crops for improvement in production, protein, and oil quality (amino and fatty acid) under saline conditions. Modern biotechnological tools such as omics approaches include genomics transcriptomics, metabolomics, proteomics, genome editing tools like CRISPR/Cas system, and speed breeding on a large scale can boost our knowledge and help scientists in developing salt tolerance ready-to-grow cotton genotypes or lines.

Conflicts of Interest

The authors declare that they have no conflicts of interest to report regarding the present study.

REFERENCES

Abrol IP, Yadav JSP, Massoud FI (1988) Salt-affected soils and their management (No. 39). *FAO soils bulletin, Food & Agriculture Organization of United Nations.* p 131

Ahmad S, Khan N, Iqbal MZ, Hussain A, Hassan M (2002) Salt tolerance of cotton *(Gossypium hirsutum* L.). *Asian J Plant Sci* 1:715–719

Ahmad G, Soleymani F, Saadatian B, Pouya M (2012) Effects of seed priming on seed germination and seedling emergence of cotton under salinity stress. *World Appl Sci J* 20:1453–1458

Ahmad Z, Shazia A, Muhammad AI, Rehman HS (2018) Foliar applied potassium enhances fiber quality, water relations and yield of cotton. *J Agric Res* 56(1): 17–25

Akhtar N, Abbas G, Hussain K, Ahmad N, Rashid S (2013) Impact of heat induced sterility on some genotypes of upland cotton under field conditions. *Int J Agric Appl Sci* 5:2–4

Alamri S, Hu Y, Mukherjee S, Aftab T, Fahad S, Raza A, Ahmad M, Siddiqui MH (2020) Silicon-induced postponement of leaf senescence is accompanied by modulation of antioxidative defense and ion homeostasis in mustard (*Brassica juncea*) seedlings exposed to salinity and drought stress. *Plant Physiol Biochem* 157:47–59

Ali A, Imtiyaz M, Adnan M, Arshad M, Rahman IU, Jamal Y, Muhammad H, Saleem N, Rahman Z (2016)

Effect of zinc activities on shoot, root biomass and phosphorus uptake in wheat gynotypes. *American-Eurasian J Agric Environ Sci* 16(1):204–208

Ali S, Abbas Z, Seleiman MF, et al (2020) Glycine betaine accumulation, significance and interests for heavy metal tolerance in plants. *Plants (Basel)* 9(7):896. Published 2020 Jul 15. doi:10.3390/plants9070896

Anagholi A, Esmaeili S, Soltani V, Khaffarian H (2013) Effects of salt stress on the growth and yield of cotton at different stages of development. *National Salinity Research Center – NSRC.* p 47 https://agris.fao.org/agris-search/search.do?recordID=IR2012039023

Ashraf M, Foolad MR (2005) Pre-sowing seed treatment-A shotgun approach to improve germination, plant growth, and crop yield under saline and non-saline conditions. *Adv Agron* 88:223–271

Ashraf M, Ozturk M., Habib-ur-Rehman A (eds) (2008) Salinity and water stress: improving crop efficiency. *Volume 44 of Tasks for Vegetation Science*, Springer Nature, p 244

Ashraf M, Akram NA, Arteca RN, Foolad MR (2010) The physiological, biochemical and molecular roles of brassinosteroids and salicylic acid in plant processes and salt tolerance. *Crit Rev Plant Sci* 29:162–190

Babu VR, Prasad SM, Rao DSK (1987) Evaluation of cotton genotypes for tolerance to saline water irrigation. *Indian J Agron* 32: 229–231.

Banerjee A, Roychoudhury A (2018) Role of beneficial trace elements in salt stress tolerance of plants. In: Hasanuzzaman M, Fujita M, Oku H, Nahar K, Hawrylak-Nowak B (eds) *Plant nutrients and abiotic stress tolerance.* Springer, Singapore. doi: 10.1007/978-981-10-9044-8_16

Bange M, Milroy S, Thongbai P (2004) Growth and yield of cotton in response to waterlogging. *Field Crops Res* 88:129–142. doi: 10.1016/j.fcr.2003.12.002.

Basal H, Dagdelen N, Unay A, Yilmaz E (2009) Effects of deficit drip irrigation ratios on cotton (*Gossypium hirsutum* L.) yield and fibre quality. *J Agron Crop Sci* 195:19–29. doi:10.1111/j.1439-037X.2008.00340.x.

Bezborodov GA, Shadmanov DK, Mirhashimov RT, Yuldashev T, Qureshi AS, Noble AD, Qadir M (2010) Mulching and water quality effects on soil salinity and sodicity dynamics and cotton productivity in Central Asia. *Agri Ecosyst Environ* 138:95–102

Bilqees R, Romman M, Subhan M, Jan S, Parvez R, Adnan M, Ikram M, Iqbal S (2019) Cardiac function profiling and antidiabetic effect of Vigna radiata in Alloxan monohydrate induced diabetic rabbits. *Pure Appl Biol* 9(1):390–395

Bozcuk S (1981) Effects of kinetin and salinity on germination of tomato, barley and cotton seeds. *Ann Bot* 48:81–84

Bradford KJ (1986) Manipulation of seed water relations via osmotic priming to improve germination under stress conditions. *Hort Sci* 21:1105–1112

Burke JJ, Wanjura DF (2009) Plant responses to temperature extremes. In: Stewart JM, Oosterhuis DM, Heitholt JJ, Mauney JR (eds) *Physiology of cotton.* Springer, New York, USA, pp 123–128

Burke J, Velten JJ, Oliver MJ (2004) In vitro analysis of cotton pollen germination. *Agron J* 96:359–368

Carmo-Silva AE, Gore MA, Andrade-Sanchez P, French AN, Hunsaker DJ, Salvucci ME (2012) Decreased CO_2 availability and inactivation of rubisco limit photosynthesis in cotton plants under heat and drought stress in the field. *Environ Exp Bot* 83:1–11 doi:10.1016/j.envexpbot.2012.04.001.

Chen W, Hou Z, Wu L, Liang Y, Wei C (2009) Effects of salinity and nitrogen on cotton growth in arid environment. *Plant Soil* 326:61–73

Cheeseman JM (2015) The evolution of halophytes, glycophytes and crops, and its implications for food security under saline conditions. *New Phytologist* 206(2):557–570

Cheng C, Zhang Y, Chen X, Song J, Guo Z, Li K, Zhang K (2018) Co-expression of AtNHX1 and TsVP improves the salt tolerance of transgenic cotton and increases seed cotton yield in a saline field. *Mol Breed* 38:19

Costa D, Azevedo V (2010) *Abiotic stress effects on physiological, agronomic and molecular parameters of 1-MCP treated cotton plants.* Doctoral dissertation, Texas A&M University

Czegeny G, Wu M, De´r A, Eriksson LA, Strid A, Hideg E (2014) Hydrogen peroxide contributes to the ultraviolet-B (280-315 nm) induced oxidative stress of plant leaves through multiple pathways. *FEBS Lett* 588:2255–2261

DeRidder BP, Salvucci M (2007) Modulation of Rubisco activase gene expression during heat stress in cotton (*Gossyoium hirsutum* L.) involves post-transcriptional mechanisms. *Plant Sci* 172:246–252

Dong H (2012a) Technology and field management for controlling soil salinity effects on cotton. *Aust J Crop Sci* 6(2):333–341

Dong H (2012b) Underlying mechanisms and related techniques of stand establishment of cotton on coastal saline-alkali soil. *Chin J Appl Ecol* 23(2):566–572

Dong H (2012c) Combating salinity stress effects on cotton with agronomic practices. *African J Agril Res* 7(34):4708–4715. doi: 10.5897/AJAR12.501

Dong H, Kong X, Luo Z, Li W, Xin C (2010a) Unequal salt distribution in the root zone increases growth and yield of cotton. *Eur J Agron* 33:285–292

Dong H, Li W, Xin C, Tang W, Zhang D (2010b) Late-planting of short-season cotton in saline fields of the Yellow River Delta. *Crop Sci* 50:292–300

Dong H, Li W, Tang W, Li Z, Zhang D (2007) Enhanced plant growth, development and fiber yield of Bt transgenic cotton by an integration of plastic mulching and seedling transplanting. *Ind Crop Prod* 26:298–306

Dong H, Li W, Tang W, Zhang D (2008) Furrow seeding with plastic mulching increases stand establishment and lint yield of cotton in a saline field. *Agron J* 100:1640–1646

Dong H, Li W, Tang W, Zhang D (2009) Early plastic mulching increases stand establishment and lint yield of cotton in saline fields. *Field Crop Res* 111:269–275

Ella MKA, Shalaby ES (1993) Cotton response to salinity and different potassium-sodium ratio in irrigation water. *J Agron Crop Sci* 170:25–31

EL Sabagh A, Hossain A, Barutçular C, Islam MS, Ratnasekera D, Kumar N, Meena RS, Gharib HS, Saneoka H, da Silva JAT (2019a) Drought and salinity stress management for higher and sustainable canola *(Brassica napus L.)* production: A critical review. *Aust. J. Crop Sci.* 13: 88–97, doi:10.21475/ajcs.19.13.01.p1284

EL Sabagh A et al (2020a) Drought and heat stress in cotton *(Gossypium hirsutum L.)*: consequences and their possible mitigation strategies. In: Hasanuzzaman M (ed) *Agronomic Crops*. Springer, Singapore. doi: 10.1007/978-981-15-0025-1_30

EL Sabagh A, Hossain A, Islam MS, Barutçular C, Ratnasekera D, Kumar N, Meena RS, Gharib HS, Saneoka H, Teixeira da Silva JA (2019b) Sustainable soybean production and abiotic stress management in saline environments: a critical review. *Aust J Crop Sci* 13(2):228–236

EL Sabagh A, Hossain A, Barutçular C, Iqbal M A, Islam MS, Fahad S,... & Erman M (2020b) Consequences of salinity stress on the quality of crops and its mitigation strategies for sustainable crop production: an outlook of arid and semi-arid regions. In: *Environment, climate, plant and vegetation growth*. Springer, Cham, pp 503–533

EL Sabagh, A, Hossain A, Barutçular C, Islam MS, Ahmad Z, Wasaya A, Meena RS (2020c) Adverse effect of drought on quality of major cereal crops: implications and their possible mitigation strategies. In: *Agronomic crops*. Springer, Singapore, pp 635–658. 10.1007/978-981-15-0025-1_31

Ezeaku PI, Ene J, Joshua AS (2015) Application of different reclamation methods on salt affected soils for crop production. *American J Exp Agric* 9:1–11

Fan X et al (2015) Transcriptome-wide identification of salt- responsive members of the WRKY gene family in Gossypium aridum. *PLoS ONE* 10:e0126148

Farooq J, Khalid M, Muhammad WA, Atiq RM, Imran J, Valentin PMI, Nawaz N (2015) High temperature stress in cotton *Gossypium hirsutum* L. *ELBA Bioflux* 7(1):34–44

Feinerman E (1983) Crop density and irrigation with saline water. *West J Agric Econ* 8:134–140

Fernandes F, Arrabaç a M, Carvalho L (2004) Sucrose metabolism in *Lupinus albus* L. Under salt stress. *Biol Plantarum* 48:31

Fowler JL, Ray LL (1977) Response of two cotton genotypes to five equidistant spacing patterns. *Agron J* 69:733–738

Foyer CH, Noctor G (2000) Oxygen processing in photosynthesis: regulation and signalling. *New Phytology* 146, 359– 388

Francois LE (1982) Narrow row cotton *(Gossypium hirsutum* L.) under saline conditions. *Irrig Sci* 3:149–156

Gao SQ, et al (2009) A cotton *(Gossypium hirsutum)* DRE-binding transcription factor gene, *GhDREB*, confers enhanced tolerance to drought, high salt, and freezing stresses in transgenic wheat. *Plant Cell Rep* 28:301–311

Gillham FEM, Bell TM, Arin T, Matthews GA, Rumeur C, Hearn AB (1995) *Cotton production prospects for the next decade*. The World Bank: Washington, DC World Bank Technical Paper

Giorgi F, Lionello P (2008) Climate change projections for the Mediterranean region. *Global Planet Change* 63:90–104. doi:10.1016/j.gloplacha.2007.09.005.

Golan-Goldhirsh A, Hankamer B, Lips S (1990) Hydroxy-proline and proline content of cell walls of sunflower, peanut and cotton grown under salt stress. *Plant Sci* 69:27–32

Gormus O, Akdag A, EL Sabagh A, Islam MS (2017a) Enhancement of productivity and fibre quality by defining ideal defoliation and harvesting timing in cotton. *Rom Agric Res* 34: 225–232

Gormus O, Kurt F, EL Sabagh A (2017b) Impact of defoliation timings and leaf pubescence on yield and fiber quality of cotton. *Journal of Agricultural Science and Technology* 19 (4): 903–915

Gouia H, Ghorbal MH, Touraine B (1994) Effects of NaCl on flows of N and mineral ions and on NO3-reduction rate within whole plants of salt-sensitive bean and salt-tolerant cotton. *Plant Physiol* 105:1409–1418

Greenway H, Munns R (1980) Mechanisms of salt tolerance in non-halophytes. *Ann Rev Plant Physiol* 31:149–190

Guo WX, Mass SJ, Bronson KF (2012) Relationship between cotton yield and soil electrical conductivity, topography, and Landsat imagery. *Preci Agron* 13(2):678–692

Guo YH, Yu YP, Wang D, Wu CA, Yang GD, Huang JG, Zheng CC (2009) GhZFP1, a novel CCCH-type zinc finger protein from cotton, enhances salt stress tolerance and fungal disease resistance in transgenic tobacco by interacting with GZIRD21A and GZIPR5. *New Phytol* 183:62–75

Guo-Wei Z, Hai-Ling L, Lei Z, Bing-Lin C, Zhi-Guo Z (2011) Salt tolerance evaluation of cotton (*Gossypium hirsutum*) at its germinating and seedling stages and selection of related indices. *Yingyong Shengtai Xuebao* 22(8): 2045–2053

Gupta S, Schillaci M, Walker R et al (2020) Alleviation of salinity stress in plants by endophytic plant-fungal symbiosis: current knowledge, perspectives and future directions. *Plant Soil (2020).* doi: 10.1007/s111 04-020-04618-w

Hafeez MB, Raza A, Zahra N, Shaukat K, Akram MZ, Iqbal S, Basra SM (2021) *Gene regulation in halophytes in conferring salt tolerance. In Handbook of bioremediation.* Academic Press, Elsevier, pp 341–370. doi: 10.1016/B978-0-12-819382-2.00022-3

Han HL, Kang FJ (2001) Experiment and study on effect of moisture coerce on cotton producing. *Trans CSAE* 17:37–40

Hasanuzzaman M, Bhuyan MHM, Zulfiqar F, Raza A, Mohsin SM, Mahmud JA, Fujita M, Fotopoulos V (2020) Reactive oxygen species and antioxidant defense in plants under abiotic stress: revisiting the crucial role of a universal defense regulator. *Antioxidants* 9:681

Hebbar KB, Gokulpure PP, Singh VV, Gotmare V, Perumal NK, Singh P (2005) Species and genotypic response of cotton (*Gossypium* species) to salinity. *Ind J Agric Sci* 75(7):441–444

Higinbotham N (1973) Electro-potentials of plant cells. *Ann Rev Plant Physiol* 24:25–46

Hossain A et al (2020a) Nutrient management for improving abiotic stress tolerance in legumes of the family Fabaceae. In: Hasanuzzaman M, Araújo S, Gill S (eds) *The plant family Fabaceae.* Springer, Singapore. doi: 10.1007/978-981-15-4752-2_15

Hossain A, Farooq M, EL Sabagh A, Hasanuzzaman M, Erman M, Islam T (2020b) Morphological, physiobiochemical and molecular adaptability of legumes of Fabaceae to drought stress, with special reference to *Medicago sativa* L. In: Hasanuzzaman M, Araújo S, Gill S (eds) *The plant family Fabaceae.* Springer, Singapore. doi: 10.1007/978-981-15-4752-2_11

Hou Z, Chen W, Li X, Xiu L, Wu L (2009) Effects of salinity and fertigation practice on cotton yield and ^{15}N recovery. *Agric Water Manage* 96:1483–1489

Hussain S, Bai Z, Huang J, Cao X, Zhu L, Zhu C, et al. (2019b) 1-Methylcyclopropene modulates physiological, biochemical, and antioxidant responses of rice to different salt stress levels. *Front Plant Sci* 10:124. doi: 10.3389/fpls.2019.00124

Hussain S, Shaukat M, Ashraf M, Zhu C, Jin Q, Zhang J (2019a) Salinity stress in arid and semi-arid climates: effects and management in field crops. In: *Climate change and agriculture.* IntechOpen. doi: 10.5772/intechopen.87982

Iqbal MA, Qaiser M, Zahoor A, Abdul MS, Sher A, Bilal A (2015) A preliminary study on plant nutrients production as combined fertilizers, consumption patterns and future prospects for Pakistan. *Am Eur J Agric Environ Sci* 15(4):588–594

Iqbal MA, Iqbal A (2015) A study on dwindling agricultural water availability in irrigated plains of Pakistan and drip irrigation as a future life line. *Am-Eur J Agric Environ Sci* 15: 184–190

Iqbal MA (2020) Ensuring food security amid novel coronavirus (COVI-19) pandemic: Global food supplies and Pakistan's perspectives. *Acta Agric Slov* 115: 1–4

Jabeen R, Ahmad R (2009) Alleviation of the adverse effects of salt stress by foliar application of sodium antagonistic essential minerals of cotton (*Gossypium hirsutum* L). *Pak J Bot* 41(5):2199–2208

Manikandan A, Sahu DK, Blaise D, Shukla PK (2019) Cotton response to differential salt stress. *Int J Agric Sci* 11(6):8059–8065

Johnson KL, Jones BJ, Bacic A, Schultz CJ (2003) The fasciclin-like arabinogalactan proteins of Arabidopsis. A multigene family of putative cell adhesion molecules. *Plant Physiol* 133:1911–1925

Katerji N, Van Hoorn JW, Hamdy A, Mastrorilli M (2000) Salt tolerance classification of crops according to soil salinity and to water stress day index. *Agric Water Manag* 43(1):99–109

Kent L, La¨uchli A (1985) Germination and seedling growth of cotton: salinity-calcium interactions. *Plant Cell Environ* 8:155–159

Keshavarz P, Norihoseini M, Malakouti MJ (2004) Effect of soil salinity on K critical level for cotton and its response to sources and rates of K Fertilizers. *IPI regional workshop on potassium and fertigation development in West Asia and North Africa*. Rabat, Morocco, pp 24–28

Khan A, Shah N, Muhammad, Khan MS, Ahmad MS, Farooq M, Adnan M, Jawad SM, Ullah H, Yousafzai AM (2016) Quantitative determination of Lethal Concentration Lc50 of Atrazine on biochemical parameters; Total protein and serum albumin of freshwater fish Grass Carp (*Ctenopharyngodonidella*). *Pol J Environ Stud* 25(4):1–7

Khan AN, Qureshi RH, Ahmad N (2004) Effect of external sodium chloride salinity on ionic composition of leaves of cotton cultivars II. Cell Sap, chloride and osmotic pressure. *Int J Agric Biol* 6:784–785

Khan MA, Ungar IA, Showalter AM (2000) Effects of salinity on growth, water relations and ion accumulation in the subtropical perennial halophyte, Atriplex griffithii var. stocksil. *Ann Bot* 85: 225–232

Khan MA, Ungar IA, Showalter AM, Dewald HD (1998b) NaClinduced accumulation of glycinebetaine in four subtropical halophytes from Pakistan. *Physiol Plant* 102:487–492

Khan AI, Khan IA, Sadaqat HA (2008) Heat tolerance is variable in cotton (*Gossypium hirsutum* L.) and can be exploited for breeding of better yielding cultivars under high temperature regimes. *Pak J Bot* 40(5):2053–2058

Khorsandi F, Anagholi A (2009) Reproductive compensation of cotton after salt stress relief at different growth stages. *J Agron Crop Sci* 195:278–283

Kong X, Luo Z, Dong H, Eneji AE, Li W (2012) Effects of non-uniform root zone salinity on water use, Na+ recirculation, and Na+ and H+ flux in cotton. *J Exp Botany* 63(5):2105–2116

Kong X, Luo Z, Zhang Y, Li W, Dong H (2017) Soaking in H_2O_2 regulates ABA biosynthesis and GA catabolism in germinating cotton seeds under salt stress. *Acta Physiol Plant* 39:2

Lee MH, et al. (2013) Divergences in morphological changes and antioxidant responses in salt-tolerant and salt-sensitive rice seedlings after salt stress. *Plant Physiol Biochem* 70:325–335

Li L, Chen J, He Q, Khan MD, Zhu S (2012) Characterization of physiological traits, yield and fiber quality in three upland cotton cultivars grown under cadmium stress. *Aust J Crop Sci* 6(11):1527–1533

Li W et al (2015) Identification of early salt stress responsive proteins in seedling roots of upland cotton (*Gossypium hirsutum* L.) employing iTRAQ-based proteomic technique. *Front Plant Sci* 6:732

Loka DA (2012) *Effect of water-deficit stress on cotton during reproductive development*. Ph.D. dissertation, University of Arkansas, Fayetteville, USA

Loka DA, Oosterhuis DM (2012) Water stress and reproductive development in cotton. In: Oosterhuis DM, Cothren JT (eds) *Flowering and fruiting in cotton*. The Cotton Foundation, Candova, TN, pp 51–58

Lokhande S, Reddy KR (2014) Reproductive and fiber quality responses of upland cotton to moisture deficiency. *Agron J* 106:1060–1069. doi:10.2134/agronj13.0537

Lu W, Chu X, Li Y, Wang C, Guo X (2013) Cotton GhMKK1 induces the tolerance of salt and drought stress, and mediates defence responses to pathogen infection in transgenic *Nicotiana benthamiana*. *PLoS ONE* 8:e68503

Lu ZM, Zeiger E (1994) Selection for higher yields and heat resistance in pima cotton has caused genetically determined changes in stomatal conductances. *Phyiol Plant* 92:273–278

Sharif I, Aleem S, Farooq J, Rizwan M, Younas A, Sarwar G, Chohan SM (2019) Salinity stress in cotton: effects, mechanism of tolerance and its management strategies. *Physiol Mol Biol Plants* 25, 807–820 (2019). doi: 10.1007/s12298-019-00676-2

Maas EV (1990) Crop salt tolerance. Chapter 13: In Tanji KK (ed) *Agricultural salinity assessment and management. ASCE Manuals and Reports on Engineering No. 71*, American Society of Civil Engineers, New York, USA, pp 262–304

Maas E, Hoffman G (1977) Crop salt tolerance: current assessment. *J Irrig Drain Div* 103:115–134

Maathuis FJM (2014) Sodium in plants, perception, signalling, and regulation of sodium fluxes. *J Exp Bot* 65:849–858

Machado RMA, Serralheiro RP (2017) Soil salinity: effect on vegetable crop growth. Management practices to prevent and mitigate soil salinization. *Horticulturae* 3:30. doi: 10.3390/horticulturae3020030

Masood A, Shah NA, Zeeshan M, Abraham G (2006) Differential response of antioxidant enzymes to salinity stress in two varieties of *Azolla* (*Azolla pinnata* and *Azolla tilieuloides*). *Env Exp Bot* 58: 216–222

Mehdi SM, Sarfraz M, Qureshi MA, Rafa HU, Ilyas M, Javed Q, et al. (2013) Management of high RSC water in salt affected conditions under rice and wheat cropping system. *Int J Sci Eng Res* 4:684–698

Meloni DA, Oliva MA, Ruiz HA, Martinez CA (2001) Contribution of proline and inorganic solutes to osmotic adjustment in cotton under salt stress. *J Plant Nutr* 24:599–612

Meloni DA, Oliva MA, Martinez CA, Cambraia J (2003) Photosynthesis and activity of superoxide dismutase, peroxidase and glutathione reductase in cotton under salt stress. *Environ Exp Bot* 49:69–76

Meng C, Cai C, Zhang T, Guo W (2009) Characterization of six novel NAC genes and their responses to abiotic stresses in *Gossypium hirsutum* L. *Plant Sci* 176:352–359

Mittler R (2002) Oxidative stress, antioxidants and stress tolerance. *Trends Plant Sci* 7:405–410

Mitysik J, Alia B, Mohanty P (2002) Molecular mechanism of quenching of reactive oxygen species by proline under stress in plants. *Current Sci* 82:525–532

Monsur M B, Ivy NA, Haque MM, Hasanuzzaman M, El Sabagh A, Rohman MM (2020) Oxidative stress tolerance mechanism in rice under salinity. *Phyton* 89(3):497

Mojid MA, Wyseure GC (2013) Implications of municipal wastewater irrigation on soil health from a study in Bangladesh. *Soil Use and Management* 29(3):384–396

Muller B, Pantin F, Génard M, Turc O, Freixes S, Piques M, et al. (2011) Water deficits uncouple growth from photosynthesis, increase C content, and modify the relationships between C and growth in sink organs. *J Exp Bot* 62:1715–1729. doi:10.1093/jxb/erq438

Munns R, Tester M (2008) Mechanism of salinity tolerance. *Ann Rev Plant Biol* 59:651–681

Munns R, James RA, Lauchli A (2006) Approaches to increasing the salt tolerance of wheat and other cereals. *J Exp Bot* 5:1025–1043

Murtaza G, Ghafoor A, Qadir M (2006) Irrigation and soil management strategies for using saline-sodic water in a cotton-wheat rotation. *Agric Water Manage* 81:98–114

Naidoo G, Naidoo Y (2001) Effects of salinity and nitrogen on growth, ion relations and proline accumulation in *Triglochin bulbosa*. *Wetl Ecol Manag* 9:491–497

Noreen Z, Ashraf M (2009) Assessment of variation in antioxidative defense system in salt-treated pea (*Pisum sativum*) cultivars and its putative use as salinity tolerance markers. *J Plant Physiol* 166:1764–1774

Noshair Khan MA, Faqir, Khan AA, Rashid A (2014) Measurement of canopy temperature for heat tolerance in upland cotton: variability and its genetic basis. *Pak J Agril Sci* 51(2):359–365

Omar AMA, El-Menshawi M, ElOkkiah S, EL Sabagh A (2018) Foliar application of organic compounds stimulate cotton (*Gossypium hirsutum L.*) to survive late sown condition. *Open Agric* 10(3): 684–697

Oosterhuis DM (2002) Day or night high temperature: a major cause of yield variability. *Cotton Grower* 46(9):8–9

Oosterhuis DM (1999) Yield response to environmental extremes in cotton. In: OosterhuisDM (ed) *Proc. cotton research meeting summary cotton research in progress* Report 193. Arkansas Agric. Exp. Stn., Fayetteville, AR, pp 30–38

Ouda S, El-Din TN, El-Enin RA, El-Baky HA (2014) In Global climate change and its impact on food & energy security in the drylands, Proceedings of the 11th International dryland development conference, 18-21 March 2013, Beijing, China, pp 253–259

Pace PF, Cralle HT, El-Halawany SHM, Cothren JT, Senseman SA (1999) Drought-induced changes in shoot and root growth of young cotton plants. *J Cotton Sci* 3(4):183–187

Peng J, et al. (2016a) Effects of soil salinity on sucrose metabolism in cotton leaves. *PLoS ONE* 11:e0156241

Peng J, et al. (2016b) Effects of soil salinity on sucrose metabolism in cotton fiber. *PLoS ONE* 11:e0156398

Peng Z, He S, Sun J, Pan Z, Gong W, Lu Y, Du X (2016c) Na^+ compartmentalization related to salinity stress tolerance in upland cotton (*Gossypium hirsutum*) seedlings. *Sci Rep* 6:34548. doi: 10.1038/srep34548

Pessarakli M (2001) Physiological responses of Cotton (*Gossypium hirsutum* L.) to salt stress. In: Pessarakli M, Dekker M (eds) *Handbook of plant and crop physiology*. New York, USA: CRC Publisher. pp 681–696

Prochazkova D, Wilhelmova N (2007) Leaf senescence and activities of the antioxidant enzymes. *Biol Plantarum* 51:401–406

Qian H, Li J, Sun L, Chen W, Sheng GD, Liu W, Fu Z (2009) Combined effect of copper and cadmium on Chlorella vulgaris growth and photosynthesis-related gene transcription. *Aquat Toxicol* 94: 56–61

Qidar M, Shams M (1997) Some agronomic and physiological aspects of salt tolerance in cotton (*Gossypium hirsutum* L.). *J Agron Crop Sci* 179:101–106

Quan RD, Shang M, Zhang H, ZhaoY ZJ (2004) Engineering of enhanced glycine betaine synthesis improves drought tolerance in maize. *Plant Biotechnol J* 2:477–486

Raza A, Razzaq A, Mehmood SS, Zou X, Zhang X, Lv Y, Xu J (2019) Impact of climate change on crops adaptation and strategies to tackle its outcome: a review. *Plants* 8:34

Raza A, Ashraf F, Zou X, Zhang X, Tosif H (2020) Plant adaptation and tolerance to environmental stresses: mechanisms and perspectives. In: *Plant ecophysiology and adaptation under climate change: mechanisms and perspectives I.* Springer, pp 117–145. doi: 10.1007/978-981-15-2156-0_5

Rafique A, Salim M, Hussain M, Gelani S (2003) Morpho-physio- logical response of cotton (*Gossypium hirsutum* L.) cultivars to variable edaphic conditions. *Pak J Life Soc Sci* 1:5–8

Rahman Mhur, Ahmad A, Wang X, Wajid A, Nasim W, Hussain M, Ahmad B, w I, Ali Z, Ishaque W, Awais M, Shelia V, Ahmad S, Fahd S, Alam M, Ullah H, Hoogenboom G (2018) Multi-model projections of future climate and climate change impacts uncertainty assessment for cotton production in Pakistan. *Agric For Meteorol* 253–254:94–113. doi: 10.1016/j.agrformet.2018.02.008

Rahman MHU, Ahmad A, Wajid A, Hussain M, Rasul F, Ishaque W, Islam MA, Shelia V, Awais M, Ullah A, Wahid A, Sultana SR, Saud S, Khan S, Fahad S, Hussain S, Nasim W (2019) Application of CSM-CROPGRO-Cotton model for cultivars and optimum planting dates: Evaluation in changing semi-arid climate. *F Crop Res* 238:139–152. doi: 10.1016/j.fcr.2017.07.007

Rahman MH, Ahmad A, Wajid A, Hussain M, Akhtar J, Hoogenboom G (2016a) Estimation of temporal variation resilience in cotton varieties using statistical models. *Pakistan J Agric Sci* 53: 787–807. doi: 10.21162/PAKJAS/16.4549

Rahman UI, Ali S, Rahman LU, Adnan M, Ibrahim M, Saleem N, Irshad M (2016b) Effect of seed priming on growth parameters of Okra (Abelmoschusesculentus L.). *Pure Appl Biol* 5(1):165–171

Rajguru SN, Banks SW, Gossett DR, Lucas MC, Fowler TE, Millhollon EP (1999) Antioxidant response to salt stress during fiber development in cotton ovules. *J Cotton Sci* 3(1):11–18

Maggio A, Reddy MP, Joly RJ (2000) Leaf gas exchange and solute accumulation in the halophyte Salvadora persica grown at moderate salinity. *Environ Exp Botany* 44(1):31–38

Rontain D, Basset G, Hanson AD (2002) Metabolic engineering of osmoprotectant accumulation in plants. *Met Eng* 4:49–56

Saifullah A, Rajput MT, Khan MA, Sial MA, Tahir SS (2015) Screening of cotton (*Gossypium hirsutum* L.) genotypes for heat tolerance. *Pak J Bot* 47(6):2085–2091

Santhosh B, Yohan Y (2019) Abiotic stress responses of cotton: a review. *Int J Chem Studies* 7(6):795–798.

Sattar S, Hussnain T, Javaid A (2010) Effect of NaCl salinity on cotton (*Gossypium arboreum* L.) grown on MS medium and in hydroponic cultures. *J Anim Plant Sci* 20:87–89

Schittenhelm S (2010) Effect of drought stress on yield and quality of maize/sunflower and maize/sorghum intercrops for biogas production. *J Agron Crop Sci* 196:253–261. doi:10.1111/j.1439-037X.201 0.00418.x.

Shah S, Houborg R, McCabe M (2017) Response of chlorophyll, carotenoid and SPAD-502 measurement to salinity and nutrient stress in wheat (*Triticum aestivum* L.). *Agron* 7:61

Shaheen HL, Shahbaz M, Ullah I, Iqbal MZ (2012) Morpho-physiological responses of cotton (*Gossypium hirsutum*) to salt stress. *Int J Agric Biol* 14:980–984

Simões WL, Calgaro M, Coelho DS, Santos DB, Souza MA (2016) Growth of sugar cane varieties under salinity. *Revista Ceres* 63(2):265–271

Singh NK, Larosa PC, Handa AK, Hasegawa PM, Bressan RA (1987) Hormonal regulation of protein synthesis associated with salt tolerance in plant cells. *Proc Natl Acad Sci USA* 84:739–743

Snider JL, Oosterhuis DM, Kawakami EM (2010) Genotypic differences in thermo tolerance are dependent upon pre-stress capacity for antioxidant protection of the photosynthetic apparatus in *Gossypium hirsutum* L. *Physiol Plant* 138:268–277

Snowden C, Ritchie G, Cave J, Keeling W, Rajan N (2013) Multiple irrigation levels affect boll distribution, yield, and fiber micronaire in cotton. *Agron J* 105:1536–1544. doi:10.2134/agronj2013.0084

Soares LA, Fernandes PD, Lima GS, Suassuna JF, Brito ME, Sá FV (2018) Growth and fiber quality of colored cotton under salinity management strategies. *Braz J Agric Environ Eng* 22(5):332–337

Sta-Baba R, Hachicha M, Mansour M, Nahdi H, Kheder MB (2010) Response of onion to salinity. *Afr J Plant Sci Biotechnol* 4:7–12

Sun XF, Liu YL (2001) Test on criteria of evaluating salt tolerance of cotton cultivars. *Acta Agron Sin* 27:794–796

Tariq M, Khan Z, Arif M, Ali K, Waqas M, Naveed K, Ali M, Khan MA, Shafi B, Adnan M (2013) Effect of nitrogen application timings on the seed yield of Brassica cultivars and associated weeds. *Pak J Weed Sci Res* 19(4):493–502

Tester M, Davenport R (2003) Na$^+$ tolerance and Na$^+$ transport in higher plants. *Ann Bot* 91:503–527

Ullah H, Khan WU, Alam M, Khalil IH, Adhikari KH, Shahwar D, Jamal Y, Jan I, Adnan M (2016) Assessment of G × E interaction and heritability for simplification of selection in spring wheat genotypes. *Can J Plant Sci* 96(3):1021–1025

Wajid A, Ahmad A, Hussain M, Rahman MH, Khaliq T, Mubeen M, Rasul F, Bashir U, Awais M, Iqbal J, Sultana SR, Hoogenboom G (2014) Modeling growth, development and seed-cotton yield for varying nitrogen increments and planting dates using DSSAT. Pakistan *J Agric Sci* 51.

Wang N, et al. (2017) Relative contribution of Na$^+$/K$^+$ homeostasis, photochemical efficiency and antioxidant defense system to differential salt tolerance in cotton (*Gossypium hirsutum* L.) cultivars. *Plant Physiol Biochem* 119:121–131

Wang Z, Kong H, Wu D (2007) Acute and chronic copper toxicity to a saltwater cladoceran Moina monogolica Daday. *Arch Environ Con Toxicol* 53:50–56

Wang X, Hou Y, Du M, Xu D, Lu H, Tian X, Li Z (2016) Effect of planting date and plant density on cotton traits as relating to mechanical harvesting in the Yellow River valley region of China. *Field Crops Res* 198:112–121

Wu CA, Yang GD, Meng QW, Zheng CC (2004) The cotton GhNHX1 gene encoding a novel putative tonoplast Na+/H+ antiporter plays an important role in salt stress. *Plant Cell Physiol* 45(5): 600–607

Xin C, Dong H, Luo Z, Tang W, Zhang D, Li W, Kong X (2010) Effects of N, P, and K fertilizer application on cotton growing in saline soil in Yellow River Delta. *Acta Agron Sin* 36(10):1698–1706

Xue ZC, Gao HY, Zhang LT (2013) Effects of cadmium on growth, photosynthetic rate and chlorophyll content in leaves of soybean seedlings. *Biol Plant* 57: 585–590

Xue T, Li X, Zhu W, Wu C, Yang G, Zheng C (2008) Cotton metallothionein GhMT3a, a reactive oxygen species scavenger, increased tolerance against abiotic stress in transgenic tobacco and yeast. *J Exp Bot* 60:339–349

Yassin M, Mekawy AM, EL Sabagh A, Islam MS, Hossain A, Barutcular C, Alharby H, Bamagoos A, Liu L, Ueda A, Saneoka H (2019) Physiological and biochemical responses of two bread wheat *(Triticum aestivum L.)* genotypes grown under salinity stress. *Appl Ecol Environ Res* 17(2):5029–5041

Zaka MA, Schmeisky H, Hussain N, Rafa HU (2009) Utilization of brackish and canal water for reclamation and crop production. In: The International conference on water conservation in Arid regions; ICWCAR 09, Qatar. pp 1–16

Zeeshan A, Khan TM, Noorka IR (2010) Diallel analysis to determine gene action for lint percentage and fibre traits in upland cotton. *Int J Agric Appl Sci* 2(1): 11–14

Zhang B, Chen X, Shu X, Lu N, Wang X, Yang X, Ye W (2018) Transcriptome analysis of *Gossypium hirsutum* L. reveals different mechanisms among NaCl, NaOH and Na2CO3 stress tolerance. *Sci Rep* 8:13527

Zhang F et al (2016) Genetic regulation of salt stress tolerance revealed by RNA-Seq in cotton diploid wild species Gossypium davidsonii. *Sci Rep* 6:20582

Zhang H, Dong H, Li W, Sun Y, Chen S, Kong X (2009) Increased glycine betaine synthesis and salinity tolerance in AhCMO transgenic cotton lines. *Mol Breed* 23:289–298

Zhang L et al (2011) A cotton group C MAP kinase gene, GhMPK2, positively regulates salt and drought tolerance in tobacco. *Plant Mol Biol* 77:17–31

Zhang L, Ma H, Chen T, Pen J, Yu S, et al. (2014) Morphological and physiological responses of cotton (*Gossypium hirsutum* L.) plants to salinity. *PLoS ONE* 9(11):e112807. doi:10.1371/journal.pone.0112807

Zhang L, Zhang G, Wang Y, Zhou Z, Meng Y, Chen B (2013) Effect of soil salinity on physiological characteristics of functional leaves of cotton plants. *J Plant Res* 126:293–304

Zhang L, Ma H, Chen T, Pen J, Yu S, Zhao X (2014). *PLoS One* 9(11): 1–14

Zhao Q, Zhang H, Wang T, Chen S, Dai S (2013) Proteomics-based investigation of salt-responsive mechanisms in plant roots. *J Proteom* 82:230–253

Zhao D, Oosterhuis, AR (1998) Responses of field-grown cotton to shade: an overview. *Reprinted from the Proc Beltwide Cotton Conf* 2:1503–1507

Ziaf K, Amjad M, Pervez MA, Iqbal Q, Rajwana IA, Ayub M (2009) Evaluation of different growth and physiological traits as indices of salt tolerance in hot pepper (*Capsicum annuum* L.). *Pakistan J Bot* 41: 1797–1809

5

Obstacle in Controlling Major Rice Pests in Asia: Insecticide Resistance and the Mechanisms to Confer Insecticide Resistance

Juel Datta, Mahmuda Binte Monsur, Panchali Chakraborty, Swapan Chakrabarty, Shah Fahad, Afsana Hossain, Md Fuad Mondal, Sharif Ahmed, Md Panna Ali, and Ayman EL Sabagh

CONTENTS

5.1 Introduction

Rice (*Oryza sativa* L.) is one of the most consumed cereal crops across the globe (Anis et al. 2019; Gaballah et al. 2021), mostly cultivated in Asian and some African countries, but the production has been hindered by several diseases and insect pests (Seck et al. 2012; Zhang et al. 2019; Jena and Kim 2020; Rahman et al. 2016; Ahmed et al. 2021). The striped stem borer (SSB) *Chilo suppressalis* Walker, and the three rice planthopper species, namely white-backed planthopper (WBPH) *Sogatella furcifera* Horváth, brown planthopper (BPH) *Nilaparvata lugens* Stål, and small brown planthopper (SBPH) *Laodelphax striatellus* Fallén, are considered as destructive rice pests in rice-growing areas (Lou et al. 2013; Du et al. 2015; Mu et al. 2016; Lu et al. 2017b).

The control of rice pests mostly depends on chemical insecticides, and this control method has been considered as the principal component in integrated pest management strategies (Min et al. 2014; Su et al. 2014; Zhang et al. 2014; Mao et al. 2019b; Wei et al. 2019; Li et al. 2020). However, the vast use of chemical compounds led to several vicious impacts on the surrounding environment and enemy insects, where the development of insecticide resistance in the insect is considered as the prime threat in insect pest management (Mao et al. 2019b; Wang et al. 2020). It has been reported that the commonly used insecticides have long been losing their efficacy to control rice pests throughout the year in China, Thailand, Japan, India, Vietnam, Cambodia, and South Korea (Yu et al. 2012; Bao et al. 2016; Matsumura et al. 2018). The SSB is one of the most important rice pests that cause damage to rice in all the growing stages, but the heavy use of insecticides caused low to high level of resistance development against diamides, organophosphates, avermectin, macrolides (Su et al. 2014; Lu et al. 2017b; Yao et al. 2017; Sun et al. 2018b; Wei et al. 2019; Huang et al. 2020). Other than SSB, the three rice hopper species WBPH, BPH, and SBPH, drastically damaging rice plant by sucking plant-sap, and transmitting viral diseases, developed resistance to commonly used insecticides (Jeong et al. 2016; Jin et al. 2017; Fujii et al. 2020; Zhang et al. 2020b). The overall investigation of insecticide resistance of rice pests implies a high risk of resistance development and foresee future control failure by commonly used insecticide classes.

The fast development of insecticide resistance in insect pests has correlated with insecticide-associated mechanisms. Four types of resistance mechanisms in response to insecticides have been reported in insects; however, target-site insensitivity (point mutation) and metabolic resistance mechanisms (increased enzyme activities and overexpressed genes) have been found to be mostly related to conferring insecticide resistance in rice pests (Gong et al. 2014; Lu et al. 2017b; Sun et al. 2018b; Mao et al. 2019b; Wei et al. 2019; Xu et al. 2019; Yang and Lai 2019). Different mechanisms can independently govern the development of insecticide resistance, as well as coexist in resistant pests. For instance, point mutation and over-production of detoxifying enzymes have been reported in diamide-resistant SSB (Sun et al. 2018b). Additionally, the metabolic resistance mechanism is another key mechanism that confers neonicotinoid resistance in rice planthopper species (Bao et al. 2016). A better understanding of distinct mechanisms has always been an essential component of pest management strategies and may help in improving insecticide resistance management (IRM) protocols and forecasting possibilities of future pest outbreaks. In this review, we summarized the condition of insecticide resistance in four rice pests in response to major insecticides. This study further discussed two major insecticide resistance mechanisms to get an obvious idea of their relation to the development of insecticide resistance.

5.2 Monitoring of Insecticide Resistance in Rice Pests

The insecticide resistance action committee (IRAC) defines insecticide resistance as the change in sensitivity of pests to chemical compounds that implies the reduced toxicity of that chemical to control certain insect pests (Dang et al. 2017). Identification and detection of resistance in rice pests showed that insecticide resistance has significantly evolved in recent years, and this phenomenon is common in various types of pests, from pests infesting different crops to mosquito vectors transmitting diseases to humans (Adesanya et al. 2020; Guedes et al. 2020; Hayd et al. 2020; Pan et al. 2020; Spitzer et al. 2020; Zhou et al. 2020). A variety of insecticides have been used to control SSB, WBPH, BPH, and SBPH in the largest part of rice-producing areas, and different biological assays have been used to detect the level of resistance in rice pests such as stem dipping, topical application methods (Table 5.1) (Masaya et al. 2009; Wen et al. 2009; Matsumura et al. 2014; Su et al. 2014; Li et al. 2020). In contrast, the susceptibility of these pests to insecticides is reducing dramatically. A regular investigation of resistance status to the commonly used insecticides in rice pests is vital to know the current situation of resistance and to predict the future resistance emergence. We hereby discuss the recently reported emergence of resistance in individual rice pests in response to several insecticides (Table 5.1).

TABLE 5.1

Recent Insecticide Resistance Status of Rice Pests Against Different Commonly Used Insecticides

Insect	Insecticide Family	Insecticides	Location[a]	Test[b]	Efficacy[c]	Reference
C. suppressalis	Diamide	Chlorantraniliprole, Flubendiamide	China	Bioassay[d]	Resistant	(Yao et al. 2017)
	Avermectin	Abamectin, Emamectin benzoate	China	Bioassay[d]	Susceptible	(Yao et al. 2017)
	Organophosphates	Chlorpyrifos	China	Bioassay[d]	Resistant	(Yao et al. 2017)
	Nereistoxin analogues	Monosultap				
	Organophosphates	Triazophos	China	Bioassay[d]	Susceptible	(Yao et al. 2017)
	Nereistoxin analogues	Monosultap	China	Bioassay[e]	Resistant	(Shuijin et al. 2017)
	Organophosphates	Triazophos				
	Diamide	Chlorantraniliprole	China	Bioassay[d]	Resistant	(Lu et al. 2017b)
	Diamide	Chlorantraniliprole	China	Bioassay[e]	Resistant	(Sun et al. 2018b)
	Avermectin	Abamectin	China	Bioassay[d]	Resistant	(Lu et al. 2017b)
	Carbohydrazide	Methoxyfenozide				
	Diamide	Chlorantraniliprole, Cyantraniliprole	China	Bioassay[d]	Resistant	(Mao et al. 2019b, Huang et al. 2020)
	Diamide	Chlorantraniliprole	China	Bioassay[d]	Resistant	(Wei et al. 2019)
	Organophosphates	Chlorpyrifos, Triazophos	China	Bioassay[f]	Resistant	(Mao et al. 2019b)
	Avermectin	Abamectin				
	Spinosyn	Spinetoram, Spinosad	China	Bioassay[d]	Susceptible	(Mao et al. 2019b)
S. furcifera	Neonicotinoids	Imidacloprid	China, Japan, Vietnam, Cambodia	Bioassay[f]	Susceptible	(Matsumura et al. 2018)
	Pyrazole	BPMC, Fipronil		Bioassay[f]	Resistant	(Matsumura et al. 2018)
	Neonicotinoids	Imidacloprid, Thiamethoxam	China	Bioassay[d]	Susceptible	(Zhang et al. 2017)
		Nitenpyram	China	Bioassay[f]	Susceptible	(Zhang et al. 2017)
		Acetamiprid	China	Bioassay[d]	Resistant	(Zhang et al. 2017)
		Imidacloprid, Thiamethoxam, Dinotefuran, Clothianidin	China	Bioassay[d]	Resistant	(Zhang et al. 2017)
		Imidacloprid, Thiamethoxam	China	Bioassay[d]	Resistant	(Jin et al. 2017)

(Continued)

TABLE 5.1 (Continued)

Recent Insecticide Resistance Status of Rice Pests Against Different Commonly Used Insecticides

Insect	Insecticide Family	Insecticides	Location[a]	Test[b]	Efficacy[c]	Reference
	Pyridines	Pymetrozine				
	Organophosphates	Chlorpyrifos				
	Carbamate	Isoprocarb				
	IGR	Buprofezin				
	Neonicotinoids	Dinotefuran	China	Bioassay[d]	Susceptible	(Mu et al. 2016)
		Cycloxaprid, Triflumezopyrim	China	Bioassay[d]	Susceptible	(Mu et al. 2016; Li et al. 2020)
		Imidacloprid, Thiamethoxam, Dinotefuran, Clothianidin, Nitenpyram, Sufoxaflor	China	Bioassay[d]	Resistant	(Li et al. 2020)
	Pyridines	Pymetrozine				
	Carbamate	Isoprocarb				
N. lugens	Neonicotinoids	Imidacloprid, Thiamethoxam	China, Japan, Vietnam, Cambodia	Bioassay[f]	Resistant	(Matsumura et al. 2018)
		Imidacloprid	Thailand	Bioassay[f]	Resistant	(Garrood et al. 2016)
		Imidacloprid	India, China	Bioassay[f]	Resistant	(Garrood et al. 2016; Yang et al. 2016b)
		Imidacloprid	Vietnam, Philippines	Bioassay[d]	Resistant	(Sanada-Morimura et al. 2019)
		Imidacloprid	China, Thailand, Vietnam	Bioassay[f]	Resistant	(Bao et al. 2016)
		Imidacloprid, Thiamethoxam, Clothianidin	China	Bioassay[d]	Resistant	(Zhang et al. 2016b)
		Imidacloprid, Thiamethoxam, Clothianidin	Japan	Bioassay[g]	Resistant	(Fujii et al. 2020)
		Imidacloprid	Taiwan, China	Bioassay[d]	Resistant	(Yang and Lai 2019)

Class	Insecticide	Location	Method	Status	Reference
Pyrethroid	Permethrin	China	Bioassay[d]	Resistant	(Zhang et al. 2020b)
Neonicotinoids	Imidacloprid, Thiamethoxam, Dinotefuran	China	Bioassay[d]	Resistant	(Liao et al. 2020)
	Imidacloprid	Taiwan, China	Bioassay[d]	Susceptible	(Yang and Lai 2019)
	Thiamethoxam, Clothianidin	China	Bioassay[d]	Resistant	(Liao et al. 2020)
Organophosphates	Chlorpyrifos	China	Bioassay[d]	Resistant	(Mu et al. 2016; Zhang et al. 2016b)
Neonicotinoids	Thiamethoxam, Dinotefuran, Nitenpyram, Clothianidin	China	Bioassay[d]	Susceptible	(Yang et al. 2016b)
Organophosphates	Chlorpyrifos	China, Japan, Vietnam, Cambodia	Bioassay[f]	Susceptible	(Matsumura et al. 2018)
Neonicotinoids	Dinotefuran	China	Bioassay[d]	Susceptible	(Yang et al. 2019)
Pyridines	Pymetrozine	China	Bioassay[d]	Susceptible	(Liao et al. 2019)
Pyrazole	Fipronil	China	Bioassay[d]	Susceptible	(Liao et al. 2017)
Avermectin	Abamectin	China	Bioassay[d]	Susceptible	(Liao et al. 2017)
Neonicotinoids	Sulfoxaflor	China	Bioassay[d]	Resistant	(Liao et al. 2017)
		China	Bioassay[d]	Resistant	(Liao et al. 2017)
		China	Bioassay[d]	Resistant	(Zhang et al. 2020b)
		China	Bioassay[d]	Resistant	(Zhang et al. 2020b)
		China	Bioassay[d]	Resistant	(Zhang et al. 2020b)
	Trifumezopyrim	China	Bioassay[f]	Susceptible	(Liao et al. 2020)
		China	Bioassay[f]	Susceptible	(Zhang et al. 2020b)
		China	Bioassay[f]	Susceptible	(Liao et al. 2020)
		China	Bioassay[f]	Susceptible	(Zhang et al. 2020b)

(Continued)

TABLE 5.1 (Continued)

Recent Insecticide Resistance Status of Rice Pests Against Different Commonly Used Insecticides

Insect	Insecticide Family	Insecticides	Location[a]	Test[b]	Efficacy[c]	Reference
L. striatellus	Pyrazole	Fipronil	South Korea	Bioassay[f]	Susceptible	(Jeong et al. 2016)
	Neonicotinoids	Imidacloprid, Thiamethoxam, Dinotefuran	South Korea	Bioassay[f]	Resistant	(Jeong et al. 2016)
	Carbamate	Carbofuran				
	Pyrethroid	Etofenprox				
	Neonicotinoids	Dinotefuran	China	Bioassay[d]	Susceptible	(Mu et al. 2016)

Notes:
[a] Collection site of insect field populations
[b] Resistance detection test
[c] Insecticide efficacy
[d] Stem dipping method
[e] Artificial diet incorporation
[f] Topical application method
[g] Topical ingestion

5.2.1 Striped Stem Borer (SSB)

A significant number of reports have pointed out the resistance problem in SSB and revealed that the pest developed a variety of resistance levels against chlorantraniliprole and other diamides (Yao et al. 2017). The rapid development of resistance in SSB to different insecticides in distinct rice-growing areas denotes the importance of regular monitoring of resistance status. In recent years, SSB that collected from different provinces of China showed the trend of moderate resistance ratios to flubendiamide and chlorantraniliprole (Yao et al. 2017). Both bioassay dipping and artificial diet incorporation methods revealed that SSB exhibited moderate to high levels of resistance to chlorantraniliprole and cyantraniliprole (Lu et al. 2017b; Sun et al. 2018b; Mao et al. 2019b; Wei et al. 2019; Huang et al. 2020). This implies that diamides lose their efficacy to control SSB and pose a risk of significant control failure in the near future. Organophosphates also showed reduced toxicity in controlling SSB collected from distinct parts of China from 2012 to 2018 (Shuijin et al. 2017; Yao et al. 2017; Mao et al. 2019b; Huang et al. 2020). Although SSB was found susceptible to triazophos in 2014, later, it showed a range of resistance levels to chlorpyrifos and triazophos (Yao et al. 2017; Mao et al. 2019b; Huang et al. 2020). This variation may have a different explanation, like the difference in field populations and the percentage of insecticide concentration during the experiment. Resistance ratio to monosultap varies from low to moderate in field populations of SSB collected from different Chinese provinces (Shuijin et al. 2017; Yao et al. 2017). Survey of insecticide resistance have found SSB susceptible to abamectin and emamectin benzoate; however, most recently, moderate to high level of resistance has been evolved to abamectin and methoxyfenozide (Lu et al. 2017b; Yao et al. 2017; Mao et al. 2019b). It has been proved that, although some of the SSB field populations showed susceptibility to a few insecticides, most of the SSB field populations evolved resistance to many commonly used insecticides.

5.2.2 White-Backed Planthopper (WBPH)

The susceptibility test of *S. furcifera* through topical application method discovered its susceptibility to imidacloprid, a first-generation neonicotinoid, while higher levels of resistance to O-sec-butylphenyl methylcarbamate (BPMC) and fipronil was found in WBPH collected from Vietnam, China, Japan, and Cambodia (Matsumura et al. 2018). Other researches also revealed that WBPH showed susceptibility to imidacloprid, thiamethoxam, nitenpyram, dinotefuran, cycloxaprid, and triflumezopyrim (Mu et al. 2016; Zhang et al. 2017; Li et al. 2020). This implies the efficacy of insecticides to control WBPH is optimistic in several rice-growing areas. However, other data showed WBPH field populations developed resistance to a number of insecticides. Chinese WBPH showed a moderate level of resistance to acetamiprid, chlorpyrifos, isoprocarb, buprofezin, nitenpyram, and pymetrozine; and moderate to high level of resistance to imidacloprid, thiamethoxam, dinotefuran, sulfoxaflor, and clothianidin (Jin et al. 2017; Zhang et al. 2017; Li et al. 2020). The reason of resistance development in WBPH may be as a result of the continuous application of common insecticides to control WBPH and other rice hopper species. Although the level of resistance is fluctuating, the overall situation of resistance status in WBPH is posing a risk of control failure by chemical compounds.

5.2.3 Brown Planthopper (BPH)

Regardless of distinct resistance detection test methods, BPH showed an increasing trend of resistance levels to imidacloprid in China, Japan, India, Thailand, Cambodia, the Philippines, and Vietnam (Bao et al. 2016; Garrood et al. 2016; Yang et al. 2016b; Zhang et al. 2016b, 2020b; Matsumura et al. 2018; Sanada-Morimura et al. 2019; Yang and Lai 2019; Fujii et al. 2020; Liao et al. 2020). This confirmed that the efficacy of these common neonicotinoids is reduced and significantly failed to control BPH in several locations for an interminable period. Additionally, thiamethoxam is another important insecticide of neonicotinoids, which is commonly used to control BPH in different places, but BPH developed significant resistance to this insecticide (Zhang et al. 2016b, 2020b; Matsumura et al. 2018; Fujii et al. 2020; Liao et al. 2020). Surprisingly, Yang and Lai (2019) reported susceptible BPH populations found in Taiwan in response to thiamethoxam, clothianidin, and chlorpyrifos (Yang and Lai 2019). On the other hand, the reduced toxicity of dinotefuran, clothianidin, chlorpyrifos, and nitenpyram was reported in several Chinese provinces and different locations of Japan (Mu et al. 2016; Zhang et al. 2016b, 2020b; Yang and Lai 2019; Liao et al. 2020). The anomalous resistance levels indicated that the toxicities of insecticides varied from one place to another but still pose a higher risk due to their migratory behaviour. Moreover, BPH showed susceptibility to pymetrozine, fipronil, abamectin, sulfoxaflor, and triflumezopyrim (Yang et al. 2016b; Liao et al. 2017, 2019, 2020; Matsumura et al. 2018; Zhang et al. 2020b). In recent years, BPH field populations collected from China showed continuity in resistance development against sulfoxaflor (Liao et al. 2017; Zhang et al. 2020b). These data represent how simultaneously BPH developed resistance to different insecticides in several locations cultivating rice. Overall, BPH already developed a higher level of resistance to most insecticide classes belong to neonicotinoids, which suggests that the mode of action needs to be changed to control BPH more successfully. Although some insecticides are still capable of controlling BPH, they need to be used in rotation as they may fail to control as the development of resistance in BPH is fast.

5.2.4 Small Brown Planthopper (SBPH)

Compared to WBPH and BPH, very few studies have been carried out to monitor the insecticide resistance status in SBPH. Bioassay test revealed that although SBPH has susceptibility to fipronil, it showed resistance to imidacloprid, thiamethoxam, dinotefuran, carbofuran, and etofenprox (Jeong et al. 2016). On the contrary, according to another report, SBPH remained susceptible to dinotefuran (Mu et al. 2016). The toxicity of major insecticide classes has been reduced to control rice planthoppers, which would be a great crisis in places where insect pest populations significantly emerged in recent years.

5.2.5 Other Rice Pests

Rice production is becoming more challenging due to the impacts of climate change, plant diseases, and pest infestations in several rice-producing countries. Other than SSB and rice hopper species, rice production has been affected by rice leaf folders, *Cnaphalocrocis medinalis* Güenée, yellow stem borer, *Tryporyza incertulas* Walker, the black rice bug, *Scotinophara lurida* Burmeister, rice water weevil, *Lissorhoptrus oryzophilus Kuschel in rice-growing regions of India, China, Korea, Japan, and South-American countries* (Cho et al. 2008; Wang et al. 2010; Kwon et al. 2012; Han et al. 2015; Giantsis et al. 2017). *Though these pest infestations are not globally widespread rather than concentrate in some specific regions, additional investigations are needed to know the insecticide resistance levels in these pests.*

5.3 Insecticide Resistance Mechanisms

Insecticide classes are divided based on their mode of action (MoA). For instances, chlorantraniliprole belong to diamides, which inhibit ryanodine receptor (RyR); in contrast, imidacloprid of neonicotinoids affect nicotinic acetylcholine receptor (nAChR) of the central nervous system in insect pests, including rice hopper species (Tomizawa and Casida 2005; Gorman et al. 2008; Jeschke and Nauen 2008; Shao et al. 2013; Guo et al. 2014; Huang et al. 2020; Sparks et al. 2020). However, distinct resistance mechanisms underlie protecting target sites from the notorious effects of insecticides, and hence, confer insecticide resistance in rice pests (Figure 5.1). The importance of resistance mechanisms is ubiquitous in many aspects of pest management strategies. Therefore, we have further discussed the most reported resistance mechanisms in rice pests to overlook the underlying correlation between resistance mechanisms and the development of resistance.

5.3.1 Target-Site Insensitivity

Target-site resistance, for instance, ryanodine receptor (RyR) mutation, has been associated with significant levels of diamides resistance in several insect pests, as well as in *C. suppressalis* (Guo et al. 2014;

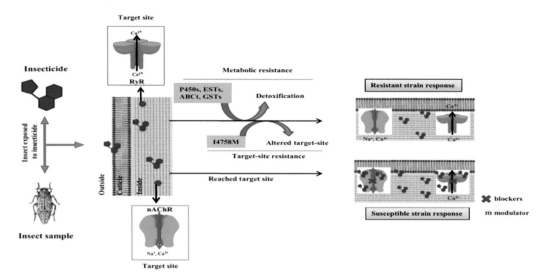

FIGURE 5.1 Two major insecticide resistance mechanisms in rice pests.

Steinbach et al. 2015; Huang et al. 2020; Zuo et al. 2020). Molecular study through RT-qPCR revealed that G4910E mutation contributes to conferring higher levels of diamide resistance in stem borer (Yao et al. 2017). Moreover, the target-site I4758M mutation has been identified in rice stem borer resistant to chlorantraniliprole, a common insecticide class of diamide group (Sun et al. 2018b; Wei et al. 2019). Hence, it implies that these mutations are associated with RyR target-site resistance and have been significantly influenced the resistance ratios of SSB to diamides. On the other hand, target-site mutations such as Y667D/C, I4758M and G4910E have been identified in chlorantraniliprole-resistant SSB (Huang et al. 2020). Similarly, I4758M and two novels Y667D and Y667C mutations have been reported in SSB that showed significant levels of resistance against diamide insecticides (Sun et al. 2018b; Hussain et al. 2016). Although I4758M mutation has been reported in all tested stem borer resistant to chlorantraniliprole, another G4910E mutation was not detected in all field populations tested; this implies that mutation has evolved independently and causes resistance development in striped rice stem borer, SSB (Wei et al. 2019; Huang et al. 2020). Further study in target-site mutation in rice pests is needed, as this type of mechanism is found to be promising in other crop pests. These results may contribute to further understanding of target-site mutations at the RyR transmembrane domains and diamide resistance in SSB.

5.3.2 Metabolic Detoxification of Xenobiotics

The metabolic resistance mechanism such as the metabolism of detoxifying enzymes is a major insecticide resistance mechanism that has been contributed to the emergence of insecticide resistance in rice insect pests (Scott 1999; Wen et al. 2009; Elzaki et al. 2017a; Xu et al. 2020; Zhang et al. 2020a; Zhou et al. 2020). A better understanding of metabolic resistance mechanisms is important to improve pest management strategies.

5.3.2.1 Metabolic Enzyme Activities

The involvement of metabolic enzymes in resistance development was measured through synergistic studies and biochemical enzymatic assays. Enhanced activities of the enzyme are much related to the higher metabolism of insecticides, which ultimately caused the evolution of insecticide resistance in rice pests. The increased enzyme activities of cytochrome P450 monooxygenases (P450), glutathione *S*-transferases (GST), uridine diphosphate glycosyltransferase (UGT), cytochrome P450-linked p-nitro O-demethylase (PNOD), MOF, and esterase/carboxylesterase (EST/CarE) have been detected in resistant SSB, WBPH and BPH field populations in response to distinct insecticide classes (Lu et al. 2017b; Yao et al. 2017; Sun et al. 2018b; Mao et al. 2019b; Xu et al. 2019; Zhao et al. 2019; Khan et al. 2016). Additionally, the antioxidant enzymes play an important role in ROS detoxification in cells (Monsur et al. 2020); the increase of the activities of three different protective enzymes, catalase (CAT), peroxidase (POD), and superoxide dismutase (SOD), have also been reported in WBPH resistant strain (Zhou et al. 2018). Further studies are needed to get a clearer picture of the specific resistance mechanism underlying the resistance development in rice pests.

5.3.2.2 Metabolic Overexpressed Genes

Functional characterization of resistance-related genes through high-throughput sequencing, RT-qPCR, and dsRNA-mediated RNA interference (RNAi) confirmed the contribution of multiple overexpressed genes encoded metabolic enzymes in the development of insecticide resistance in rice pests. Insect ATP-Binding Cassette (ABC) transporters are reported to be associated with insecticide resistance (Zhou et al. 2018; Wu et al. 2019). Three ABC transporter genes, *CsABCC8, CsABCG1C,* and *CsABCH1*, were significantly up-regulated in diamides resistant SSB (Meng et al. 2020). This implies the importance of transporter systems in the evolution of insecticide resistance in SSB. Similarly, multiple ABC transporter genes were overexpressed in WBPH strains showed resistance to neonicotinoids, insect growth regulators, and avermectin (Table 5.2) (Yang et al. 2016a; Zhou et al. 2018). Little research has been carried out on the metabolic UGTs; however, the functional characterization of *CsUGT40AL1, CsUGT33AG3* confirmed their role in the emergence of diamide resistance in SSB (Zhao et al. 2019). Moreover, CYP450s contribution in conferring insecticide resistance has been widely found in several insect pests (Daborn et al. 2002; Chen et al. 2019a; Adnan et al. 2018). The overexpression of two P450 genes, *CYP6ER1* and *CYP6AY1*, play both an individual and combined role in conferring a higher level of resistance against neonicotinoids in BPH (Bao et al. 2016; Garrood et al. 2016; Pang et al. 2016; Zhang et al. 2016a, 2016c; Sun et al. 2018a; Jin et al. 2019; Liao et al. 2019, 2020; Mao et al. 2019a).

It has been proved in several studies that the involvement of single and multiple P450 genes enhanced the metabolism of insecticides in resistant SSB, WBPH, SBPH, and BPH (Zhang et al. 2016c; Elzaki et al. 2017b; Wang et al. 2019; Xu et al. 2019). The P450s contribution in rice pests to metabolize insecticides is also found in relation with diamides, insect growth regulator, organophosphates, and pyrethroids (Table 5.2) (Elzaki et al. 2017a, 2017b; Sun et al. 2017; Ali et al. 2019; Xu et al. 2019; Liao et al. 2020). Zhang et al. (2018) investigated the role of nAChR genes side by side P450 genes in neonicotinoid-resistant BPH and found a higher level of expression of a number of P450s and three nAChR genes (Zhang et al. 2018). Although these efforts gave fundamentals of metabolic mechanisms that contributed to resistance development in rice pests, further studies are necessary for a better understanding of specific insecticide resistance mechanisms. Induction of *CarE* gene expression also found to be linked with resistance development, such as overexpression of *NlCarE* was found in BPH showing higher levels of resistance to organophosphates (Lu et al. 2017a). The findings revealed that single and multiple resistance mechanisms govern a range of insecticide resistance ratios in rice pests.

5.4 Conclusion and Future Challenges

The monitoring of insecticide resistance status in rice pests showed control failure and reduced toxicity of different insecticide classes. Although regular investigations of resistance development possess enormous importance, there is still a lack of regular and prompt monitoring of resistance in rice-growing areas. The insecticide resistance may spread in wide areas cultivating rice and may cause heavy yield losses-even in places where resistance has never been reported as most of the pests can migrate a long distance. Collecting and compiling data from the areas producing rice, using modern tools and techniques in monitoring such as artificial intelligence, geographical information system, and improving resistance detection test might be the resolutions that can prompt research on the evolution of insecticide resistance. Detection of insecticide resistance and reporting information of the actual condition of the resistance ratio of a particular region is considered as the

TABLE 5.2

Resistance Mechanisms of Rice Pests in Response to Different Insecticides

Species	Methods	Metabolic Enzyme	Key Findings — Mutation, Overexpressed Genes	Insecticide Group	Reference
SSB	Enzyme assay	P450, GST, EST	–	Diamide	(Mao et al. 2019b)
	RT-qPCR	–	CsABCC8, CsABCG1C, CsABCH1	Diamide	(Meng et al. 2020)
	RT-qPCR	EST	G4910E mutation	Diamide	(Yao et al. 2017)
	RT-qPCR	–	I4758M mutation	Diamide	(Wei et al. 2019)
	Enzyme assay	EST	–	Diamide	(Lu et al. 2017b)
	Synergism, enzyme assay, RT-qPCR	P450, EST	I4758M, Y4667C, Y4667D mutation	Diamide	(Sun et al. 2018b)
	Genotyping	–	Y4667D/C, I4758M, G4910E mutation	Diamide	(Huang et al. 2020)
	Synergism, enzyme assay, RT-qPCR, RNAi	P450	CYP6CV5, CYP9A68, CYP321F3, CYP324A12	Diamide	(Xu et al. 2019)
	Enzyme assay	CarE, GST, PNOD	–	Pyrazole	(Xiao et al. 2017)
	Synergism, RT-qPCR, RNAi	UGT	CsUGT40AL1, CsUGT33AG3	Diamide	(Zhao et al. 2019)
WBPH	Enzyme assay, RT-qPCR, RACE	CarE, GST, MFO, CAT,[a] POD,[a] SOD[a]	sfABCG9	Neonicotinoids	(Zhou et al. 2018)
			sfABCG2, sfABCG12, sfABCG13	Neonicotinoids, Insect growth regulator	
			sfABCG1, sfABCG3, sfABCG4, sfABCG5, sfABCG6, sfABCG8, sfABCG10, sfABCG11, sfABCG14	Neonicotinoids, Insect growth regulator, Avermectin	
	Enzyme assay	EST	–	Neonicotinoids	(Li et al. 2020)
	Enzyme assay	CarE, GST, MFO	–	Insect growth regulator	(Chang et al. 2015)
	RNA-Seq, RT-qPCR	–	P450, GST, CCE and ABC transporter genes	Neonicotinoids	(Yang et al. 2016a)
	Synergism, RNA-Seq, RT-qPCR, RNAi	P450	CYP6FD1, CYP6FD2	Neonicotinoids	(Wang et al. 2019)
	Synergism, enzyme assay, RT-qPCR	P450, GST	CYP302A1, CYP304H1	Insect growth regulator	(Ali et al. 2019)

(Continued)

TABLE 5.2 (Continued)

Resistance Mechanisms of Rice Pests in Response to Different Insecticides

Species	Methods	Key Findings		Insecticide Group	Reference
		Metabolic Enzyme	Mutation, Overexpressed Genes		
SBPH	RT-qPCR	–	*CYP353D1v2, CYP439A1v3, CYP306A2v2, CYP6FU1, CYP4C72, CYP439A1v3*	Organophosphates, Pyrethroid, Neonicotinoids	(Elzaki et al. 2017b)
	qPCR	–	*CYPs*[16]*, Lsα1, Lsβ1, Lsβ3*	Neonicotinoids	(Zhang et al. 2018)
	–	–	*CYP353D1v2*	Insect growth regulator	(Elzaki et al. 2017a)
BPH	RT-qPCR	–	*CYP6ER1*	Neonicotinoids	(Garrod et al. 2016)
	Synergism, enzyme assay, RT-qPCR, RNAi	P450	*CYP6ER1, CYP6AY1*	Neonicotinoids	(Bao et al. 2016)
	Enzyme assay, RT-qPCR, RNAi	P450	*CYP6ER1*	Neonicotinoids	(Sun et al. 2018a)
	Synergism, RT-qPCR, RNAi	–	*CYP6ER1, CYP6AY1, CYP6CE1, CYP6CW1*	Neonicotinoids	(Zhang et al. 2016c)
	Synergism, RT-qPCR	–	*CYP6ER1, CYP6AY1, CYP6CS1*	Neonicotinoids	(Zhang et al. 2016a)
	RT-qPCR, RNAi	–	*CYP6ER1*	Neonicotinoids	(Pang et al. 2016)
	Synergism, enzyme assay, RT-qPCR, RNAi	P450, CarE	*CYP6ER1*	Neonicotinoids	(Liao et al. 2019; Mao et al. 2019a)
	Synergism, enzyme assay, RT-qPCR, RNAi	CarE	*NlCarE*	Organophosphates	(Lu et al. 2017a)
	Synergism, enzyme assay, RT-qPCR, RNAi	P450	*CYP6ER1*	Neonicotinoids	(Jin et al. 2019)
	Synergism, RT-qPCR, RNAi	–	*CYP6FU1, CYP425A1, CYP6AY1*	Pyrethroid	(Sun et al. 2017)
	RT-qPCR	–	*CYP6AY1, CYP6ER1*	Neonicotinoids, Insect growth regulator	(Liao et al. 2020)

Notes:
[a] protective enzyme
CCE carboxyl
MFO mixed-function oxidase

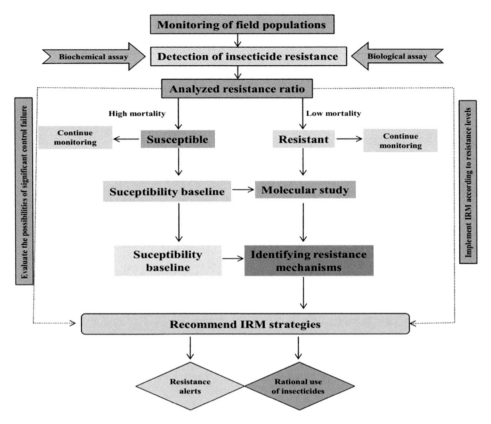

FIGURE 5.2 The flowchart represents an anticipated process of successful insecticide resistance management (IRM) protocol.

initial steps in insecticide resistance management (IRM) protocol (Figure 5.2). Implementation of successful IRM is a noteworthy choice to contain resistance development in pests and to overcome the problem (Sparks et al. 2020; Alam et al. 2016). Several settings and approaches could build a successful IRM protocol as we depicted one such IRM protocol (Figure 5.2). The regular monitoring of insecticide resistance would help to implement particular contents of IRM, rotation use of common insecticides, reducing/increasing concentration of insecticide doses, use of non-chemical control measures, and introducing new chemicals, would be some alternatives to reduce the pace of insecticide resistance development.

Target-site binding, increased activities of CYP450, EST, and GST enzymes have been significantly contributed to the evolution of insecticide resistance in rice pests, and studies have been emphasized for improving our understandings of these mechanisms. However, we are still looking for characterizing more resistance mechanisms such as multi-resistance mechanisms, a specific number of resistance genes, identification of regulatory mechanisms of resistance genes, transcription factors. Advanced molecular technology would be a better choice for further research, for example, in plants (Jun et al. 2019; Monsur et al. 2020) and in model insects, e.g. *Drosophila* (Douris et al. 2020), scientists are remarkably using state-of-the-art genome editing tools such as CRISPR/Cas9, base editing, and RNAi more effectively to understand particular molecular mechanisms (Taning et al. 2017; Chen et al. 2019b; Zhu and Palli 2020). In contrast, in the study of insecticide resistance mechanisms of rice pests, the use of the most advanced tools and techniques is still insufficient. The combination of biological, biochemical, and molecular studies would provide a better solution for resistance problems in rice pests. When commonly used insecticides cannot control target pests, choosing new insecticide classes might promise to be an essential element in integrated pest management strategies. But developing and introducing a new chemical compound will entail a great deal of cost. To avoid this problem, we have to use more creative tools to improve pest management strategies and also adapt the IRM processes to delay resistance development in pests.

Funding

No specific funding has been received to support this work.

Conflict of Interest

The authors declare that they have no conflict of interest.

Ethics Approval

The author did not perform any study with animals or human participants to prepare this manuscript.

REFERENCES

Adesanya AW, Waters TD, Lavine MD, Walsh DB, Lavine LC, Zhu F (2020) Multiple insecticide resistance in onion thrips populations from Western USA. *Pestic Biochem Physiol* 165:104553

Adnan M, Shah Z, Sharif M, Rahman H (2018) Liming induces carbon dioxide (CO_2) emission in PSB inoculated alkaline soil supplemented with different phosphorus sources. *Environ Sci Pollut Res* 25:9501–9509

Ahmed S, Alam MJ, Hossain A, Islam Akmm, Awan TH, Soufan W, Qahtan AA, Okla MK, El Sabagh A (2021) Interactive effect of weeding regimes, rice cultivars, and seeding rates influence the rice-weed competition under dry direct-seeded condition. *Sustainability* 13(1):317

Alam A, Baig J, Din SU, Arshad M, Adnan M, Khan A (2016) Relative abundance of benthic macro-invertebrates in relation to abiotic environment in Hussainbadnallah, Hunza, Gilgit Baltistan, Pakistan. *Int J Biosci* 9(3):185–193

Ali E, Mao K, Liao X, Jin R, Li J (2019) Cross-resistance and biochemical characterization of buprofezin resistance in the white-backed planthopper, *Sogatella furcifera* (Horvath). *Pestic Biochem Physiol* 158:47–53

Anis G, Hassan H, El-Sherif A, Saneoka H, EL Sabagh A (2019) Evaluation of new promising rice hybrid and its parental lines for floral, agronomic traits and genetic purity assessment. *Pak J Agric Sci* 56(3): 567–576

Bao H, Gao H, Zhang Y, Fan D, Fang J, Liu Z (2016) The roles of *CYP6AY1* and *CYP6ER1* in imidacloprid resistance in the brown planthopper: expression levels and detoxification efficiency. *Pestic Biochem Physiol* 129:70–74

Chang X, Yuan Y, Zhang T, Wang D, Du X, Wu X, Chen H, Chen Y, Jiao Y, Teng H (2015) The toxicity and detoxifying mechanism of cycloxaprid and buprofezin in controlling *Sogatella furcifera* (Homoptera: Delphacidae). *J Insect Sci* 15(1). doi:10.1093/jisesa/iev077

Chen C, Shan T, Liu Y, Shi X, Gao X (2019a) Identification of a novel cytochrome P450 *CYP3356A1* linked with insecticide detoxification in *Bradysia odoriphaga*. *Pest Manag Sci* 75:1006–1013

Chen K, Wang Y, Zhang R, Zhang H, Gao C (2019b) CRISPR/Cas genome editing and precision plant breeding in agriculture. *Annu Rev Plant Biol* 70:667–697

Cho JR, Lee M, Kim H S, Boo KS (2008) Effect of photoperiod and temperature on reproductive diapause of *Scotinophara lurida* (Burmeister) (Heteroptera: Pentatomidae). *J Asia-Paci Entomol* 11:53–57

Daborn P, Yen J, Bogwitz M, Le Goff G, Feil E, Jeffers S, Tijet N, Perry T, Heckel D, Batterham P (2002) A single P450 allele associated with insecticide resistance in *Drosophila*. *Science* 297:2253–2256

Dang K, Doggett SL, Veera Singham G, Lee CY (2017) Insecticide resistance and resistance mechanisms in bed bugs, *Cimex* spp. (Hemiptera: Cimicidae). *Parasit Vectors* 10:318

Douris V, Denecke S, Van Leeuwen T, Bass C, Nauen R, Vontas J (2020) Using CRISPR/Cas9 genome modification to understand the genetic basis of insecticide resistance: *Drosophila* and beyond. *Pestic Biochem Physiol* 167:104595

Du B, Wei Z, Wang Z, Wang X, Peng X, Du B, Chen R, Zhu L, He G (2015) Phloem-exudate proteome analysis of response to insect brown plant-hopper in rice. *J Plant Physiol* 183: 13–22

Elzaki MEA, Miah MA, Han Z (2017a) Buprofezin is metabolized by *CYP353D1v2*, a cytochrome P450 associated with imidacloprid resistance in *Laodelphax striatellus*. *Int J Mol Sci* 18:2564

Elzaki MEA, Zhu J, Pu Y, Zhang W, Sun H, Wu M, Han Z (2017b) Cross-resistance among common insecticides and its possible mechanism in *Laodelphax striatellus* Fallén (Hemiptera: Delphacidae). *Orient Insects* 52:2–15

Fujii T, Sanada-Morimura S, Oe T, Ide M, Van Thanh D, Van Chien H, Van Tuong P, Loc PM, Cuong LQ, Liu ZW, Zhu ZR, Li JH, Wu G, Huang SH, Estoy GF, Sonoda JS, Matsumura M (2020) Long-term field insecticide susceptibility data and laboratory experiments reveal evidence for cross resistance to other neonicotinoids in the imidacloprid-resistant brown planthopper *Nilaparvata lugens*. *Pest Manag Sci* 76:480–486

Gaballah MM, Metwally AM, Skalicky M, Hassan MM, Brestic M, EL Sabagh A, Fayed AM (2021) Genetic diversity of selected rice genotypes under water stress conditions. *Plants* 10(1):27

Garrood WT, Zimmer CT, Gorman KJ, Nauen R, Bass C, Davies TG (2016) Field-evolved resistance to imidacloprid and ethiprole in populations of brown planthopper *Nilaparvata lugens* collected from across South and East Asia. *Pest Manag Sci* 72:140–149

Giantsis IA, Castells Sierra J, Chaskopoulou A (2017) The distribution of the invasive pest, rice water weevil *Lissorhoptrus oryzophilus* Kuschel (*Coleoptera: Curculionidae*), is expanding in Europe: first record in the Balkans, confirmed by CO1 DNA barcoding. *Phytoparasitica* 45:147–149

Gong W, Yan HH, Gao L, Guo YY, Xue CB (2014) Chlorantraniliprole resistance in the diamondback moth (Lepidoptera: Plutellidae). *J Econ Entomol* 107:806–814

Gorman K, Liu Z, Denholm I, Bruggen KU, Nauen R (2008) Neonicotinoid resistance in rice brown planthopper, *Nilaparvata lugens*. *Pest Manag Sci* 64:1122–1125

Guedes RNC, Beins K, Costa DN, Coelho GE, Bezerra H (2020) Patterns of insecticide resistance in *Aedes aegypti*: meta-analyses of surveys in Latin America and the Caribbean. *Pest Manag Sci* 76:2144–2157

Guo L, Wang Y, Zhou X, Li Z, Liu S, Pei L, Gao X (2014) Functional analysis of a point mutation in the ryanodine receptor of *Plutella xylostella* (L.) associated with resistance to chlorantraniliprole. *Pest Manag Sci* 70:1083–1089

Han Y, Lei W, Wen L, Hou M (2015) Silicon-mediated resistance in a susceptible rice variety to the rice leaf folder, *Cnaphalocrocis medinalis* Guenée (Lepidoptera: Pyralidae). *PLoS One* 10:e0120557

Hayd RLN, Carrara L, Lima JDM, de Almeida NCV, Lima JBP, Martins AJ (2020) Evaluation of resistance to pyrethroid and organophosphate adulticides and kdr genotyping in *Aedes aegypti* populations from Roraima, the northernmost Brazilian State. *Parasit Vectors* 13:264

Huang JM, Rao C, Wang S, He LF, Zhao SQ, Zhou LQ, Zhao YX, Yang FX, Gao CF, Wu SF (2020) Multiple target-site mutations occurring in lepidopterans confer resistance to diamide insecticides. *Insect Biochem Mol Biol* 121:103367

Hussain I, Alam SS, Khan I, Shah B, Naeem A, Khan N, Ullah W, Iqbal B, Adnan M, Junaid K, Shah SRA, Ahmed N, Iqbal M (2016) Study on the biological control of fusarium wilt of tomato. *J Ento & Zool Studies* 4(2):525–528

Jena KK, Kim SR (2020) Genomics, biotechnology and plant breeding for the improvement of rice production. In: Gosal SS, Wani SH (eds) *Accelerated plant breeding, volume 1: cereal crops*. Springer International Publishing, Cham, pp 217–232

Jeong IH, Lee SW, Choi BR, Lee SH, Kwon DH (2016) Monitoring and evaluation of differential insecticide resistance profiles in the immigrant vs. indigenous populations of the small brown planthopper (*Laodelphax striatellus* Fallén) in Korea. *J Asia-Paci Entomol* 19:247–252

Jeschke P, Nauen R (2008) Neonicotinoids-from zero to hero in insecticide chemistry. *Pest Manag Sci* 64:1084–1098

Jin JX, Jin DC, Li WH, Cheng Y, Li FL, Ye ZC (2017) Monitoring trends in insecticide resistance of field populations of *Sogatella furcifera* (Hemiptera: Delphacidae) in Guizhou Province, China, 2012-2015. *J Econ Entomol* 110:641–650

Jin R, Mao K, Liao X, Xu P, Li Z, Ali E, Wan H, Li J (2019) Overexpression of *CYP6ER1* associated with clothianidin resistance in *Nilaparvata lugens* (Stål). *Pestic Biochem Physiol* 154:39–45.

Jun R, Xixun H, Kejian W, Chun W (2019) Development and application of CRISPR/Cas system in rice. *Rice Science* 26:69–76

Khan N, Shah Z, Adnan M, Ali M, Khan B, Mian IA, Ali A, Zahoor M, Roman M, Ullah L, Khaliq A, Khan WA, Alam A (2016) Evaluation of soil for important properties and chromium concentration in the basin of chromite hills in lower Malakand. *Adv Environ Biol* 10(7):141–147

Kwon YS, Chung N, Bae MJ, Li F, Chon TS, Park YS (2012) Effects of meteorological factors and global warming on rice insect pests in Korea. *J Asia-Paci Entomol* 15:507–515

Li W, Mao K, Liu C, Gong P, Xu P, Wu G, Le W, Wan H, You H, Li J (2020) Resistance monitoring and assessment of the control failure likelihood of insecticides in field populations of the whitebacked planthopper *Sogatella furcifera* (Horváth). *Crop Prot* 127:104973

Liao X, Jin R, Zhang X, Ali E, Mao K, Xu P, Li J, Wan H (2019) Characterization of sulfoxaflor resistance in the brown planthopper, *Nilaparvata lugens* (Stål). *Pest Manag Sci* 75:1646–1654

Liao X, Mao K, Ali E, Zhang X, Wan H, Li J (2017) Temporal variability and resistance correlation of sulfoxaflor susceptibility among Chinese populations of the brown planthopper *Nilaparvata lugens* (Stål). *Crop Prot* 102:141–146

Liao X, Xu PF, Gong PP, Wan H, Li JH (2020) Current susceptibilities of brown planthopper *Nilaparvata lugens* to triflumezopyrim and other frequently used insecticides in China. *Insect Sci* 28(1): 115–126

Lou YG, Zhang GR, Zhang WQ, Hu Y, Zhang J (2013) Biological control of rice insect pests in China. *Biol Control* 67:8–20

Lu K, Wang Y, Chen X, Zhang Z, Li Y, Li W, Zhou Q (2017a) Characterization and functional analysis of a carboxylesterase gene associated with chlorpyrifos resistance in *Nilaparvata lugens* (Stål). *Comp Biochem Physiol C Toxicol Pharmacol* 203:12–20

Lu Y, Wang G, Zhong L, Zhang F, Bai Q, Zheng X, Lu Z (2017b) Resistance monitoring of *Chilo suppressalis* (Walker) (Lepidoptera: Crambidae) to chlorantraniliprole in eight field populations from east and central China. *Crop Prot* 100:196–202

Mao K, Li W, Liao X, Liu C, Qin Y, Ren Z, Qin X, Wan H, Sheng F, Li J (2019b) Dynamics of insecticide resistance in different geographical populations of *Chilo suppressalis* (Lepidoptera: Crambidae) in China 2016-2018. *J Econ Entomol* 112:1866–1874

Mao K, Zhang X, Ali E, Liao X, Jin R, Ren Z, Wan H, Li J (2019a) Characterization of nitenpyram resistance in *Nilaparvata lugens* (Stål). *Pestic Biochem Physiol* 157:26–32

Masaya M, Takeuchi H, Satoh M, Sachiyo SM, Akira O, Tomonari W, Van TD (2009) Current status of insecticide resistance in rice planthoppers in Asia. In: Heong KL and Hardy B (eds) *Planthoppers: new threats to the sustainability of intensive rice production systems in Asia.* International Rice Research Institute, Los Baños, Philippines, pp 233–244

Matsumura M, Sanada-Morimura S, Otuka A, Ohtsu R, Sakumoto S, Takeuchi H, Satoh M (2014) Insecticide susceptibilities in populations of two rice planthoppers, *Nilaparvata lugens* and *Sogatella furcifera*, immigrating into Japan in the period 2005-2012. *Pest Manag Sci* 70:615–622

Matsumura M, Sanada-Morimura S, Otuka A, Sonoda S, Van Thanh D, Van Chien H, Van Tuong P, Loc PM, Liu ZW, Zhu ZR, Li JH, Wu G, Huang SH (2018) Insecticide susceptibilities of the two rice planthoppers *Nilaparvata lugens* and *Sogatella furcifera* in East Asia, the Red River Delta, and the Mekong Delta. *Pest Manag Sci* 74:456–464

Meng X, Yang X, Wu Z, Shen Q, Miao L, Zheng Y, Qian K, Wang J (2020) Identification and transcriptional response of ATP-binding cassette transporters to chlorantraniliprole in the rice striped stem borer, *Chilo suppressalis*. *Pest Manag Sci* 76(11):3626–3635

Min S, Lee SW, Choi BR, Lee SH, Kwon DH (2014) Insecticide resistance monitoring and correlation analysis to select appropriate insecticides against *Nilaparvata lugens* (Stål), a migratory pest in Korea. *J Asia-Paci Entomol* 17:711–716

Monsur MB, Shao G, Lv Y, Ahmad S, Wei X, Hu P, Tang S (2020) Base editing: the ever expanding Clustered Regularly Interspaced Short Palindromic Repeats (CRISPR) tool kit for precise genome editing in plants. *Genes (Basel)* 11(4):466

Monsur MB, Ivy NA, Haque MM, Hasanuzzaman M, EL Sabagh A et al (2020) Oxidative stress tolerance mechanism in rice under salinity. *Phyton-Int J Exp Bot* 89(3):497–517

Mu XC, Zhang W, Wang LX, Zhang S, Zhang K, Gao CF, Wu SF (2016) Resistance monitoring and cross-resistance patterns of three rice planthoppers, *Nilaparvata lugens*, *Sogatella furcifera* and *Laodelphax striatellus* to dinotefuran in China. *Pestic Biochem Physiol* 134:8–13

Pan Y, Zeng X, Wen S, Gao X, Liu X, Tian F, Shang Q (2020) Multiple ATP-binding cassette transporters genes are involved in thiamethoxam resistance in *Aphis gossypii* glover. *Pestic Biochem Physiol* 167:104558

Pang R, Chen M, Liang Z, Yue X, Ge H, Zhang W (2016) Functional analysis of *CYP6ER1*, a P450 gene associated with imidacloprid resistance in *Nilaparvata lugens*. *Sci Rep* 6:4992

Rahman I, Ali S, Alam M, Adnan M, Basir A, Ullah H, Malik MFA, Shah AS, Ibrahim M (2016) Effect of seed priming on germination performance and Yield of okra (*Abelmoschus esculentus* L.). *Pak J Agric Res* 29(3):253–262

Sanada-Morimura S, Fujii T, Chien HV, Cuong LQ, Estoy GF, Matsumura JM (2019) Selection for imida-cloprid resistance and mode of inheritance in the brown planthopper, *Nilaparvata lugens*. *Pest Manag Sci* 75:2271–2277

Scott JG (1999) Cytochrome P450 and insecticide resistance. *Insect Biochem Mol Biol* 29:757–777

Seck PA, Diagne A, Mohanty S, Wopereis MCS (2012) Crops that feed the world 7: rice. *Food Secur* 4:7–24

Shao X, Liu Z, Xu X, Li Z, Qian X (2013) Overall status of neonicotinoid insecticides in China: production, application and innovation. *J Pesti Sci* 38:1–9

Shuijin H, Qiong C, Wenjing Q, Yang S, Houguo Q (2017) Resistance monitoring of four insecticides and a description of an artificial diet incorporation method for *Chilo suppressalis* (Lepidoptera: Crambidae). *J Econ Entomol* 110:2554–2561

Sparks TC, Crossthwaite AJ, Nauen R, Banba DS, Cordova F, Earley U, Kintscher E, Fujioka S, Hirao A, Karmon D, Kennedy R, Nakao, Popham HJR, Salgado V, Watson GB, Wedel BJ, Wessels FJ (2020) Insecticides, biologics and nematicides: updates to IRAC's mode of action classification - a tool for resistance management. *Pestic Biochem Physiol* 167:104587

Spitzer T, Bílovský J, Matušinsky P (2020) Changes in resistance development in pollen beetle (*Brassicogethes aeneus* F.) to lambda-cyhalothrin, etofenprox, chlorpyrifos-ethyl, and thiacloprid in the Czech Republic during 2013–2017. *Crop Prot* 135:105224

Steinbach D, Gutbrod O, Lummen P, Matthiesen S, Schorn C, Nauen R (2015) Geographic spread, genetics and functional characteristics of ryanodine receptor based target-site resistance to diamide insecticides in diamondback moth, *Plutella xylostella*. *Insect Biochem Mol Biol* 63:14–22

Su J, Zhang Z, Wu M, Gao C (2014) Changes in insecticide resistance of the rice striped stem borer (Lepidoptera: Crambidae). *J Econ Entomol* 107:333–341

Sun H, Yang B, Zhang Y, Liu Z (2017) Metabolic resistance in *Nilaparvata lugens* to etofenprox, a non-ester pyrethroid insecticide. *Pestic Biochem Physiol* 136:23–28

Sun X, Gong Y, Ali S, Hou M (2018a) Mechanisms of resistance to thiamethoxam and dinotefuran compared to imidacloprid in the brown planthopper: roles of cytochrome P450 monooxygenase and a P450 gene *CYP6ER1*. *Pestic Biochem Physiol* 150:17–26

Sun Y, Xu L, Chen Q, Qin W, Huang S, Jiang Y, Qin H (2018b) Chlorantraniliprole resistance and its biochemical and new molecular target mechanisms in laboratory and field strains of *Chilo suppressalis* (Walker). *Pest Manag Sci* 74:1416–1423

Taning CNT, Van Eynde B, Yu N, Ma S, Smagghe G (2017) CRISPR/Cas9 in insects: applications, best practices and biosafety concerns. *J Insect Physiol* 98:245–257

Tomizawa M, Casida JE (2005) Neonicotinoid insecticide toxicology: mechanisms of selective action. *Annu Rev Pharmacol Toxicol* 45:247–268

Wang LX, Zhang YC, Tao S, Guo D, Zhang Y, Jia YL, Zhang S. Khan C, Gao CF, Wu SF (2020) Pymetrozine inhibits reproductive behavior of brown planthopper *Nilaparvata lugens* and fruit fly *Drosophila melanogaster*. *Pestic Biochem Physiol* 165:104548

Wang XG, Ruan YW, Gong C, Xiang W, Xu XX, Zhang YM, Shen LT (2019) Transcriptome analysis of *Sogatella furcifera* (Homoptera: Delphacidae) in response to sulfoxaflor and functional verification of resistance-related P450 genes. *Int J Mol Sci* 20:1–20

Wang Y, Zhang G, Du J, Liu B, Wang M (2010) Influence of transgenic hybrid rice expressing a fused gene derived from *cry1Ab* and *cry1Ac* on primary insect pests and rice yield. *Crop Prot* 29:128–133

Wei Y, Yan R, Zhou Q, Qiao L, Zhu G, Chen M (2019) Monitoring and mechanisms of chlorantraniliprole resistance in *Chilo suppressalis* (Lepidoptera: Crambidae) in China. *J Econ Entomol* 112:1348–1353

Wen Y, Liu Z, Bao H, Han Z (2009) Imidacloprid resistance and its mechanisms in field populations of brown planthopper, *Nilaparvata lugens* Stål in China. *Pestic Biochem Physiol* 94:36–42

Wu C, Chakrabarty S, Jin M, Liu K, Xiao Y (2019) Insect ATP-binding cassette (ABC) transporters: roles in xenobiotic detoxification and Bt insecticidal activity. *Int J Mol Sci* 20(11):2829

Xiao C, Luan S, Xu Z, Lang J, Rao W, Huang Q (2017) Tolerance potential of *Chilo suppressalis* larvae to fipronil exposure via the modulation of detoxification and GABA responses. *J Asia-Paci Entomol* 20:1287–1293

Xu L, Luo G, Sun Y, Huang S, Xu GD, Xu Z, Gu Z, Zhang Y (2020) Multiple down-regulated cytochrome P450 monooxygenase genes contributed to synergistic interaction between chlorpyrifos and imidacloprid against *Nilaparvata lugens*. *J Asia-Paci Entomol* 23:44–50

Xu L, Zhao J, Sun Y, Xu D, Xu G, Xu X, Zhang Y, Huang S, Han Z, Gu Z (2019) Constitutive overexpression of cytochrome P450 monooxygenase genes contributes to chlorantraniliprole resistance in *Chilo suppressalis* (Walker). *Pest Manag Sci* 75:718–725

Yang H, Zhou C, Jin DC, Gong MF, Wang Z, Long GY (2019) Sublethal effects of abamectin on the development, fecundity, and wing morphs of the brown planthopper *Nilaparvata lugens*. *J Asia-Paci Entomol* 22:1180–1186

Yang Y, Huang L, Wang Y, Zhang Y, Fang S, Liu Z (2016b) No cross-resistance between imidacloprid and pymetrozine in the brown planthopper: status and mechanisms. *Pestic Biochem Physiol* 130:79–83

Yang Y, Zhang Y, Yang B, Fang J, Liu Z (2016a) Transcriptomic responses to different doses of cycloxaprid involved in detoxification and stress response in the whitebacked planthopper, *Sogatella furcifera*. *Entomol Exp Appl* 158:248–257

Yang YY, Lai CT (2019) Synergistic effect and field control efficacy of the binary mixture of permethrin and chlorpyrifos to brown planthopper (*Nilaparvata lugens*). *J Asia-Paci Entomol* 22:67–76

Yao R, Zhao DD, Zhang S, Zhou LQ, Wang X, Gao CF, Wu SF (2017) Monitoring and mechanisms of insecticide resistance in *Chilo suppressalis* (Lepidoptera: Crambidae), with special reference to diamides. *Pest Manag Sci* 73:1169–1178

Yu YL, Huang LJ, Wang LP, Wu JC (2012) The combined effects of temperature and insecticide on the fecundity of adult males and adult females of the brown planthopper *Nilaparvata lugens* Stål (Hemiptera: Delphacidae). *Crop Prot* 34:59–64

Zhang BZ, Su X, Xie LF, Zhen CA, Hu GL, Jiang K, Huang ZY, Liu RQ, Gao YF, Chen XL, Gao XW (2020a) Multiple detoxification genes confer imidacloprid resistance to *Sitobion avenae* Fabricius. *Crop Prot* 128:105014

Zhang J, Zhang Y, Wang Y, Yang Y, Cang X, Liu Z (2016a) Expression induction of P450 genes by imidacloprid in *Nilaparvata lugens*: a genome-scale analysis. *Pestic Biochem Physiol* 132:59–64

Zhang K, Zhang W, Zhang S, Wu SF, Ban LF, Su JY, Gao CF (2014) Susceptibility of *Sogatella furcifera* and *Laodelphax striatellus* (Hemiptera: Delphacidae) to six insecticides in China. *J Econ Entomol* 107:1916–1922

Zhang X, Liao X, Mao K, Yang P, Li D, Ali E, Wan H, Li J (2017) Neonicotinoid insecticide resistance in the field populations of *Sogatella furcifera* (Horváth) in Central China from 2011 to 2015. *J Asia-Paci Entomol* 20:955–958

Zhang X, Liao X, Mao K, Zhang K, Wan H, Li J (2016b) Insecticide resistance monitoring and correlation analysis of insecticides in field populations of the brown planthopper *Nilaparvata lugens* (stål) in China 2012-2014. *Pestic Biochem Physiol* 132:13–20

Zhang X, Yin F, Xiao S, Jiang C, Yu T, Chen L, Ke X, Zhong Q, Cheng Z, Li W (2019) Proteomic analysis of the rice (*Oryza officinalis*) provides clues on molecular tagging of proteins for brown planthopper resistance. *BMC Plant Biol* 19:30

Zhang Y, Liu B, Zhang Z, Wang L, Guo H, Li Z, He P, Liu Z, Fang J (2018) Differential expression of P450 Genes and nAChR subunits associated with imidacloprid resistance in *Laodelphax striatellus* (Hemiptera: Delphacidae). *J Econ Entomol* 111:1382–1387

Zhang Y, Yang Y, Sun H, Liu Z (2016c) Metabolic imidacloprid resistance in the brown planthopper, *Nilaparvata lugens*, relies on multiple P450 enzymes. *Insect Biochem Mol Biol* 79:50–56

Zhang YC, Feng ZR, Zhang S, Pei XG, Zeng B, Zheng C, Gao CF, Xu Y (2020b) Baseline determination, susceptibility monitoring and risk assessment to triflumezopyrim in *Nilaparvata lugens* (Stål). *Pestic Biochem Physiol* 167:104608

Zhao J, Xu L, Sun Y, Song P, Han Z (2019) UDP-glycosyltransferase genes in the striped rice stem borer, *Chilo suppressalis* (Walker), and their contribution to chlorantraniliprole resistance. *Int J Mol Sci* 20:1064. doi:10.3390/ijms20051064

Zhou C, Yang H, Wang Z, Long GY, Jin DC (2018) Protective and detoxifying enzyme activity and ABCG subfamily gene expression in *Sogatella furcifera* under insecticide stress. *Front Physiol* 9:1890

Zhou CS, Cao Q, Li GZ, Ma DY (2020) Role of several cytochrome P450s in the resistance and cross-resistance against imidacloprid and acetamiprid of *Bemisia tabaci* (Hemiptera: Aleyrodidae) MEAM1 cryptic species in Xinjiang, China. *Pestic Biochem Physiol* 163: 209–215

Zhu KY, Palli SR (2020) Mechanisms, applications, and challenges of insect RNA interference. *Annu Rev Entomol* 65:293–311

Zuo YY, Ma HH, Lu WJ, Wang XL, Wu SW, Nauen R, Wu YD, Yang YH (2020) Identification of the ryanodine receptor mutation I4743M and its contribution to diamide insecticide resistance in *Spodoptera exigua* (Lepidoptera: Noctuidae). *Insect Sci* 27:791–800

6

Role of Nanotechnology for Climate Resilient Agriculture

Afifa Younas, Madiha Rashid, and Sajid Fiaz

CONTENTS

6.1 Introduction

Our planet is figuratively 'on fire.' There is a significant increase in the world's temperature. The Intergovernmental Panel on Climate Change (IPCC) has reported that the temperature has risen and is expected to rise further in the near future, in case the action is not taken for it (IPCC 2001, 2007). A number of views have been given by the scientists to manifest climate change. One significant way to control this condition is through improved agricultural systems. The condition is especially problematic for South Asia due to its high population and dependence on agriculture for basic needs. One important plan of the Action on Climate Today (ACT) was to allow the Climate Resilient Agriculture (CRA) in that particular area (Adnan et al. 2018a, 2019; Akram et al. 2018a, 2018b; Aziz et al. 2017a, 2017b; Habib et al. 2017;

DOI: 10.1201/9781003160717-6

Hafiz et al. 2016, 2019; Kamran et al. 2017; Muhammad et al. 2019; Sajjad et al. 2019; Saud et al. 2013, 2014, 2016, 2017, 2020; Shah et al. 2013; Qamar et al. 2017; Wajid et al. 2017; Yang et al. 2017; Zahida et al. 2017; Depeng et al. 2018; Hussain et al. 2020; Hafiz et al. 2020a, 2020b; Shafi et al. 2020; Wahid 2020; Subhan et al. 2020; Zafar-ul-Hye et al. 2020a, 2020b; Adnan et al. 2020; Ilyas et al. 2020; Saleem 2020; Farhat et al. 2020; Wu et al. 2020; Mubeen 2020).

Agriculture is one of the global warming contributors. The forestry, agriculture, and other land use areas are emitting one-fourth of the total global anthropogenic greenhouse gases (GHG). The other important contributors are enteric fermentation (methane from livestock), fertilizers (organic & synthetic), manure management, swamp rice production, burning biomass, and organic residues in soil. This overview did not involve land-use reform, forestry, and peat drainage or burning, which is contributing almost the same towards greenhouse gas emission. Also, it has great consequences on agriculture and its associated political, social, and physical areas (Lamboll et al. 2017; Khan et al. 2017). The interrelationship of various causes and effects of climate change has shown the insufficiency of technical solutions. Apart from these solutions, cultural, social, institutional, economic, and political feedbacks are also needed. Along with these issues, climate change is also influencing the agriculture sector. Global warming is enhancing food security issues by increasing temperature, unpredictable seasons, rise in sea levels, and increasing the extremity of weather events (Wu et al. 2019; Ahmad et al. 2019; Baseer et al. 2019; Hafiz et al. 2018; Tariq et al. 2018; Fahad and Bano 2012; Fahad et al. 2013, 2014a, 2014b, 2015a, 2015b, 2016a, 2016b, 2016c, 2016d, 2017, 2018, 2019a, 2019b; Hesham and Fahad 2020; Iqra et al. 2020; Akbar et al. 2020; Mahar et al. 2020; Noor et al. 2020; Bayram et al. 2020; Amanullah et al. 2020; Rashid et al. 2020; Arif et al. 2020; Amir et al. 2020; Saman et al. 2020; Muhammad et al. 2020; Md Jakirand and Allah 2020; Farah et al. 2020; Sadam et al. 2020; Unsar et al. 2020; Fazli et al. 2020; Md Enamul et al. 2020; Gopakumar et al. 2020; Zia-ur-Rehman 2020; Ayman et al. 2020; Mohammad et al. 2020a, 2020b; Senol 2020; Amjad et al. 2020; Ibrar et al. 2020; Sajid et al. 2020).

Nanotechnology may have wide-ranging applications in agriculture by opening up new prospects like the manufacturing of nanosensors to protect food from pests and pathogens. It could be done by identifying pathogens to control the increasing incidence of its attack due to changes in the climate. Advanced molecular and cell biology tools related to genetic engineering and molecular nanodevices can help in controlling this situation. According to El-Beltagy (2008), nanotechnology has also been used to stabilize sand dunes the Norwegian scientists. Nanotechnology is now at a rate in various parts of the world, including several developed countries. The application of nanotechnology in agriculture has started and will continue to evolve with major impacts on the various fields of agriculture and the food industry. It would definitely encourage scientists to build strategies to improve people's ability to deal with the detrimental impacts of climate change.

This chapter is based on exploring agriculture and climate change through a study on projected impacts, the opportunities for mitigation and adaptation strategies, and the importance of an approach to climate change innovation systems. Most of the data and projections in the review are global in nature for agriculture sector.

6.2 Nanotechnology and Climate Adaptive Agriculture

Even at the early stage of the nano revolution, current and upcoming nanotechnology applications in agriculture, food, and water, have already been presenting great potential for the poor. In this section, we have concentrated on these innovations by discussing their potential benefits (risk avoidance, etc.) and why they are likely to increase benefits for the poor. Currently, in many developing countries, few nanotechnology projects are already targeting specifically the needs of the poor (OECD and Allianz 2008).

Nanotechnology has been proved as a new scale-neutral technology, applicable to both large-scale commercial agriculturists and small-scale (poor in resources) farmers (Lal 2007). Although many applications of nanotechnology are of great interest to the agriculture sector, the technique has not widely been used in this sector, even in developed countries. The technique has currently been recognized in the food production chain, including nanosensors and nano-agrochemicals. Its use for soil cleaning and nanopore filters in the food production chain has also been reported (Bouwmeester et al. 2009).

The use of nanoceramic devices and silver nanoparticles has also been reported in the food production and processing phase of its production chain (Bouwmeester et al. 2009). Over the next 5–10 years, a growing array of nanotechnology applications is expected in developing countries for food and agricultural use (USDA 2003). These technologies included nanosensors, nano-delivery systems, nanocoatings, nanofilms, nanoparticles, and quantum dots (Scott 2005; FAO-WHO 2009). Nanosensors are able to detect extremely small quantities of chemical contaminants, viruses, and bacteria in food, water, and environmental media (Scott 2005). Nano delivery systems can carry medicines or nutrients to the location where they are required at the time they are needed and have the ability to reduce the usage of the products they distribute.

Nanotechnology technologies targeting low-use efficiencies of inputs (such as fertilizer, irrigation water, and pesticides) and drought stress, and elevated soil temperature are of special interest to developing countries. Nanoscale agrichemical formulations have increased the efficiency of use and reduce environmental losses (Lal 2007). More effective distribution of nutrients can be expected to result in improved yields (Joseph and Morrison 2006). Nanoporous materials capable of retaining water and steadily increasing it during periods of drought could also be expected to increase yields. The use of nanotechnology to reduce the impact of aflatoxin would raise the weight of food animals, resulting in increased availability of meat (Shi et al. 2006).

6.3 Natural Resources and Nanotechnology

Natural resources, including those related to soil, water, plant and animal diversity, vegetation cover, sustainable energy supplies, climate and ecological services, are central to the development and operation of agricultural systems and the social and environmental sustainability of life on earth. Historically, the direction of global agricultural growth has been primarily based on improved production rather than on a more systemic convergence of NRM (Natural Resource Management) with food and nutrition conservation. A systemic or system-oriented approach is preferred as it can solve the challenging problems involved with the complexities of food and other production systems in diverse ecologies, environments, and cultures. The most significant potential prospects and approaches for the application of nanotechnology are to increase the productivity of the use of the required inputs (light, water, soil) for crop farming and to better handle biotic and abiotic stresses (Lowry et al. 2019).

6.3.1 Water

Water is being scarce not only in arid and drought-prone regions but also in the areas where precipitation is ample. The scarcity of water concerns the quantity and quality of the available resource because the use of degraded water resources for more stringent water use is considered useless. Therefore, a goal for agriculture in water shortage regions would therefore be efficient water usage – protection of the land, the friendly climate, technical adequacy, economic survival, and social acceptability for growth concerns (Periera 2003).

The lowest per capita water supplies are available in the WANA area. Some countries are faced with extreme water shortages (e.g. Jordan and Palestine), and some are predicted in the very near future to come into the same group. Water is the only product that can radically alter the capacity of a low-rainfall environment if accessible at a low cost (El-Beltagy 2008). Water continues to be misused in spite of its scarcity in WANA. The water recovering capacity above the water recovery level has been improved by emerging technologies, and aquifers have been depleted to exhaustion. Modern infrastructure can promote desertification. The convenient transportation has made it possible for farmers to move water for their cattle to grassland, encouraging overgrazing in the same area, resulting in wind erosion and biodiversity destruction. These are some of the main problems that have led to the degradation, global warming, starvation, and poverty of millions of people.

The use of nanomaterial is an innovative approach with improved tolerance, capability, and selectivity for heavy metals and other pollutants. Higher reactivity, greater surface contact, and better waste disposal are the advantages of using these materials. There are some examples of nanoparticles and nanomaterials which could also be used for water remediation, including CNTs (carbon nanotubes), zeolites, zero-valent-iron nanoparticles (ZVI), zero-valent-iron monolayer nanoparticles, biopolymers, and single enzyme nanoparticles. Nanoparticles can give a combined treatment of inorganic and organic chemical contaminants besides removing pathogenic microorganisms. Nanoparticles of polymer have different applications, such as sunscreen and water treatment. Polymeric nanoparticles are based on a similar concept to surfactant micelles with having amphiphilic properties and their hydrophobic and hydrophilic parts in each element. As a remediation agent, nanoparticles of amphiphilic polyurethane (APU) have strong prospects (Tungittiplakorn et al. 2004).

The efficiency of water filtration membranes can be increased by nanofibres and nanobiocides (Marguerite du Plessis 2011). For membrane fouling caused by water quality reduction bacteria, surface-modified nanofibres can induce inhibition of these bacteria. Nanofiltration is a form of filtration, which is powered by pressure. Multivalent ions, pesticides, and heavy metals have a greater thrust or rejection of nanofiltration membranes than standard approaches. Today, this technology has become the main and most modern water treatment system and is available in your house, enterprise, or manufacturing facility for practical use. Others include self-assembled mono-layers on mesophorosic silica, dendrimers or dendritic polymers, single nanoparticle enzyme (SEN), tunable biopolymers, nanocrystalline zeolites. There are also many other examples of nanotechnology uses for water treatment.

6.3.2 Land

More than 900 million inhabitants are now directly and negatively impacted by land loss in about 100 countries. Without slowing and reversing the degradation rate, humanity will be threatened with food security and preventing poorer nations from increasing their wealth by improving productivity. Now land destruction can be seen in all agro-climate areas on all continents. Although climatic conditions like drought and flooding are also contributing to the destruction, human activities are affecting it badly. Land destruction is a local phenomenon in many areas, but at the international and global levels, it does have systemic consequences. The most badly affected are developed countries, particularly in arid and semi-arid areas.

Land destruction is an all-encompassing challenge. We must never misinterpret it, however, if the influence of the land loss is immediately obvious and most drastic, restricted to agricultural land or farm livelihoods. Land depletion is a symbol of underdevelopment in the developed world. It is the combined result of social and economic causes such as poverty and the uneven allocation of land resources, and inefficient land usage, and methods of cultivation. In dry regions, the atmosphere and the fragility of habitats intensify these causes. Since agriculture is the main employer of workers, the consequences of the destruction of the soil are always catastrophic, leading to starvation and civil instability (El-Beltagy 2006).

The presence of toxic composites in the natural soil system contributes to soil pollution. Popular soil contaminants are heavy metals, which naturally exist in the soil but rarely at toxic amounts and which are key sources of polluted soil, especially industrial waste (such as paint residues, batteries, electric waste, etc.) or municipal or industrial waste sites. Heavy metals should be regarded as one of the most difficult soil contaminants since they are non-degradable and can linger soon as they are added into the environment; mercury and selenium are the only exceptions because they can be converted and volatilized by microorganisms (Figure 6.1). Due to the sorption-desorption reactions with other soil elements, we used a wide variety of modification agents to control the bioavailability of heavy metals and to avoid their diffusion into the soil by various sorting processes: adsorption of mineral surfaces, creation of organic ligand stable compounds, ion exchange and surface precipitation.

Nanoscale particles have become essential for soil immobilization of heavy metals in recent years (Figure 6.1). As nanoparticles tend easy to assemble into micro-millimetre aggregates easily, their distinctive characteristics such as high specific surface area and soil deliverability have been lost. Organic polymers such as starch and carboxymethyl cellulose (CMC) are also added as stabilizers to the nanoparticles in order to avoid the agglomeration of nanoparticles by steric and/or electrostatic stabilization systems, as well as to increase the physical stability and mobility of the soils and of a wider particular surface region. Compounds of phosphate could be used as useful substances for the in situ immobilization of polluted soils of heavy metals, as seen by the immobilization of lead (Pb) since phosphate has been used on soil either in soluble form (phosphoric acid) or solid forms such as synthesized apatite, natural phosphate, and even fish bones (the useful composition of apatite) (Yang et al. 2017). Furthermore, these could also be used in the development of a stable compound known as lead phosphate, also called pyromorphite.

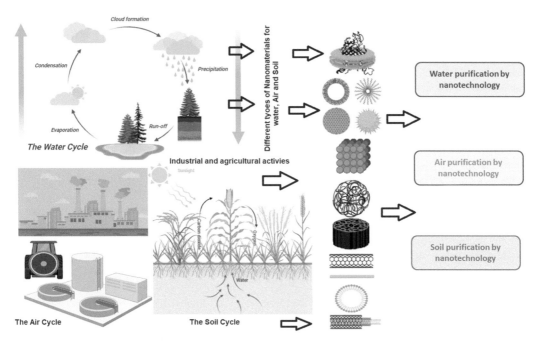

FIGURE 6.1 Application of nanotechnology and natural resources purification.

A few researches have focused on the use of naturally occurring iron-rich nanomaterials for the adsorption of toxic metals in soils. In order to avoid Pb in polluted agricultural soil, iron-rich synthesized nanomaterials (Fe and Fe oxides) and natural iron-rich materials (NRE) have been used. For heavy metals immobilization and maintenance of biological soil products, the NRE addition is recommended (Almaroai et al. 2014). FeO nanoparticles (also include some Fe-oxide nanoparticles) have been commonly used to treat a range of other kinds of contaminants, including anions (perchloride and nitrogen and dichromate), heavy metals (nickel, chromium, lead, and mercury), and radionuclides (uranium dioxide) (Johnson and Hill 1999; Wilson et al. 2008). The polyelectrolyte multilayer method can be used to synthesize FeO nanoparticles of similar sizes.

6.3.3 Air

The direct and indirect effects of climate change on agriculture have been explored (Figure 6.1). It illustrated how agriculture and land use transition lead to emissions of greenhouse gases or global warming, and then solutions have been given for supply and demand reduction in farming and food systems. A number of examples for adaptation and challenges to adaptation have been addressed in order to adapt agriculture to climate change. In agriculture, such as climate-smart agriculture, important principles and approaches to combat climate change are being taken into account. Finally, it is discussed as to how agricultural innovation structures can be improved in order to address climate change under the headings of public policies and regulatory context, knowledge management, capability improvements, the private sector and economy position, weather financing and agriculture, and risk-increasing management, inadequacy, and conflict-related management. It is concluded with suggestions for ways to respond to climate change in agricultural systems (Lamboll et al. 2017).

Nuisance odours generated from waste and wastewater treatment plants have been a cause of public discomfort. This situation is affecting air quality and has become a growing problem for social and public health, particularly in developing countries. In a wastewater treatment plant, various modelling approaches for biological treatment, as well as physio-chemical processes, have been developed and implemented successfully. A very complex problem is still seen in the mathematical modelling of the odour generation process, mainly because of olfactory disturbance caused by many chemical compounds, and the perception of odours is impaired by subjective limits. Moreover, dynamic atmospheric dispersion processes have a strong influence on air quality from odour sources (Carrera-Chapela et al. 2014).

Examples

The design methods and formulations have given the candidature for the elimination of a variety of pollutants. These included odoriferous compounds containing unwanted odours, toxins from the body, such as mold and mildew, algae, spores, etc. The nanocomposites, which constitute mixed clathrates and methods for using such compositions to capture guest bands such as malodors, molds, and cells (reactive nano-component fumed synthetic faujasite and amorphous synthetic mordenite) have been disclosed (Pronovost and Hickey 2014).

The creation of NOx (NO and NO_2) emissions out of the combustion of fossil fuel was a big endeavour. The typical adsorbents used for low-temperature NOx removal include zeolites from ion exchange, active carbon and FeOOH spread over active carbon fibre. Due to the reactivity of surface functional groups, NO can effectively be adsorbed into the activated carbon while it is still little in the number of adsorbed species. CNTs can be used to delete NO as an adsorbent. At room temperature, the CNTs were successful in absorbing NOx, SO_2, and CO_2 (Long and Yang 2001).

Several systems have been tested for CO2 capture, including absorption, adsorption, cryogenics, membrane, etc. (Aaron and Tsouris 2005). Of these innovations, the most evolved method was the adsorption-regeneration system. That is the mechanism for the absorption of amine or ammonia. CNTs would have a good ability to accumulate greenhouse gas carbon dioxide through chemical alteration. Generally, with elevated relative humidity, the efficiency of CO_2 adsorption on modified CNTs increases but decreases as the temperature increases.

Nanotechnology can also purify harmful gases in natural air. The method of CNTs and adsorption of gold particles for toxic gas purification is an example of nanotechnology. CNTs are special macromolecules with a uniform structure, thermal stability, and excellent chemical characteristics, as well as single-wall nanotubes (SWNTs) and multiple-wall nanotubes (MWNTs). These nanomaterials have shown a high degree of ability as superior adsorbents for the elimination in air currents and in the water of different forms of organic or inorganic contaminants. It has been stated that SWNTs are NO2 and NH3 chemical sensors. Nanomaterials are based on metal-oxide with improved sensing characteristics, fast recovery, and low limit detection. The variation of modern metal oxide nanotechnology is synthesized to track NOx, SO2, and formaldehyde in the sinks of the main outdoor and indoor greenhouse gases (Lin et al. 2019).

Dioxin and its similar compounds are soluble and particularly toxic chemicals, such as polychlorinated dibenzofuran, polychlorinated biphenyls. Dioxin compounds are produced primarily by waste incineration combustion of organic compounds. The interaction of dioxins with active carbon is about three times greater than CNTs. These adsorbents included silica, activated carbon, single-wall nanotubes, zeolites, and molecular nanoporous silica baskets.

Nanosize semiconductors in air remediation such as titanium dioxide (TiO_2), zinc oxide (ZnO) (Fe_2O_3), and tungsten oxide (WO_3) could be used as photocatalysts. Photocatalysts can oxidize chemical compounds into non-toxic products in relation to the atmosphere and water remediation (Watlington 2005). Probably, it is due to low toxicity, high photoconductivity, high photostability, and readily accessible and economical material that TiO_2 is being used as advanced methods of photochemical oxidation for water remediation. Furthermore, via nanotechnology development, the reactivity and selectivity of semiconductor photocatalysts have been altered.

6.4 Synthesized and Natural Nanoparticles and Climate Resilient Agriculture

Nanotechnology is promoting nanometre-size technologies (Feynman 1991). Biology and material science are merged in the practice of nanomaterials in biotechnology. Nanoparticles provide a very versatile tool that demonstrates specific properties with a wide range of applications (Murray et al. 2000).

6.4.1 Synthesized Nanoparticles

Owing to the various incentives for the use of nanoparticle synthesis in biological systems, some research groups have exploited the use of biological systems (Table 6.1). Nanoparticles are remarkable in their properties and utility in different ways, given the identical size of nanoparticles and biomolecules, such as proteins and polynucleic acids (Ferrari 2005).

TABLE 6.1

Application of Chemically, Physically Synthesized, and Naturally Synthesized Nanoparticles

Sr. No.	Nature of Nanoparticle	Name of Nanoparticle	Function	References
1	Synthesized nanoparticle	Zinc oxide	Disease management	Nandhini et al. (2019)
2		Nanocapsules	Biocides, repellents, attractants, or plant growth regulators	Camara et al. (2019)
3		Maghemite (γ-Fe_2O_3) nanoparticles	Acts as nanozymes and fertilizer, improving growth and abiotic stress tolerance	Palmqvist et al. (2017)
4	Chemical synthesis of nanoparticles	chitosan-copper nanoparticles	Disease management e.g. phytopathogens, *Rhizoctonia solani*, and *Pythium aphanidermatum*	Vanti et al. (2020)
5		Titanium dioxide nanoparticles (TiO_2 NPs)	Improved all agronomic traits and increased antioxidant enzyme activity compared with plants grown under salinity	Gohari et al. (2020)
		Synthesized TiO_2-NPs	Synthesis and characterization of titanium dioxide nanoparticles by chemical and green methods and their antifungal activities against wheat rust	Irshad et al. (2020)
6	Physical Synthesis of Nanoparticles	SiO_2 NPs	SiO_2 NPs could maintain the integrity of the cell, increase the thickness of the cell wall and the ratio of As in the pectin	Cui et al. (2020)
7		Silica-coated Fe_3O_4 nanoparticles	Synthesis of silica-coated Fe_3O_4 nanoparticles by microemulsion method and evaluation of antimicrobial activity	Asab et al. (2020)
8		Fe_3O_4 doped ZrO_2 nanoparticles	Microwave-assisted synthesis and antimicrobial activity of Fe_3O_4-doped ZrO_2 nanoparticles	Imran et al. (2019)
9	Natural synthesized nanoparticles	*Gossypium hirsutum*-derived silver nanoparticles	Plant pathogens *Xanthomonas axonopodis* pv. *malvacearum* and *Xanthomonas campestris* pv. *campestris*	Vanti et al. (2019)
10		Green synthesis of zinc oxide nanoparticles (ZnONPs) using plant extract of chamomile flower (*Matricaria chamomilla* L.), olive leaf (*Olea europaea*), and red tomato fruit (*Lycopersicon esculentum* M.)	*Xanthomonas oryzae* pv. oryzae	Ogunyemi et al. (2019)
11		Biogenic silver nanoparticles	Disease management *S. sclerotiorum*	Guilger et al. (2017)
12	Biological synthesis of nanoparticles	Green synthesis of silver nanoparticles	Improve seed germination, plant development, and photosynthetic efficiency	Soliman et al. (2020)
13		Green source microalgae Nanoparticle synthesis	Biosensing, environmental pollution detection properties	Jacob et al. (2020)
14		Magnesium hydroxide nanoparticles	Improved seed germination and growth	Shinde et al. (2020)

6.4.1.1 Chemical Synthesis of Nanoparticles

The chemical synthesis process is useful since the synthesis of large amounts of nanoparticles requires only a short amount of time. However, capping agents for stabilizing nanoparticles in size are required under this process. Three chemical methods have been reported for the synthesis of nanoparticles:

- Dispersion of preformed polymers
- Polymerization of monomers
- Ionic gelation or coacervation of hydrophilic polymers

6.4.1.2 Physical Synthesis of Nanoparticles

The above approach in physical processes is hardly ever used. Metal nanoparticles are synthesized by evaporation – condensation, which may occur using a tube furnace at ambient pressure. The raw material inside the ship is vaporized into a carrier gas in the furnace. Nanoparticles of various materials such as Ag, Au, PbS and fullerene were previously manufactured using evaporation/condensation techniques (Kruis et al. 2000). Initially, two-dimensional arrays of colloidal Au particles have been developed and identified, and later Grabar reported that a new loom was used for Au colloid production by surface-enhanced Raman scattering (SERS) substrates. In the colloid monolayers, they are very desirable for simple and functional applications, with uniform roughness, good stability, and bio-compatibility (Grabar et al. 1995).

6.4.2 Natural Synthesized Nanoparticles

The biogenically synthesized nanoparticles demonstrated strong polydispersion, dimensions, and stability. The nanoparticles have been synthesized using physical, chemical, and biological processes (Chen et al. 2008). Among them, physical and chemical processes are very costly (Li et al. 1999).

6.4.2.1 Biological Synthesis of Nanoparticles

The biological methods of the synthesis of nanoparticles would help to eliminate harsh conditions of processing by facilitating synthesis at specific pH, temperature, time, pressure, and simultaneously at a low price. Many microorganisms, whether intracellular or extracellular, have been found to be competent to synthesize inorganic nanoparticles. In recent years, nanoparticles have been exceptional in many areas, including electricity, medicine, climate, agriculture, etc., due to unpredictable properties (Table 6.1). Nanoparticles preparation could be established by either of two ways (Raveendran et al. 2003):

  i. Nanoparticles synthesis
  ii. Processing of nanomaterials into nanostructure particles

6.5 Impact of Climate Change on Cash Crops and Nanotechnology

The key aspect of climate change, one of the biggest environmental issues in the world, is global warming. It has also become a matter of general interest for the science world, policymakers, and the social public. A variety of issues, not least nature's shifts in the world itself, have been brought on by climate change that poses a significant challenge to human life and the global economy. Agriculture is one of the main and most important sector impacted by climate change, particularly for crop production and food security. Therefore, the influence of climate change is still one of the most critical problems against agricultural development.

The economic condition and development in climate change studies have been systematically outlined with respect to agricultural production in China. It has been introducing the modelling methodology, advances in the experiment on greenhouse gas accumulation in atmospheric effects on crops, the impact on agricultural climate resources and potential developments of climate change, probable influence on agricultural crop growth and yield of climate change, impacts on crop plants and varieties to enhance the utilization of agricultural climate resources.

On the grounds of this, an issue was also proposed for climate change impact assessment in agriculture. A more focused analysis into the uncertainties of possible climate change projections, model simulation and evaluation approaches must be taken with a view to enhancing the accurate and rational risk assessment for climate change in agriculture. More research is also needed on the effects of climate change extremes on agricultural development, the impacts on agricultural plant diseases and insect pests, the impacts on cash crops, fruit, animal husbandry, and the farmland environment (Jianping 2015).

In the past, traditional rice cultivation practices were in use. Highly critical in traditional agriculture was to maintain crop production, land structure, and fertility. The traditional practices described in this chapter are modern agriculture, inorganic chemical fertilizer, ecological, and Sri Lankan agricultural systems. These traditional rice crop growing activities have demonstrated lower soil fertility and have a more detrimental effect on habitats. These traditional approaches enhance the risk of global warming and decrease successful operations in agriculture. Nanotechnology is one of the most exciting methods of revolutionizing the traditional food science and technologies to achieve the necessary food production in the last few decades. Twenty-first-century science is the nanotechnology era. This new discipline included nanopesticides and nanofertilizers, i.e., plant-based nanosystems for fostering growth and development. Nanotechnology is providing the production of nano-based food products, smart nutrient supplies, and bioactive products in nano-food processing. Younas et al. (2020) also focused on nano-agrochemicals, plant pathogen detection, and nano-food processing as emerging nanotechnology approaches in the agricultural and food industry.

The use of nanomaterials has also reduced the number of sprayed agrochemicals by the efficient provision of active substances, minimizing fertilizer nutrient losses, and increasing yield by optimized management of water and nutrients (Figure 6.2). Nanosensors could also increase the efficiency of water, nutrients, and chemical use. It is, therefore, an eco-friendly, cost-effective method. In the programme of plant improvement and genomic transformation, nanotechnology-led innovations are also used. New biomass nanoparticles such as extremely porous lignocellulose fibre jute nanocarbons provide surprisingly high value for agricultural goods and raw materials. Therefore, for sustainable crop production a system under the growing world, the advanced research-based potential use of nanotechnology is required (Hossain et al. 2020).

6.5.1 Nano Growth Promoting Systems

New research on improving the yield and quality of agricultural production is of the utmost significance, despite climate change scenarios. Researchers have shown interest in recent years in the use of nanomaterials and plant growth-enhancing bacteria (PGPB). In the future, the development of new nanobiotechnological innovations by using nanomaterials along with bacteria would be more beneficial than traditional methods (Figure 6.2). There is also a decrease in the amount of fertilizers used on agricultural fields, giving out maximum efficiency with the lowest input values. The sum of chemical inputs in agricultural zones will be decreased, and efficient protection will be given against different stress factors in plants. It is expected that mineral abundance in farmlands can be improved according to chemical formulations by manufacturing nano-bioactive materials. The obtained results certainly support agriculture and nature effectively. Nanobiotechnological methods with new opportunities for study are also essential for scientific researchers (Cinsli et al. 2020).

The method not only catalyses waste degradation and toxic materials directly but also leads to increasing the performance of microorganisms in waste and toxic materials degradation. Bioremediation uses living organisms to break down and extract toxins and toxic materials from

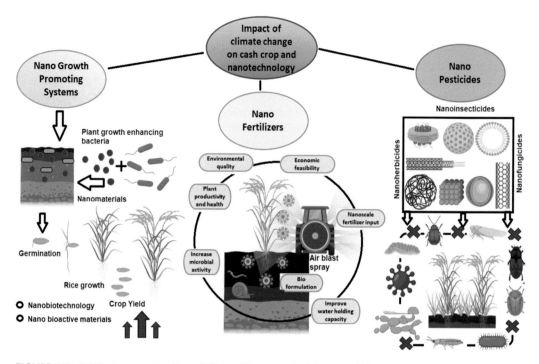

FIGURE 6.2   Brief scheme on the effect of Climate Change on Cash Crops and Nanotechnology.

agricultural soil and water. Other terms like bioremediation (through beneficial microbes), phytoremediation, and mycoremediation (fungi and mushrooms) are also commonly used. Thus, microorganisms can naturally and effectively extract heavy metals from soil and water with the bioremediation process (Dixit et al. 2015).

6.5.2 Nanopesticides

Nanomaterials and pesticides are introduced in the field, so the plants must be safe from chemical, abiotic and biotic, threats. Nanopesticides to replace classical pesticides have recently been created. The wide use of nanomaterials (and possible residues) in farm practices contributes to its deposition in the natural environment.

Nanotechnology may be effective at the time of climate change in order to combat climate change by storing carbon in land-based pools. The nanomaterials are shown to improve carbon stability and their potential sequestration in soil due to their special nanoscale properties. However, one of the key reasons for limiting the extensive use of this technique in order to mitigate climate change is that the conflicting studies on the possible effects of nanomaterial on soil microorganisms have inconsistent consequences. Continued efforts are, however, required to explore the potential of nanotechnology for carbon sequestration without losing ecosystem productivity level to build smart agriculture for the environment. This chapter has focused on the potential of nanomaterials for better soil carbon management and possible research implications for soil carbon study (Pramanik et al. 2020).

The consistency of products from the required chemicals supply to the target biological process is produced by microencapsulation-like nanoencapsulation. Any goods from Syngenta in Switzerland, like Karate ZEON, Subdue MAXx, Ospray's Chyella, Penncap-M, or microencapsulated BASF items will combat fit nanoscale. Lately, a range of chemicals firms have been encouraging the selling of nanoscale plagiarisms as microencapsulated pesticides. Such goods, including Primo MAXX, Banner MAXX, Subdue MAXX, etc. are also sold in the Australian market. While they are classified in the industry as microemulsions, they are actually emulsions in the nanoscale. A very narrow interface between the words microemulsion and nanoemulsions has been confirmed (Gouin 2004).

6.5.3 Nanofertilizers

Recent advances on nanotechnology applications in agriculture, including cropping with focus on contaminated-soil nanofertilizers. Nanomaterials play a significant role in soil pollutant fate, stability, and toxicity and are considered as central components in multiple biotic and abiotic remediation strategies. The utility and fate of nanomaterials depend heavily on their properties and interactions with soil components which are also addressed critically in this analysis. Study into remediation and the fate of nanoparticles in soil remains sparse and is mainly restricted to lab studies.

When nanomaterials enter the soil environment, their soil content and plant growth may be influenced by the effects of nutrient discharge in target nutrients, soil biota, organic carbon, and plant morphological and physiological responses. There was also a description of the processes of nanomaterial absorption and translocation in plants and related defence mechanisms (Usman et al. 2020). Global population is rising on daily basis, so the key challenge is how precisely the increasing demands with limited resources (less cost of production, less land) could be met and environmental threats be reduced. Therefore, disease management is most often considered to be the main task (Younas et al. 2020). Over the last decade, the industry is freely using nanofertilizers, but the largest chemical industries continue to dominate farm fertilizers, in particular. The nanofertilizers can include nanozinc, silica, iron, and titanium dioxide, ZnCd series/ZnS core shells, QDs InP/ZnS core shells, QDs in Mn/ZnSe, QDs in gold, QDs core shell, etc. Uptake tests of many metal oxide NPs biological fate and toxicity, viz. Al_2O_3, TiO_2, CeO2, FeO, and ZnONP have been intensively carried out for agricultural production (Dimkpa 2014; Zhang et al. 2016).

6.6 Nanotechnology for Biotic and Abiotic Stress Management

Sustainable farming is a crucial factor in meeting an increasingly growing global population's rising demand for food. A promising method for sustainable agriculture is nanobiotechnology. However, certain nanoparticles with special physicochemical characteristics naturally improve plant culture and stress resistance rather than acting as nanocarriers. Nanoparticles have this biological function in terms of physicochemical properties, the method of application (foliar distribution, hydroponics, soil), and the concentration applied. Here, the impacts of different forms, properties, and concentrations of nanoparticles have been discussed on plant growth and various abiotic (salinity, drought, sun, heavy metal, and light) and biotic (pathogens and herbivores) stresses (Figure 6.3) (Khan et al. 2016; Zhao et al. 2020).

6.6.1 Biotic Stress Management and Nanotechnology

Pathogens and herbivores are the main biotic stresses. Nanoparticles are often considered to be capable of promoting plant growth by having beneficial effects on seed germination, root or leaf growth, and biomass or grain yield. The provided knowledge would allow for a better selection of nanoparticles for starting materials within and outside the nanobiotechnology sector in agricultural applications. Finally, a switch to a new nanoparticle design focused on agriculture needs from testing/use of existing nanoparticles would allow the use of nanotechnology in sustainable agriculture (Zhao et al. 2020; Zahoor et al. 2016).

Climate change would adversely affect crop production by increasing the occurrence of diseases and reducing the impact of traditional approaches to disease control. Nanotechnology is an innovative new approach for the management of plant diseases, which has many advantages over traditional products, such as increased performance, lower input requirements, and less eco-toxicity. Results have been encouraging from studies of crop plants using nanomaterials (NMs) as protective agents. This review has focused on the use of NMs, in three different ways, in disease management:

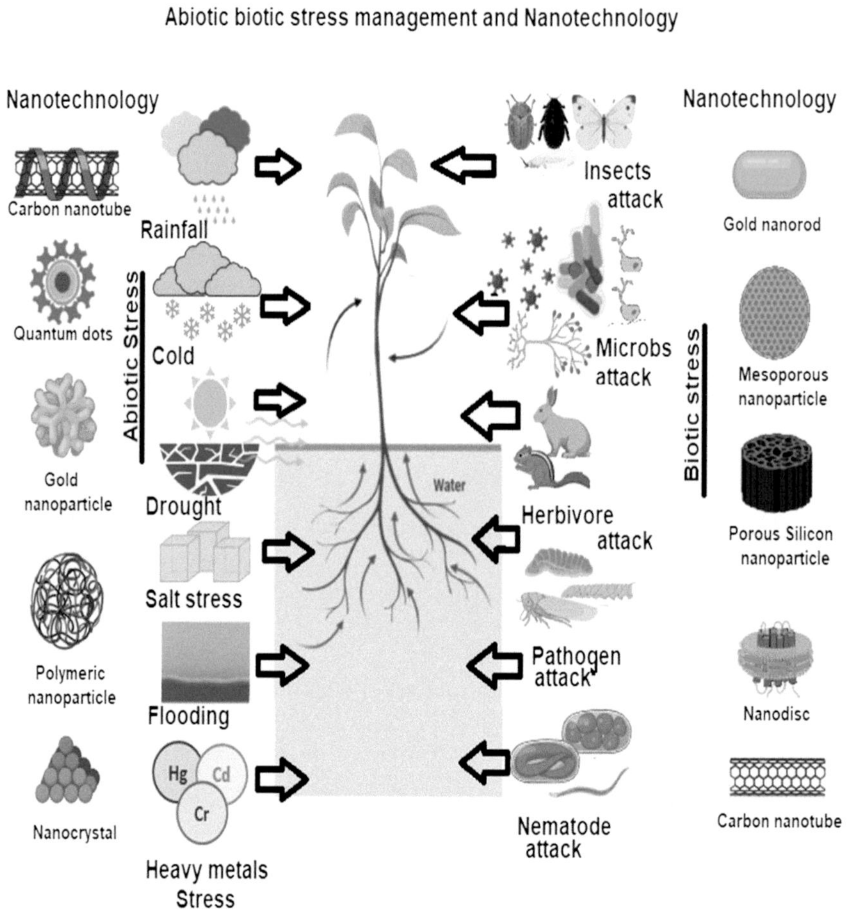

FIGURE 6.3 Effect of nanotechnology on biotic and abiotic stress.

    i. as antimicrobial agents

    ii. as biostimulants that may induce plant innate immunity

    iii. as carriers for active ingredients such as pesticides, micronutrients, and elicitors.

The possible benefits of nanotechnology have been taken into account, along with the role of NMs in future disease control measures and crop adaptation actions (Lin et al. 2020).

6.6.2 Abiotic Stress Management and Nanotechnology

The application of nanoparticles (NP) in the relief of various abiotic stresses is very effective and encouraging. These stress conditions are contributing towards oxidative damage and ions by producing the reactive oxygen species (ROS) and the cytosolic aggregation of toxic ions. These ROSs are highly reactive and interact with many processes of plant physiology by acting on plant growth and various abiotic (salinity, droughts, humidity, high light intensity, and heavy metals) stresses of different nature, properties, and concentrations of nanoparticles.

All abiotic stresses mainly reduce photosynthesis through chlorophyll degradation and disruption to other chloroplast ultrastructures. The NPs will mitigate these effects by improving the safety against antioxidants and osmoregulation to offset these harmful effects. Numerous NPs replicate the function of antioxidants and help retain redox potential under various abiotic stresses. In addition, these NPs improve photosynthetic effectiveness by increasing the amount of chlorophyll, light absorption, Fv/Fm, and other characteristics. For future studies, the molecular pathways of the NPs that have contributed to an increase in growth under various abiotic stresses must be investigated. By simulating the seeds and plants, chlorophyll levels and photosynthesis may also be decreased, and oxidative damage may be induced (Tanveer et al. 2020). This chapter has addressed the beneficial role of NPs in redox regulation, ionic homeostasis, and photosynthesis under various conditions of abiotic stresses.

The special physiological properties of nanoparticles (NPs) naturally improve plant growth and stress resistance. The biological function of NPs depends on the properties and the method of use (foliar application, hydroponics, or soil) of the nanoparticles and their relevant concentration. Different forms, characteristics, and concentrations of nanoparticles affect plant growth in different abiotic stresses like salinity, droughts, humidity, high light intensity, and heavy metals. Nanoparticles are also taken into account in promoting plant development through beneficial effects on seed germination, root and spring development, and biomass or grain yields. The carbon dots treated with rice demonstrated outstanding tolerance to abiotic stress with greater antioxidant ability and adequate nutrients. This report suggested that carbon dots have a strong ability to support plants from abiotic stresses (Li et al. 2020).

6.7 Opportunities and Challenges for Nanotechnology in the Agri-tech Revolution

Nanotechnology manages a leading agricultural control mechanism with its miniature component in particular. Furthermore, the implementation of nanotechnology allows for a resonant weight loss with various potential benefits, such as increased efficiency and protection, reductions in agricultural inputs, and enhanced soil absorption with nanoscale nutrients. The challenges such as sustainability, vulnerability, public wellbeing, and a safe life include agriculture, food, and natural resources. Nanomaterials in agriculture seek to minimize spreading pesticides, minimize fertilizer nutrient shortages and improve yields through the management of pesticide products and nutrients. Nanotechnology aims at boosting agriculture and food quality with innovative nano-instruments for monitoring rapid diagnoses of diseases, including increasing plant capacity to absorb nutrients. Relevant uses such as nanopesticides to trial products and nutrients to improve production in cultivation without decontamination of soils, rivers, and defence against many insects, pests, and microbial diseases have core interest in using nanotechnologies are the major priorities of the agricultural industry. Nanotechnology can serve as sensors to control the soil quality of agricultural land, and thus, protect the health of farm plants (Prasad et al. 2017).

Current farming practices introduced during the green revolution, particularly against climate change and the increasing populations, are becoming unsustainable. A main factor in the imminent agritech revolution will be nanotechnology, which will offer a more sustainable, productive, and robust agriculture system, thus encouraging food safety. Here, the discussion is on the most incredible new prospects and approaches for the use of nanotechnology in order to maximize the efficiency of use of the required inputs (light, water, soil) in crop agriculture and to enhance biotic and abiotic stress management. Potential obstacles towards development and execution are addressed to highlight the need for a framework approach to the design of proposed nanotechnologies.

The application of nanotechnologies in agriculture may also generate various threats for the human body. Nurkiewicz et al. (2008) found that ENM systemic circulation in rats has been accomplished with inhalation of Nanosized titanium dioxide engineered nanotechnology products, although it was not tested for human skin (Monteiro-Riviere and Rasmussen 2007). However, the data from a study conducted by Tinkle et al. (2003) showed that, during continuous flexing, latex particles below 1 μm reach the outer layers of the skin sample. Moreover, some experiments have shown that nanostructured particles cannot penetrate stable, intact tissue of skin (Tsuji et al. 2006).

It was demonstrated by Nurkiewicz et al. (2008) about the inhalation exposure of rats showing that even with low concentration exposure with nanosized titanium dioxide could enhance the microvascular dysfunction. Radomski et al. (2005) reported the accumulation of the platelet and vascular thrombosis due to the exposure of nanoparticles. More research on nanomaterials and nanoparticles' impact on blood and vessels is, therefore, needed because ENM may potentially lead to adverse effects in any body organ. The well-vascularized organ may affect organs or enter the brain cells and interfere with the metabolism of the foetus. Similarly, the human brain and blood vessels may also be the organs that could get affected adversely by nanotechnology (Elder et al. 2009).

6.8 Recommendations

The ecological agriculture of food chains and their associated flow of resources must be considered as an ecosystem process in which abiotic-biotic living beings function in compliance with organized stability. For optimizing agricultural productivity, emerging technology, innovation, expanded nano-chemical applications, specialization policies, and government policy should be adapted. In order to solve this condition, new technologies in the food industry must be developed. Thus, nanotechnology is the new and emerging technology that has very unique properties in the food supply chain (from the field to the table: cultivation, use of agricultural products such as nanopesticides, nanoherbicides, na-nofertilizers, etc., specific growing methods, smart foods, improved food consistency and texture, im-proved bioavailability/nutritional values, packaging and labelling, etc.) around the whole agricultural sector. New environmental and protection schemes for the distribution of specific materials, minerals, plants, and so on, could also have the potential for pharmaceutical applications.

In insect/pest control and agriculture food products, the biosensors related to nanotechnology have an important role. Consumers can always receive actual information about the condition of such products through intelligent nanosensor food packaging. Nanomaterial properties, including size, dosage, exposure duration, surface composition, structures, immune response, accumulation, retained time, etc. should be closely inspected, and all other effects should be investigated. For identifying, verifying, and achieving the effects of each nanomaterial/nano-food on entire environments, new analytical methods are needed to be developed. Life cycle of nanomaterials/nano-foods could also be studied. In order to leverage this in-formation, it is required to develop a vast database and international cooperation in governance, innovations, and regulation. The authorities should therefore set out specific recommendations and roadmaps to reduce the risks raised by the use of nanotechnological goods.

New networks and forums of collaboration should be opened in order to address the impacts of this technology on human society, economy, and research, with the involvement of various groups, such as consumers, experts, policymakers, industry, and other related sectors. This research could also provide worldwide creative and cost-effective production routes for human nutrition.

6.9 Conclusion

In order to fulfil different agricultural demands, nanotechnology has benefited through control over their composition, morphology, and functionality by engineered nanomaterials. For climate-sensitive cultivation, the mixture of nanomaterials and other organic and inorganic substances can further expand the use of nanomaterials to solve existing and emerging global problems such as chemical detection, pollution emission control, decontamination, and anti-terrorism activities. Since there are also some concerns about the toxicity of nanomaterials themselves, therefore, future recommendations for research to encourage sustainable nano-enabled agriculture research have been established. Finding and production of new materials could create a better environment for nanotechnological applications in the fields of chemistry, material science, biology, environmental sciences, chemical engineering, and a number of other relevant fields.

REFERENCES

Aaron, D and Tsouris, D (2005). Separation of CO2 from flue gases: a review. *Separat Sci Technol* 40: 321–348.

Adnan M, Fahad S, Khan IA, Saeed M, Ihsan MZ, Saud S, Riaz M, Wang D, Wu C (2019). Integration of poultry manure and phosphate solubilizing bacteria improved availability of Ca bound P in calcareous soils. *3 Biotech* 9(10):368.

Adnan M, Fahad S, Muhammad Z, Shahen S, Ishaq AM, Subhan D, Zafar-ul-Hye M, Martin LB, Raja MMN, Beena S, Saud S, Imran A, Zhen Y, Martin B, Jiri H, Rahul D (2020) Coupling phosphate-solubilizing bacteria with phosphorus supplements improve maize phosphorus acquisition and growth under lime induced salinity stress. *Plants* 9(900). doi:10.3390/plants9070900

Adnan M, Shah Z, Sharif M, Rahman H (2018a) Liming induces carbon dioxide (CO_2) emission in PSB inoculated alkaline soil supplemented with different phosphorus sources. *Environ Sci Pollut Res Int* 25(10):9501–9509.

Adnan M, Zahir S, Fahad S, Arif M, Mukhtar A, Imtiaz AK, Ishaq AM, Abdul B, Hidayat U, Muhammad A, Inayat-Ur R, Saud S, Muhammad ZI, Yousaf J, Amanullah, Hafiz MH, Wajid N (2018b) Phosphate-solubilizing bacteria nullify the antagonistic effect of soil calcification on bioavailability of phosphorus in alkaline soils. *Sci Rep* 8:4339. doi:10.1038/s41598-018-22653-7

Ahmad S, Kamran M, Ding R, Meng X, Wang H, Ahmad I, Fahad S, Han Q (2019) Exogenous melatonin confers drought stress by promoting plant growth, photosynthetic capacity and antioxidant defense system of maize seedlings. *PeerJ* 7:e7793. doi:10.7717/peerj.7793

Akbar H, Timothy JK, Jagadish T, Golam M, Apurbo KC, Muhammad F, Rajan B, Fahad S, Hasanuzzaman M (2020) Agricultural land degradation: processes and problems undermining future food security. In: Fahad S, Hasanuzzaman M, Alam M, Ullah H, Saeed M, Khan AK, Adnan M (eds) *Environment, climate, plant and vegetation growth*. Springer Publ Ltd, Springer Nature Switzerland AG. Part of Springer Nature, pp 17–62. doi:10.1007/978-3-030-49732-3

Akram R, Turan V, Hammad HM, Ahmad S, Hussain S, Hasnain A, Maqbool MM, Rehmani MIA, Rasool A, Masood N, Mahmood F, Mubeen M, Sultana SR, Fahad S, Amanet K, Saleem M, Abbas Y, Akhtar HM, Waseem F, Murtaza R, Amin A, Zahoor SA, ul Din MS, Nasim W (2018a) Fate of organic and inorganic pollutants in paddy soils. In Hashmi MZ and Varma A (eds) *Environmental pollution of paddy soils, soil biology*. Springer International Publishing AG, Cham, Switzerland,
pp 197–214.

Akram R, Turan V, Wahid A, Ijaz M, Shahid MA, Kaleem S, Hafeez A, Maqbool MM, Chaudhary HJ, Munis, MFH, Mubeen M, Sadiq N, Murtaza R, Kazmi DH, Ali S, Khan N, Sultana SR, Fahad S, Amin A, Nasim W (2018b) Paddy land pollutants and their role in climate change. In: Hashmi MZ and Varma A (eds) *Environmental pollution of paddy soils, soil biology*. Springer International Publishing AG, Cham, Switzerland, pp 113–124.

Almaroai, YA, Vithanage, M, Rajapaksha, AU, Lee, SS, Dou, X, Lee, YH, … Ok YS (2014). Natural and synthesized iron-rich amendments for As and Pb immobilisation in agricultural soil. *Chem Ecol* 30(3):267–279.

Amanullah SK, Imran, Hamdan AK, Muhammad A, Abdel RA, Muhammad A, Fahad S, Azizullah S, Brajendra P (2020) Effects of climate change on irrigation water quality. In: Fahad S, Hasanuzzaman M, Alam M, Ullah H, Saeed M, Khan AK, Adnan M (eds) *Environment, climate, plant and vegetation growth*. Springer Publ Ltd, Springer Nature Switzerland AG. Part of Springer Nature, pp 123–132. doi:10.1007/978-3-030-49732-3

Amir M, Muhammad A, Allah B, Sevgi Ç, Haroon ZK, Muhammad A, Emre A (2020) Biofortification under climate change: the fight between quality and quantity. In: Fahad S, Hasanuzzaman M, Alam M, Ullah H, Saeed M, Khan AK, Adnan M (eds) *Environment, climate, plant and vegetation growth*. Springer Publ Ltd, Springer Nature Switzerland AG. Part of Springer Nature, pp 173–228. doi:10.1007/978-3-03 0-49732-3.

Amjad I, Muhammad H, Farooq S, Anwar H (2020) Role of plant bioactives in sustainable agriculture. In: Fahad S, Hasanuzzaman M, Alam M, Ullah H, Saeed M, Khan AK, Adnan M (eds) *Environment, climate, plant and vegetation growth*. Springer Publ Ltd, Springer Nature Switzerland AG. Part of Springer Nature, pp 591–606. doi:10.1007/978-3-030-49732-3

Arif M, Talha J, Muhammad R, Fahad S, Muhammad A, Amanullah, Kawsar A, Ishaq AM, Bushra K, Fahd R (2020) Biochar; a remedy for climate change. In: Fahad S, Hasanuzzaman M, Alam M, Ullah H, Saeed M, Khan AK, Adnan M (eds) *Environment, climate, plant and vegetation growth*. Springer Publ Ltd, Springer Nature Switzerland AG. Part of Springer Nature, pp 151–172. doi:10.1 007/978-3-030-49732-3

Asab G, Zereffa EA, Abdo Seghne T (2020) Synthesis of silica-coated Fe_3O_4 nanoparticles by microemulsion method: characterization and evaluation of antimicrobial activity. *Int J Biomater* 2020.

Aziz K, Daniel KYT, Fazal M, Muhammad ZA, Farooq S, Fan W, Fahad S, Ruiyang Z (2017a) Nitrogen nutrition in cotton and control strategies for greenhouse gas emissions: a review. *Environ Sci Pollut Res* 24:23471–23487. doi:10.1007/s11356-017-0131-y

Aziz K, Daniel KYT, Muhammad ZA, Honghai L, Shahbaz AT, Mir A, Fahad S (2017b) Nitrogen fertility and abiotic stresses management in cotton crop: a review. *Environ Sci Pollut Res* 24:14551–14566. doi:10.1007/s11356-017-8920-x

Baseer M, Adnan M, Fazal M, Fahad S, Muhammad S, Fazli W, Muhammad A, Jr. Amanullah, Depeng W, Saud S, Muhammad N, Muhammad Z, Fazli S, Beena S, Mian AR, Ishaq AM (2019) Substituting urea by organic wastes for improving maize yield in alkaline soil. *J Plant Nutrition*. doi:10.1080/01904167.2 019.1659344

Bayram AY, Seher Ö, Nazlican A (2020) Climate change forecasting and modeling for the year of 2050. In: Fahad S, Hasanuzzaman M, Alam M, Ullah H, Saeed M, Khan AK, Adnan M (eds) *Environment, climate, plant and vegetation growth*. Springer Publ Ltd, Springer Nature Switzerland AG. Part of Springer Nature, pp 109–122. doi:10.1007/978-3-030-49732-3

Bouwmeester H, Dekkers S, Noordam M, Hagens W, Bulder A, De Heer C, ten Voorde S, Wijnhoven S, Marvin H, Sips A (2009) Review of health safety aspects of nanotechnologies in food production. *Regulat Toxicol Pharmacol* 53:52–62.

Camara MC, Campos EVR, Monteiro RA, Santo Pereira ADE, de Freitas Proença PL, Fraceto LF (2019) Development of stimuli-responsive nano-based pesticides: emerging opportunities for agriculture. *J Nanobiotechnol* 17(1):100.

Carrera-Chapela F, Donoso-Bravo A, Souto JA, Ruiz-Filippi G (2014) Modeling the odor generation in WWTP: an integrated approach review. *Water Air Soil Pollut* 225(6):1932.

Chen H, Roco MC, Li X, Lin Y (2008) Trends in nanotechnology patents. *Nat Nanotechnol* 3:123–125.

Cui J, Li Y, Jin Q, Li F (2020) Silica nanoparticles inhibit arsenic uptake into rice suspension cells via improving pectin synthesis and the mechanical force of the cell wall. *Environ Sci Nano* 7(1):162–171.

Depeng W, Fahad S, Saud S, Muhammad K, Aziz K, Mohammad NK, Hafiz MH, Wajid N (2018) Morphological acclimation to agronomic manipulation in leaf dispersion and orientation to promote "Ideotype" breeding: evidence from 3D visual modeling of "super" rice (*Oryza sativa* L.). *Plant Physiol Biochem* 135:499–510. doi:10.1016/j.plaphy.2018.11.010

Dimkpa CO (2014) Can nanotechnology deliver the promised benefits without negatively impacting soil microbial life? *J Basic Microbiol* 54(9):889–904.

Dixit R, Malaviya D, Pandiyan K, Singh UB, Sahu A, Shukla R, … Paul D (2015) Bioremediation of heavy metals from soil and aquatic environment: an overview of principles and criteria of fundamental processes. *Sustainability* 7(2):2189–2212.

El-Beltagy A (2006) Land degradation: a global and regional problem. In: van Ginkel H, Barrett B, Court J, Velasquez J (eds) *Human development and the environment challenges for the United Nations in the new millennium*. United Nations University Press, Tokyo, New York, Paris, pp 245–263.

El-Beltagy A (2008) Enhancing resilience in dry areas to cope with vagaries of climate change. Proceedings of the 9th International Conference on the Development of Drylands: November 7–10, 2008; IDCC, Alexandria, Egypt, pp 69–76.

Elder L, Grieger CR, Gatti G (2009) Human health risks of engineered nanomaterials: critical knowledge gaps in nanomaterials risk assessment. In: Linkov I, Steevens J (eds) *Nanomaterials: risks and benefits*. Springer, Dordrecht, pp 3–29.

Fahad S, Adnan M, Hassan S, Saud S, Hussain S, Wu C, Wang D, Hakeem KR, Alharby HF, Turan V, Khan MA, Huang J (2019b) Rice responses and tolerance to high temperature. In: Hasanuzzaman M, Fujita M, Nahar K, Biswas JK (eds) *Advances in rice research for abiotic stress tolerance*. Woodhead Publ Ltd, Cambridge, UK, pp 201–224.

Fahad S, Bajwa AA, Nazir U, Anjum SA, Farooq A, Zohaib A, Sadia S, NasimW, Adkins S, Saud S, Ihsan MZ, Alharby H, Wu C, Wang D, Huang J (2017) Crop production under drought and heat stress: plant responses and management options. *Front Plant Sci* 8:1147. doi:10.3389/fpls.2017.01147

Fahad S, Bano A (2012) Effect of salicylic acid on physiological and biochemical characterization of maize grown in saline area. *Pak J Bot* 44:1433–1438.

Fahad S, Chen Y, Saud S, Wang K, Xiong D, Chen C, Wu C, Shah F, Nie L, Huang J (2013) Ultraviolet radiation effect on photosynthetic pigments, biochemical attributes, antioxidant enzyme activity and hormonal contents of wheat. *J Food Agric Environ* 11(3&4):1635–1641.

Fahad S, Hussain S, Bano A, Saud S, Hassan S, Shan D, Khan FA, Khan F, Chen Y, Wu C, Tabassum MA, Chun MX, Afzal M, Jan A, Jan MT, Huang J (2014a) Potential role of phytohormones and plant growth-promoting rhizobacteria in abiotic stresses: consequences for changing environment. *Environ Sci Pollut Res* 22(7):4907– 4921. doi:10.1007/s11356-014-3754-2

Fahad S, Hussain S, Matloob A, Khan FA, Khaliq A, Saud S, Hassan S, Shan D, Khan F, Ullah N, Faiq M, Khan MR, Tareen AK, Khan A, Ullah A, Ullah N, Huang J (2014b) Phytohormones and plant responses to salinity stress: a review. *Plant Growth Regul* 75(2):391–404. doi: 10.1007/s10725-014-0013-y

Fahad S, Hussain S, Saud S, Hassan S, Chauhan BS, Khan F et al (2016a) Responses of rapid viscoanalyzer profile and other rice grain qualities to exogenously applied plant growth regulators under high day and high night temperatures. *PLoS One* 11(7):e0159590. doi:10.1371/journal.pone.0159590

Fahad S, Hussain S, Saud S, Hassan S, Ihsan Z, Shah AN, Wu C, Yousaf M, Nasim W, Alharby H, Alghabari F, Huang J (2016c) Exogenously applied plant growth regulators enhance the morphophysiological growth and yield of rice under high temperature. *Front Plant Sci* 7:1250. doi:10.3389/fpls.2016.01250

Fahad S, Hussain S, Saud S, Hassan S, Tanveer M, Ihsan MZ, Shah AN, Ullah A, Nasrullah KF, Ullah S, Alharby H, Nasim W, Wu C, Huang J (2016d) A combined application of biochar and phosphorus alleviates heat-induced adversities on physiological, agronomical and quality attributes of rice. *Plant Physiol Biochem* 103:191–198.

Fahad S, Hussain S, Saud S, Khan F, Hassan S, Amanullah, Nasim W, Arif M, Wang F, Huang J (2016b) Exogenously applied plant growth regulators affect heat-stressed rice pollens. *J Agron Crop Sci* 202:139–150.

Fahad S, Hussain S, Saud S, Tanveer M, Bajwa AA, Hassan S, Shah AN, Ullah A, Wu C, Khan FA, Shah F, Ullah S, Chen Y, Huang J (2015a) A biochar application protects rice pollen from high-temperature stress. *Plant Physiol Biochem* 96:281–287.

Fahad S, Muhammad ZI, Abdul K, Ihsanullah D, Saud S, Saleh A, Wajid N, Muhammad A, Imtiaz AK, Chao W, Depeng W, Jianliang H (2018) Consequences of high temperature under changing climate optima for rice pollen characteristics-concepts and perspectives. *Archives Agron Soil Sci*. doi:10.1080/03650340.2018.1443213

Fahad S, Nie L, Chen Y, Wu C, Xiong D, Saud S, Hongyan L, Cui K, Huang J (2015b) Crop plant hormones and environmental stress. *Sustain Agric Rev* 15:371–400.

Fahad S, Rehman A, Shahzad B, Tanveer M, Saud S, Kamran M, Ihtisham M, Khan SU, Turan V, Rahman MHU (2019a) Rice responses and tolerance to metal/metalloid toxicity. In: Hasanuzzaman M, Fujita M, Nahar K, Biswas JK (eds) *Advances in rice research for abiotic stress tolerance*. Woodhead Publ Ltd, Cambridge, UK, pp 299–312.

FAO-WHO (2009) FAO/WHO joint expert meeting on the application of nanotechnologies in the food and agriculture sectors: potential food safety applications. Meeting Report. Rome.

Farah R, Muhammad R, Muhammad SA, Tahira Y, Muhammad AA, Maryam A, Shafaqat A, Rashid M, Muhammad R, Qaiser H, Afia Z, Muhammad AA, Muhammad A, Fahad S (2020) Alternative and non-conventional soil and crop management strategies for increasing water use efficiency. In: Fahad S, Hasanuzzaman M, Alam M, Ullah H, Saeed M, Khan AK, Adnan M (eds) *Environment, climate, plant and vegetation growth*. Springer Publ Ltd, Springer Nature Switzerland AG. Part of Springer Nature, pp 323–338. doi:10.1007/978-3-030-49732-3

Farhana G, Ishfaq A, Muhammad A, Dawood J, Fahad S, Xiuling L, Depeng W, Muhammad F, Muhammad F, Syed AS (2020) Use of crop growth model to simulate the impact of climate change on yield of various wheat cultivars under different agro-environmental conditions in Khyber Pakhtunkhwa, Pakistan. *Arabian J Geosci* 13:112. doi:10.1007/s12517-020-5118-1

Farhat A, Hafiz MH, Wajid I, Aitazaz AF, Hafiz FB, Zahida Z, Fahad S, Wajid F, Artemi C (2020) A review of soil carbon dynamics resulting from agricultural practices. *J Environ Manage* 268(2020):110319.

Fazli W, Muhmmad S, Amjad A, Fahad S, Muhammad A, Muhammad N, Ishaq AM, Imtiaz AK, Mukhtar A, Muhammad S, Muhammad I, Rafi U, Haroon I, Muhammad A (2020) Plant-microbes interactions and functions in changing climate. In: Fahad S, Hasanuzzaman M, Alam M, Ullah H, Saeed M, Khan AK, Adnan M (eds) *Environment, climate, plant and vegetation growth*. Springer Publ Ltd, Springer Nature Switzerland AG. Part of Springer Nature, pp 397–420. doi:10.1007/978-3-030-49732-3

Ferrari M (2005) Cancer nanotechnology: opportunities and challenges. *Nat Rev Cancer* 5:61–171.

Feynman R (1991) There's plenty of room at the bottom. *Science* 29:1300–1301.

Gohari G, Mohammadi A, Akbari A, Panahirad S, Dadpour MR, Fotopoulos V, Kimura S (2020) Titanium dioxide nanoparticles (TiO$_2$ NPs) promote growth and ameliorate salinity stress effects on essential oil profile and biochemical attributes of *Dracocephalum moldavica*. *Sci Rep* 10(1):1–14.

Gopakumar L, Bernard NO, Donato V (2020) Soil microarthropods and nutrient cycling. In: Fahad S, Hasanuzzaman M, Alam M, Ullah H, Saeed M, Khan AK, Adnan M (eds) *Environment, climate, plant and vegetation growth*. Springer Publ Ltd, Springer Nature Switzerland AG. Part of Springer Nature, pp 453–472. doi:10.1007/978-3-030-49732-3

Gouin S (2004) Microencapsulation: industrial appraisal of existing technologies and trends. *Trends Food Sci Technol* 15(7-8):330–347.

Grabar KC, Freeman RG, Hommer MB, Natan MJ (1995) Preparation and characterization of Au colloid monolayers. *Anal Chem* 67:735–743.

Guilger M, Pasquoto-Stigliani T, Bilesky-Jose N et al (2017) Biogenic silver nanoparticles based on *Trichoderma harzianum*: synthesis, characterization, toxicity evaluation and biological activity. *Sci Rep* 7:44421. doi:10.1038/srep44421

Habib R, Ashfaq A, Aftab W, Manzoor H, Fahd R, Wajid I, Md Aminul I, Vakhtang S, Muhammad A, Asmat U, Abdul W, Syeda RS, Shah S, Shahbaz K, Fahad S, Manzoor H, Saddam H, Wajid N (2017) Application of CSM-CROPGRO-Cotton model for cultivars and optimum planting dates: evaluation in changing semi-arid climate. *Field Crops Res*. doi:10.1016/j.fcr.2017.07.007

Hafiz MH, Abdul K, Farhat A, Wajid F, Fahad S, Muhammad A, Ghulam MS, Wajid N, Muhammad M, Hafiz FB (2020a) Comparative effects of organic and inorganic fertilizers on soil organic carbon and wheat productivity under arid region. *Commun Soil Sci Plant Anal*. doi:10.1080/00103624.2020.1763385

Hafiz MH, Farhat A, Ashfaq A, Hafiz FB, Wajid F, Carol Jo W, Fahad S, Gerrit H (2020b) Predicting kernel growth of maize under controlled water and nitrogen applications. *Int J Plant Prod*. doi:10.1007/s42106-020-00110-8

Hafiz MH, Farhat A, Shafqat S, Fahad S, Artemi C, Wajid F, Chaves CB, Wajid N, Muhammad M, Hafiz FB (2018) Offsetting land degradation through nitrogen and water management during maize cultivation under arid conditions. *Land Degrad Dev* 1–10. doi:10.1002/ldr.2933

Hafiz MH, Muhammad A, Farhat A, Hafiz FB, Saeed AQ, Muhammad M, Fahad S, Muhammad A (2019) Environmental factors affecting the frequency of road traffic accidents: a case study of sub-urban area of Pakistan. *Environ Sci Pollut Res*. doi:10.1007/s11356-019-04752-8

Hafiz MH, Wajid F, Farhat A, Fahad S, Shafqat S, Wajid N, Hafiz FB (2016) Maize plant nitrogen uptake dynamics at limited irrigation water and nitrogen. *Environ Sci Pollut Res* 24(3):2549–2557. doi:10.1007/s11356-016-8031-0

Hesham FA, Fahad S (2020) Melatonin application enhances biochar efficiency for drought tolerance in maize varieties: modifications in physio-biochemical machinery. *Agron J* 112(4):1–22.

Hossain A, Kerry RG, Farooq M, Abdullah N, Islam, MT (2020) Application of nanotechnology for sustainable crop production systems. In: *Nanotechnology for food, agriculture, and environment*. Springer, Cham, pp 135–159.

Hussain MA, Fahad S, Rahat S, Muhammad FJ, Muhammad M, Qasid A, Ali A, Husain A, Nooral A, Babatope SA, Changbao S, Liya G, Ibrar A, Zhanmei J, Juncai H (2020) Multifunctional role of brassinosteroid and its analogues in plants. *Plant Growth Regul*. doi:10.1007/s10725-020-00647-8

Ibrar K, Aneela R, Khola Z, Urooba N, Sana B, Rabia S, Ishtiaq H, Ur Rehman M, Salvatore M (2020) Microbes and environment: global warming reverting the frozen zombies. In: Fahad S, Hasanuzzaman M, Alam M, Ullah H, Saeed M, Khan AK, Adnan M (eds) *Environment, climate, plant and vegetation growth*. Springer Publ Ltd, Springer Nature Switzerland AG. Part of Springer Nature, pp 607–634. doi:10.1007/978-3-030-49732-3

Ilyas M, Mohammad N, Nadeem K, Ali H, Aamir HK, Kashif H, Fahad S, Aziz K, Abid U (2020) Drought tolerance strategies in plants: a mechanistic approach. *J Plant Growth Regulation*. doi:10.1007/s00344-020-10174-5

Imran M, Riaz S, Sanaullah I, Khan U, Sabri AN, Naseem S (2019) Microwave assisted synthesis and antimicrobial activity of Fe3O4-doped ZrO2 nanoparticles. *Ceram Int* 45(8):10106–10113.

Intergovernmental Panel on Climate Change (IPCC): Summary for Policymakers (2007) *Climate change 2007: impacts, adaptation and vulnerability: contribution of Working Group II to the fourth assessment report of the Intergovernmental Panel on Climate Change.* Edited by: Parry ML, Canziani OF, Palutiko JP, van der Linden PJ, Hanson CE. Cambridge University Press, Cambridge.

Intergovernmental Panel on Climate Change (IPCC): Synthesis Report 2001. *Contribution of Working Group I, II, and III to the third assessment report of the Intergovernmental Panel on Climate Change.* Cambridge University Press, Cambridge.

Iqra M, Amna B, Shakeel I, Fatima K, Sehrish L, Hamza A, Fahad S (2020) Carbon cycle in response to global warming. In: Fahad S, Hasanuzzaman M, Alam M, Ullah H, Saeed M, Khan AK, Adnan M (eds) *Environment, climate, plant and vegetation growth.* Springer Publ Ltd, Springer Nature Switzerland AG. Part of Springer Nature, pp 1–16. doi:10.1007/978-3-030-49732-3

Irshad MA, Nawaz R, Ur Rehman MZ, Imran M, Ahmad MJ, Ahmad S, … Ali S (2020). Synthesis and characterization of titanium dioxide nanoparticles by chemical and green methods and their antifungal activities against wheat rust. *Chemosphere*, 127352.

Jacob JM, Ravindran R, Narayanan M, Samuel SM, Pugazhendhi A, Kumar G (2020). Microalgae: a prospective low cost green alternative for nanoparticle synthesis. *Curr Opin Environ Sci Health.*

Jan M, Muhammad Anwar-ul-Haq, Adnan NS, Muhammad Y, Javaid I, Xiuling L, Depeng W, Fahad S (2019) Modulation in growth, gas exchange, and antioxidant activities of salt-stressed rice (*Oryza sativa* L.) genotypes by zinc fertilization. *Arabian J Geosci* 12:775. doi:10.1007/s12517-019-4939-2

Jianping G (2015) Advances in impacts of climate change on agricultural production in China. *J Appl Meteorol Sci* 26(1):1–11.

Johnson RP, Hill, CL (1999) Polyoxometalate oxidation of chemical warfare agent simulants in fluorinated media. *J Appl Toxicol* 19(S1):S71–S75.

Joseph T, Morrison M (2006) Nanotechnology in agriculture and food. Woodrow Wilson International Center for Scholars. Project on Emerging Nanotechnologies. www.nanoforum.org. Accessed 20 May 2010.

Kamran M, Wenwen C, Irshad A, Xiangping M, Xudong Z, Wennan S, Junzhi C, Shakeel A, Fahad S, Qingfang H, Tiening L (2017) Effect of paclobutrazol, a potential growth regulator on stalk mechanical strength, lignin accumulation and its relation with lodging resistance of maize. *Plant Growth Regul* 84:317–332. doi:10.1007/s10725-017-0342-8

Khan A, Shah N, Gul A, Sahar NU, Ismail A, Muhammad Aziz F, Farooq M, Adnan M, Rizwan M (2016) Comparative study of toxicological impinge of glyphosate and atrazine (herbicide) on stress biomarkers; blood biochemical and hematological parameters of the freshwater common carp (*Cyprinus carpio*). *Pol J Environ Stud* 25(5):1993–1999.

Khan MA, Basir A, Adnan M, Shah AS, Noor M, Khan A, Shah JA, Ali Z, Rehman A (2017) Wheat phenology and density and fresh and dry weight of weeds as affected by potassium sources levels and tillage practices. *Pak J Weed Sci Res* 23(4):451–462.

Kruis FE, Fissan H, Rellinghaus B (2000) Sintering and evaporation characteristics of gas-phase synthesis of size-selected PbS nanoparticles. *Mater Sci Eng B* 69:329–334.

Lal R (2007) Ushering soil science into the 21 century. President's Message. Soil Science Society of America, 7 November 2007, Madison Wisconsin. https://www.soils.org/about-society/presidents-message/archive/16. Accessed 22 May 2010.

Lamboll R, Stathers T, Morton J (2017) Climate change and agricultural systems. In: *Agricultural Systems.* Academic Press, pp 441–490.

Li Y, Duan X, Qian Y, Yang L, Liao H (1999) Nanocrystalline silver particles: synthesis, agglomeration, and sputtering induced by electron beam. *J Colloid Interface Sci* 209:347–349.

Li Y, Gao J, Xu X, Wu Y, Zhuang J, Zhang X, … Hu C 2020. Carbon dots as a protective agent alleviating abiotic stress on rice (*Oryza sativa* L.) through promoting nutrition assimilation and the defense system. *ACS Appl Mater Interfaces* 12(30):33575–33585.

Lin F, Wang Z, Dhankher OP, Xing B (2020) Nanotechnology as a new sustainable approach for controlling crop diseases and increasing agricultural production. *J Exp Bot* 71(2):507–519. doi:10.1093/jxb/erz314

Lin C, Xu W, Yao Q, Wang X (2019) Nanotechnology on toxic gas detection and treatment. In: *Novel nanomaterials for biomedical, environmental and energy applications.* Elsevier, pp 275–297.

Long RQ, Yang RT (2001) Carbon nanotubes as a superior sorbent for nitrogen oxides. *Ind Eng Chem Res* 40:4288–4291. doi:10.1021/ie000976k

Lowry GV, Avellan A, Gilbertson LM (2019). Opportunities and challenges for nanotechnology in the agri-tech revolution. *Nat Nanotechnol* 14(6):517–522.

Mahar A, Amjad A, Altaf HL, Fazli W, Ronghua L, Muhammad A, Fahad S, Muhammad A, Rafiullah, Imtiaz AK, Zengqiang Z (2020) Promising technologies for Cd-contaminated soils: drawbacks and possibilities. In: Fahad S, Hasanuzzaman M, Alam M, Ullah H, Saeed M, Khan AK, Adnan M (eds) *Environment, climate, plant and vegetation growth*. Springer Publ Ltd, Springer Nature Switzerland AG. Part of Springer Nature, pp 63–92. doi:10.1007/978-3-030-49732-3

Marguerite du Plessis D (2011) *Fabrication and characterization of anti-microbial and biofouling resistant nanofibers with silver nanoparticles and immobilized enzymes for application in water filtration*. Master thesis, University of Stellenbosch.

Md Jakir H, Allah B (2020) Development and applications of transplastomic plants; a way towards eco-friendly agriculture. In: Fahad S, Hasanuzzaman M, Alam M, Ullah H, Saeed M, Khan AK, Adnan M (eds) *Environment, climate, plant and vegetation growth*. Springer Publ Ltd, Springer Nature Switzerland AG. Part of Springer Nature, pp 285–322. doi:10.1007/978-3-030-49732-3

Md Enamul H, Shoeb AZM, Mallik AH, Fahad S, Kamruzzaman MM, Akib J, Nayyer S, Mehedi AKM, Swati AS, Md Yeamin A, Most SS (2020) Measuring vulnerability to environmental hazards: qualitative to quantitative. In: Fahad S, Hasanuzzaman M, Alam M, Ullah H, Saeed M, Khan AK, Adnan M (eds) *Environment, climate, plant and vegetation growth*. Springer Publ Ltd, Springer Nature Switzerland AG. Part of Springer Nature, pp 421–452. doi:10.1007/978-3-030-49732-3

Mohammad AW, Ahmad M, Usman ARA, Akanji M, Rafique MI (2020a) Advances in pyrolytic technologies with improved carbon capture and storage to combat climate change. In: Fahad S, Hasanuzzaman M, Alam M, Ullah H, Saeed M, Khan AK, Adnan M (eds) *Environment, climate, plant and vegetation growth*. Springer Publ Ltd, Springer Nature Switzerland AG. Part of Springer Nature, pp 535–576. doi:10.1007/978-3-030-49732-3

Mohammad AW, Abdelazeem S, Munir A, Khalid E, Adel RAU (2020b) Extent of climate change in Saudi Arabia and its impacts on agriculture: a case study from Qassim region. In: Fahad S, Hasanuzzaman M, Alam M, Ullah H, Saeed M, Khan AK, Adnan M (eds) *Environment, climate, plant and vegetation growth*. Springer Publ Ltd, Springer Nature Switzerland AG. Part of Springer Nature, pp 635–658. doi:10.1007/978-3-030-49732-3

Monteiro-Riviere I, Rasmussen R (2007) Dermal effects of nanomaterials. In: Monteiro-Riviere NA, Tran CL (eds) *Nanotoxicology – characterization, dosing and health effects*. Informa Healthcare, New York.

Mubeen M, Ashfaq A, Hafiz MH, Muhammad A, Hafiz UF, Mazhar S, Muhammad SD, Asad A, Amjed A, Fahad S, Wajid N (2020) Evaluating the climate change impact on water use efficiency of cotton-wheat in semi-arid conditions using DSSAT model. *J Water Clim Change*. doi:10.2166/wcc.2019.179/622035/jwc2019179.pdf

Muhammad TQ, Amna F, Amna B, Barira Z, Xitong Z, Ling-Ling C (2020) Effectiveness of conventional crop improvement strategies vs. omics. In: Fahad S, Hasanuzzaman M, Alam M, Ullah H, Saeed M, Khan AK, Adnan M (eds) *Environment, climate, plant and vegetation growth*. Springer Publ Ltd, Springer Nature Switzerland AG. Part of Springer Nature, pp 253–284. doi:10.1007/978-3-030-49732-3

Muhammad Z, Abdul MK, Abdul MS, Kenneth BM, Muhammad S, Shahen S, Ibadullah J, Fahad S (2019) Performance of *Aeluropus lagopoides* (mangrove grass) ecotypes, a potential turfgrass, under high saline conditions. *Environ Sci Pollut Res*. doi:10.1007/s11356-019-04838-3

Murray CB, Kagan CR, Bawendi MG (2000) Synthesis and characterisation of monodisperse nanocrystals and close-packed nanocrystal assemblies. *Annu Rev Mater Res* 30:545–610.

Nandhini M, Rajini SB, Udayashankar AC, Niranjana SR, Lund OS, Shetty HS, Prakash HS (2019) Biofabricated zinc oxide nanoparticles as an eco-friendly alternative for growth promotion and management of downy mildew of pearl millet. *Crop Prot* 121:103–112.

Noor M, Ur Rehman N, Ajmal J, Fahad S, Muhammad A, Fazli W, Saud S, Hassan S (2020) Climate change and coastal plant lives. In: Fahad S, Hasanuzzaman M, Alam M, Ullah H, Saeed M, Khan AK, Adnan M (eds) *Environment, climate, plant and vegetation growth*. Springer Publ Ltd, Springer Nature Switzerland AG. Part of Springer Nature, pp 93–108. doi:10.1007/978-3-030-49732-3

Nurkiewicz TR, Porter DW, Hubbs AF, Cumpston JL, Chen BT, Frazer DG, Castranova V (2008) Nanoparticle inhalation augments particle-dependent systemic microvascular dysfunction. *Part Fibre Toxicol* 5:1. doi:10.1186/1743-8977-5-1

OECD and Allianz (2008) Sizes that matter: opportunities and risks of nanotechnologies. Report in Cooperation with the OECD International Futures Programme. http://www.oecd.org/dataoecd/32/1/441 08334.pdf. Accessed 20 May 2010.

Ogunyemi SO, Abdallah Y, Zhang M, Fouad H, Hong X, Ibrahim E, … Li B (2019) Green synthesis of zinc oxide nanoparticles using different plant extracts and their antibacterial activity against *Xanthomonas oryzae* pv. oryzae. *Artif Cells Nanomed Biotechnol* 47(1):341–352.

Palmqvist NM, Seisenbaeva GA, Svedlindh P et al (2017) Maghemite nanoparticles acts as nanozymes, improving growth and abiotic stress tolerance in *Brassica napus*. *Nanoscale Res Lett* 12:631.

Periera L (2003) Irrigation demand management to cope with drought and water scarcity. Water Science and Technology Library. In: *Tools for drought mitigation in Mediterranean regions*, vol. 44, pp 19–33. doi:10.1007/978-94-010-0129-8_2

Pramanik P, Ray P, Maity A, Das S, Ramakrishnan S, Dixit P (2020) Nanotechnology for improved carbon management in soil. In: *Carbon management in tropical and sub-tropical terrestrial systems*. Springer, Singapore, pp 403–415.

Prasad R, Bhattacharyya A, Nguyen QD (2017) Nanotechnology in sustainable agriculture: recent developments, challenges, and perspectives. *Front Microbiol* 8:1014.

Pronovost AD, Hickey ME (2014) U.S. Patent No. 8871186. U.S. Patent and Trademark Office, Washington, DC.

Qamar Z, Zubair A, Muhammad Y, Muhammad ZI, Abdul K, Fahad S, Safder B, Ramzani PMA, Muhammad N (2017) Zinc biofortification in rice: leveraging agriculture to moderate hidden hunger in developing countries. *Arch Agron Soil Sci* 64:147–161. doi:10.1080/03650340.2017.1338343

Radomski A, Jurasz P, Alonso-Escolano D, Drews M, Morandi M, Malinski T, Radomski MW (2005) Nanoparticle-induced platelet aggregation and vascular thrombosis. *Brit J Pharmacol* 146(6):882–893.

Rashid M, Qaiser H, Khalid SK, Mohammad AW, Zhang A, Muhammad A, Shahzada SI, Rukhsanda A, Ghulam AS, Shahzada MM, Sarosh A, Muhammad FQ (2020) Prospects of biochar in alkaline soils to mitigate climate change. In: Fahad S, Hasanuzzaman M, Alam M, Ullah H, Saeed M, Khan AK, Adnan M (eds) *Environment, climate, plant and vegetation growth*. Springer Publ Ltd, Springer Nature Switzerland AG. Part of Springer Nature, pp 133–150. doi:10.1007/978-3-030-49732-3

Raveendran P, Fu J, Wallen SL (2003) Completely "green" synthesis and stabilization of metal nanoparticles. *J Am Chem Soc* 125:13940–13941.

Rehman M, Fahad S, Saleem MH, Hafeez M, Ur Rahman MH, Liu F, Deng G (2020) Red light optimized physiological traits and enhanced the growth of ramie (*Boehmeria nivea* L.). *Photosynthetica* 58(4):922–931.

Sabagh AEL, Hossain A, Barutçular C, Iqbal MA, Islam MS, Fahad S, Sytar O, Çig F, Meena RS, Erman M (2020) Consequences of salinity stress on the quality of crops and its mitigation strategies for sustainable crop production: an outlook of arid and semi-arid regions. In: Fahad S, Hasanuzzaman M, Alam M, Ullah H, Saeed M, Khan AK, Adnan M (eds) *Environment, climate, plant and vegetation growth*. Springer Publ Ltd, Springer Nature Switzerland AG. Part of Springer Nature, pp 503–534. doi:10.1007/978-3-030-49732-3

Sadam M, Muhammad TQ, Ghulam M, Muhammad SK, Faiz AJ (2020) Role of biotechnology in climate resilient agriculture. In: Fahad S, Hasanuzzaman M, Alam M, Ullah H, Saeed M, Khan AK, Adnan M (eds) *Environment, climate, plant and vegetation growth*. Springer Publ Ltd, Springer Nature Switzerland AG. Part of Springer Nature, pp 339–366. doi:10.1007/978-3-030-49732-3

Sajid H, Jie H, Jing H, Shakeel A, Satyabrata N, Sumera A, Awais S, Chunquan Z, Lianfeng Z, Xiaochuang C, Qianyu J, Junhua Z (2020) Rice production under climate change: adaptations and mitigating strategies. In: Fahad S, Hasanuzzaman M, Alam M, Ullah H, Saeed M, Khan AK, Adnan M (eds) *Environment, climate, plant and vegetation growth*. Springer Publ Ltd, Springer Nature Switzerland AG. Part of Springer Nature, pp 659–686. doi:10.1007/978-3-030-49732-3

Sajjad H, Muhammad M, Ashfaq A, Waseem A, Hafiz MH, Mazhar A, Nasir M, Asad A, Hafiz UF, Syeda RS, Fahad S, Depeng W, Wajid N (2019) Using GIS tools to detect the land use/land cover changes during forty years in Lodhran district of Pakistan. *Environ Sci Pollut Res*. doi:10.1007/s11356-019-06072-3

Saleem MH, Fahad S, Adnan M, Mohsin A, Muhammad SR, Muhammad K, Qurban A, Inas AH, Parashuram B, Mubassir A, Reem MH (2020a) Foliar application of gibberellic acid endorsed phytoextraction of copper and alleviates oxidative stress in jute (*Corchorus capsularis* L.) plant grown in highly copper-contaminated soil of China. *Environ Sci Pollution Res*. doi: 10.1007/s11356-020-09764-3

Saleem MH, Fahad S, Shahid UK, Mairaj D, Abid U, Ayman ELS, Akbar H, Analía L, Lijun L (2020c) Copper-induced oxidative stress, initiation of antioxidants and phytoremediation potential of flax (*Linum usitatissimum* L.) seedlings grown under the mixing of two different soils of China. *Environ Sci Poll Res*. doi: 10.1007/s11356-019-07264-7

Saleem MH, Rehman M, Fahad S, Tung SA, Iqbal N, Hassan A, Ayub A, Wahid MA, Shaukat S, Liu L, Deng G (2020b) Leaf gas exchange, oxidative stress, and physiological attributes of rapeseed (*Brassica napus* L.) grown under different light-emitting diodes. *Photosynthetica* 58(3):836–845.

Saleh J, Maftoun M (2008) Interactive effect of NaCl levels and Zinc sources and levels on the growth and mineral composition of rice. *J Agric Sci Tech* 10:325–336.

Saman S, Amna B, Bani A, Muhammad TQ, Rana MA, Muhammad SK (2020) QTL mapping for abiotic stresses in cereals. In: Fahad S, Hasanuzzaman M, Alam M, Ullah H, Saeed M, Khan AK, Adnan M (eds) *Environment, climate, plant and vegetation growth*. Springer Publ Ltd, Springer Nature Switzerland AG. Part of Springer Nature, pp 229–252. doi: 10.1007/978-3-030-49732-3

Saud S, Chen Y, Fahad S, Hussain S, Na L, Xin L, Alhussien SA (2016) Silicate application increases the photosynthesis and its associated metabolic activities in Kentucky bluegrass under drought stress and post-drought recovery. *Environ Sci Pollut Res* 23(17):17647–17655. doi: 10.1007/s11356-016-6957-x

Saud S, Chen Y, Long B, Fahad S, Sadiq A (2013) The different impact on the growth of cool season turf grass under the various conditions on salinity and drought stress. *Int J Agric Sci Res* 3:77–84.

Saud S, Fahad S, Cui G, Chen Y, Anwar S (2020) Determining nitrogen isotopes discrimination under drought stress on enzymatic activities, nitrogen isotope abundance and water contents of Kentucky bluegrass. *Sci Rep* 10:6415. doi: 10.1038/s41598-020-63548-w

Saud S, Fahad S, Yajun C, Ihsan MZ, Hammad HM, Nasim W, Amanullah Jr, Arif M, Alharby H (2017) Effects of nitrogen supply on water stress and recovery mechanisms in Kentucky bluegrass plants. *Front Plant Sci* 8:983. doi: 10.3389/fpls.2017.00983

Saud S, Li X, Chen Y, Zhang L, Fahad S, Hussain S, Sadiq A, Chen Y (2014) Silicon application increases drought tolerance of Kentucky bluegrass by improving plant water relations and morph physiological functions. *Sci World J* 2014:1–10. doi: 10.1155/2014/368694

Scott NR (2005) Nanotechnology and animal health. *Rev Sci Tech Off Int Epiz* 24(1):425–432.

Senol C (2020) The effects of climate change on human behaviors. In: Fahad S, Hasanuzzaman M, Alam M, Ullah H, Saeed M, Khan AK, Adnan M (eds) *Environment, climate, plant and vegetation growth*. Springer Publ Ltd, Springer Nature Switzerland AG. Part of Springer Nature, pp 577–590. doi: 10.1007/ 978-3-030-49732-3

Shafi MI, Adnan M, Fahad S, Fazli W, Ahsan K, Zhen Y, Subhan D, Zafar-ul-Hye M, Martin B, Rahul D (2020) Application of single superphosphate with humic acid improves the growth, yield and phosphorus uptake of wheat (*Triticum aestivum* L.) in calcareous soil. *Agron* (10):1224. doi: 10.3390/ agronomy10091224

Shah F, Lixiao N, Kehui C, Tariq S, Wei W, Chang C, Liyang Z, Farhan A, Fahad S, Huang J (2013) Rice grain yield and component responses to near 2°C of warming. *Field Crop Res* 157:98–110

Sh YH, Xu ZR, Feng JL, Wang CZ (2006) Efficacy of modified montmorillonite nanocomposite to reduce the toxicity of aflatoxin in broiler chicks. *Animal Feed Sci Technol* 129:138–148.

Shinde S, Paralikar P, Ingle AP, Rai M (2020) Promotion of seed germination and seedling growth of Zea mays by magnesium hydroxide nanoparticles synthesized by the filtrate from *Aspergillus niger*. *Arab J Chem* 13(1):3172–3182.

Soliman M, Qari SH, Abu-Elsaoud A et al (2020) Rapid green synthesis of silver nanoparticles from blue gum augment growth and performance of maize, fenugreek, and onion by modulating plants cellular antioxidant machinery and genes expression. *Acta Physiol Plant* 42:148.

Subhan D, Zafar-ul-Hye M, Fahad S, Saud S, Martin B, Tereza H, Rahul D (2020) Drought stress alleviation by ACC Deaminase Producing *Achromobacter xylosoxidans* and *Enterobacter cloacae*, with and without timber waste biochar in maize. *Sustain* 12(6286). doi: 10.3390/su12156286

Tanveer M, Shahzad B, Ashraf U (2020) Nanoparticle application and abiotic-stress tolerance in plants. In: *Plant life under changing environment*. Academic Press, pp 627–641.

Tariq M, Ahmad S, Fahad S, Abbas G, Hussain S, Fatima Z, Nasim W, Mubeen MH, Khan MA, Adnan M (2018) The impact of climate warming and crop management on phenology of sunflower-based cropping systems in Punjab, Pakistan. *Agric Forest Met* 15(256):270–282.

Tinkle A, Rich R, Salmen D, Adkins (2003) Skin as a route of exposure and sensitisation in chronic beryllium disease. *Environ Health Perspect* 111(9):1202–1208.

Tsuji M, Howard (2006) Research strategies for safety evaluation of nanomaterials, part IV: risk assessment of nanoparticles. *Toxicol Sci* 89(1):42–50.

Tungittiplakorn W, Lion LW, Cohen C, Kim JY (2004) Engineered polymeric nanoparticles for soil remediation. *Environ Sci Technol* 38:1605–1610. doi:10.1021/es0348997

United States Department of Agriculture (USDA) (2003) Nanoscale science and engineering for agriculture and food systems. National Planning Workshop, 18–19 November 2002. A Report Submitted to the Cooperative State Research and Extension Service, Washington, DC. 68 p.

Unsar NU, Muhammad R, Syed HMB, Asad S, Mirza AQ, Naeem I, Muhammad HR, Fahad S, Shafqat S (2020) Insect pests of cotton crop and management under climate change scenarios. In: Fahad S, Hasanuzzaman M, Alam M, Ullah H, Saeed M, Khan AK, Adnan M (eds) *Environment, climate, plant and vegetation growth*. Springer Publ Ltd, Springer Nature Switzerland AG. Part of Springer Nature, pp 367–396. doi:10.1007/978-3-030-49732-3

Usman M, Farooq M, Wakeel A, Nawaz A, Cheema SA, Ur Rehman H, … Sanaullah M (2020) Nanotechnology in agriculture: current status, challenges and future opportunities. *Sci Total Environ* 137778.

Vanti GL, Masaphy S, Kurjogi M, Chakrasali S, Nargund VB (2020) Synthesis and application of chitosan-copper nanoparticles on damping off causing plant pathogenic fungi. *Int J Biol Macromol* 156: 1387–1395.

Vanti GL, Nargund VB, Vanarchi R, Kurjogi M, Mulla SI, Tubaki S, Patil RR (2019) Synthesis of *Gossypium hirsutum*-derived silver nanoparticles and their antibacterial efficacy against plant pathogens. *Appl Organomet Chem* 33(1):e4630.

Wahid F, Fahad S, Subhan D, Adnan M, Zhen Y, Saud S, Manzer HS, Martin B, Tereza H, Rahul D (2020) Sustainable management with mycorrhizae and phosphate solubilizing bacteria for enhanced phosphorus uptake in calcareous soils. *Agriculture* 10(334). doi:10.3390/agriculture10080334

Wajid N, Ashfaq A, Asad A, Muhammad T, Muhammad A, Muhammad S, Khawar J, Ghulam MS, Syeda RS, Hafiz MH, Muhammad IAR, Muhammad ZH, Muhammad HR, Veysel T, Fahad S, Suad S, Aziz K, Shahzad A (2017) Radiation efficiency and nitrogen fertilizer impacts on sunflower crop in contrasting environments of Punjab. *Pak Environ Sci Pollut Res* 25:1822–1836. doi:10.1007/s11356-017-0592-z

Watlington K (2005) *Emerging nanotechnologies for site remediation and wastewater treatment*. National Network for Environmental Management Studies Fellow, North Carolina State University.

Wilson EF, Abbas H, Duncombe BJ, Streb C, Long DL, Cronin L (2008) Probing the self-assembly of inorganic cluster architectures in solution with cryospray mass spectrometry: growth of polyoxomolybdate clusters and polymers mediated by silver (I) ions. *J Am Chem Soc* 130(42):13876–13884.

Wu C, Kehui C, She T, Ganghua L, Shaohua W, Fahad S, Lixiao N, Jianliang H, Shaobing P, Yanfeng D (2020) Intensified pollination and fertilization ameliorate heat injury in rice (*Oryza sativa* L.) during the flowering stage. *Field Crops Res* 252:107795.

Wu C, Tang S, Li G, Wang S, Fahad S, Ding Y (2019) Roles of phytohormone changes in the grain yield of rice plants exposed to heat: a review. *PeerJ* 7:e7792. doi:10.7717/peerj.7792

Yang Z, Zhang Z, Zhang T, Fahad S, Cui K, Nie L, Peng S, Huang J (2017) The effect of season-long temperature increases on rice cultivars grown in the central and southern regions of China. *Front Plant Sci* 8:1908. doi:10.3389/fpls.2017.01908

Younas A, Yousaf Z, Rashid M, Riaz N, Fiaz S, Aftab A, Haung S (2020) Nanotechnology and plant disease diagnosis and management. In: *Nanoagronomy*. Springer, Cham, pp 101–123.

Younas A, Yousaf Z, Riaz N, Rashid M, Razzaq Z, Tanveer M, Huang S (2020) Role of nanotechnology for enhanced rice production. In: *Nutrient dynamics for sustainable crop production*. Springer, Singapore, pp 315–350.

Zafar-ul-Hye M, Muhammad N, Subhan D, Fahad S, Rahul D, Mazhar A, Ashfaq AR, Martin B, Jiˇrí H, Zahid HT, Muhammad N (2020a) Alleviation of cadmium adverse effects by improving nutrients uptake in bitter gourd through cadmium tolerant rhizobacteria. *Environments* 7(54). doi:10.3390/environments7080054

Zafar-ul-Hye M, Muhammad TH, Muhammad A, Fahad S, Martin B, Tereza D, Rahul D, Subhan D (2020b) Potential role of compost mixed biochar with rhizobacteria in mitigating lead toxicity in spinach. *Sci Rep* 10:12159. doi:10.1038/s41598-020-69183-9

Zahida Z, Hafiz FB, Zulfiqar AS, Ghulam MS, Fahad S, Muhammad RA, Hafiz MH, Wajid N, Muhammad S (2017) Effect of water management and silicon on germination, growth, phosphorus and arsenic uptake in rice. *Ecotoxicol Environ Saf* 144:11–18.

Zahoor M, Afzal M, Ali M, Mohammad W, Khan N, Adnan M, Ali A, Saeed M (2016) Effect of organic waste and NPK fertilizer on potato yield and soil fertility. *Pure Appl Biol* 5(3):439–445.

Zhang Q, Han L, Jing H, Blom DA, Lin Y, Xin HL, Wang H (2016) Facet control of gold nanorods. *ACS Nano* 10(2):2960–2974.

Zhao L, Lu L, Wang A, Zhang H, Huang M, Wu H, Ji R (2020) Nano-biotechnology in agriculture: use of nanomaterials to promote plant growth and stress tolerance. *J Agric Food Chem* 68(7):1935–1947.

Zia-ur-Rehman M (2020) Environment, climate change and biodiversity. In: Fahad S, Hasanuzzaman M, Alam M, Ullah H, Saeed M, Khan AK, Adnan M (eds) *Environment, climate, plant and vegetation growth*. Springer Publ Ltd, Springer Nature Switzerland AG. Part of Springer Nature, pp 473–502. doi:10.1007/978-3-030-49732-3

7

Elevated CO_2 in Combination with Heat Stress Influences the Growth and Productivity of Cereals: Adverse Effect and Adaptive Mechanisms

Ayman EL Sabagh, Akbar Hossain, Mohammad Sohidul Islam, Sharif Ahmed, Ali Raza, Muhammad Aamir Iqbal, Allah Wasaya, Disna Ratnasekera, Adnan Arshad, Arpna Kumari, Subhan Danish, Paul Ola Igboji, Rahul Datta, Sytar Oksana, Skalický Milan, Marian Brestic, Kulvir Singh, Muhammad Ali Raza, and Shah Fahad

CONTENTS

DOI: 10.1201/9781003160717-7

7.1 Introduction

The global picture on climate change has been terrifying. Apart from the controversial ozone layer depletion, there is increased radiation, global warming, incidences of excessive or sparse rainfall, flood, drought, desertification, melting of arctic ice, rise in sea level, tsunami, and possible armageddon, etc. being experienced with loss of agricultural and water resources (Raza et al. 2019b, 2020). The unabated loss of global agricultural and water resources poses a significant threat to world peace and togetherness. Yet, these losses cannot be made up when the world is languishing under the effects of hunger and starvation, especially in Sub-Saharan Africa, Asia, and Latin America, where most of the world's citizens reside. Most of the time, what breeds civil conflicts and unrest is a struggle for scarce natural resources, especially land, food, clothing, and shelter. As the saying goes, 'A hungry man is an angry man' (Igboji et al. 2018). Changing climate is possibly the most nagging and imperative environmental issue on the world's priority list. The earth's systems of air, water, and land have always been dynamic as studies of ancient climates show that there have been alternating periods of global warmth and global chill at various times. However, it's now that ample proofs are testifying to the fact that human actions are altering or possibly rushing climate change (Raza et al. 2019).

The global temperature level has increased by 0.6°C in the previous 130 years. Whereas carbon dioxide (CO_2) concentrations in the air have increased by about 25% in the past 200 years, a cumulating effect from about 280–356 ppm. Methane (CH_4) concentrations in the air have crumpled during the past 100 years. Collectively taken, CO_2, CH_4 and N_2O are greenhouse smokes/gases which comprise reradiation of heat energy released from the ground's outward and which is making the earth hotter. The gas levels are increasing mostly due to human activities associated with energy production, transportation, and farming. The fossil fuel-linked emissions account for 65% of the extra transience degree attributable to atmospheric contamination and 70% of anthropogenic atomizers' environment preservation (Lelieveld et al. 2019). The rank of the gas ordered by percentage of emission due to human-induced universal heating is CO_2 (70%), CH_4 (20%), and N_2O plus other gases (10%). It is supposed that particles in the air, for instance, from engineering activities or volcanic eruptions, imitate rays and harvest a chilling outcome. A doubling of CO_2 levels would theoretically lead to an average global temperature rise of 1–2 °C even if all other factors remain the same (Anderson et al. 2016). But in reality, other factors will also change in response to the rising temperature and may show some negative (bad) and positive (good) fallouts. For example, water vapour in the atmosphere increases as temperature rises (Hudson Institute 1999). Information on the opposing influences of climate alteration on the growth and production of crops because of the elevated CO_2 (eCO_2) concentration and elevated temperature (eT) is obstinately illuminating the necessity for the advancement of tools to measure the response of crops owing to climate alteration aspects (Mina et al. 2019). Most of the physiological traits of crops are adversely affected by changing precipitation, a higher level of CO_2 and rise in temperature (Lawlor and Mitchell 2000; Raza et al. 2019; EL Sabagh et al. 2020). Such adverse effects on crops are usually referred to as the changing environment induced stresses (Raza 2020). It is reported that the increased temperature to 30 °C during floret development can cause sterility in wheat (Saini and Aspinall 1982). Schlenker and Roberts (2009) reported that maize yield could be increased up to 29°C but beyond that it decreases. It is established by Lobell et al. (2011) that the increase of temperature by every 1°C negatively impacts the yield of maize. The model projects carried out in India show that a rise of 2°C in temperature stops the optimistic signs of crop performance, like crop quality, from the greater level of CO_2. For example, the yield of wheat is anticipated to be lessened by 19.04%, consequent upon the mutual effect of rising temperature and CO_2 concentration to 4.5°C and 390–630 ppm, respectively (Verma and Misra 2018). Therefore, this chapter discussed the importance of eCO_2 and temperature on cereal crops' physiological traits, so as to understand the subject matter and accelerate work on addressing the critical challenges.

7.2 Impact of Climate Change on Escalating Temperature and eCO$_2$

The projected rise in average temperatures anywhere up to 3.7 degrees (IPCC 2007) by the end of the present era might have a substantial influence on crops vernalization (Adhikari et al. 2016; Ali et al. 2016). Rising temperature is also affecting barley phenology particularly pollination, flowering time (Amasino 2010), and grain filling by both disturbing the degree of growth and expansion precisely and persuading eCO$_2$ beneficial offsets (Robredo ct al. 2007; Smith et al. 2011; Moore and Lobell 2014). Studies of Wallwork et al. (1998a) andSingh et al. (2008) evidenced the product degradation of endosperm storage when the high temperature (up to 35 degrees) for five days was applied to malting barley plants during grain-filling. Furthermore, a sign of endosperm cell wall and crushed cell layer (CCL) progress were highly sensitive to maximum heat, and the compact width of the CCL caused incomplete hydrolysis of b-glucans. Contraction of 11–75% in the movement of enzymes under study following exposure to extreme temperatures is evident (Wallwork et al. 1998b). Meanwhile, the development of the embryo significantly increased in heat-treated seeds than that in the normal condition seeds.

Higher CO$_2$ (>1000 ppm) might potentially progress yield of important crops (Körner 2006), like amongst species of C$_3$ plants which are likely to inherit such an advantage. The shared effect on flowering time and grain filling needs systematic evaluation (Fuhrer 2003; Lee 2011). Lee (2011) reported that by rising of temperature by 4°C, the time of flowering was advanced by 31 and 50 days for *Setaria viridis* and *Chenopodium album*, respectively.

Stomatal conductance unerringly alters plant water contents and photosynthesis. Its effects also improved with growing temperature notwithstanding the decline in leaf water probable, upsurge in transpiration, and upsurge in intercellular CO$_2$ meditation, and was decoupled by photosynthesis (Urban et al. 2017). It was observed that control on stomatal inaugural under eCO$_2$ was not altered by temperature; however, their amalgamation meaningfully enhanced entire-plant functioning (Habermann et al. 2019).

Although rising atmospheric CO$_2$ levels are identified with increased photosynthesis and production of biomass, the signs of crop performance, like crop quality, on plants' chemical composition requires study (Erbs et al. 2015). The conclusion reached from the FACE (free air CO$_2$ enrichment) facility in California, USA, suggested that when eCO$_2$ touched 680 ppm on a grassland environment, forbs flowering was accelerated by 2–4 days. Whereas, in the dominant grasslands, flowering time of grasses were delayed by 2–6 days (Cleland et al. 2007). However, the peak inspected concentration of CO$_2$ at 3000 ppm (ambient +4°C-day/5°C-night) significantly stimulated biomass production of barley (Juknys et al. 2012). The maximum biomass accumulation was observed under the shared effect of temperature and eCO$_2$ for tomato and barley.

Ko et al. (2019) had run the DSSAT CERES-barley crop model to simulate the yields potential of multiple barley cultivars, and reported reasonable increases under demonstrative attentiveness pathways 4.5 (RCP 4.5) and quick increases under the RCP 8.5 scenario. In the context of future scenarios, it is very important to develop barley varieties adaptive to forthcoming environments with calls for better-quality yield under weather warming and eCO$_2$ scenarios. Augmented temperature may defend the nutritional excellence of crops under upcoming eCO$_2$ meditations (Kohler et al. 2019). The mixture of eCO$_2$ with eT usually reinstated seed iron and zinc meditations gained under usual CO$_2$ and temperature environments, signifying that the possible danger to human nourishment by growing CO$_2$ meditation may not be complete.

7.3 Influence of eCO$_2$ and Temperature on Physiological Traits of Cereal Crops

Every passing decade is becoming more warmer than the previous one since 1850 (IPCC 2013). Melting of ice, increasing sea level, and continuous greenhouse gas emissions are making the situation worse (Nastis et al. 2012). Under such circumstances, global food security has become a complex challenge, especially in developing countries (Tyfield 2011; Mendelsohn 2014; De Laurentiis et al. 2016).

Changing climatic conditions are directly associated with the crop's productivity (Jones 2013), especially cereals crops that face unpredictable challenges due to increasing levels of CO_2 in the air and increased temperature of the earth (Raza et al. 2019, 2020; Raza 2020). Most of the physiological traits of crops are adversely affected by changing precipitation patterns, a higher level of CO_2, and increased temperature (Lawlor and Mitchell 2000). Such adverse effects on crops are usually referred to as the changing environment induced stresses. Some of the negative effects of changing climate on cereal crops are provided in Figure 7.1.

For a better understanding of crop response towards eCO_2 and eT, characterization of stomata is a key physiological trait (Lobell et al. 2013). These stomata are directly in contact with atmospheric CO_2 (aCO_2) level, which is negatively associated with rice's stomata density. The stomata size has shown positive correlation with eCO_2 concentration in the atmosphere (Franks et al. 2009). Mesophyll cells and inner lateral walls of epidermal guard cells of stomata are specialized in sensing the CO_2 concentration (Tubiello and Ewert 2002; Attri and Rathore 2003). The higher atmospheric temperature usually caused stomatal closure in cereal crops when water evaporation is increased (Figure 7.2). Such a cereal plant strategy played a decisive role in leaf moisture conservation (Bernacchi et al. 2007). Lower latent heat loss due to stomatal closure also increased the leaf's surface temperature in conditions of eCO_2 (Kimball and Bernacchi 2006). It has been observed that low conductance of stomata also disturbs the photosynthesis rate, which is a key factor for low carbohydrate accumulation in the C_4 cereal seeds and poor productivity of crops. Xu et al. (2011) and

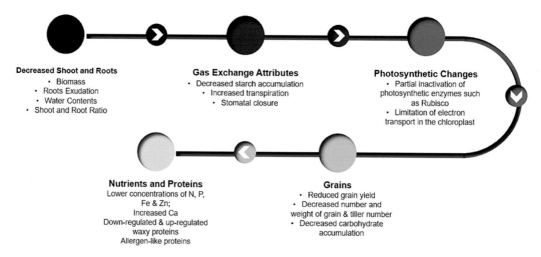

FIGURE 7.1 Different negative possessions of eCO_2 and temperature on cereals crops.

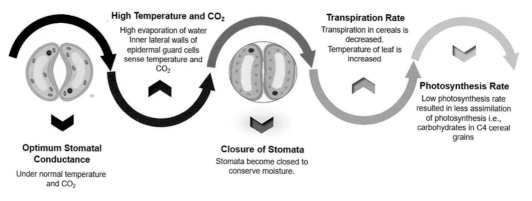

FIGURE 7.2 Stomatal responses of cereals towards eCO_2 and temperature.

Lobell et al. (2013) observed a significant reduction in maize plants' stomata conductance under temperature conditions seen to hover in the range of 36–42°C.

At the same time, Chavan et al. (2019) found that eCO_2 improves the adverse influence of heat stress on wheat functioning, nonetheless without affecting grain harvest. In non-heat stressed plants, eCO_2 enhances CO_2 adaptation rates (+36%) evaluated at growth CO_2, notwithstanding having down-regulated the photosynthetic potential. Heat stress compact assimilation rates (–42%) in atmospheric carbon dioxide (aCO_2) but not in eCO_2 resulted in full-fledged plants since CO_2 threatened photo-synthesis via rising ribulose bisphosphate rejuvenation scope and decreasing photochemical injury under heat stress (Chavan et al. 2019).

7.3.1 C_3 Cereal Plants and eCO_2

If there is short-term exposure to CO_2, crops have been observed to display an increase in the rate of photosynthesis (Moore et al. 1998; Xu and Shen 2002; Von Caemmerer and Furbank 2003). However, long-term exposure to high CO_2 concentration may acclimatize the crop to its environment with prolonged growth at elevated atmospheric CO_2 or they may become insensitive (Delucia et al. 1985; Arp 1991; Farage et al. 1991; Lin et al. 1997). A higher level of CO_2 in the air increased the accumulation of carbohydrates in C_3 cereals as a result of occurrence of successive acclimation of photosynthetic capacity (Myers et al. 2014). In contrast, during prolonged exposure of crop to elevated CO_2, Rubisco becomes deactivated, indicating that cereal crops do not necessarily behave in predictable ways like showing an ideal acclimation response towards a high level of CO_2 (Makino 1994).

7.3.2 C_4 Cereal Plants and eCO_2

A wide range of adaptation mechanisms have been documented so far in C_4 cereal plants vis-à-vis exposure to a higher level of CO_2 viz. (a) Rubisco in the bundle sheath cells (Ziska and Bunce 1997), (b) CO_2 saturation effect (Ziska and Bunce 1997), (c) Adjustment of direct CO_2 and bundle sheath leakiness (Watling et al. 2000), (d) C_3-like photosynthesis in early C_4 leaves (Ziska et al. 1999) and (e) Lower stomatal conductance (Ainsworth and Rogers 2007).

Upsurge in the respiration of plants as compared to photosynthesis caused a decline in a net gain of carbon under high temperature (Valentini et al. 2000). In C_3 cereal crops, i.e., rice and wheat, the rate of photosynthesis becomes maximum when the temperature ranges between 20 and 32°C. The respiration rate becomes non-linearly rapid from 15 to 40°C and, then, eventually becomes linearly decreased (Porter and Semenov 2005). In chloroplast, limited activity of electron transport chain and deactivation of rubisco activase have been seen to decrease the potential photosynthesis of cereal crops in the face of heat levels increasing due to high temperature (Joshi and Karan 2013). Modification in kinetic constants of rubisco have been noted to disturb the CO_2/O_2 ratio in cereal plants with results including decreased solubilization of CO_2 and increased photorespiration (Crafts-Brandner and Law 2000). Such conditions also minimized the carboxylation in plants at eT (Andersson and Backlund 2008; Zahoor et al. 2016). The photosystem II of cereal crops exhibited maximum sensitivity towards high temperature-induced stress, as it does not show signs of being disturbed until the temperature of leaf touches 40°C or more (Markelz et al. 2014). Depending upon the cultivar of cereal crops, an increase of 0.5–1.7°C in leaf temperature has been docu-mented under conditions of twice the concentration of CO_2 as normal. The increase in leaf tem-perature can also be a high of 3°C under specified weather conditions. In addition to the above facts, the longevity of proteins and chlorophyll in crops is also curtailed under high atmospheric temperature which also caused a significant decrease in the high photosynthetic capacity of cereal crops (Paul et al. 2014).

Under elevated temperature and CO_2 induced stress, the accumulation of toxic compounds i.e., reactive oxygen species (ROS), is increased. High ROS-induced oxidative damage in lipids, DNA, and proteins which are synthesized in peroxisomes, chloroplasts, and mitochondria is well docu-mented (Hasanuzzaman et al. 2020). These ROS caused peroxidation of lipid membrane in cereal crops at the cellular level, especially in wheat and rice (Király and Czövek 2002; Shah et al. 2011). Antioxidants malondialdehyde (MDA), ascorbic acid (AsA), and glutathione (GSH) contents are also increased under high-temperature stress which played a vital role in detoxification of ROS. Synthesis of scavenging enzymes, i.e., superoxide dismutase (SOD), catalase (CAT), ascorbate peroxidase (APX), peroxiredoxin (PrxR), and glutathione peroxidase (GPX), also acts as an imperative signalling molecule in minimizing the toxic effects of ROS (Savicka and Škute 2010; Hasanuzzaman et al. 2013, 2020).

Furthermore, accumulation of endogenous stress ethylene in cereal crops under abiotic stress i.e., high temperature, is a well-known fact. Stress conditions stimulate the methionine to convert into S-adenosyl-Met. Activation of enzyme ACC synthase changes into S-adenosyl-Met into ACC. This ACC is then catalyzed by ACC oxidase, with resulting biosynthesized stress generating ethylene (Glick et al. 1998). Higher biosynthesis and accumulation of ethylene ultimately induced negative changes in the crops' development phases and decreased plant growth (Arshad et al. 2008). Higher accumulation of stress ethylene in plant roots plays a decisive role in decreasing the root elongation. It promotes the thickness of plant roots via accumulation of dead cells in the root cortex. Such accumulation of dead cells in the root cortex results in the formation of lysigenous aerenchyma. In the region of hypocotyls, stress ethylene also decreases the cell division. Low cell division eventually results in poor elongation of roots and shoots in plants (Skirycz et al. 2011).

Matile et al. (1997) also observed a similar reduction in the synthesis of chlorophyll in plants as a result of higher biosynthesis of ethylene under stress, and they suggested that the outburst of ethylene under stress condition degraded the lipid which resulted in the loss of chloroplast cell membrane integrity. In chloroplast, the chlorophyllase (chlase) gene becomes stimulated by higher ethylene (Figure 7.3) accumulation which starts degradation of the lipid as it comes in contact with chlorophyll (Matile et al. 1997). However, scientists have observed that a low level of ethylene in plants also played an important role in decreasing the adverse signs of ROS under temperature trauma (Overmyer et al. 2003).

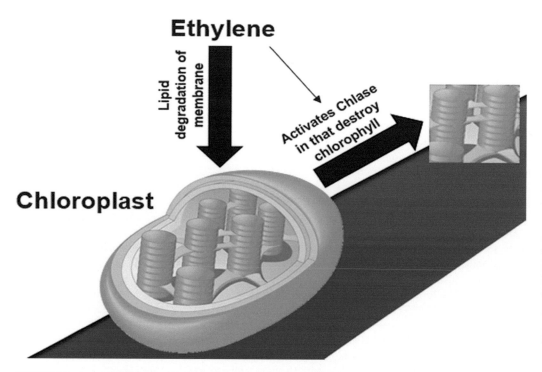

FIGURE 7.3   Stress ethylene and degradation of chlorophyll under high temperature induced stress.

7.4 Effect of eCO$_2$ and Temperature on Cereal Crops: Seed Germination and Seedling Growth

The effects of eT on physiological traits of the root, stem, and leaf area of three cereal crops, i.e., maize, rice, and sorghum at two weeks after germination, were monitored by Iloh et al. (2014) in Nigeria. They observed a reduction in germination speed as temperature regimes were raised. The shoot and root lengths of those crops were decreased with increasing temperature. However, the maize root length is decreased from 13.10±1.5 cm to 3.80±0.5 cm with increasing temperature from 40 to 50°C, respectively. Rice had a decrease in leaf length of 6.70±1.0 cm to 4.20±1.5 cm at 50°C. Iloh et al. (2014) reported that sorghum was severely affected by the elevated temperature among the crops studied. Their findings suggest growth stagnation of rice, maize, and sorghum with rises in temperature (Table 7.1).

Mina et al. (2019) evaluated the effects of eT and eCO$_2$ on maize genotypes' health index and generated crop health index (CHI) for maize cultivars (PEHM5 and CM119) in response to eT and eCO$_2$. In an open field experiment, in open-top chambers, PEHM5 and CM119 maize cultivars during their growth were subjected to two levels of CO$_2$ (400 ppm) which is ambient and eCO$_2$ (550±20 ppm), and three temperatures (ambient, ambient + 1.5°C and ambient + 3°C). These workers monitored the reply of maize plants to temperature and eCO$_2$ using 13 stress pointer constraints– morphological, physiological, and biochemical at three growth phases – vegetative, tasselling, and dent. The scientists' data to define minimum dataset (MDS) receptive to mixtures of temperature and eCO$_2$ actions via PCA is shown in Table 7.2. According to these findings, out of 13 pointers, total dry weight, relative leaf water content, and photosynthetic degree had advanced incidences for MDS at all the three phases. After generating CHI results, it was reported that CHI for maize plants under changed handlings was wide-ranging in 0.14–0.93. The normal CHI under different handlings were meaningfully connected to the harvest of both maize genotypes with R^2 of 0.82 and 0.90, correspondingly. The authors found that upsurge in temperature had a damaging consequence on CM119 and PEHM5 maize plants, with least mean CHI of 0.20–0.53 (Mina et al. 2019).

TABLE 7.1

The Effects of Elevated Temperature on Maize after 2 Weeks of Exposure (Adapted from Iloh et al. 2014)

Physiological Parameter (cm)	RCO	RCI	RCII	RCIII	RCIV
Rice					
Length of root	8.30±1.0	7.10±1.5	7.60±2.9	6.50±1.5	5.80±0.1
Length of leaf	8.20±0.1	6.70±1.0	5.80±0.6	4.90±3.2	4.20±1.5
Width of leaf	0.70±0.1	0.50±0.1	0.50±0.1	0.20±0.1	0.20±0.1
Length of stem	6.80±0.3	5.90±0.5	5.30±0.5	4.0±0.5	3.00±0.1
Sorghum	SCO	SEI	SEII	SEIII	SEIV
Length of root	8.90±2.3	6.30±1.3	6.10±1.9	5.50±1.6	1.50±0.1
Length of leaf	11.10±0.6	8.80±2.2	7.70±0.4	6.90±2.5	5.80±1.1
Width of leaf	0.50±0.1	0.20±0.0	0.03±0.1	0.30±0.1	0.10±0.1
Length of stem	8.60±2.2	5.00±0.4	4.20±0.9	4.0±2.3	3.40±0.5
Maize	MCO	MEI	MEII	MEIII	MEIV
Length of root	16.3±1.0	13.10±1.5	11.6±2.9	8.50±1.5	3.80±0.5
Length of leaf	19.90±1.6	15.40±3.6	11.60±1.7	9.70±3.2	5.80±1.1
Width of leaf	1.30±0.1	1.20±0.1	1.00±0.1	0.90±0.1	0.50±0.1
Length of stem	7.80±0.3	6.90±0.8	5.30±0.8	4.0±0.4	4.0±0.7

Note: MCO, control experiment at 37°C, MEI experiment at 40°C heat stress; MEII experiment at 42°C heat stress; MEIII experiment at 45°C heat stress; MEIV at 50°C heat stress; SCO, control experiment at 37°C, SEI experiment at 40°C heat stress; SEII experiment at 42°C heat stress; SEIII experiment at 45°C heat stress; SEIV at 50°C heat stress; RCO, control experiment at 37°C, RCI experiment at 40°C heat stress; RCII experiment at 42°C heat stress; RCIII experiment at 45°C heat stress; RCIV at 50°C heat stress.

TABLE 7.2

Principal Components/Indicators Parameters Response to eCO_2 and Temperature on the Vegetative Stage in PEHM5 Maize Genotype (Aadapted from Mina et al. 2019)

PCs	PC1	PC2	PC3
Eigen value	4.162	2.256	1.805
% variance	34.687	18.796	15.042
Cumulative %	34.687	53.483	68.526
Factors loading: Eigenvector			
Plant height	0.727	0.313	−0.259
Stem girth	0.777	−0.394	0.042
Leaf Area	0.398	0.036	−0.206
Relative Leaf Water Content (RLWC)	−0.085	0.083	0.750
Photosynthetic rate	−0.368	0.549	−0.154
Stomatal conductance	0.845	0.234	0.295
Transpiration rate	0.826	0.410	0.116
Total dry biomass	0.926	0.447	0.337
Total chlorophyll content	0.109	−0.550	0.632

7.5 Influence of eCO_2 and High Temperature Stress during Flowering and Grain Filling Periods in Rice

7.5.1 High Temperature Stress

High temperature stress is a serious threat to sustaining yield and grain quality of rice. Increasing eCO_2 is prophesied to improve photosynthesizing efficiency ensuring higher biological and grain yield. Contrary to this, both high-temperature stress and a higher level of eCO_2 are believed to decline the rice grain quality. Though, the interaction between eCO_2 and high-temperature stress on rice grain harvest, quality characters, and predominantly inorganic configuration under normal field environment in cool environmental conditions are not yet fully disclosed (Chaturvedi et al. 2017). The progressive ineluctability of haphazardous atmospheric conditions leading to regularization of stress incidents among recent world climatic models averring in, for example, growing global temperature level causing an augmented regularity of heat stress incidents (Battisti and Naylor 2009). Rice is susceptible to high-temperature stress (Jagadish et al. 2010) and the forewarned variations in the present and upcoming climate profiles would adversely influence worldwide rice productivity (Teixeira et al. 2013) and grain quality (Lin et al. 2010) and, consequently, decreasing its commercial and nutritional status (Wang et al. 2011; Lyman et al. 2013). It was proposed that further research must be devoted to investigate high temperature stress differences among various types of rice varieties. The revolutionary rice-planting methods based on planting system changes in rice planting regions with extreme high temperature stress. At the same time, high temperature detection and indication systems must be upgraded to gain optimal high temperature stress management efficiencies (Wang et al. 2019). Contrarily, high-temperature stress revelation throughout reproduction and grain formation stages is known to decrease rice harvest by reducing the amount of fertile spikelets (Jagadish et al. 2010) and reduced grain formation period (Ahmed et al., 2015) along with damage of sink efficiency (Kim et al. 2011). Furthermore, leaf sensation, leading to reduction in net photosynthetic proficiency (Kim et al. 2011) and starch alteration enzymes action are evident under heat stress which caused deterioration in final grain yield in rice (Ahmed et al. 2015) and other major crops like maize and wheat (Hawker and Jenner 1993; Cheikh and Jones 2006).

7.5.2 Effect of Elevated CO_2 on Rice

Rice is a C_3 crop, therefore, it shows a positive response against eCO_2 and produces maximum yield in rice (Shimono et al. 2009). Yang et al. (2009) presented evidence in favour of eCO_2 at the rate 580 ppm having improved the rice yield up to 24–34% when all other environmental factors like irrigation, light, and temperature were kept optimum. Mittler and Blumwald (2010) and Jagadish et al. (2016) revealed that environment and climatic issues such as eCO_2 and eT might differ significantly and interrelate in fully open-air field environments upsetting several growth and developmental phases of crops. Yang et al. (2007) documented that eCO_2 harms the proportion of rice chopping in relation to head rice recovery because of accelerated crop health. Myers et al. (2014) observed a close relationship between eCO_2 and reduced grain protein content, as well as changes inside the composition of grain minerals. However, it was observed that wheat and rice grown in eCO_2 have reduced grain nutrient contents (Myers et al. 2014). Among the researchers, a few reported that eCO_2 reduces photorespiration, improves photosynthesis, and WUE and, thereby, increases rice grain yield (Bowes 1996; Leakey et al. 2009a; Ziska et al. 2012). Ainsworth and Bush (2011) revealed that encouragingly, photosynthetic efficiency and primary production at eCO_2 are unerringly correlated to the absorbing efficacy when using or storing additional non-structural carbohydrates (NSC), which otherwise could lead to the adaptation of photosynthesizing activity in the original tissue. Consequently, varieties with an increased absorber size and durability for storing or using photo assimilation will find this useful when exposed to eCO_2, reducing the adjustment of photosynthetic efficiency and maintaining maximum production at the leaf, side by side (Zhu et al. 2015).

7.5.3 Effects of Elevated CO_2 and Heat Stress on Rice

Climate change will have an impact on agricultural production. Though high temperatures can damage crops, the maximum CO_2 concentration will have a positive influence on crop evolution and productivity (Raza et al. 2019, 2020a, 2020b). Reports exist that vouch to the fact that the maximum concentration of eCO_2 in the atmosphere maximizes the potential yield of C_3 crops, especially of rice in high latitudes (Taylor et al. 2018). Rice is a necessary food crop with half of the biosphere's population being the staple food in the everyday diet where rice is taken every day (Maclean et al. 2002). It is also the staple food across Asia, where about half of the biosphere's poor people live, and the rest in Africa and Latin America. Different studies on rice have also shown that eCO_2 frequently increases the photosynthesis rate, plant weight, and grain harvest (Horie et al. 2000; Chakrabarti et al. 2012).

Elevated temperature causes a decrease in total dry matter, tiller impermanence, reduced number of panicles and grains per panicle, floret sterility, and grain weight which ultimately reduces the crop yield. The high-temperature stress of 3.9°C significantly reduces the grain and biomass yield of rice. An increase in daily average temperature from 28°C to 32°C significantly reduces the total dry biomass, root dry biomass, root extent, leaf area, and specific leaf part of rice (Rankoth and De Costa 2013; Raj et al. 2016). Cao et al. (2009) recognized that increase in temperature at flowering stage and early grain filling phase of various rice varieties showed lower photosynthetic efficiency in the crop. Moreover, eCO_2-facilitated variations in material sugar levels and a straight CO_2-ambitious controlling pathway connecting a key flowering gene (*MFT*) are strong indications that eCO_2 plays an important role in the flowering period regulation in crops. Increasing temperature and eCO_2 are main climate alteration influences that could disturb plant suitability and flowering-related procedures (Jagadish et al. 2016). Hussain et al. (1999) revealed that heat stress throughout grain filling primes will decrease application of extra NSC in the sink in spite of amplified integrate amount from leaves in response to eCO_2. For instance, high temperature condensed grain mass or grain starch level, even with developed sucrose source from basis tissue, has been found to exist exposed to below-eCO_2 in rice. Temperature and photoperiod are the two main powerful forces for crop growth. The optimal temperature for standard rice development is 27–32°C (Yin et al. 1996). Wahid et al. (2007) studied that elevated temperature disturbs nearly all the growth phases of rice from budding to maturing and gathering. The developmental phase at which the plant is exposed to heat stress regulates the harshness of the likely injury to the crop.

Ziska and Teramura (1992) documented that the aCO_2 level is likely to enhance in the forthcoming years, which will have long-term effects on numerous plants. As a C_3 plant, rice will surely have advantage from this enhancement in CO_2, mostly through condensed photorespiration. A positive role of elevated CO_2 has also been delineated for biomass accretion, tillering, panicles per plant, and grain yield of rice. Maximum CO_2 levels will impact stomatal conductance positively by decreasing moisture loss through transpiration efficiency, thus improving WUE (Wassmann et al. 2009b). Therefore, it is evident that predictable increase in CO_2 concentration will be beneficial in several ways for rice development. However, the general impact is adverse when rise in CO_2 and temperature are taken into consideration simultaneously (Wassmann et al. 2009a). Kim et al. (1996) studied that rising of aCO_2 level is considered to exclusively exacerbate heat persuaded spikelet sterility because it causes stomatal conductance under increased aCO_2 owing to heat stress.

Anjum et al. (2008) revealed that high-temperature stress harmfully influences pollen cell microspore progressing into male barrenness. However, heat stress of above 30°C during floral growth can cause whole sterility reliant on genes (Kaur and Behl 2010). Heat stress adversely affects the grain formation stage that ultimately reduces grain yield (Farooq et al. 2011). Elevated temperature stress usually increases the degree of grain-filling and curtails the grain-filling time period (Dias and Lidon 2009a). But, the grain development rate and period over which it happened reduced in crops having dissimilar grain weight constancy (Vijayalakshmi et al. 2010).

Flowering has been recognized as the phase most susceptible to heat stress, and the dominant ambient temperatures throughout anthesis have been correlated to reproductive offsets (Jagadish et al. 2007). At the plant level, a maximum CO_2 level improves photosynthetic rate, growth period, developmental phases, and yield of several growing crops, including rice (Ainsworth and Rogers 2007). So far, most of the researches with eCO_2 and rice have verified conservatively bred japonica (Yang et al. 2007) and Indica cultivars (De Costa et al. 2007). Nevertheless, recent hybrid rice genotypes display developed seedling vigour, degree of tillering, comparatively advanced growth degree, and advanced yield probable as compared to conventional inbred cultivars (Xie et al. 1996). Rise in temperature during flower initiation and grain formation stage is stated to decrease grain harvest and improve superiority in rice quality (Usui et al., 2014; Bahuguna et al. 2015). Elevated temperature stress-tolerant rice cultivar (NL-44) upheld advanced seed-set and pliability with starch metabolism enzymes on exposure to eCO_2 and high temperature simultaneously. Establishing rice genotypes with eCO_2 exposure integrated with greater tolerance to heat throughout the flowering and grain filling by givers such as NL-44 is capable of lessening the adverse effects of temperature, and support higher food availability efficiency in a global context, advancing from CO_2-rich surroundings (Chaturvedi et al. 2017).

7.6 Elevated CO₂ and Heat Stress Effect on Maize

Elevated CO_2 is a significant abiotic stress issue and leaves noteworthy fertilization signals on various crops. Early wide-ranging investigations on this concern for crops have described that eCO_2 suggestively enhanced WUE, reduced transpiration frequency, curtailed maize development retro and augmented plant height, leaf number, leaf zone, development rate, and harvest. In addition, increasing CO_2 disturbs rainfall stability, which can alter the periodic precipitation delivery (Easterling et al. 2000). It has been projected that this result would carry about 10% upsurge or decline in water resources at numerous zones (Wallace 2000). Elevated temperature causes heat damage and physiological syndromes resulting in condensed harvest (Johkan et al. 2011). Raised temperature due to eCO_2 also had a principal effect on food grain harvest reliant on the sites. With temperature upsurge by 1.0–2.0°C in tropical and subtropical states, such as India, food grain production is expected to decline by 30% (Johkan et al. 2011).

The C_4 grass maize is the third most important food crop globally because of harvest and its yield is presaged to jump by 45% from 1997 to 2020 (Young and Long 2000). Research on maize yield responses to eCO_2 displayed changing signals of growth extending from no stimulus of biomass (Hunt et al. 1991) to 50% stimulation (Rogers and Dahlman 1993). High temperature unfavourably affects growth, yield, and crop quality of maize plants (Pathak et al. 2012). At the same time, maize does not respond without major signs to high temperature stress. The plant transduces the abscisic acid (ABA) signal to direct heat shock proteins (HSPs), which are molecular chaperones that contribute in protein refolding and deprivation triggered by high temperature. The HSPs can steady selective protein arrangements and tortuously provision constancy of thylakoid membrane construction, electron transportation, and secure carbon adjustment which may lead to greater photosynthesis (Tao et al. 2016). There is a lot of information on the effect of eCO_2 on maize yield varying from little optimistic consequence (Leakey et al. 2009a), no consequence (Kim et al. 2007) to upsurged harvest by 50% (Vanaja et al. 2015). Though there are several studies on the influences of eCO_2 and eT on maize yield, that used precise atmosphere amenities such as phytotron and plant growth cavities or crop growth reproduction copies (Pathak et al. 2012), still there are an inadequate number of investigations on the effects of eCO_2 level and temperature on this important crop under field environments. Consequently, there is a necessity to develop or to recognize cultivars which can accomplish healthier growth and productivity under presaged climatic variations(EL Sabagh et al 2020a). Further, there is a need to characterize maize genotypes that differ in heat tolerance at a molecular level and identification of candidate genes for heat stress tolerance in maize with the potential to assist maize breeding programmes.

7.7 Elevated CO_2 and Heat Stress Effect on Wheat

Rising CO_2 concentration is linked to the increased surface mean temperatures worldwide. As a result of changing climate, warm days are becoming warmer threatening global wheat production. Heat stress (HS) means that as the daily temperature rises above the threshold levels for plants (Wahid et al. 2007), it drastically affects agricultural productivity, (Asseng et al. 2015; Raza et al. 2019a; Salim and Raza 2020). Heat stress bounds the growth and development of plants through altering physio-biochemical processes (Wahid et al. 2007; Yildirim et al. 2018; EL Sabagh et al. 2019) and hindering photosynthesis (Sage and Kubien 2007; Farooq et al. 2011; Barutcular et al. 2017). Plants may adapt an HS-induced oxidative damage by triggering several survival mechanisms (Pan et al. 2018; Zhang et al. 2018).

The maximum degree of ribulose bisphosphate (Ri_{BP}) carboxylation (Vcmax) is an exact reaction to temperatures of 40°C, but the highest degree of Ri_{BP} rejuvenation or electron transportation (E_{Trs}) usually causes shrinkages at lower temperatures of 33°C (Medlyn et al. 2002). The comparison of eCO_2 vis-à-vis remaining photosynthesis is superior at HS because of the conquest of photorespiration (Li et al. 2013). The eCO_2 may also upsurge the OTT of photosynthesis (Alonso et al. 2008; Ghannoum et al. 2010). At eCO_2, the reply of photosynthesis to eT conversions is progressively imperfect by E_{Trs} and Rubisco initiation (Sage and Kubien 2007). The OTT of photosynthesis reproduces that of E_{Trs} in plants grown at eCO_2. Beyond OTT, acclimation of photosynthesis to HS is linked with augmented electron transportation and/or eT constancy of Rubisco initiation (Sage and Kubien 2007).

Therefore, for the development of wheat varieties suitable to forthcoming climate changes, it is imperative to gain knowledge on how eCO_2 and HS interactively influence wheat yields. For confirmation of the hypothesis, Chavan et al. (2019) conducted a glasshouse trial with a modern, HYV wheat cultivar, 'Scout' at 419 µl l^{-1} of aCO_2 or 654 µl l^{-1} of eCO_2 through maintaining day/night temperature at 22/15°C, respectively. Around 50% of the wheat plants at anthesis stage were exposed to 40/24°C for 5 days for inducing by HS. Results of the study revealed that in control plants, eCO_2 improved (+36%) aCO_2 acclimatization rates measured at growth aCO_2 despite photosynthetic capacity being downregulated. HS condensed AsRat (–42%) in aCO_2- nonetheless not in eCO_2 fully fledged plants since eCO_2 sheltered photosynthesis by enhancing the capacity of Ri_{BP} rejuvenation and dropping photochemical injury under HS. The eCO_2 reduces the stomatal conductance and upsurges the photosynthesis by encouraging carboxylation and Rubisco oxygenation, recognized as photorespiration (Ainsworth and Rogers 2007; Leakey et al. 2009a).

7.8 Elevated CO₂ and Heat Stress Effect on Sorghum

Sorghum [*Sorghum bicolor* (L.) Moench.], a C_4 plant, is a tall cereal grain that is an important food and feed crop on a global scale grown particularly in the USA, Mexico, Sudan, and some developing countries of Africa and South Asia. It is a hardy crop that is best adapted to warm climes and is resilient to drought and heat because its large fibrous roots can reach far down into the soil to extract water. Growing temperature and eCO_2 are the main climate alteration factors that may adversely affect growth, development, and grain yield of sorghum. Future predictable variations in temperature could harshly impact sorghum yield worldwide. Several reproduction models predicted 7–17% yield reduction in sorghum in Asia and Africa (Butt et al. 2005; Tingem et al. 2008; Srivastava et al. 2010; Sultan et al. 2014).

The average optimal temperature for seed germination of sorghum ranges from 21–35°C, where optimal temperatures are 26–34°C for vegetative development and 25–28°C for reproductive development (Maiti 1996; Prasad et al. 2006, 2008). Higher temperature than optimum affects both vegetative and reproductive development of sorghum, and the influence of eT on the generative growth phase is more noticeable compared to vegetative phase (Prasad and Staggenborg, 2008; Nguyen et al. 2013; Singh et al. 2015). Previous studies have shown that increase in temperature enhances crop development rate resulting in earlier flowering, augmented leaf number, and leaf appearance rate, but with no result on leaf size. Though, there was a noteworthy reduction in plant height, pollen feasibility, grain number, individual grain heaviness, and grain yield (Prasad and Staggenborg, 2008; Nguyen et al. 2013). The result of eT (36–40°C) on sorghum leaf number plant^{-1} seems to be slight, as reduction in leaf number is remunerated by an enlarged leaf amount axis^{-1} (Jain et al. 2007; Van Oosterom et al. 2011).

Prasad et al. (2006, 2008) stated that sorghum is frequently cultivated in the regions where day/night temperatures are more than 32/22°C, as high-temperature throughout pre-anthesis (sporogenesis) reduces pollen feasibility and number of pollen grains, that results in reduced seedset. They also concluded that high temperature throughout the grain-filling phase declined separate grain extent owing to reduction of grain filling period that reduced the source and finally reduced grain size and yield. The eT>33/28°C throughout the initial phases of panicle growth persuade floret and embryo abortion (Downes 1972). Raised temperature throughout floret formation changes pollen morphology with adverse consequences on flowering, having an irregular exile wall, deterioration of tapetum cells, and sheath injury, foremost to pollen barrenness (Prasad and Staggenborg 2008; Djanaguiraman et al. 2014). Prasad et al. (2015) reported that anthesis period (5 days before and 5 days after) is very sensitive to elevated temperature stress producing supreme diminution in floret fruitfulness of sorghum. High temperature stress throughout floret expansion or grain filling phase of sorghum reduced floral fertility, grain mass, and grain mass per panicle. CO_2 is an important element that affects plant development mostly by its direct effect on photosynthesis and stomatal composition (Uprety et al. 2002; Shimono et al. 2013). The rising of CO_2 concentration due to climate change is extensively likely to enhanced photosynthesis resulting in increased yield and crop output in crop plants (Kant et al. 2012; Hasegawa et al. 2013; Hussain et al. 2016). The growth and developmental responses greatly vary with the growth stage of plants, e.g., a better response in early growth phases and a reduced response as plants enter the adult stage (Kramer 1981; Geiger et al. 1999). Chaudhuri et al. (1986) reported that sorghum grown at eCO_2 produced extra roots and shoots as compared to plants grown with ambient CO_2. eCO_2 may improve crop productivity and alter N level in different plant tissues of sorghum (Torbert et al. 2004).

Watling et al. (2000) conducted a study on sorghum using CO_2 at the rate 350 (ambient) or 700 (raised) μmol mol^{-1} on key rudiments of the C4 pathway, and they observed a 2-fold decline in the width of the package sheath cell wall in plants grown at raised CO_2 relative to ambient CO_2. They also reported both carboxylation proficiency and CO_2-soaked degree of photosynthesis which was lesser in plants fully-fledged at raised CO_2 relative to ambient CO_2. 49% decrease was observed in phosphoenol pyruvate carboxylase (PEPC) content of leaves (area basis) in the plant developed in raised CO_2 against eCO_2, but no alteration in Rubisco was found. There was a 3-fold upsurge in C-isotope taste in leaves of plants grown in eCO_2, and bundle sheath leakiness was projected to be 24 and 33%, for the ambient and eCO_2-grown plants, respectively.

During the anthesis phase, eCO_2 has been described to increase tissue temperature by dropping the dangerous temperature verge, subsequent in advanced spikelet infertility in sorghum (Prasad et al. 2006). The positive response of leaf carbon exchange rate (CER) to eCO_2 was better in young leaves as compared to that in the old leaves. In young leaves, eCO_2 improved Rubisco movement at 30/20°C and 36/26°C, while PEPC action was not exaggerated by eCO_2 at 30/20°C but was slightly improved at 36/26°C. Leaf CER and Rubisco action were improved by eCO_2 at the early phases of leaf ontogeny for the sorghum (Prasad et al. 2009).

Prasad et al. (2009) stated that vegetative growth and dry matter production of sorghum were significant at high day/night temperatures (36/26°C) as compared to a cooler temperature (30/20°C) with eCO_2, however, economic yield losses were greater with the combination of high temperatures with eCO_2 than at ambient CO_2. The improvement in CER and up-regulation of Rubisco action of the midsection of leaf rudiments of sorghum were found to be better at early phases of leaf growth which may underwrite to the better nonsexual development and dry mass construction experiential at eCO_2 and high temperatures. Prasad et al. (2006) investigated the communication of eCO_2 at dissimilar levels of temperatures on sorghum, and found that no substantial changes occur in seed figures or seed extent at ambient (350 μmol mol^{-1}) or eCO_2 (700 μmol mol^{-1}) in augmented day/night temperatures of 36/26°C. However, the percent seed-set was decreased under eCO_2 than ambient CO_2 at either 32/22 or 36/26°C resulting in decreased yield by 40–100% and of harvesting index by 24%. In the semi-arid tropics, where sorghum is presently fully fledged throughout the rainy period, the cruel crop-season temperatures were previously approaching or had crossed over these optimal temperatures (Prasad et al. 2006; Singh et al. 2014).

7.9 Elevated CO_2 and Heat Stress Effect on Barley

Barley (*Hordeum vulgare* L.) is a more or less climate-adaptable crop cultivated worldwide and with a global yield recently recorded as reaching 149 million tonnes. It is used in multiple food preparations, beverages, animal feeds, and has other uses (Calvo et al. 2017; EL Sabagh et al. 2019). On the other hand, increasing atmospheric CO_2 levels is the main reason for rising average surface temperatures worldwide as well as the main reason behind the prolonged duration of, rate of recurrence in, and intensity of a heatwave. Elevated CO_2 and rising temperatures are the key climate change aspects that might distress plant suitability, flowering grain filling, and yield of barley (Jagadish et al. 2016). Excessively higher temperature in the warmer regions of the world during the crop growth period is threatening the yield potential of barley globally. (Savin et al. 1996; Passarella et al. 2005). Flowering and grain filling stages are critical determinants for plant reproductive outcome and seed-set. Numerous studies have examined the response of barley to eCO_2 (Pritchard et al. 1999; Smith et al. 2011; Jagadish et al. 2016). Increasing the temperature by 2°C can improve the agronomic parameters (plant height, dry matter quality and ear length) of highland barley, but no significant impact of temperature on barley harvest was observed as a result of eCO_2 while raising the temperature by 4°C will seriously disturb the normal growth and harvest of highland barley (Liu et al. 2018). Globally, the daily temperature is considered as the most important factor upsetting grain quality (Taub et al. 2008) in the majority of cereal growing regions. Research (Senghor et al. 2017) findings revealed that the short-term periods of extreme temperature throughout flowering and grain filling can reduce grain size, weight, yield, and quality of crop (Passarella et al. 2005, 2008; Senghor et al. 2017). Flowering and grain filling periods are controlled by a compound signalling system including the modulation of ecological spurs (i.e., temperature, photoperiod) and endogenous genetic behaviours (Passarella et al. 2002). A study of Clausen et al. (2011) has revealed that CO_2 levels above the existing ambient concentration normally enhanced the growth and yield of C_3 crop species. Further, similar trends could stimulate future harvest, except snowballing climate change effects such as escalating temperatures, erratic rainfalls, and atmospheric interference. Existing variation in the stimulation in the harvest at eCO_2 concentrations has been detected in many cereals, including barley (Clausen et al. 2011). Although, field trials of Chavan et al. (2019) revealed that

eCO_2 alleviated the damaging impacts of heat stress at the anthesis stage on wheat photosynthesis. However, grain harvest was condensed by high temperature in both eCO_2 handlings.

7.10 Adaptations to Elevated CO_2

The changes in climate, including eCO_2, increasing temperature, and changed rainfall have extreme influences on vegetation function and crop yields affecting sustainable food production (Lobell et al. 2011; Ruiz-Vera et al. 2013; Lavania et al. 2015) and causing exacerbated impacts when above climate vicissitudes are joined with other ecological boundary conditions, such as limited or toxic nutrition status (Peñuelas et al. 2012; Xu et al. 2013a, 2013b; Wang et al. 2015).

Rising aCO_2 has a significant influence on almost all important biological procedures such as photosynthesis, respiration, hormone signalling, and antioxidant defence, as well as other significant tributary metabolic processors in crop plants (Matros et al. 2006; Peñuelas et al. 2013; Singh and Agrawal 2015). CO_2 has direct impacts on plant metabolism predominantly through its role in photosynthesis, while it indirectly influences plant performance by its effects on atmospheric temperature and ground water status. It is well understood that the adaptations to eCO_2 are mainly due to changes in primary metabolic processors and associated physiological functions.

7.10.1 Adaptations of Photosynthesis to Elevated CO_2

The elevated aCO_2 concentration accelerates the rate of photoassimilation in the leaves of many plants, which enhances plant growth leading to increase in crop yield (Long et al. 2004, 2006). Contrary to this, rising CO_2-mediated photoassimilation stimulates transcriptional upregulation of genes associated with respirational pathways (Fukayama et al. 2011; Markelz et al. 2014). However, photosynthetic down-regulation was detected with long-term exposure to eCO_2 (Darbah et al. 2010). The long-term impact of CO_2 enrichment has been more pronounced on C_3 species as compared to C_4 species. This photosynthetic down-regulation is said to be due to limited ATP production, which is not sufficient for RuBP regeneration, as ATP produced in respiration processes and consumed more with eCO_2 (Watanabe et al. 2014) causes an undesirable reaction to result during photosynthesis. This shows the close connection among the photosynthetic and respiratory processes (Moroney et al. 2013; Watanabe et al. 2014).

Increased aCO_2 concentrations depicted increased levels of soluble sugars and starch in plant leaves, indicating that photoassimilation increases under eCO_2 concentrations (Teng et al. 2006). However, the extent of the response to eCO_2 differs with the plant functional types and plant groups, for example, extreme response for trees and C_3 grasses, reasonable for C_3 shrubs, and minimum for C_4 grasses (Ainsworth and Long 2005; Ainsworth and Rogers 2007) and their environmental conditions such as nutrient and water availability.

The eCO_2 inhibits the oxygenation of RuBP by down-regulating the affinity of O_2 to Rubisco while up-regulating the carboxylation of RuBP by facilitating CO_2 affinity resulting in stimulation of photosynthesis (Kane et al. 2013; Moroney et al. 2013). Reports showed photorespiration caused approximately 30% loss of the carbon fixed by photosynthesis with the rising temperature. Nonetheless, the CO_2 fixing can be increased >50% when suppressing the oxygenation reaction, as the carboxylation reaction occurs (Long et al. 2006) showing high adaptations to increased CO_2 concentrations.

The restricted photorespiration at eCO_2 reduces the H_2O_2 productions, which protects the chloroplasts by oxidative stress (Watanabe et al. 2014; Zinta et al. 2014) implying photorespiration has a defensive role contrary to photooxidation (Zinta et al. 2014). Photosynthetic responses to eCO_2 may be to weaken when interacting with other abiotic factors such as water deficit (Xu et al. 2014), N deficit (Markelz et al. 2014) and raised temperature (Ruiz-Vera et al. 2013). Under CO_2-enriched environment, more carbohydrate accumulates due to increase in size of starch granules and number of chloroplasts (Teng et al. 2006, 2009). However, high carbohydrate accumulation damages chloroplasts which reduces photosynthetic capacity (Aranjuelo et al. 2011) under CO_2-enriched environment.

7.10.2 Adaptations of Antioxidant Systems to Elevated CO_2

Rising CO_2 up-regulates the antioxidant pathways such as polyphenols, ascorbate, alkaloids, and antioxidant defence activities including CAT and SOD which results in decreased cellular ROS levels (Mishra and Agrawal 2014; Zinta et al. 2014; Raza et al. 2020; Hasanuzzaman et al. 2020). Kumari et al. (2013) reported that ascorbate and phenol levels were increased in *Beta vulgaris* under eCO_2. The up-regulated ascorbate, GSH activities, and their redox position were reported in *Lolium perenne* and *Medicago lupulina* (Farfan-Vignolo and Asard 2012). The up-regulation of ascorbate synthesis was activated by enhanced photoassimilation at eCO_2 (Zinta et al. 2014; Ali et al. 2016). The declined oxidative stress at eCO_2 was detected in *Zingiber officinale* (Ghasemzadeh et al. 2010), *Catharanthus roseus* (Singh and Agrawal 2015), *Caragana microphylla* (Xu et al. 2014), *Vigna radiata* (Mishra and Agrawal 2014), and *Arabidopsis thaliana* plants (Zinta et al. 2014).

The plant antioxidant defence systems are closely related with carbon metabolism (Smirnoff and Wheeler 2000) where ascorbate synthesis has been drastically activated by enhanced carbohydrate biosynthesis because of eCO_2 (Smirnoff and Wheeler 2000; Zinta et al. 2014). The degree of mitigation of oxidative stress, extent of response, and magnitude of damage may vary with crop genotypes, varieties, growth stages, abiotic and biotic factors, and their interactions (Hodges and Forney 2000; Gill and Tuteja 2010; AbdElgawad and Asard 2013; Kumari et al. 2013; Zinta et al. 2014). Zinta et al. (2014) stated the constant transcription expression levels of ROS regulatory enzymes in *Arabidopsis* at increased CO_2 were below optimum water status, while O_3 partially suppressed the transcription expression levels of antioxidant metabolism at eCO_2 in soybean (Gillespie et al. 2012).

The impacts of oxidative damages caused by severe abiotic stresses such as drought (Xu et al. 2014; AbdElgawad et al. 2015), high temperature (Zinta et al. 2014; AbdElgawad et al. 2015), O_3 pollution (Kumari et al. 2013), and salinity (Pérez-López et al. 2009) could be mitigated by CO_2 fertilization. The CO_2-mediated up regulation of sugar production and sugar-derived reactive carbonyls were triggered in wheat plants under extreme environmental conditions (Takagi et al. 2014).

7.10.3 Changers in Key Metabolites, Hormones, and Gene Expressions under Elevated CO_2

Since early times, it has been reported that eCO_2 can influence primary and secondary metabolic processes and their compositions in plants (Lavola and Julkunen-Tiitto 1994; Lavola et al. 2013). At eCO_2, up-regulated accretion of carbon compounds is linked with up-regulation of phosphoglyceratemutase, and down-regulation of the adenosine diphosphate glucose pyrophosphatase protein which is responsible for the equilibrium among the carbon sink and basis in wheat plants (Aranjuelo et al. 2011). The decreased N compounds, increased total non-structural carbohydrates (starch and sugar) (Luo et al. 2004; Markelz et al. 2014), and steady whole structural carbohydrates (including cellulose, lignin, and lipids) (Markelz et al. 2014) were detected as present under CO_2 fertilization. However, Ribeiro et al. (2012) detected enhanced nitrogen assimilation in CO_2-rich atmosphere, while Guo et al. (2013) reported increased phloem amino acid contents in *Medicago truncatula* with eCO_2.

Secondary metabolites (phenylpropanoids, tannins, triterpenoids, phenolic acids, and alkaloids) were reported to increase with rising CO_2 (Matros et al. 2006; Ghasemzadeh et al. 2010; Lavola et al. 2013). For instance, huge accumulation of phenylpropanoids, chlorogenic acid, and scopoletin coumarins in tobacco leaves (Matros et al. 2006)-triggered levels of kaempferol and fisetin in ginger (Ghasemzadeh et al. 2010) were explored under CO_2 fertilization. CO_2 fertilization triggered the salicylic acid (SA) (Zavala et al. 2013) and brassinosteroids (BRs) levels (Jiang et al. 2012), but decreased jasmonates (JA) and ethylene meditations (Zavala et al. 2013; Vaughan et al. 2014). Enhanced auxin levels, decreased ABA, and constant cytokinins were observed in *Pinus tabulaeformis* disclose to enriched CO_2 environments, counteracting impacts of O_3 thereby stabilizing damage (Li et al. 2011). Ribeiro et al. (2012) found incorporating carbohydrate and inorganic nitrogen metabolism was associated with gibberellic acids with eCO_2. Thus, CO_2 fertilization is useful in replacing certain metabolic bioprocesses in the low-gibberellin crops.

The changed expressions of different genes in various crops have been reported with eCO_2. The overexpression of Rubiscoactivase in rice showed the reduced photoassimilation with CO_2-rich environments (Fukayama et al. 2011). The diminished Rubisco gene expressions were reported in N-deficient environments at eCO_2 (Leakey et al. 2009b; Markelz et al. 2014). The gene expression of PS II, D1 protein has been down-regulated with eCO_2 in high temperature environments in certain rice cultivars but not in others, indicating genotypic dependence (Gesch et al. 2003).

Mature wheat plants at senescence stage showed up-regulation of nitrogen remobilization genes and down-regulation of carbon remobilization genes (Buchner et al. 2015). The down-regulated photosynthetic gene expressions due to extreme weather events can be repaired by eCO_2 (Zinta et al. 2014). Winter wheat (*Triticumaestivum*) at low temperatures exhibited higher expressions of many genes with eCO_2 (Kane et al. 2013). A knockout mutant *bou-2* encoding mitochondrial carrier gene, *BOUT DE SOUFFLE* (*BOU*) in *Arabidopsis* showed halted growth at ambient CO_2 but not with eCO_2 (Eisenhut et al. 2013). High expressions of respiratory genes have been reported in deficient and optimum N environments with eCO_2 in *A. thaliana* plants (Markelz et al. 2014).

7.10.4 Effect of eCO_2 on Stomatal Development

Elevated CO_2 decreases both stomatal thickness and catalogue in a number of species, although cultivar or species-specific responses are also observed (Bettarini et al. 1998; Luomala et al. 2005; Field et al. 2015; Caldera et al. 2016). Stomatal development under eCO_2 is controlled by *HIC* (*High Carbon Dioxide*) that encodes a putative 3-ketoacyl coenzyme A synthase (KCS) (Gray et al. 2000). The *HIC* mutant showed an increase in stomatal density with eCO_2, indicating that *HIC* is a negative regulator of CO_2-dependent stomatal development. In addition to the transcriptional level regulation, post-transcriptional regulatory components are also involved in stomatal development. For instance, a naturally occurring Arabidopsis mutant, *28y*, showed an augmented stomatal thickness and index. Through map-based cloning, *ARGONAUTE1* (*AGO1*) was recognized as the causal mutation and it acts downstream of *TMM* (Yang et al. 2014). AGO1 is a RNA-binding protein and its role in RNA silencing and post-transcriptional gene regulation is well established. In addition to AGO1, other mechanisms of the microRNA pathway, such as HYL1 and HEN1, were also identified to participate in the establishment of stomatal patterns (Jover-Gil et al. 2012). Overall, these studies suggest that transcriptional and post-transcriptional gene regulatory factors control stomatal development and patterning. However, their role in controlling stomatal development under high CO_2 still requires further investigation. Jasmonic acid (JA) deficient mutant *fad-4* and ethylene insensitive mutant *ein-2* were found to be impaired in stomatal development at high CO_2 (Lake et al. 2002). As JA and ethylene are mainly involved in defence responses to pathogens, it could be interpreted that CO_2 signalling and defence signalling pathways cross-talk. Further, systemic signalling from mature to young leaves control stomatal density in the developing leaves under high CO_2, although the nature and identity of such signals are not known (Lake et al. 2001). Differentiation and distribution of stomata are controlled differentially at the adaxial and abaxial surfaces, as surface-specific molecular regulators can also influence stomatal responses to high CO_2 (Driscoll et al. 2006; Soares et al. 2008).

7.10.5 Effect of Elevated CO_2 on Stomatal Conductance

The impact of eCO_2 on stomatal conductance (gs) is widely studied using a wide range of experimental systems such as growth rooms, open top chambers (OTC), and FACE (free-air CO_2 enrichment) (Bunce 2001a; Seneweera et al. 2001; Uprety et al. 2002; Long et al. 2004; Ainsworth and Rogers 2007). Although gs response varies between experimental systems, high CO_2-induced stomatal closure is frequently observed. The duration of CO_2 exposure also influences stomatal responses, where short-term exposure to eCO_2 has a larger impact. In contrast, plants exposed to eCO_2 for an extended period of time shows a certain degree of stomatal acclimation. The exact mechanisms underlying this stomatal acclimation to high CO_2 are yet to be fully understood. Stomatal conductance also relies on the aerodynamic properties, and changes in the stomatal conductance unequivocally translate to a change in transpiration. Besides CO_2, aerial environmental factors such as temperature, photosynthetic photon flux density (PPFD), and water vapour pressure difference (VPD) also affect stomatal conductance (Bunce 2004). The unique combinations of these factors influence stomatal conductance in a species-dependent manner (Bunce 2001b). The reduction of stomatal conductance caused by eCO_2 is greater in monocots (~ 42%) compared to dicots (~28%), and C_4 species have faster gs response compared to C_3 species (Vodnik et al. 2013). Early reports highlighted the association among temperature-dependent photosynthetic potential and stomatal conductance (Wong et al. 1979). Further studies have shown that stomatal conductance is not firmly related to the photosynthetic potential, and decreases in Rubisco gratified condensed both guard cell and mesophyll photosynthesis with no seeming result on stomatal conductance (Von Caemmerer et al. 2004). This is contrary to the usual detected association between photosynthetic potential and stomatal conductance when the photosynthetic capacity is manipulated via antisense RNA technology. Also, CO_2-induced oscillations of gs are known to be triggered by sudden changes in other environmental factors (Kaiser and Kappen 2001). Thus, it is of particular interest to know how the relative reduction in stomatal conductance at eCO_2 for a given species varies with other environmental factors.

7.10.6 High CO_2 Improves Water-Use Efficiency

Loss of water from aerial parts of plants during CO_2 uptake is inevitable. The proportion of the rate of left CO_2 assimilation (An) to transpiration (E) (An/E) for a given period of time throughout photosynthesis is usually demarcated as plant water-use efficiency (WUE). Plants with greater WUE fix more carbon per unit water lost, which is advantageous for survival in dry or drought-stressed environments. An improved formula for WUE was described later as a function of rate of An to leaf stomatal conductance (g_w), that is An/g_w, referred as physiological water-use efficiency (Feng 1999). Several lessons have designated an important upsurge in WUE in increasingly lean water seasons due to the soaring atmospheric CO_2 concentration from the year 1900 to present (Mccarroll and Loader 2004; Gagen et al. 2011). Examination of isotope timetables in tree jewels indicated an extensive upsurge in tree WUE over 40 years and found that WUE augmented by 20.5%, on an average, though stem development did not show any significant change (Peñuelas et al. 2011). Studies on angiosperm and conifer tree species showed a dynamic reduction of water loss in angiosperms, while conifers were found to be insensitive when exposed to eCO_2 (Brodribb and Mcadam 2013). A meta-analysis of statistics from 13 lasting (>one year) field-created investigations on European forest tree species with eCO_2 indicated an important reduction (21%) of stomatal conductance, and the response to CO_2 was significantly greater in new deciduous trees compared to coniferous trees. This response is extremely consistent in trees under water stressed conditions in long-term studies (Medlyn et al. 2001). Overall, high CO_2 improves WUE in a broad range of species, suggesting that it is a generally observed phenomenon.

7.11 Challenges to Improving Production with Elevated CO_2 and Heat Stress

Global climate change projections expect a rise in atmospheric CO_2 concentration up to 700 ppm and that in temperature by approximately 4.8°C by the end of the 21st century, respectively (Sultana et al. 2009). The aggravated levels of atmospheric CO_2 have outperformed during the pre-industrial phase, i.e., approximately 40% (IPCC 2013). Besides, it is believed that in the second half of the century, it can increase twice over the existing level (Jia et al. 2015). The preeminent CO_2 amount in the environment received a lot of attention because of its detrimental impacts on food crops. Similarly, the other constraint, i.e., heat stress, has also induced innumerable perturbations in the form of morpho-physiological responses that cannot even be ignored due to the consequent productivity loss. Therefore, this section is particularly based on the challenges that are faced frequently when researchers are exploring productivity improvement possibilities.

7.11.1 Strategies and Challenges for Improving Crop Production

After the perusal of literature, the details of strategies adopted for the improvement of crop productivity and challenges with enhanced CO_2 concentration and incidents of heat stress are summarized in the following subsections.

7.11.1.1 With Elevated CO_2

The relevant and recent reports on elevated CO_2 level are comprehensively addressed as below:

- According to the viewpoints of Hatfield et al. (2011), high amount of CO_2 has positive effects on plant growth and development along with the improvement in water use efficiency. In contrast, it is a challenge to predict whether these positive faces of elevated CO_2 can persist even with other changing environmental factors like temperature, precipitation, and interaction with biotic stress factors (weeds, insects, diseases, etc.).

- Due to the limited efficiency of the C_3 photosynthetic pathway, the growth and development of crop plants are adversely affected. In this context, the existence of CCM is reported in some algae along with micro-autotrophs and cyanobacteria (Beardall and Raven 2017). In this mechanism, Rubisco accelerates the process of carbon assimilation via highly proficient CCM that enhances the CO_2 concentration around Rubisco. However, innumerable contravenes are present, for example, comprehension of bicarbonate transporters' activation and energization processes, issues related to C_3 chloroplast, carboxysomal Rubiscos' chaperone requirements, etc. (Price et al. 2011; Breuers et al. 2012).

- As the source and sink association regulates biomass accumulation and crop yield, so it was recently chronicled that a thermostable Rubisco can be engineered via modifying Rubisco's small and large subunits (Peterhansel and Offermann 2012).

- As per the adjudications of Erb and Zarzycki (2016), it was revealed that this concern can be overcome via improving Rubisco efficiency, applying CCM, instituting bypasses' of photorespiration, etc.

- It was reported that C_3 crop plants can be benefited from the enhanced CO_2 because of their photosynthesis responsiveness to CO_2. Besides, there are other challenges, and the major one is the requirements of greater sink capacity (Dingkuhn et al. 2020).

- The outcomes from the study of Eshete et al. (2020) indicated that the aggravated levels of CO_2 negatively impacted the production of both traded and non-traded crops. Therefore, the Ethiopian Government has inaugurated a strategy to relegate the emission of CO_2, i.e., climate-resilient green economy (CRGE).

7.11.1.2 Under Heat Stress

The temperature beyond the physiologically optimal limit that disturbs normal plant growth and development is referred to as high temperature, whereas extremely high temperatures are referred to as heat stress. Many conventional and modern strategies were employed to resolve heat stress-induced damages to environment. On contrary, there are several challenges that are yet to be resolved and addressed to deal with heat stress as summarized below:

- There are numerous loopholes in the complete mechanism of heat tolerance, thereby, the number of management strategies is limited to precise phenotyping techniques. Furthermore, the emphasis should be laid on crop diversity, identification, and selection of superior adaptive characters. The integration of molecular breeding, functional genomics, and transgenic techniques can make a difference, so that, reinforced with the facilities on next-generation phenomics (Jha et al. 2014).

- The incidences of high-temperature variations are reported to be associated with considerable floret-fertility loss that ultimately causes yield reduction. The prompt elevation in night-time temperature is reported to mediate the narrowing of diurnal temperature amplitude that is another budding challenge to sustainable crop productivity (Prasad et al. 2017).

- An inclusive account of conventional as well as modern approaches to deal with heat stress has been addressed by Fahad et al (2017). It was reported that the identification of lines with superior characters via applying breeding techniques is the easiest option (Ehlers and Hall 1998). Heat tolerant plants are characterized via *the* minimization of photosynthetic damages along with the biosynthesis of protective compounds (Bita and Gerats 2013). Furthermore, the reproductive phase of plants is reported to be more vulnerable, therefore, many challenges are still there viz., improved photosynthetic rate, fruit setting, thermostability imparting to membranes, etc. (Fahad et al. 2017). Apart from this, the effectiveness of any method and/or strategy cannot be applied to all crops as it is under study.

- Heat stress is recorded to affect wheat plants adversely along with grain filling which, subsequently, causes a reduction in crop yield. Thus, a deep understanding of metabolic dynamics is essential to achieve heat stress-mediated consequences (Abdelrahman et al. 2020).

7.12 Conclusion

It can be easily inferred that climate change is a complex phenomenon that needs patience to solve. When the going gets tough, only the tough get going. Hence, this chapter is a welcome call to understand threats on cereal crops that feed more than 90% of world citizens with over 90% of them residing in Africa, Asia, and Latin America whose lands are severely threatened by rising temperature, as carbon dioxide exacerbates global warming and climate change. Constant research and applications are needed to address these issues in agriculture and the environment. Therefore, this research work discussed the importance of eCO_2 and temperature on cereal crops' physiological traits so as to understand the problem in hand and accelerate efforts at addressing the critical challenges facing us – not only they who must daily deal with it but also they who are responsible for solving it. It is well understood that the adaptations to eCO_2 are mainly due to changes in primary metabolic processors and associated physiological functions.

REFERENCES

AbdElgawad H, Asard H (2013) Elevated CO_2 attenuates oxidative stress caused by drought and elevated temperature in four C3 plant species. *Biotechnologia* 94:156–202.

AbdElgawad H, Farfan-Vignolo ER, de Vos D, Asard H, (2015) Elevated CO_2 mitigates drought and temperature-induced oxidative stress differently in grasses and legumes. *Plant Sci* 231:1–10.

AbdElgawad H, Zinta G, Beemster GTS, Janssens IA, Asard H (2016) Future climate CO_2 levels mitigate stress impact on plants: increased defense or decreased challenge? *Front Plant Sci.* doi:10.3389/fpls.2016.00556

Abdelrahman M, Burritt DJ, Gupta A, Tsujimoto H, Tran LSP (2020) Heat stress effects on source–sink relationships and metabolome dynamics in wheat. *J Exp Bot* 71(2):543–554.

Adhikari P, Ale S, Bordovsky JP, et al (2016) Simulating future climate change impacts on seed cotton yield in the Texas High Plains using the CSM-CROPGRO-Cotton model. *Agric Water Manag* 164. doi:10.1016/j.agwat.2015.10.011

Agüera E, De la Haba P (2018) Leaf senescence in response to elevated atmospheric CO_2 concentration and low nitrogen supply. *Biol Plantar* 62(3):401–408.

Ahmed N, et al (2015). Effect of high temperature on grain filling period, yield, amylose content and activity of starch biosynthesis enzymes in endosperm of basmati rice. *J Sci Food Agric* 95:2237–2243.

Ainsworth EA, Bush DR (2011) Carbohydrate export from the leaf: a highly regulated process and target to enhance photosynthesis and productivity. *Plant Physiol* 155:64–69.

Ainsworth EA, Long SP (2005) What have we learned from 15 years of free-air CO_2 enrichment (FACE)? A meta-analytic review of the responses of photosynthesis, canopy properties and plant production to rising CO_2. *New Phytol* 165:351–372. doi:10.1111/j.1469-8137.2004.01224.x

Ainsworth EA, Rogers A (2007) The response of photosynthesis and stomatal conductance to rising [CO_2]: mechanisms and environmental interactions. *Plant Cell Environ* 30:258–270. doi:10.1111/j.1365-304 0.2007.01641.x

Ali J, Muhammad H, Ullah I, Rashid JA, Adnan M, Ali M, Ahmad W, Rehman A, Khan J (2016) Mango seed germination in different media at different depth. *J Nat Sci Res* 6(1):56–59.

Ali M, Ullah Z, Mian IA, Khan M, Khan N, Adnan M, Saeed M (2016) Response of maize to nitrogen levels and seed priming. *Pure Appl Biol* 5(3):578–587.

Alonso A, Pérez P, Morcuende R, Martinez-Carrasco R (2008) Future CO_2 concentrations, though not warmer temperatures, enhance wheat photosynthesis temperature responses. *Physiol Plant* 132:102–112.

Amasino R (2010) Seasonal and developmental timing of flowering. *Plant J*. doi:10.1111/j.1365-313X.201 0.04148.x

Anderson TR, Hawkins E, Jones PD (2016) CO_2, the greenhouse effect and global warming: from the pioneering work of Arrhenius and Callendar to today's Earth System Models. *Endeavour* 40(3):178–187. doi:10.1016/j.endeavour.2016.07.002

Andersson I, Backlund A (2008) Structure and function of Rubisco. *Plant Physiol Biochem* 46:275–291. doi:1 0.1016/j.plaphy.2008.01.001

Anjum F, Wahid A, Javed F, Arshad M (2008) Influence of foliar applied thiourea on flag leaf gas exchange and yield parameters of bread wheat (*Triticum aestivum*) cultivars under salinity and heat stresses. *Int J Agric Biol* 10:619–626.

Apel K, Hirt H (2004) Reactive oxygen species: metabolism, oxidative stress, and signal transduction. *Ann Rev Plant Biol* 55:373–399.

Aranjuelo I, Cabrera-Bosquet L, Morcuende R, Avice JC, Nogués S, Araus JL, et al (2011) Does ear C sink strength contribute to overcoming photosynthetic acclimation of wheat plants exposed to elevated CO_2? *J Exp Bot* 62:3957–3969. doi:10.1093/jxb/err095

Aranjuelo I, Sanz-Sáez Á, Jauregui I, Irigoyen JJ, Araus JL, Sánchez-Díaz M, Erice G (2013) Harvest index, a parameter conditioning responsiveness of wheat plants to elevated CO_2. *J Exp Bot* 64(7):1879–1892.

Arp WJ (1991) Effects of source-sink relations on photosynthetic acclimation to elevated CO_2. *Plant Cell Environ* 14:869–875. doi:10.1111/j.1365-3040.1991.tb01450.x

Arshad M, Shaharoona B, Mahmood T (2008) Inoculation with *Pseudomonas* spp. containing ACC-Deaminase partially eliminates the effects of drought stress on growth, yield, and ripening of pea (*Pisum sativum* L.). *Pedosphere* 18:611–620. doi:10.1016/S1002-0160(08)60055-7

Asada K (1999) The water-water cycle in chloroplasts: scavenging of active oxygens and dissipation of excess photons. *Annu Review Plant Biol* 50(1):601–639.

Asseng S, Ewert F, Martre P, et al (2015) Rising temperatures reduce global wheat production. *Nat Climate Change* 5:143–147.

Assmann SM, Jegla T (2016) Guard cell sensory systems: recent insights on stomatal responses to light, abscisic acid, and CO_2. *Curr Opin Plant Biol* 33:157–167.

Attri SD, Rathore LS (2003) Simulation of impact of projected climate change on wheat in India. *Int J Climatol* 23:693–705. doi:10.1002/joc.896

Bahuguna RN, Jha J, Pal M, Shah D, Lawas LM, Khetarpal S, Jagadish SVK (2015) Physiological and biochemical characterization of NERICA-L-44: a novel source of heat tolerance at the vegetative and reproductive stages in rice. *Physiol Plant* 154:543–559.

Barutcular C, EL Sabagh A, Koç M, Ratnasekera D (2017) Relationships between grain yield and physiological traits of durum wheat varieties under drought and high temperature stress in Mediterranean conditions. *Fres Environ Bull* 26(4):4282–4291.

Bates GW, Rosenthal DM, Sun J, Chattopadhyay M, Peffer E, Yang J, Ort DR, Jones AM (2012) A comparative study of the *Arabidopsis thaliana* guard-cell transcriptome and its modulation by sucrose. *PLoS One* 7(11).

Battisti DS, Naylor RL (2009) Historical warnings of future food insecurity with unprecedented seasonal heat. *Science* 323:240–244.

Beardall J, Raven JA (2017) Cyanobacteria vs green algae: which group has the edge?. *J Exp Bot* 68(14): 3697–3699.

Bernacchi CJ, Kimball BA, Quarles DR, Long SP, Ort DR (2007) Decreases in stomatal conductance of soybean under open-air elevation of [CO_2] are closely coupled with decreases in ecosystem evapotranspiration. *Plant Physiol* 143:134–144. doi:10.1104/pp.106.089557

Bettarini I, Vaccari FP, Miglietta F (1998) Elevated CO_2 concentrations and stomatal density: observations from 17 plant species growing in a CO_2 spring in central Italy. *Glob Change Biol* 4:17–22.

Bhattacharyya P, Roy KS, Neogi S, Dash PK, Nayak AK, Mohanty S, Rao KS (2013) Impact of elevated CO_2 and temperature on soil C and N dynamics in relation to CH_4 and N_2O emissions from tropical flooded rice (*Oryza sativa* L.). *Sci Total Environ* 461:601–611.

Bita C, Gerats T (2013) Plant tolerance to high temperature in a changing environment: scientific fundamentals and production of heat stress-tolerant crops. *Front plant Sci* 4:273. doi:10.3389/fpls.2013.00273

Booker FL, Reid CD, Brunschön-Harti S, Fiscus EL, Miller JE (1997) Photosynthesis and photorespiration in soybean [*Glycine max* (L.). Merr.] chronically exposed to elevated carbon dioxide and ozone. *J Exp Bot* 48:1843–1852.

Bowes G (1996) Photosynthetic responses to changing atmospheric carbon dioxide concentrations. In: Baker NA (ed) *Advances in photosynthesis: photosynthesis and the environment*, vol. 5. Kluwer Academic: Dordrecht, The Netherlands, pp 387–407.

Brestic M, Zivcak M, Hauptvogel P, Misheva S, Kocheva K, Yang X, Allakhverdiev SI (2018) Wheat plant selection for high yields entailed improvement of leaf anatomical and biochemical traits including tolerance to non-optimal temperature conditions. *Photosyn Res* 136(2):245–255.

Breuers FK, Bräutigam A, Geimer S, Welzel UY, Stefano G, Renna L, Weber AP (2012) Dynamic remodeling of the plastid envelope membranes–a tool for chloroplast envelope in vivo localizations. *Front Plant Sci* 3:7. doi:10.3389/fpls.2012.00007

Brodribb TJ, Mcadam SA (2013) Unique responsiveness of angiosperm stomata to elevated CO_2 explained by calcium signalling. *PLoS One* 8:e82057.

Buchner P, Tausz M, Ford R, Leo A, Fitzgerald GJ, Hawkesford MJ, et al (2015) Expression patterns of C- and N-metabolism related genes in wheat are changed during senescence under elevated CO_2 in dry-land agriculture. *Plant Sci* 236:239–249. doi:10.1016/j.plantsci.2015.04.006

Bunce JA (2001a) Direct and acclimatory responses of stomatal conductance to elevated carbon dioxide in four herbaceous crop species in the field. *Glob Change Biol* 7:323–331.

Bunce JA (2001b) Direct and acclimatory responses of stomatal conductance to elevated carbon dioxide in four herbaceous crop species in the field. *Glob Change Biol* 7:323–331.

Bunce JA (2004) Carbon dioxide effects on stomatal responses to the environment and water use by crops under field conditions. *Oecologia* 140:1–10.

Butt TA, McCarl BA, Angerer J, Dyke PT, Stuth JW (2005) The economic and food security implications for climate change in Mali. *Climate Change* 68:355–378. doi:10.1007/s10584-005-6014-0

Cai C, Yin X, He S, Jiang W, Si C, Struik PC, Xiong Y (2016) Responses of wheat and rice to factorial combinations of ambient and elevated CO_2 and temperature in FACE experiments. *Glob Change Biol* 22(2):856–874.

Cairns JE, Sonder K, Zaidi PH, Verhulst N, Mahuku G, et al (2012) Maize production in a changing climate: impacts, adaptation, and mitigation strategies. *Adv Agron* 114:1–65.

Caldera HIU, De Costa W, Ian Woodward F, Lake JA, Ranwala SM (2016) Effects of elevated carbon dioxide on stomatal characteristics and carbon isotope ratio of *Arabidopsis thaliana* ecotypes originating from an altitudinal gradient. *Physiol Plantarum* 149:74–92.

Calvo OC, Franzaring J, Schmid I, et al (2017) Atmospheric CO_2 enrichment and drought stress modify root exudation of barley. *Glob Change Biol*. doi:10.1111/gcb.13503

Calzadilla A, Zhu T, Rehdanz K, Tol RSJ, Ringler C (2013) Economy wide impacts of climate change on agriculture in Sub-Saharan Africa. *Ecol Econ* 93:150–165.

Cao YY, Duan H, Yang LN, Wang ZQ, Liu LJ, Yang JC (2009) Effect of high temperature during heading and early filling on grain yield and physiological characteristics in Indica rice. *Acta Agron Sin* 35: 512–521.

Carter R, Woolfenden H, Baillie A, Amsbury S, Carroll S, Healicon E, Sovatzoglou S, Braybrook S, Gray JE, Hobbs J, Morris RJ (2017) Stomatal opening involves polar, not radial, stiffening of guard cells. *Current Biol* 27(19):2974–2983.

Cassia R, Nocioni M, Correa-Aragunde N, Lamattina L (2018) Climate change and the impact of greenhouse gasses: CO_2 and NO, friends and foes of plant oxidative stress. *Front Plant Sci* 9:273. doi:10.3389/fpls.2018.00273

Chakrabarti B, Singh SD, Kumar SN, Aggarwal PK, Pathak H, Nagarajan S (2012) Low-cost facility for assessing impact of carbon dioxide on crops. *Curr Sci* 102:1035–1040.

Chaturvedi AK, Bahuguna RN, Pal M, Shah D, Maurya S, Jagadish KSV (2017) Elevated CO_2 and heat stress interactions affect grain yield, quality and mineral nutrient composition in rice under field conditions. *Field Crops Res* 206:149–157.

Chaudhuri UN, Burnett RB, Kirkham MB, Kanemasu ET (1986) Effect of carbon dioxide on sorghum yield, root growth, and water use. *Agric Forest Meteorol* 37:109–122.

Chavan SG, Duursma RA, Tausz M, and Ghannoum O. (2019) Elevated CO_2 alleviates the negative impact of heat stress on wheat physiology but not on grain yield. *J Exp Bot* 70(21):6447–6459. doi:10.1093/jxb/erz386

Cheikh N, Jones RJ (2006) Heat stress effects on sink activity of developing maize kernels grown in vitro. *Physiol Plant* 95:59–66.

Cho JY, Lee HJ, Kim GA, Kim GD, Lee YS, Shin SC, Park KY, Moon JA (2012) Quantitative analyses of individual g-oryzanol (steryl ferulates) in conventional and organic brown rice (*Oryza sativa L.*). *J Cereal Sci* 55:337–343.

Clausen SK, Frenck G, Linden LG, et al (2011) Effects of single and multifactor treatments with elevated temperature, CO_2 and ozone on oilseed, rape and barley. *J Agron Crop Sci*. doi:10.1111/j.1439-037X.2011.00478.x

Cleland EE, Chuine I, Menzel A, et al (2007) Shifting plant phenology in response to global change. *Trends Ecol Evol* 22(7):357–365.

Collins M, Knutti R, Arblaster J, Dufresne JL, Fichefet T, Friedlingstein P, Krinner G (2013) Long-term climate change: projections, commitments and irreversibility. In: *Climate change 2013-the physical science basis: contribution of Working Group I to the fifth assessment report of the Intergovernmental Panel on Climate Change*. Cambridge University Press, pp 1029–1136.

Crafts-Brandner SJ, Law RD (2000) Effect of heat stress on the inhibition and recovery of the ribulose-1,5-bisphosphate carboxylase/oxygenase activation state. *Planta* 212:67–74.

Cunniff J, Charles M, Jones G, Osborne CP (2016) Reduced plant water status under sub-ambient p CO_2 limits plant productivity in the wild progenitors of C3 and C4 cereals. *Ann Bot* 118(6):1163–1173.

Daloso D, Daloso M, Anjos LD, Fernie AR (2016) Roles of sucrose in guard cell regulation. *New Phytol* 211(3):809–818. doi:10.1111/nph.13950

Darbah JNT, Kubiske ME, Nelson N, Kets K, Riikonen J, Sober A, Rouse L, Karnosky DF (2010). Will photosynthetic capacity of aspen trees acclimate after long-term exposure to elevated CO_2 and O3? *Environ Pollut* 158:983–991.

De Costa Wajm, Weerakoon WMW, Chinthaka KGR, Herath Hmlk, Abeywardena RMI (2007) Genotypic variation in the response of rice (*Oryza sativa* L.) to increased atmospheric carbon dioxide and its physiological basis. *J Agron Crop Sci* 193:117–130.

De Laurentiis V, Hunt DV, Rogers CD (2016) Overcoming food security challenges within an energy/water/food nexus (EWFN) approach. *Sustain* 8:95. doi:10.3390/su8010095

Delucia EH, Sasek TW, Strain BR (1985) Photosynthetic inhibition after long-term exposure to elevated levels of atmospheric carbon dioxide. *Photosynth Res* 7:175–184.

Dias AS, Lidon FC (2009a) Evaluation of grain filling rate and duration in bread and durum wheat under heat stress after anthesis. *J Agron Crop Sci* 195:137–147.

Dietzen CA, Larsen KS, Ambus PL, Michelsen A, Arndal MF, Beier C, Reinsch S, Schmidt IK (2019) Accumulation of soil carbon under elevated CO_2 unaffected by warming and drought. *Glob Change Biol* 25(9):2970–2977.

Dingkuhn M, Luquet D, Fabre D, Muller B, Yin X, Paul MJ (2020) The case for improving crop carbon sink strength or plasticity for a CO_2-rich future. *Curr Opin Plant Biol* 56:159–272.

Di Toppi LS, Marabottini R, Badiani M, Raschi A (2002) Antioxidant status in herbaceous plants growing under elevated CO_2 in mini-FACE rings. *J Plant Physiol* 159(9):1005–1013.

Djanaguiraman M, Prasad PVV, Murugan M, Perumal M, Reddy UK (2014) Physiological differences among sorghum (*Sorghum bicolor* L. Moench) genotypes under high temperature stress. *Environ Exp Bot* 100:43–54.

Dong H, Bai L, Zhang Y, Zhang G, Mao Y, Min L, Xiang F, Qian D, Zhu X, Song CP (2018) Modulation of guard cell turgor and drought tolerance by a peroxisomal Acetate–Malate Shunt. *Mol Plant* 11(10):1278–1291.

Downes RW (1972) Effect of temperature on the phenology and grain yield of *Sorghum bicolor*. *Aust J Agric Res* 23:585–594.

Dreccer MF, Wockner KB, Palta JA, McIntyre CL, Borgognone MG, Bourgault M, Reynolds M, Miralles DJ (2014) More fertile florets and grains per spike can be achieved at higher temperature in wheat lines with high spike biomass and sugar content at booting. *Fun Plant Biol* 41:482–495.

Driscoll S, Prins A, Olmos E, Kunert K, Foyer C (2006) Specification of adaxial and abaxial stomata, epidermal structure and photosynthesis to CO_2 enrichment in maize leaves. *J Exp Bot* 57:381–390.

Easterling DR, Evans JL, Groisman PY, Karl TR, Kunkel KE, et al (2000) Observed variability and trends in extreme climate events: a brief review. *Bull Amer Meteor Soc* 81:417–425.

Ehlers JD, Hall AE (1998) Heat tolerance of contrasting cowpea lines in short and long days. *Field Crops Res* 55(1–2):11–21.

Eisenhut M, Planchais S, Cabassa C, Guivarćh A, Justin AM, Taconnat L, et al (2013) Arabidopsis a bout de souffle is a putative mitochondrial transporter involved in photorespiratory metabolism and is required for meristem growth at ambient CO_2 levels. *Plant J* 73:836–849. doi: 10.1111/tpj.12082.

EL Sabagh A, Hossain A, Islam MS, Iqbal MA, Raza A, Karademir Ç, Karademir E, Rehman A, Rahman MA, Singhal RK, Llanes A, Raza MA, Mubeen M, Nasim W, Barutçular C, Meena RS, Saneoka H (2020) Elevated CO_2 concentration improves heat-tolerant ability in crops. In: *Abiotic stress in plants*. IntechOpen.

EL Sabagh A, Hossain A, Barutçular C, Islam MS, Awan SI, Galal A, et al (2019) Wheat (*Triticum aestivum l.*) production under drought and heat stress–adverse effects, mechanisms and mitigation: a review. *Appl Ecol Environ Res* 17(4):8307–8332. doi:10.15666/aeer/1704_83078332

EL Sabagh A, Hossain A, Iqbal MA, Barutçular C, Islam MS, Çiğ F, Erman M, Sytar O, Brestic M, Wasaya A, Jabeen T, Bukhari MA, Mubeen M, Athar HR, Azeem F, Akdeniz H, Konuşkan Ö, Kizilgeci F, Ikram M, Sorour S, Nasim W, Elsabagh M, Rizwan M, Meena RS, Fahad S, Ueda A, Liu L, Saneoka H (2020a) Maize adaptability to heat stress under changing climate. *Plant Stress Physiology*. IntechOpen. doi:10.5772/intechopen.92396

EL Sabagh A, Hossain A, Islam MS, Barutcular C, Hussain S, Hasanuzzaman M, Kumar N (2019) Drought and salinity stresses in barley: Consequences and mitigation strategies. *Aust J Crop Sci* 13(6):810

Erb TJ, Zarzycki J (2016) Biochemical and synthetic biology approaches to improve photosynthetic CO_2-fixation. *Curr Opin Chem Biol* 34:72–79.

Erbs M, Manderscheid R, Jansen G, et al (2015) Elevated CO_2 (FACE) Affects Food and Feed Quality of Cereals (Wheat, Barley, Maize): Interactions with N and Water Supply. *Procedia Environ Sci*. 10.1016/j.proenv.2015.07.155

Fahad S, Bajwa AA, Nazir U, Anjum SA, Farooq A, Zohaib A, Ihsan MZ (2017) Crop production under drought and heat stress: plant responses and management options. *Front Plant Sci* 8:1147.

Fair P, Tew J, Cresswell C (1973) Enzyme activities associated with carbon dioxide exchange in illuminated leaves of Hordeum vulgare L. II. Effects of external concentrations of carbon dioxide and oxygen. *Ann Bot* 37:1035–1039.

Farage PK, Long SP, Lechner EG, Baker NR (1991) The sequence of change within the photosynthetic apparatus of wheat following short-term exposure to ozone. *Plant Physiol* 95:529–535. 10.1104/pp.95.2.529

Farfan-Vignolo ER, Asard H (2012) Effect of elevated CO_2 and temperature on the oxidative stress response to drought in Lolium perenne L. and Medicago sativa L. *Plant Physiol Biochem* 59:55–62.doi: 10.1016/j.plaphy.2012.06.014

Farooq M, Bramley H, Palta JA, Siddique KHM (2011) Heat stress in wheat during reproductive and grain-filling phases. *Crit Rev Plant Sci* 30:491–507.

Feng X (1999) Trends in intrinsic water-use efficiency of natural trees for the past 100–200 years: a response to atmospheric CO_2 concentration. *Geochim Cosmochim Ac* 63:1891–1903.

Field KJ, Duckett JG, Cameron DD, Pressel S (2015) Stomatal density and aperture in non-vascular land plants are non-responsive to above-ambient atmospheric CO_2 concentrations. *Ann Bot* 115(6):915–922.

Franks PJ, Cowan IR, Farquhar GD (1998) A study of stomatal mechanics using the cell pressure probe. *Plant Cell Environ* 21:94–100.

Franks PJ, Drake PL, Beerling DJ (2009) Plasticity in maximum stomatal conductance constrained by negative correlation between stomatal size and density: An analysis using Eucalyptus globulus. *Plant, Cell Environ* 32:1737–1748. 10.1111/j.1365-3040.2009.002031.x

Fuhrer J (2003) Agroecosystem responses to combinations of elevated CO_2, ozone, and global climate change. *Agric Ecosyst Environ* 97:1–20.

Fukayama H, Sugino M, Fukuda T, Masumoto C, Taniguchi Y, Okada M, et al (2011). Gene expression profiling of rice grown in free air CO_2 enrichment (FACE) and elevated soil temperature. *Field Crops Res* 121:195–199. doi:10.1016/j.fcr.2010.11.018

Gagen M, Mccarroll D, Loader NJ, Robertson I (2011) *Stable isotopes in dendroclimatology: moving beyond 'potential' in dendroclimatology*. Springer, pp 147–172.

Gálusová T, Piršelová B, Rybanský Ľ, Krasylenko Y, Mészáros P, Blehová A, Matušíková I (2020) Plasticity of soybean stomatal responses to Arsenic and Cadmium at the whole plant level. *Polish J Environ Stud*. doi:10.15244/pjoes/116444

Geiger M, Haake V, Ludewig F, Sonewald U, Stitt M (1999) The nitrate and ammonium nitrate supply have a major influence on the response of photosynthesis, carbon metabolism, nitrogen metabolism, and growth to elevated carbon dioxide in tobacco. *Plant Cell Environ* 22:1177–1199.

Gesch RW, Kang IH, Gallo-Meagher M, Vu JCV, Boote KJ, Allen LH, et al (2003) Rubisco expression in rice leaves is related to genotypic variation of photosynthesis under elevated growth CO_2 and temperature. *Plant Cell Environ* 26:1941–1950. doi:10.1046/j.1365-3040.2003.01110.x

Ghannoum O (2009) C4 photosynthesis and water stress. *Ann Bot* 103:635–644.

Ghannoum O, Phillips NG, Sears MA, Logan BA, Lewis JD, Conroy JP, Tissue DT (2010) Photosynthetic responses of two eucalypts to industrial-age changes in atmospheric [CO_2] and temperature. *Plant Cell Environ* 33:1671–1681.

Ghasemzadeh A, Jaafar HZE, Rahmat A (2010) Elevated carbon dioxide increases contents of flavonoids and phenolic compounds, and antioxidant activities in Malaysian young ginger (*Zingiber officinale* roscoe.) varieties. *Molecules* 15:7907–7922. doi:10.3390/molecules15117907

Gill SS, Tuteja N (2010) Reactive oxygen species and antioxidant machinery in abiotic stress tolerance in crop plants. *Plant Physiol Biochem* 48:909–930. doi:10.1016/j.plaphy.2010.08.016

Gillespie KM, Xu F, Richter KT, McGrath JM, Markelz RC, Ort DR, et al (2012) Greater antioxidant and respiratory metabolism in field-grown soybean exposed to elevated O3 under both ambient and elevated CO_2. *Plant Cell Environ* 35:169–184. doi:10.1111/j.1365-3040.2011.02427.x

Glick B, Penrose D, Li J (1998) A model for the lowering of plant ethylene concentrations by plant growth-promoting bacteria. *J Theor Biol* 190:63–68. doi:10.1006/jtbi.1997.0532

Gonzalez-Meler MA, Ribas-Carbó M, Siedow JN, Drake BG (1996) Direct inhibition of plant mitochondrial respiration by elevated CO_2. *Plant Physiol* 112:1349–1355.

Goufo P, Pereira J, Moutinho-Pereira J, Correia CM, Figueiredo N, Carranca C, Trindade H (2014) Rice (*Oryza sativa* L.) phenolic compounds under elevated carbon dioxide (CO_2) concentration. *Environ Exp Bot* 99:28–37.

Goufo P, Trindade H (2014) Rice antioxidants: phenolic acids, flavonoids, anthocyanins, proanthocyanidins, tocopherols, tocotrienols, γ-oryzanol, and phytic acid. *Food Sci Nutr* 2(2):75–104.

Gray JE, Holroyd GH, Van Der Lee, FM, Bahrami AR, Sijmons PC, Woodward FI, Schuch W, Hetherington AM (2000) The HIC signalling pathway links CO_2 perception to stomatal development. *Nature* 408:713–716.

Guo H, Sun Y, Li Y, Tong B, Herris M, Zhu-Salzman K, et al (2013) Pea aphid promotes amino acid metabolism both in *Medicago truncatula* and bacteriocytes to favor aphid population growth under elevated CO_2. *Global Change Biol* 19:3210–3223. doi: 10.1111/gcb.12260

Habermann E, Dias de Oliveira EA, Contin DR, San Martin JAB, Curtarelli L, Gonzalez-Meler MA, Martinez CA (2019) Stomatal development and conductance of a tropical forage legume are regulated by elevated [CO_2] under moderate warming. *Front Plant Sci* 10:609. doi:10.3389/fpls.2019.00609.

Hammer GL, Broad IJ (2003) Genotype and environment effects on dynamics of harvest index during grain filling in sorghum. *Agron J* 95:199–206.

Hasanuzzaman M, Bhuyan MHM, Zulfiqar F, Raza A, Mohsin SM, Mahmud JA, Fotopoulos V (2020) Reactive oxygen species and antioxidant defense in plants under abiotic stress: revisiting the crucial role of a universal defense regulator. *Antioxidants* 9(8):681. doi:10.3390/antiox9080681

Hasanuzzaman M, Nahar K, Gill SS, Fujita M (2013) Drought stress responses in plants, oxidative stress, and antioxidant defense. In: *Climate Change and plant abiotic stress tolerance*, pp 209–250. doi:10.1002/9783527675265.ch09

Hasegawa T, Sakai H, Tokida T, Nakamura H, Zhu C, Usui Y, Yoshimoto M, Fukuoka M, Wakatsuki H, Katayanagi N, Matsunami T, Kaneta Y, Sato T, Takakai F, Sameshima R, Okada M, Mae T, Makino A

(2013) Rice cultivar responses to elevated CO_2 at two free-air CO_2 enrichment (FACE) sites in Japan. *Funct Plant Bio* 140:148–159.

Hatfield JL, Boote KJ, Kimball B, Ziska L, Izaurralde RC, Ort D, Wolfe D (2011). Climate impacts on agriculture: implications for crop production. *Agron J* 103(2):351–370.

Hawker JS, Jenner CF (1993) High temperature affects the activity of enzymes in the committed pathway of starch synthesis in developing wheat endosperm. *Aust J Plant Physiol* 20:197–209.

Hodges DM, Forney CF (2000) The effects of ethylene, depressed oxygen and elevated carbon dioxide on antioxidant profiles of senescing spinach leaves. *J Exp Bot* 51:645–655. doi: 10.1093/jexbot/51.344.645

Horie T, Baker JT, Nakagawa H, Matsui T, Kim HY (2000) Crop ecosystem responses to climatic change: rice. In: Reddy, KR, Hodges, HF (eds) *Climate change and global crop productivity*. CABI Publishing, Wallingford, pp 81–106.

Horvat D, Šimić G, Drezner G, Lalić A, Ledenčan T, Tucak M, Zdunić Z (2020) Phenolic acid profiles and antioxidant activity of major cereal crops. *Antioxidants* 9(6):527. doi:10.3390/antiox9060527

Hudson Institute. 1999. *Global warming – boon for mankind*. www.hudson.org/American

Hunt R, Hand D, Hannah M, Neal A (1991) Response to CO_2 enrichment in 27 herbaceous species. *Funct Ecol* 5:410–421.

Hussain H, Raziq F, Khan I, Shah B, Altaf M, Attaullah, Ullah W, Naeem A, Adnan M, Junaid K, Shah SRA, Iqbal M (2016) Effect of Bipolarismaydis (Y. Nisik & C. Miyake) shoemaker at various growth stages of different maize cultivars. *J Entomol & Zool Stud* 4(2):439–444.

Hussain MW, Allen LHJ, Bowes G (1999) Up-regulation of sucrose phosphate synthase in rice grown under elevated CO_2 and temperature. *Photosynth Res* 60:199–208.

Hymus GJ, Baker NR, Long SP (2001) Growth in elevated CO_2 can both increase and decrease photo-chemistry and photoinhibition of photosynthesis in a predictable manner. *Dactylis glomerata* grown in two levels of nitrogen nutrition. *Plant Physiol* 127(3):1204–1211.

Igboji P, Okey N, Udeh A (2018) Agricultural products fumigation poses risk of food contamination in Abakaliki, southeastern Nigeria. *Int J Agric Resour Gov Ecol* 14:91–103.

Igboji PO (2000) *Developing the agricultural potentials of ebonyi state*, 2nd edition. Ebonyi Dove Magazine.

Igboji PO (2015) Agriculture in peril. *Elixir Agric* 82:32329–32335.

Iloh, AC, Omatta, G, Ogbadu, GH, Onyenekwe PC (2014) Effects of elevated temperature on seed germi-nation and seedling growth on three cereal crops in Nigeria. *Sci Res Essays* 9:806–813. doi:10.5897/sre2014.5968

IPCC (Intergovernmental Panel on Climate Change) (2013) Summary for policymakers. In: *Climate change 2013: the physical science basis. Contribution of Working Group I to the fifth assessment report of the Intergovernmental Panel on Climate Change*. Cambridge University Press, Cambridge, UK and New York, p 9.

IPCC (2007) Summary for policy makers. In: Solomon SD, Qin M, Manning Z, Chen M, Marquis M, Avery KB, Tignor M, Miller HL (eds) *Climate change 2007: the physical science basis. Contribution of Working Group I to the fourth assessment report of the Intergovernmental Panel on Climate Change*. Cambridge University Press, UK.

Isner JC, Begum A, Nuehse T, Hetherington AM, Maathuis FJ (2018). KIN7 kinase regulates the vacuolar TPK1 K+ channel during stomatal closure. *Curr Biol* 28(3):466–472.

Jagadish SVK, Bahuguna RN, Djanaguiraman M, Gamuyao R, Prasad PV, Craufurd PQ (2016). Implications of high temperature and elevated CO_2 on flowering time in plants. *Front Plant Sci* 7:913. doi:10.3389/fpls.2016.00913.

Jagadish SVK, Craufurd PQ, Wheeler TR (2007). High temperature stress and spikelet fertility in rice (*Oryza sativa* L.). *J Exp Bot* 58:1627–1635.

Jagadish SVK, Muthurajan R, Oane R, Wheeler TR, Heuer S, Bennett J, Craufurd PQ (2010) Physiological and proteomic approaches to address heat tolerance during anthesis in rice (*Oryza sativa* L.). *J Exp Bot* 61:143–156.

Jain M, Prasad PVV, Boote KJ, Hartwell AL, Chourey PS (2007) Effects of season-long high temperature growth conditions on sugar-to-starch metabolism in developing microspores of grain sorghum (*Sorghum bicolor* L. Moench). *Planta* 227:67–79. doi:10.1007/s00425007-0595-y

Jayawardena DM, Heckathorn SA, Bista DR, Mishra S, Boldt JK, Krause CR (2017) Elevated CO_2 plus chronic warming reduce nitrogen uptake and levels or activities of nitrogen-uptake and assimilatory proteins in tomato roots. *Physio Plant* 159(3):354–365.

Jezek M, Blatt MR (2017) The membrane transport system of the guard cell and its integration for stomatal dynamics. *Plant Physiol* 174(2):487–519. doi:10.1104/pp.16.01949

Jha UC, Bohra A, Singh NP (2014) Heat stress in crop plants: its nature, impacts and integrated breeding strategies to improve heat tolerance. *Plant Breeding* 133(6):679–701.

Jia G, Zhang MQ, Wang XW, Zhang WJ (2015) A possible mechanism of mineral responses to elevated atmospheric CO_2 in rice grains. *J Integr Agric* 14(1):50–57.

Jiang YP, Cheng F, Zhou YH, Xia XJ, Shi K, Yu JQ (2012) Interactive effects of CO_2 enrichment and brassinosteroid on CO_2 assimilation and photosynthetic electron transport in *Cucumis sativus*. *Environ Exp Bot* 75:98–106. doi:10.1016/j.envexpbot.2011.09.002

Johkan M, Oda M, Maruo T, Shinohara Y (2011) Crop production and global warming impacts, case studies on the economy, human health, and on urban and natural environments. http://www.intechopen.com

Jones HG (2013) *Plants and microclimate: a quantitative approach to environmental plant physiology*, 3rd edition. Cambridge University Press. doi:10.1017/CBO9780511845727

Joshi R, Karan R (2013) Physiological, biochemical and molecular mechanisms of drought tolerance in plants. In: *Molecular approaches in plant abiotic stress*, pp 209–231.

Jover-Gil S, Candela H, Robles P, Aguilera V, Barrero JM, Micol JL, Ponce MR (2012) The microRNA pathway genes AGO1, HEN1 and HYL1 participate in leaf proximal–distal, venation and stomatal patterning in Arabidopsis. *Plant Cell Physiol* 53:1322–1333.

Juknys R, Januškaitienė I, Dikšaitytė A, Šliumpaitė I (2012) Impact of warming climate on barley and tomato growth and photosynthetic pigments. *Biologija*. doi:10.6001/biologija.v58i2.2490

Kadam NN, Xiao G, Melgar RJ, Bahuguna RN, Quinones C, Tamilselvan A, Prasad PVV, Jagadish KS (2014) Agronomic and physiological responses to high temperature, drought, and elevated CO_2 interactions in cereals. *Adv Agron* 127:111–156.

Kaiser H, Kappen L (2001) Stomatal oscillations at small apertures: indications for a fundamental insufficiency of stomatal feedback-control inherent in the stomatal turgor mechanism. *J Exp Bot* 52:1303–1313.

Kane K, Dahal KP, Badawi MA, Houde M, Hüner NP, Sarhan F (2013) Long-term growth under elevated CO_2 suppresses biotic stress genes in non-acclimated, but not cold-acclimated winter wheat. *Plant Cell Physiol* 54:1751–1768. doi:10.1093/pcp/pct116

Kang Y, Outlaw WHJ, Andersen PC, Fiore GB (2007) Guard-cell apoplastic sucrose concentration – a link between leaf photosynthesis and stomatal aperture size in the apoplastic phloem loader *Vicia faba* L. *Plant Cell Environ* 30:551–558.

Kant S, Seneweera S, Rodin J, Materne M, Burch D, Rothstein SJ, Spangenberg G (2012) Improving yield potential in crops under elevated CO_2: integrating the photosynthetic and nitrogen utilization efficiencies. *Front Plant Sci*. doi:10.3389/fpls.2012.00162

Kaur V, Behl RK (2010) Grain yield in wheat as affected by short periods of high temperature, drought and their interaction during pre- and post-anthesis stages. *Cereal Res Commun* 38:514–520.

Kim HY, Horie T, Nakagawa H Wada K (1996) Effect of elevated CO_2 and high temperature on growth and yield of rice II. The effect on yield and its components of Akihikari rice. *Jpn J Crop Sci* 65:644–651.

Kim J, et al (2011) Relationship between grain filling duration and leaf senescence of temperate rice under high temperature. *Field Crops Res* 122:207–213.

Kim SH, Gitz DC, Sicher RC, Baker JT, Timlin DJ (2007) Temperature dependence of growth development, and photosynthesis in maize under elevated CO_2. *Environ Exp Bot* 61:224–236.

Kimball BA, Bernacchi CJ (2006) Evapotranspiration, canopy temperature, and plant water relations. In: *Managed ecosystems and CO_2*. Springer, Berlin, Heidelberg, pp 311–324.

Király I, Czövek P (2002) Changes of MDA level and O_2 scavenging enzyme activities in wheat varieties as a result of PEG treatment. *Acta Biol Szeged* 46:105–106.

Ko J, Ng CT, Jeong S, et al (2019) Impacts of regional climate change on barley yield and its geographical variation in South Korea. *Int Agrophys*. doi:10.31545/intagr/104398

Kohler IH, Huber SC, Bernacchi CJ, Baxter IR (2019) Increased temperatures may safeguard the nutritional quality of crops under future elevated CO_2 concentrations. *Plant J* 97:872–886.

Körner C (2006) Plant CO_2 responses: an issue of definition, time and resource supply. *New Phytol*.

Kramer PJ (1981) Carbon dioxide concentration, photosynthesis, and dry matter production. *Bio Sci* 31:29–33.

Kumar A, Nayak AK, Sah RP, Sanghamitra P, Das BS (2017) Effects of elevated CO_2 concentration on water productivity and antioxidant enzyme activities of rice (*Oryza sativa* L.) under water deficit stress. *Field Crops Res* 212:61–72.

Kumari S, Agrawal M, Tiwari S (2013) Impact of elevated CO_2 and elevated O3 on *Beta vulgaris* L.: pigments, metabolites, antioxidants, growth and yield. *Environ Poll* 174:279–288.

Lake J, Quick W, Beerling DJ, Woodward FI (2001) Plant development: signals from mature to new leaves. *Nature* 411:154.

Lake JA, Woodward FI, Quick WP (2002) Long-distance CO_2 signalling in plants. *J Exp Bot* 53:183–193.

Landete JM (2013) Dietary intake of natural antioxidants: vitamins and polyphenols. *Crit Rev Food Sci Nutri* 53(7):706–721.

Lavana D, Dhingra A, Siddiqui MH, Al-Whaibi MH, Grover A (2015) Current status of the production of high temperature tolerant transgenic crops for cultivation in warmer climates. *Plant Physiol Biochem* 86:100–108. doi:10.1016/j.plaphy.2014.11.019

Lavania D, Dhingra A, Siddiqui MH, Al-Whaibi MH, Grover A (2015) Current status of the production of high temperature tolerant transgenic crops for cultivation in warmer climates. *Plant Physiol Biochem* 86:100–108. doi:10.1016/j.plaphy.2014.11.019

Lavola A, Julkunen-Tiitto R (1994) The effect of elevated carbon dioxide and fertilization on primary and secondary metabolites in birch, *Betula pendula* (Roth). *Oecologia* 99:315–321. doi:10.1007/BF00627744

Lavola A, Nybakken L, Rousi M, Pusenius J, Petrelius M, Kellomaki S, et al (2013) Combination treatment of elevated UVB radiation, CO_2 and temperature has little effect on silver birch (*Betula pendula*) growth and phytochemistry. *Physiol Plant* 149:499–514. doi:10.1111/ppl.12051

Lawlor DW, Mitchell RA (2000) Crop ecosystem responses to climatic change: wheat. In: Reddy KR, Hodges HF (eds) *Climate change and global crop productivity*. CAB International, Wallingford, UK, pp 57–80.

Leakey A, Uribelarrea M, Ainsworth E, Naidu S, Rogers A, et al (2006) Photosynthesis, productivity and yield of maize are not affected by open-air elevation of CO_2 concentration in the absence of drought. *Plant Physiol* 140(2):779–790.

Leakey ADB, Ainsworth EA, Bernacchi CJ, Rogers A, Long SP, Ort DR (2009a) Elevated CO_2 effects on plant carbon, nitrogen, and water relations: six important lessons from FACE. *J Exp Bot* 60:2859–2876.

Leakey ADB, Xu F, Gillespie KM, McGrath JM, Ainsworth EA, Ort DR (2009b) Genomic basis for stimulated respiration by plants growing under elevated carbon dioxide. *Proc Natl Acad Sci USA* 106:3597–3602. doi:10.1073/pnas.0810955106

Lee JS (2011) Combined effect of elevated CO_2 and temperature on the growth and phenology of two annual C3 and C4 weedy species. *Agric Ecosyst Environ*. doi:10.1016/j.agee.2011.01.013

Lelieveld J, Klingmüller K, Pozzer A, Burnett RT, Haines A, Ramanathan V (2019) Effects of fossil fuel and total anthropogenic emission removal on public health and climate. *Proc Natl Acad Sci USA* 116(15):7192–7197. doi:10.1073/pnas.1819989116

Li B, Feng Y, Zong Y, Zhang D, Hao X, Li P (2020a) Elevated CO_2-induced changes in photosynthesis, antioxidant enzymes and signal transduction enzyme of soybean under drought stress. *Plant Physiol Biochem* 154:105–114.

Li J, et al (2008) Effects of elevated CO_2 on growth, carbon assimilation, photosynthate accumulation and related enzymes in rice leaves during sink-source transition. *J Integr Plant Biol* 50:723–732.

Li M, Li Y, Zhang W, Li S, Gao Y, Ai X, Zhang D, Liu B, Li Q (2018) Metabolomics analysis reveals that elevated atmospheric CO_2 alleviates drought stress in cucumber seedling leaves. *Anal Biochem* 559:71–85.

Li S, Li Y, Gao Y, He X, Zhang D, Liu B, Li Q (2020b) Effects of CO_2 enrichment on non-structural carbohydrate metabolism in leaves of cucumber seedlings under salt stress. *Sci Hortic* 265:109275.

Li W, Zhu Q, Wang Y, Wang S, Chen X, Zhang D, Wang W (2013) *The relationships between physiological and biochemical indexes and the yield characteristics of rice under high temperature stress*, 29(9).

Li X, Ahammed G, Zhang Y, Zhang G, Sun Z, Zhou J, Zhou Y, Xia X, Yu J, Shi K (2014). Carbon dioxide enrichment alleviates heat stress by improving cellular redox homeostasis through an ABA-independent process in tomato plants. *Plant Biol* 17:81–89.

Li XM, Zhang LH, Ma LJ, Li YY (2011) Elevated carbon dioxide and/or ozone concentrations induce hormonal changes in *Pinus tabulaeformis*. *J Chem Ecol* 37:779–784. doi:10.1007/s10886-011-9975-7

Li Y, Zhang Y, Zhang X, Korpelainen H, Berninger F, Li C (2013) Effects of elevated CO_2 and temperature on photosynthesis and leaf traits of an understory dwarf bamboo in subalpine forest zone, *China Physiol Plantar 148*(2):261–272.

Lin CJ, Li CY, Lin SK, Yang FH, Huang JJ, Liu YH, Lur HS (2010) Influence of high temperature during grain filling on the accumulation of storage proteins and grain quality in rice (*Oryza sativa* L.). *J Agric Food Chem* 58:10545–10552.

Lin JS, Wang GX (2002) Doubled CO_2 could improve the drought tolerance better in sensitive cultivars than in tolerant cultivars in spring wheat. *Plant Sci* 163:627–637.

Lin W, Ziska LH, Namuco OS, Bai K (1997) The interaction of high temperature and elevated CO_2 on photosynthetic acclimation of single leaves of rice in situ. *Physiol Plant* 99:178–184. 10.1111/j.1399-3054.1997.tb03446.x

Liu GY, Xie YC, Hou YH, Li X, Zhang HG, Wan YF (2018) Effects of elevated temperature and carbon dioxide concentration on the growth of highland barley. *Chin J Agrometeorol* 39(9):567–574.

Lobell DB, Bänziger M, Magorokosho C, Vivek B (2011) Nonlinear heat effects on African maize as evidenced by historical yield trials. *Nat Clim Chang 1*:42. doi:10.1038/nclimate1043.

Lobell DB, Gourdji SM (2012) The influence of climate change on global crop productivity. *Plant Physiol* 160:1686–1697.

Lobell DB, Hammer GL, McLean G, Messina C, Roberts MJ, Schlenker W (2013) The critical role of extreme heat for maize production in the United States. *Nat Clim Chang* 3:497–501. 10.1038/nclimate1832

Lobell DB, Schlenker W, Costa-Roberts J (2011) Climate trends and global crop production since 1980. *Science* 333:616–620. doi:10.1126/science.1204531

Long SP, Ainsworth EA, Rogers A, Ort DR (2004) Rising atmospheric carbon dioxide: plants FACE the future. *Annu Rev Plant Biol* 55:591–628.

Long SP, Zhu XG, Naidu S, Ort DR (2006) Can improvement in photosynthesis increase crop yields? *Plant Cell Environ* 29:315–330. doi:10.1111/j.1365-3040.2005.01493.x

Luo Y, Su B, Currie WS, Dukes JS, Finzi A, Hartwig U, et al (2004) Progressive nitrogen limitation of ecosystem responses to rising atmospheric carbon dioxide. *BioScience* 54:731–739. doi:10.1641/0006-3568(2004)054

Luomala E, Laitinen K, Sutinen S, Kellomäki S, Vapaavuori E (2005) Stomatal density, anatomy and nutrient concentrations of Scots pine needles are affected by elevated CO_2 and temperature. *Plant Cell Environ* 28:733–749.

Lyman NB, Jagadish KSV, Nalley LL, Dixon BL, Siebenmorgen T (2013) Neglecting rice milling yield and quality underestimates economic losses from high-temperature stress. *PLoS One* 8:e72157.

Maclean JL, Dawe DC, Hardy B, Hettel GP (2002). *Rice Almanac*, 3rd edition. IRRI, WARDA, CIAT and FAO, Philippines.

Madan P, Jagadish S, Craufurd P, Fitzgerald M, Lafarge T, Wheeler T (2012) Effect of elevated CO_2 and high temperature on seed-set and grain quality of rice. *J Exp Bot* 63(10):3843–3852.

Maiti RK (1996) *Sorghum science*. Science Publishers, Lebanon, 352 p.

Maity PP, Chakrabarti B, Bhatia A, Purakayastha TJ, Saha ND, Jatav RS, Sharma A, Bhowmik A, Kumar V, Chakraborty D (2019) Effect of elevated CO_2 and temperature on growth of rice crop. *Int J Curr Microbiol Appl Sci* 8:1906–1911.

Makino A (1994) Biochemistry of C3-photosynthesis in high CO_2. *J Plant Res* 107:79–84. doi:10.1007/BF02344533

Markelz RJC, Lai LX, Vosseler LN, Leakey ADB (2014) Transcriptional reprogramming and stimulation of leaf respiration by elevated CO_2 concentration is diminished, but not eliminated, under limiting nitrogen supply. *Plant Cell Environ* 37:886–898. doi:10.1111/pce.12205

Matile P, Schellenberg M, Vicentini Γ (1997) Planta localization of chlorophyllase in the chloroplast envelope. *Planta* 201:96–99. doi:10.1007/BF01258685

Matros A, Amme S, Kettig B, Buck-Sorlin GH, Sonnewald U, Mock HP (2006) Growth at elevated CO_2 concentrations leads to modified profiles of secondary metabolites in tobacco cv. SamsunNN and to increased resistance against infection with potato virus Y. *Plant Cell Environ* 29:126–137. doi:10.1111/j.1365-3040.2005.01406.x

Mccarroll D, Loader NJ (2004) Stable isotopes in tree rings. *Quater Sci Rev* 23:771–801.

McLachlan DH, Lan J, Geilfus CM, Dodd AN, Larson T, Baker A, H~orak H, Kollist H, He Z, Graham I, et al (2016) The breakdown of stored triacylglycerols is required during light-induced stomatal opening. *Curr Biol* 26:707–712.

Medlyn B, Barton C, Broadmeadow M, Ceulemans R, De Angelis P, Forstreuter M, Freeman M, Jackson S, Kellomäki S, Laitat E (2001) Stomatal conductance of forest species after long-term exposure to elevated CO_2 concentration: a synthesis. *New Phytol* 149:247–264.

Medlyn BE, Dreyer E, Ellsworth D, et al (2002) Temperature response of parameters of a biochemically based model of photosynthesis. II. A review of experimental data. *Plant Cell Environ* 25:1167–1179.

Mendelsohn R (2014) The impact of climate change on agriculture in Asia. *J Integr Agric* 13:660–665. doi:10.1016/S2095-3119(13)60701-7

Min MK, Kim R, Moon SJ, Lee Y, Han S, Lee S, Kim BG (2020) Selection and functional identification of a synthetic partial ABA agonist, S7. *Sci Rep* 10(1):1–10.

Mina U, Kumar R, Gogoi R, Bhatia A, Harit RC, Singh D, Kumar A, Kumar A (2019) Effect of elevated temperature and carbon dioxide on maize genotypes health index. *Ecol Indic* 105:292–302. doi:10.1016/j.ecolind.2017.08.060

Mishra AK, Agrawal SB (2014) Cultivar specific response of CO_2 fertilization on two tropical mung bean (*Vigna radiata* L.) cultivars: ROS generation, antioxidant status, physiology, growth, yield and seed quality. *J Agron Crop Sci* 20:273–289. doi:10.1111/jac.12057

Misra BB, Acharya BR, Granot D, Assmann SM, Chen S (2015) The guard cell metabolome: functions in stomatal movement and global food security. *Front Plant Sci* 6:334. doi:10.3389/fpls.2015.00334

Misra BB, Reichman SM, Chen S (2019) The guard cell ionome: understanding the role of ions in guard cell functions. *Progr Biophys Mol Biol* 146:50.

Mittler, R, Blumwald, E, (2010). Genetic engineering for modern agriculture: challenges and perspectives. *Annu Rev Plant Biol* 61:443–462.

Mittler R, Vanderauwera S, Gollery M, Van Breusegem F (2004) Reactive oxygen gene network of plants. *Trends Plant Sci* 9(10):490–498.

Moore BD, Cheng SH, Rice J, Seemann JR (1998) Sucrose cycling, Rubisco expression, and prediction of photosynthetic acclimation to elevated atmospheric CO_2. *Plant Cell Environ* 21:905–915. doi:10.1046/j.1365-3040.1998.00324.x

Moore FC, Lobell DB (2014) Adaptation potential of European agriculture in response to climate change. *Nat Clim Change*. doi:10.1038/nclimate2228

Moroney JV, Jungnick N, DiMario RJ, Longstreth DJ (2013) Photorespiration and carbon concentrating mechanisms, two adaptations to high O2, low CO_2 conditions. *Photosyn Res* 117:121–131. doi:10.1007/s11120-013-9865-7

Mphande W, Nicolas ME, Seneweera S, Bahrami H (2016) Dynamics and contribution of stem water-soluble carbohydrates to grain yield in two wheat lines contrasted under drought and elevated CO_2 conditions. *J Plant Physiol* 214(2):1037–1058.

Mubeen M, et al (2019) Evaluating the climate change impact on crop water requirement of cotton- wheat in semi-arid conditions using DSSAT model. doi:10.2166/wcc.2019.179

Munemasa S, Hauser F, Park J, Waadt R, Brandt B, Schroeder JI (2015) Mechanisms of abscisic acid-mediated control of stomatal aperture. *Curr Opin Plant Biol* 28:154–162.

Myers SS, Zanobetti A, Kloog I, Huybers P, Leakey ADB, Bloom AJ, Carlisle E, Dietterich LH, Fitzgerald G, Hasegawa T, Holbrook NM, Nelson RL, Ottman MJ, Raboy V, Sakai H, Sartor KA, Schwartz J, Seneweera S, Tausz M, Usui Y (2014) Increasing CO_2 threatens human nutrition. *Nature* 510:139–142. doi:10.1038/nature13179

Nastis SA, Michailidis A, Chatzitheodoridis F (2012) Climate change and agricultural productivity. *Afri J Agric Res* 7:4885–4893. doi:10.5897/ajar11.2395

Nguyen TC, Singh V, van Oosterom EJ, Chapman SC, Jordan DR, Hammer GL (2013) Genetic variability in high temperature effects on seed-set in sorghum. *Funct Plant Biol* 40:439–448.

Ohama N, Sato H, Shinozaki K, Yamaguchi-Shinozaki K (2017) Transcriptional regulatory network of plant heat stress response. *Trends Plant Sci* 22(1):53–65.

Oliveira MFD, Marenco RA (2019) Gas exchange, biomass allocation and water-use efficiency in response to elevated CO_2 and drought in andiroba (*Carapa surinamensis*, Meliaceae). *iForest Biogeosci Foresty* 12(1):61–68.

Osakabe Y, Watanabe T, Sugano SS, Ueta R, Ishihara R, Shinozaki K, Osakabe K (2016) Optimization of CRISPR/Cas9 genome editing to modify abiotic stress responses in plants. *Sci Rep* 6:26685.

Overmyer K, Brosché M, Kangasjärvi J (2003) Reactive oxygen species and hormonal control of cell death. *Trends Plant Sci* 8:335–342.

Pan C, Ahammed GJ, Li X, Shi K (2018). Elevated CO_2 improves photosynthesis under high temperature by attenuating the functional limitations to energy fluxes, electron transport and redox homeostasis in tomato leaves. *Front Plant Sci* 9:1739.

Parvin S, Uddin S, Tausz-Posch S, Armstrong R, Fitzgerald G, Tausz M (2019) Grain mineral quality of dryland legumes as affected by elevated CO_2 and drought: a FACE study on lentil (*Lens culinaris*) and faba bean (*Vicia faba*). *Crop Past Sci* 70(3):244–253.

Passarella VS, Savin R, Slafer GA (2002) Grain weight and malting quality in barley as affected by brief periods of increased spike temperature under field conditions. *Aust J Agric Res.* doi:10.1071/AR02096

Passarella VS, Savin R, Slafer GA (2005) Breeding effects on sensitivity of barley grain weight and quality to events of high temperature during grain filling. *Euphytica.* doi:10.1007/s10681-005-5068-4

Passarella VS, Savin R, Slafer GA (2008) Are temperature effects on weight and quality of barley grains modified by resource availability? *Aust J Agric Res.* doi:10.1071/AR06325

Pastore D, Fratianni A, Di Pede S, Passarella S (2000) Effects of fatty acids, nucleotides and reactive oxygen species on durum wheat mitochondria. *FEBS Lett* 470:88–92.

Pathak H, Aggarwal PK, Singh SD (2012) *Climate change impacts, adaptations and mitigation in agriculture: methodology for assessment and application.* Indian Agricultural Research Institute, New Delhi, pp 302.

Paul MM, Rai A, Kumar S (2014) Classification of cereal proteins related to abiotic stress based on their physicochemical properties using support vector machine. *Curr Sci* 107:1283–1289.

Peñuelas J, Canadell JG, Ogaya R (2011) Increased water-use efficiency during the 20th century did not translate into enhanced tree growth. *Global Ecol Biogeogr* 20:597–608.

Peñuelas J, Sardans J, Estiarte M, Ogaya R, Carnicer J, Coll M, et al (2013) Evidence of current impact of climate change on life: a walk from genes to the biosphere. *Global Change Biol* 19:2303–2338. doi: 10.1111/gcb.12143

Peñuelas J, Sardans J, Rivas-Ubach A, Janssess I (2012) The human-induced imbalances between C, N and P in Earth's life system. *Global Change Biol* 18:3–6. doi:10.1111/j.1365-2486.2011.02568.x

Pérez-López U, Robredo A, Lacuesta M, Sgherri C, Mena-Petite A, Navari-Izzo F, Muñoz-Rueda A (2010) Lipoic acid and redox status in barley plants subjected to salinity and elevated CO_2. *Physiol Plantar* 139(3):256–268.

Pérez-López U, Robredo A, Lacuesta M, Sgherri C, Muñoz-Rueda A, Navari-Izzo F, et al (2009) The oxidative stress caused by salinity in two barley cultivars is mitigated by elevated CO_2. *Physiol Plantar* 135:29–42. doi:10.1111/j.1399-3054.2008.01174.x

Peterhansel C, Offermann S (2012) Re-engineering of carbon fixation in plants–challenges for plant biotechnology to improve yields in a high-CO_2 world. *Curr Opin Biotechnol* 23(2):204–208.

Porter JR, Semenov MA (2005) Crop responses to climatic variation. *Philos Trans R Soc B Biol Sci*, 2021–2035.

Prasad PV, Bheemanahalli R, Jagadish SK (2017) Field crops and the fear of heat stress—opportunities, challenges and future directions. *Field Crops Res* 200:114–121.

Prasad PV, Boote KJ, Allen JLH, Thomas JM (2002) Effects of elevated temperature and carbon dioxide on seed-set and yield of kidney bean (*Phaseolus vulgaris* L.). *Global Change Biol* 8(8):710–721.

Prasad PVV, Boote KJ, Allen LH (2006) Adverse high temperature effects on pollen viability, seed-set, seed yield and harvest index of grain-sorghum (*Sorghum bicolor* L. Moench) are more severe at elevated carbon dioxide due to higher tissue temperatures. *Agr For Meteorol* 139:237–251.

Prasad PVV, Djanaguiraman M (2014) Response of floret fertility and individual grain weight of wheat to high temperature stress: sensitive stages and thresholds for temperature and duration. *Funct Plant Biol* 41:1261–1269.

Prasad PVV, Djanaguiraman M, Perumal R, Ciampitti IA (2015) Impact of high temperature stress on floret fertility and individual grain weight of grain sorghum: sensitive stages and thresholds for temperature and duration. *Front Plant Sci* 6:820. doi:10.3389/fpls.2015.00820

Prasad PVV, Staggenborg SA (2008) *Drought and/or heat stress on physiological, developmental, growth, and yield processes of crop plants' response to limited water: understanding and modelling water stress effects on plant growth processes*. Advances in Agricultural Systems Modeling Series 1. Madison, USA, pp 301–355.

Prasad PVV, Vu JCV, Boote KJ, Allen LH (2009) Enhancement in leaf photosynthesis and upregulation of Rubisco in the C4 sorghum plant at elevated growth carbon dioxide and temperature occur at early stages of leaf ontogeny. *Funct Plant Biol* 36:761–769.

Price GD, Badger MR, von Caemmerer S (2011) The prospect of using cyanobacterial bicarbonate transporters to improve leaf photosynthesis in C3 crop plants. *Plant Physiol* 155(1):20–26.

Pritchard SG, Rogers HH, Prior SA, Peterson CM (1999) Elevated CO_2 and plant structure: a review. *Glob Change Biol* 5(7):808–837.

Rae BD, Long BM, Förster B, Nguyen ND, Velanis CN, Atkinson N, McCormick AJ (2017) Progress and challenges of engineering a biophysical CO_2-concentrating mechanism into higher plants. *J Exp Bot* 68(14):3717–3737.

Raj A, Chakrabarti B, Pathak H, Singh SD, Mina U, Mittal R (2016) Growth, yield components and grain yield response of rice to temperature and nitrogen levels. *J Agrometeorol* 18(1):1–6.

Rankoth LM, De Costa M (2013) *Response of growth, biomass partitioning and nutrient uptake of lowland rice to elevated temperature at the vegetative stage*. Book of Abstracts of the Peradeniya University Research Sessions, Sri Lanka, 2012, 17, p 6.

Rao MV, Hale BA, Ormrod DP (1995) Amelioration of ozone-induced oxidative damage in wheat plants grown under high carbon dioxide (role of antioxidant enzymes). *Plant Physiol* 109:421–432.

Raza A (2020) Eco-physiological and biochemical responses of rapeseed (*Brassica napus* L.) to abiotic stresses: consequences and mitigation strategies. *J Plant Growth Regul*. doi:10.1007/s00344-020-10231-z

Raza A, Mehmood SS, Tabassum J, Batool R (2019a) Targeting plant hormones to develop abiotic stress resistance in wheat. In: *Wheat production in changing environments*. Springer, Singapore, pp 557–577.

Raza A, Razzaq A, Mehmood SS, Zou X, Zhang X, Lv Y, Xu J (2019b) Impact of climate change on crops adaptation and strategies to tackle its outcome: a review. *Plants* 8(2):34. doi:10.3390/plants8020034

Reddy KR, Vara Prasad P, Kakani VG (2005) Crop responses to elevated carbon dioxide and interactions with temperature: cotton. *J Crop Improv* 13(1-2):157–191.

Ribeiro DM, Araujo WL, Fernie AR, Schippers JHM, Mueller-Roeber B (2012) Action of Gibberellins on growth and metabolism of Arabidopsis plants associated with high concentration of carbon dioxide. *Plant Physiol* 160:1781–1794. doi:10.1104/pp.112.204842

Robredo A, Pérez-López U, de la Maza HS, et al (2007) Elevated CO_2 alleviates the impact of drought on barley improving water status by lowering stomatal conductance and delaying its effects on photosynthesis. *Environ Exp Bot*. doi:10.1016/j.envexpbot.2006.01.001

Rodrigues WP, Martins MQ, Fortunato AS, Rodrigues AP, Semedo JN, Simões-Costa MC, et al (2016) Long-term elevated air [CO_2] strengthens photosynthetic functioning and mitigates the impact of supra-optimal temperatures in tropical *Coffea arabica* and *C. canephora* species. *Glob Change Biol* 22:415–431. doi:10.1111/gcb.13088

Roger HH, Bingham GE, Cure JD, Smith JM, Surano KA (1983) Responses of selected plant species to elevated carbon dioxide in the field. *J Environ Qua* 12:569–574.

Rogers HH, Dahlman RC (1993) Crop responses to CO_2 enrichment. CO_2 and biosphere. *Adv Vegeta Sci* 14:117–131.

Rosenzweig C, Elliott J, Deryng D, Ruane AC, Müller C, Arneth A, Khabarov N (2014) Assessing agricultural risks of climate change in the 21st century in a global gridded crop model intercomparison. *Proc Nati Acad Sci* 111(9):3268–3273.

Roy S, Banerjee A, Mawkhlieng B, Misra AK, Pattanayak A, Harish GD, Bansal KC (2015) Genetic diversity and population structure in aromatic and quality rice (*Oryza sativa* L.) landraces from North-Eastern India. *PLoS One*, 10(6):e0129607.

Ruiz-Vera UM, Siebers M, Gray SB, Drag DW, Rosenthal DM, Kimball BA, et al (2013) Global warming can negate the expected CO_2 stimulation in photosynthesis and productivity for soybean grown in the Midwestern United States. *Plant Physiol* 162:410–423. doi:10.1104/pp.112.211938

Sage RF, Kubien DS (2007) The temperature response of C3 and C4 photosynthesis. *Plant Cell & Environ* 30:1086–1106.

Saini H, Aspinall D (1982) Abnormal sporogenesis in wheat (*Triticum aestivum* L.) induced by short periods of high temperature. *Ann Bot* 49:835–846. doi:10.1093/oxfordjournals.aob.a086310.

Sakai H, Tokida T, Usui Y, Nakamura H, Hasegawa T (2019) Yield responses to elevated CO_2 concentration among Japanese rice cultivars released since 1882. *Plant Prod Sci* 22(3):352–366.

Salazar-Parra C, Aguirreolea J, Sánchez-Díaz M, Irigoyen JJ, Morales F (2012) Climate change (elevated CO_2, elevated temperature and moderate drought) triggers the antioxidant enzymes' response of grapevine cv. Tempranillo, avoiding oxidative damage. *Physiol Plantar* 144:99–110.

Salim N, Raza A (2020) Nutrient use efficiency (NUE) for sustainable wheat production: a review. *J Plant Nutri* 43(2):297–315.

Satapathy BS, Duary B, Saha S, Pun KB, Singh T (2017) Effect of weed management practices on yield and yield attributes of wet direct seeded rice under lowland ecosystem of Assam. ORYZA Inter *J Rice* 54(1):29–36.

Savicka M, Škute N (2010) Effects of high temperature on malondialdehyde content, superoxide production and growth changes in wheat seedlings (*Triticum aestivum* L.). *Ekologija* 56:26–33. doi:10.2478/v1 0055-010-0004-x

Savin R, Stone PJ, Nicolas ME (1996) Responses of grain growth and malting quality of barley to short periods of high temperature in field studies using portable chambers. *Aust J Agric Res*. doi:10.1071/AR9960465

Schlenker W, Roberts MJ (2009) Nonlinear temperature effects indicate severe damages to US crop yields under climate change. *Proc Natl Acad Sci USA* 106:15594–15598. doi:10.1073/pnas.0906865106

Scranton MI, de Angelis MA (2001) Gas exchange in estuaries. In: Steele JH (ed) *Encyclopedia of ocean sciences*, 2nd edition. Academic Press, pp 1–8. doi:10.1016/B978-012374473-9.00080-1

Seneweera S, Ghannoum O, Conroy JP (2001) Root and shoot factors contribute to the effect of drought on photosynthesis and growth of the C4 grass *Panicum coloratum* at elevated CO_2 partial pressures. *Funct Plant Biol* 28:451–460.

Senghor A, Dioh RMN, Müller C, Youm I (2017) Cereal crops for biogas production: a review of possible impact of elevated CO_2. *Renew Sustain Energy Rev* 71:548–554.

Sgherri C, Pérez-López U, Micaelli F, Miranda-Apodaca J, Mena-Petite A, Muñoz-Rueda A, Quartacci MF (2017) Elevated CO_2 and salinity are responsible for phenolics-enrichment in two differently pigmented lettuces. *Plant Physiol Biochem* 115:269–278.

Shah F, Huang J, Cui K, Nie L, Shah T, Chen C, Wang K (2011) Impact of high-temperature stress on rice plant and its traits related to tolerance. *J Agric Sci* 149:545–556. doi:10.1017/S0021859611000360

Shanmugam S, Kjaer KH, Ottosen CO, Rosenqvist E, Kumari Sharma D, Wollenweber B (2013) The alleviating effect of elevated CO_2 on heat stress susceptibility of two wheat (*Triticum aestivum* L.) cultivars. *J Agron Crop Sci* 199:340–350.

Sharma P, Jha AB, Dubey RS, Pessarakli M (2012). Reactive oxygen species, oxidative damage, and antioxidative defense mechanism in plants under stressful conditions. *J Bot* 2012:217037.

Sharma S, Kaur R, Singh A (2017) Recent advances in CRISPR/Cas mediated genome editing for crop improvement. *Plant Biotechnol Rep* 11(4):193–207.

Shimono H, Nakamura H, Hasegawa T, Okada M (2013) Lower responsiveness of canopy evapotranspiration rate than of leaf stomatal conductance to open-air CO_2 elevation in rice. *Global Change Biol* 19:2444–2453.

Shimono H, Okada M, Yamakawa Y, Nakamura H, Kobayashi K, Hasegawa T (2009) Genotypic variation in rice yield enhancement by elevated CO_2 relates to growth before heading, and not to maturity group. *J Exp Bot* 60:523–532.

Shtein I, Shelef Y, Marom Z, Zelinger E, Schwartz A, Popper ZA, Bar-On B, Harpaz-Saad S (2017) Stomatal cell wall composition: distinctive structural patterns associated with different phylogenetic groups. *Ann Bot* 119(6):1021–1033.

Singh A, Agrawal M (2015). Effects of ambient and elevated CO_2 on growth, chlorophyll fluorescence, photosynthetic pigments, antioxidants, and secondary metabolites of Catharanthusroseus (L.) G Don. grown under three different soil N levels. *Environ Sci Poll Res Int* 22:3936–3946. doi:10.1007/s11356-014-3661-6

Singh P, Nedumaran S, Traore PCS, Boote KJ, Rattunde HFW, Prasad PVV, et al (2014). Quantifying potential benefits of drought and heat tolerance in rainy season sorghum for adapting to climate change. *Agric Forest Meteorol* 185:37–48. doi:10.1016/j.agrformet.2013.10.012

Singh S, Asthir B, Bains NS (2008) High temperature tolerance in relation to carbohydrate metabolism in Barley. *Ecol Environ Cons* 14:55–59.

Singh SD, Chakrabarti B, Muralikrishna KS, Chaturvedi AK, Kumar V, Mishra S, Harit R (2013) Yield response of important field crops to elevated air temperature and CO. *Ind J Agric Sci* 83(10): 1009–1012.

Singh V, Nguyen TC, van Oosterom EJ, Chapman SC, Jordan DR and Hammer GL (2015) Sorghum genotypes differ in high temperature responses for seed set. *Field Crops Res* 171:32–40.

Skirycz A, Claeys H, de Bodt S, Oikawa A, Shinoda S, Andriankaja M, Maleux K, Eloy NB, Coppens F, Yoo SD, Saito K, Inzé D (2011) Pause-and-stop: the effects of osmotic stress on cell proliferation during early leaf development in Arabidopsis and a role for ethylene signaling in cell cycle arrest. *Plant Cell* 23:1876–1888. doi:10.1105/tpc.111.084160

Slesak I, Libik M, Karpinska B, Karpinski S, Miszalski Z (2007) The role of hydrogen peroxide in regulation of plant metabolism and cellular signalling in response to environmental stresses. *Acta Biochim Polon* 54(1):39–50.

Smirnoff N, Wheeler GL (2000) Ascorbic acid in plants: biosynthesis and function. *Crit Rev Plant Sci* 19:267–290. doi:10.1016/S0735-2689(00)80005-2

Smith DL, Dijak M, Bulman P, et al (2011) *Barley: physiology of yield*. Crop Yield.

Smith CJ, Forster PM, Allen M, et al (2019) Current fossil fuel infrastructure does not yet commit us to 1.5 °C warming. *Nat Commun* 10:101. doi:10.1038/s41467-018-07999-w

Soares AS, Driscoll SP, Olmos E, Harbinson J, Arrabaça MC, Foyer CH (2008) Adaxial/abaxial specification in the regulation of photosynthesis and stomatal opening with respect to light orientation and growth with CO_2 enrichment in the C4 species *Paspalum dilatatum*. *New Phytol* 177:186–198.

Srivastava A, Kumar SN, Aggrawal PK (2010) Assessment on vulnerability of sorghum to climate change in India. *Agric Ecosyst Environ* 138:160–169. doi:10.1016/j.agee.2010.04.012

Sultan B, Guan K, Kouressy M, Biasutti M, Piani C, Hammer GL, et al (2014) Robust features of future climate change impacts on sorghum yields in West Africa. *Environ Res Lett* 9:104006. doi:10.1088/174 8-9326/9/10/104006

Sultana H, Ali N, Iqbal MM, Khan AM (2009) Vulnerability and adaptability of wheat production in different climatic zones of Pakistan under climate change scenarios. *Clim Change* 94(1-2):123–142.

Takagi D, Inoue H, Odawara M Shimakawa G, Miyake C (2014) The Calvin cycle inevitably produces sugar-derived reactive carbonyl methylglyoxal during photosynthesis: a potential cause of plant diabetes. *Plant Cell Physiol* 55:333–340. doi:10.1093/pcp/pcu007

Tan K, Zhou G, Lv X, Guo J, Ren S (2018) Combined effects of elevated temperature and CO_2 enhance threat from low temperature hazard to winter wheat growth in North China. *Sci Rep* 8(1):1–9.

Tao Zh-Q, Chen Y-Q, Li Ch, Zou J-X, Yan P, Yuan Sh-F, Wu X, Sui P (2016) The causes and impacts for heat stress in spring maize during grain filling in the North China Plain—a review. *J Int Agric* 15(12):2677–2687. doi:10.1016/S2095-3119(16)61409-0

Taub DR, Miller B, Allen H (2008) Effects of elevated CO_2 on the protein concentration of food crops: a meta-analysis. *Glob Change Biol*. doi:10.1111/j.1365-2486.2007.01511.x

Tausz-Posch S, Borowiak K, Dempsey RW, Norton RM, Seneweera S, Fitzgerald GJ, Tausz M (2013) The effect of elevated CO_2 on photochemistry and antioxidative defence capacity in wheat depends on environmental growing conditions–a FACE study. *Environ Exp Bot* 88:81–92.

Taylor SH, Aspinwall MJ, Blackman CJ, Choat B, Tissue DT, Ghannoum O (2018) CO_2 availability influences hydraulic function of C3 and C4 grass leaves. *J Exp Bot* 69(10):2731–2741.

Teixeira EI, Fischer G, van Velthuizen H, Walter C, Ewert F (2013) Global hot-spots of heat stress on agricultural crops due to climate change. *Agric Forest Meteorol* 170:206–215.

Teng N, Jin B, Wang Q, Hao H, Ceulemans R, Kuang T, et al (2009) No detectable maternal effects of elevated CO_2 on *Arabidopsis thaliana* over 15 generations. *PLoS One* 4:e6035. doi:10.1371/journal. pone.0006035

Teng N, Wang J, Chen T, Wu X, Wang Y, Lin J (2006) Elevated CO_2 induces physiological, biochemical and structural changes in leaves of *Arabidopsis thaliana*. *New Phytol* 172:92–103. doi:10.1111/j.1469-813 7.2006.01818.x

Tingem M, Ravington M, Bellocchi G, Azam-Ali S, Collis J (2008) Effect of climate change on crop production in Cameroon. *Clim Res* 36:65–77. doi:10.3354/cr00733

Torbert HA, Prior SA, Rogers HH, Runion GB (2004) Elevated atmospheric CO_2 effects on N fertilization in grain sorghum and soybean. *Field Crops Res* 88(1):57–67.

Toum L, Torres PS, Gallego SM, Benavídes MP, Vojnov AA, Gudesblat GE (2016) Coronatine inhibits stomatal closure through guard cell-specific inhibition of NADPH oxidase-dependent ROS production. *Front Plant Sci* 7:1851.

Tubiello FN, Ewert F (2002) Simulating the effects of elevated CO_2 on crops: approaches and applications for climate change. *Eur J Agron*, 57–74.

Tyfield D (2011) Food systems transition and disruptive low carbon innovation: Implications for a food security research agenda. *J Exp Bot* 62:3701–3706. doi:10.1093/jxb/err123

Uprety DC, Dwivedi N, Jain V, Mohan R (2002) Effect of elevated carbon dioxide concentration on the stomatal parameters of rice cultivars. *Photosynthetica* 40:315–319.

Urban J, Ingwers M, McGuire MA, Teskey RO (2017). Stomatal conductance increases with rising temperature. *Plant Signal Behav* 12(8):e1356534. doi:10.1080/15592324.2017.1356534.

Usui Y, Sakai H, Tokida T, Nakamura H, Nakagawa H, Hasegawa T (2014) Heat-tolerant rice cultivars retain grain appearance quality under free-air CO_2 enrichment. *Rice* 7:6.

Vaidya AS, Helander JD, Peterson FC, Elzinga D, Dejonghe W, Kaundal A, Park SY, Xing Z, Mega R, Takeuchi J, Khanderahoo B (2019) Dynamic control of plant water use using designed ABA receptor agonists. *Science* 366(6464):eaaw8848. doi:10.1126/science.aaw8848.

Valentini R, Matteucci G, Dolman AJ, Schulze ED, Rebmann C, Moors EJ, Granier A, Gross P, Jensen NO, Pilegaard K, Lindroth A, Grelle A, Bernhofer C, Grünwald T, Aubinet M, Ceulemans R, Kowalski AS, Vesala T, Rannik Ü, Berbigier P, Loustau D, Guomundsson J, Thorgeirsson H, Ibrom A, Morgenstern K, Clement R, Moncrieff J, Montagnani L, Minerbi S, Jarvis PG (2000) Respiration as the main determinant of carbon balance in European forests. *Nature* 404:861–865. doi:10.1038/35009084

Vanaja M, Maheswari M, Lakshmi NJ, Sathish P, Yadav SK, Salini K, Vagheera P, Kumar GV, Razak A (2015) Variability in growth and yield response of maize genotypes at elevated CO_2 concentration. *Adv Plants Agric Res* 2:42. doi:10.15406.apar.2015.02.00042

Vanaja M, Yadav SK, Archana G, Lakshmi NJ, Reddy PR, Vagheera P, Razak SA, Maheswari M, Venkateswarlu B (2011) Response of C4 (maize) and C3 (sunflower) crop plants to drought stress and enhanced carbon dioxide concentration. *Plant Soil Environ* 57(5):207–215.

Vandegeer RK, Powell KS, Tausz M (2016) Barley yellow dwarf virus infection and elevated CO_2 alter the antioxidants ascorbate and glutathione in wheat. *J Plant Physiol* 199:96–99.

Van Oosterom EJ, Borrell AK, Deifel KS, Hammer GL (2011) Does increased leaf appearance rate enhance adaptation to postanthesis drought stress in sorghum? *Crop Sci* 51:2728–2740. doi:10.2135/cropsci2 011.01.0031

Vaughan MM, Huffaker A, Schmelz EA, Dafoe NJ, Christensen S, Sims J, et al (2014) Effects of elevated [CO_2] on maize defence against mycotoxigenic Fusarium verticillioides. *Plant Cell Environ* 37:2691–2706. doi:10.1111/pce.12337

Verma M, Misra AK (2018) Effects of elevated carbon dioxide and temperature on crop yield: a modeling study. *J Appl Math Comput* 58:503–526. doi:10.1007/s12190-017-1154-8

Vicente R, Bolger AM, Martinez-Carrasco R, Pérez P, Gutiérrez E, Usadel B, Morcuende R (2019) De novo transcriptome analysis of durum wheat flag leaves provides new insights into the regulatory response to elevated CO_2 and high temperature. *Front Plant Sci* 10:1605.

Vijayalakshmi K, Fritz AK, Paulsen GM, Bai G, Pandravada S, Gill BS (2010) Modeling and mapping QTL for senescence-related traits in winter wheat under high temperature. *Mol Breed* 26:163–175.

Visioli F, Lastra Cadl, Andres-Lacueva C, Aviram M, Calhau C, Cassano A, Llorach R (2011). Polyphenols and human health: a prospectus. *Crit Rev Food Sci Nutr* 51(6):524–546.

Vodnik D, Hladnik J, Vrešak M, Eler K (2013) Interspecific variability of plant stomatal response to step changes of [CO_2]. *Environ Exp Bot* 88:107–112.

Von Caemmerer S, Furbank RT (2003) The C4 pathway: an efficient CO_2 pump. *Photosyn Res* 77:191–207.

Von Caemmerer S, Lawson T, Oxborough K, Baker NR, Andrews TJ, Raines CA (2004) Stomatal conductance does not correlate with photosynthetic capacity in transgenic tobacco with reduced amounts of Rubisco. *J Exp Bot* 55:1157–1166.

Wahid A, Gelani S, Ashraf M, Foolad MR (2007) Heat tolerance in plants: an overview. *Env Exp Bot* 61:199–223.

Walker BJ, VanLoocke A, Bernacchi CJ, Ort DR (2016) The costs of photorespiration to food production now and in the future. *Annu Rev Plant Biol* 67:107–129.

Wallace JS (2000) Increasing agricultural water use efficiency to meet future food production. *Agric Ecosyst Environ* 82(1):105–119.

Wallwork MAB, Jenner CF, Logue SJ, Sedgley M (1998a) Effect of high temperature during grain-filling on the structure of developing and malted barley grains. *Ann Bot*. doi:10.1006/anbo.1998.0721

Wallwork MAB, Logue SJ, MacLeod LC, Jenner CF (1998b) Effect of high temperature during grain filling on starch synthesis in the developing barley grain. *Aust J Plant Physiol*. doi:10.1071/PP97084

Wang D, Heckathorn SA, Barua D, Joshi P, Hamilton EW, LaCroix JJ (2008) Effects of eCO_2 on the tolerance of photosynthesis to acute heat stress in C3, C4, and CAM species. *Am J Bot* 95(2):165–176. doi:10.3 732/ajb.95.2.165

Wang RS, Pandey S, Li S, Gookin TE, Zhao Z, Albert R, Assmann SM (2011) Common and unique elements of the ABA-regulated transcriptome of Arabidopsis guard cells. *BMC Genom* 12(1):216.

Wang X, Taub DR, Jablonski LM (2015) Reproductive allocation in plants as affected by elevated carbon dioxide and other environmental changes: a synthesis using meta-analysis and graphical vector analysis. *Oecologia* 177:1075–1087. doi:10.1007/s00442-014-3191-4

Wang Y, Frei M, Song Q, Yang L (2011) The impact of atmospheric CO_2 concentration enrichment on rice quality- a research review. *Acta Ecol Sin* 31:277–282.

Wang Y, Wang L, Zhou J, Hu S, Chen H, Xiang J, Zhang Y, Zeng Y, Shi Q, Zhu D, Zhang Y (2019) Research progress on heat stress of rice at flowering stage. *Rice Sci* 26(1):1–10. doi:10.1016/j.rsci.201 8.06.009

Wassmann R, Jagadish SVK, Heuer S, Ismail A, Redona E, Serraj R, Singh RK, Howell G, Pathak H, Sumfleth K (2009a) Climate change affecting rice production: the physiological and agronomic basis for possible adaptation strategies. *Adv Agron* 101:59–122.

Wassmann R, Jagadish SVK, Sumfleth K, Pathak H, Howell G, Ismail A, Serraj R, Redona E, Singh RK, Heuer S (2009b) Regional vulnerability of climate change impacts on Asian rice production and scope for adaptation. *Adv Agron* 102:91–133.

Watanabe CK, Sato S, Yanagisawa S, Uesono Y, Terashima I, Noguchi K (2014) Effects of elevated CO_2 on levels of primary metabolites and transcripts of genes encoding respiratory enzymes and their diurnal patterns in Arabidopsis thaliana: possible relationships with respiratory rates. *Plant Cell Physiol* 553:41–357. doi:10.1093/pcp/pct185

Watling JR, Press MC, Quick WP (2000) Elevated CO_2 induces biochemical and ultrastructural changes in leaves of the C4 cereal sorghum. *Plant Physiol* 123:1143–1152. doi:10.1104/pp.123.3.1143

Wong SC, Cowan LR, Farquhar GD (1979) Stomatal conductance correlates with photosynthetic capacity. *Nature* 282:424–426.

Xie HA, Zheng JT, Zhang SG, Lin MJ (1996) Breeding theory and practice of 'Shanyou 63', the variety with the largest cultivated area in China. *J Fujian Acad Agric Sci* 11:1–6.

Xie X, He Z, Chen N, Tang Z, Wang Q, Cai Y (2019) The roles of environmental factors in regulation of oxidative stress in plant. *BioMed Res Inter*, 9732325. doi:10.1155/2019/9732325

Xu DQ, Shen YK (2002) Photosynthetic efficiency and crop yield. In: Pessarakli M (ed) *Handbook of plant and crop physiology*. CRC Press, Boca Raton.

Xu Z, Jiang Y, Jia B, Zhou G (2016) Elevated-CO_2 response of stomata and its dependence on environmental factors. *Front Plant Sci* 7:657. doi:10.3389/fpls.2016.00657

Xu Z, Jiang Y, Zhou G (2015) Response and adaptation of photosynthesis, respiration, and antioxidant systems to elevated CO_2 with environmental stress in plants. *Front Plant Sci* 6:701.

Xu Z, Shimizu H, Yagasaki Y, Ito S, Zheng Y, Zhou G (2013) Interactive effects of elevated CO_2, drought, and warming on plants. *J Plant Growth Regulator* 32:692–707.

Xu Z, Zhou G, Han G, Li Y (2011) Photosynthetic potential and its association with lipid peroxidation in response to high temperature at different leaf ages in maize. *J Plant Growth Regul* 30:41–50. doi:10.1 007/s00344-010-9167-7

Xu ZZ, Shimizu H, Ito S, Yagasaki Y, Zou CJ, Zhou GS, et al (2014) Effects of elevated CO_2, warming and precipitation change on plant growth, photosynthesis and peroxidation in dominant species from North China grassland. *Planta* 239:421–435. doi:10.1007/s00425-013-1987-9

Xu ZZ, Shimizu H, Yagasaki Y, Ito S, Zheng YR, Zhou GS (2013a) Interactive effects of elevated CO_2, drought, and warming on plants. *J Plant Growth Regul* 32:692–707. doi:10.1007/s00344-013-9337-5

Xu ZZ, Yu ZW, Zhao JY (2013b) Theory and application for the promotion of wheat production in china: past, present and future. *J Sci Food Agric* 93:2339–2350. doi:10.1002/jsfa.6098

Yang K, Jiang M, Le J (2014) A new loss-of-function allele 28y reveals a role of ARGONAUTE1 in limiting asymmetric division of stomatal lineage ground cell. *J Integr Plant Biol* 56:539–549.

Yang L, Huang J, Yang H, Dong G, Liu H, Liu G, Zhu J, Wang Y (2007) Seasonal changes in the effects of free-air CO_2 enrichment (FACE) on nitrogen (N) uptake and utilization of rice at three levels of N fertilization. *Field Crops Res* 100:189–199.

Yang L, Liu H, Wang Y, Zhu J, Huang J, Liu G, Dong G, Wang Y (2009). Impact of elevated CO_2 concentration on inter-specific hybrid rice cultivar Liangyoupeijiu under fully open-air field conditions. *Field Crop Res* 112:7–15.

Yang L, Wang Y, Dong G, Gu H, Huang J, Zhu J, Yang H, Liu G, Han Y (2007) The impact of free-air CO_2 enrichment (FACE) and nitrogen supply on grain quality of rice. *Field Crop Res* 102: 128–140.

Ye Y, Zhou L, Liu X, Liu H, Li D, Cao M, Chen H, Xu L, Zhu JK, Zhao Y (2017) A novel chemical inhibitor of ABA signaling targets all ABA receptors. *Plant Physiol* 173(4):2356–2369.

Yildirim M, Barutcular C, Koc M, Dizlek H, EL Sabagh A, Hossain A, Islam MS, Toptas I, Basdemir F, Albayrak O, Akinci C (2018) Assessment of the grain quality of wheat genotypes grown under multiple environments using GGE biplot analysis. *Fres Environ Bull* 27(7):4830–4837.

Yin X, Kroff MJ, Goudriaan J (1996) Differential effects of day and night temperature on development to flowering in rice. *Ann Bot* 77:203–213.

Young KJ, Long SP (2000) Crop ecosystem responses to climatic change: maize and sorghum. In: Reddy KR, Hodges HF (eds) *Climate change and global crop productivity*. CABI International, Oxon, UK, pp 107–131.

Yuhui W, Denghua Y, Junfeng W, Yi D, Xinshan S (2017) Effects of elevated CO_2 and drought on plant physiology, soil carbon and soil enzyme activities. *Pedosphere* 27(5):846–855.

Zacharias M, Singh SD, Naresh Kumar S, Harit RC, Aggarwal PK (2010) Impact of elevated temperature at different phenological stages on the growth and yield of wheat and rice. *Indian J Plant Physiol* 15(4):350.

Zahoor M, Khan N, Ali M, Saeed M, Ullah Z, Adnan M, Ahmad B (2016) Integrated effect of organic waste and NPK fertilizers on nutrients uptake in potato crop and soil fertility. *Pure Appl Biol* 5(3):601–607.

Zavala JA, Nabity PD, Delucia EH (2013) An emerging understanding of mechanisms governing insect herbivory under elevated CO_2. *Annu Rev Entomol* 58:79–97. doi:10.1146/annurev-ento-12 0811-153544

Zhang TJ, Feng L, Tian XS, Yang CH, Gao J (2015) Use of chlorophyll fluorescence and P700 absorbance to rapidly detect glyphosate resistance in goosegrass (*Eleusine indica*). *J Integr Agric* 14:714–723.

Zhang X, Högy P, Wu X, Schmid I, Wang X, Schulze WX, Jiang D, Fangmeier A (2018) Physiological and proteomic evidence for the interactive effects of post-anthesis heat stress and elevated CO_2 on wheat. *Proteomics* 18:1800262.

Zhu C, Xu X, Wang D, Zhu J, Liu G (2015) An Indica rice genotype showed a similar yield enhancement to that of hybrid rice under free air carbon dioxide enrichment. *Sci Rep* 5:12719. doi:10.1038/srep12719

Zhu M, Dai S, McClung S, Yan X, Chen S (2009) Functional differentiation of Brassica napus guard cells and mesophyll cells revealed by comparative proteomics. *Mol Cell Proteom* 8(4):752–766. doi:10.1074/mcp.M800343-MCP200

Zinta G, AbdElgawad H, Domagalska MA, Vergauwen L, Knapen D, Nijs I, et al (2014) Physiological, biochemical, and genome-wide transcriptional analysis reveals that elevated CO_2 mitigates the impact of combined heat wave and drought stress in *Arabidopsis thaliana* at multiple organizational levels. *Global Change Biol* 20:3670–3685. doi:10.1111/gcb.12626

Zinta G, AbdElgawad H, Peshev D, Weedon JT, Van den Ende W, Nijs I, Janssens IA, Beemster GT, Han A (2018) Dynamics of metabolic responses to combined heat and drought spells in *Arabidopsis thaliana* under ambient and rising atmospheric CO_2. *J Exp Bot* 69(8):2159–2170.

Ziska LH, Bunce JA (1997) Influence of increasing carbon dioxide concentration on the photosynthetic and growth stimulation of selected C4 crops and weeds. *Photosyn Res* 54:199–208. doi:10.1023/A:100594 7802161

Ziska LH, Bunce JA, Shimono H, Gealy DR, Baker JT, Newton PC, Howden M (2012) Food security and climate change: on the potential to adapt global crop production by active selection to rising atmospheric carbon dioxide. *Proc R Soc B: Biol Sci* 279(1745):4097–4105.

Ziska LH, Sicher RC, Bunce JA (1999) The impact of elevated carbon dioxide on the growth and gas exchange of three C4 species differing in CO_2 leak rates. *Physiol Plantar* 105:74–80. doi:10.1034/j.13 99-3054.1999.105112.x

Ziska LH, Teramura AH (1992) Intraspecific variation in the response of rice (*Oryza sativa* L.) to increased CO_2 – photosynthetic, biomass and reproductive characteristics. *Physiol Plantar* 84:269–274.

8

Molecular Mechanisms of Stress Tolerance in Plants

Abdul Saboor Khan, Muhammad Adnan, Aamir Hamid, and Adnan Akbar

CONTENTS

8.1 Main

Biotic stress such as pathogen, virus, or bacterial attack, or, abiotic stress conditions such as drought, salinity, metal toxicity, extreme temperatures, low nutrient availability, and high ultraviolet radiation exposures impose restrictions on plant development (Adnan et al. 2018a; Adnan et al. 2019; Akram et al. 2018a, 2018b; Aziz et al. 2017a, 2017b; Habib et al. 2017; Hafiz et al. 2016; Hafiz et al. 2019; Kamaran et al. 2017; Muhmmad et al. 2019; Sajjad et al. 2019; Saud et al. 2013; Saud et al. 2014; Saud et al. 2017; Saud et al. 2016; Shah et al. 2013; Saud et al. 2020; Qamar et al. 2017; Wajid et al. 2017; Yang et al. 2017; Zahida et al. 2017; Depeng et al. 2018; Hussain et al. 2020; Hafiz et al. 2020a, 2020b; Shafi et al. 2020; Wahid et al. 2020; Subhan et al. 2020; Zafar-ul-Hye 2020a, 2020b; Adnan et al. 2020; Ilyas et al. 2020; Saleem 2020a, 2020b, 2020c; Rehman 2020; Farhat et al. 2020; Wu et al. 2020; Mubeen et al. 2020; Farhana 2020; Jan et al. 2019; Wu et al. 2019; Ahmad et al. 2019; Baseer et al. 2019; Hafiz et al. 2018; Tariq et al. 2018; Fahad and Bano 2012; Fahad et al. 2017; Fahad et al. 2013; Fahad et al. 2014a, 2014b; Fahad et al. 2016a, 2016b, 2016c, 2016d; Fahad et al. 2015a, 2015b; Fahad et al. 2018; Fahad et al. 2019a, 2019b; Hesham and Fahad 2020; Iqra et al. 2020; Akbar et al. 2020; Mahar et al. 2020; Noor et al. 2020; Bayram et al. 2020; Amanullah et al. 2020; Rashid et al. 2020; Arif et al. 2020; Amir et al. 2020; Saman et al. 2020; Muhammad Tahir et al. 2020; Md Jakirand Allah 2020; Farah et al. 2020; Sadam et al. 2020; Unsar et al. 2020; Fazli et al. 2020; Md. Enamul et al. 2020; Gopakumar et al. 2020; Zia-ur-Rehman 2020; Al-Wabel et al. 2020a, 2020b; Senol 2020; Amjad et al. 2020; Ibrar et al. 2020; Sajid et al. 2020). This aspect also constrains the output and timing of agricultural harvests. In their perseverance to temperatures, living spaces occupied by plants growing in contrasting natural environments often differ, ranging from freezing temperatures to temperatures more than 60°C (Nahar et al. 2015). These have an impact in unusual ways on molecular events that are already well understood. Greater understanding is required to interpret plants' distribution into several natural plant communities or opposition between species concerning temperature (Nahar et al. 2015; Fahad et al. 2016a, 2016b, 2016c, 2016d; Fahad et al. 2018; Fahad et al. 2019a, 2019b). The temperature effects on plant life are generally at the optimum range, but it is important to research how high

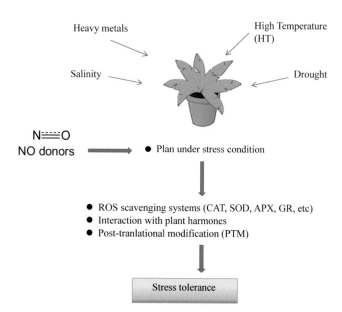

FIGURE 8.1 Schematic representation of the general mechanism of actions of NO donors on plants under different abiotic stress conditions leading to plant stress tolerance.

temperatures lead to some plants' death. High temperature (HT) means temperatures that are sufficiently high to harm metabolic processes and influence plant growth and development (Balla et al. 2009; Fahad et al. 2016a, 2016b, 2016c, 2016d; Fahad et al. 2018; Fahad et al. 2019a, 2019b). High-temperature stresses are considered chronic exposures beyond the ideal range that lasts for days or weeks, whereas those that last for shorter durations are referred to as acute exposures. The incidents arising from this differ substantially (Kappen 1981). One of the most detrimental stresses that threaten greater plant production and survival worldwide is HT stress. Drought-stressed plants with nitric oxide (NO) is likely to lead to plant mechanistic behaviour for water deficit mitigation. The activity of NO enhances antioxidant enzyme activity (CAT), dismutase (SOD), ascorbate peroxidase (APX), reductase (GPX), and peroxidase (POD), for instance. CAT is located in the active ROS metabolism of plant peroxisomes. O2- to H2O2 and O2 are catalyzed by SOD. Simultaneously, H_2O_2 and other peroxides are made from CAT, APX, and GPX (Farooq et al. 2009). Figure 8.1 shows the general mechanism of NO donor actions on the plants seen in a schematic representation. However, the nitrogen content of some species is reported to be decreased. In contrast, others' nitrogen content continues to decrease when the growth tempo is increasing, implying a lower nitrogen demand to satisfy photosynthetic functions (Hakeem et al. 2011a, 2011b). Although higher plants which are sturdierdevelop defensive strategies to combat HT stress, they are often not enough not to cause severe damage. Plants' metabolism is modified in response to HT stress, and helps to defend them against stress that is modulated with antimetabolisms, secondary metabolites, hormonal, osmoprotectants, and many other major biomolecules (Nahar et al. 2015). High temperatures cause photosynthetic membrane damage, which eventually results in the swelling of grana piles and the aberrant piling, which increases leaf cell ion leakages considerably and changes photosystem energy flow (Allakhverdiev et al. 2008). In Jatropha curcas, electrolyte leakage (EL) increases by 28%, and lipid peroxidation is predicted to increase by 50% when exposed to HT at 43°C, due to the concentration of thiobarbituric acid-reactive compounds (Silva et al. 2010).

HT (38°C) in wheat has been reported to result in oxidative stress, resulting in the growth of lipid membrane peroxidation and change in malondialdehyde (MDA) levels by 76 and 144%, respectively, over 24 and 48 hours, compared to control plants grown under normal stressfree conditions. Multi-isogeneous (Betl/Bet1) and defective (Betl/Bet1) F maize producing homozygotes had been exposed to HT (Hasanuzzaman et al. 2012). When the leaf or leaf segments were exposed to 45°C of temperature, both lines of maize had differential membrane integrity. The line Bet l/Bet 1 has been less affected by diaphragms than the near-isogenic bet l/Bet1 sister line. Bet l/bet l lines were seen as the threshold HT causing damage *in vitro* to the catastrophic membrane, with a mean temperature difference of 2°C. This resistance has been attributable to the one gene that conducts glycine betaine growth, which protects leaf plasma membranes destabilized in HT (Yang et al. 1996). The cotton, sorghum, and soybean were also documented to cause heat stress lipid chronic inflammation and severe membrane damage (Djanaguiraman et al. 2009; Khan et al. 2017; Djanaguiraman et al. 2011).

Over the last several decades, the usage of intracellular phytoprotectants such as osmolytes, plant growth regulators, flavonoids, signage particles, organic compounds, and polyamins has given substantial defence against HT-induced plant harm. Nevertheless, the exact defence and signal transduction processes are still unknown. It is important to analyze the correct dosage and period of care of exogenous protectants and the protectants' best application methods. More studies into molecular approaches are required to expose the defence mechanisms under extreme stress upon the protectants' application. Engineering plants for the synthesis of these appropriate compounds can be an effective way of improving heat tolerance in valuable crop plants, which certainly and wondrously reflect a potentially exciting field of study (Nahar et al. 2015).

The cotton species, Gossypium hirsutum L., is mainly cultivated in areas that often reach 48–50°C in summer. It was documented that vegetative growth and boll development were seriously disrupted in certain Gossypium hirsutum genotypes due to HT in June, July, and early August high temperatures (Taha et al. 1981). Similar findings by Hoffman and Rawlins (1970) show a small boll set in 38°C in cotton. The crop typically grows in early June in the majority of countries where high temperatures will adversely affect cotton plant germination, production, and vegetative growth.

8.2 High-temperature Yield Loss

Crops usually grow under specific intermediate temperature ranges. Excessively cold or hot temperatures interrupt physiological processes and, consequently, hurt rates of output. The frequency and duration are expected to increase in high-temperature events of climate change. Temperature changes are expected to adversely affect crop yields (Porter and Gawith 1999; Ottman et al. 2012); regional temperature shifts are expected to be more accurate than rainfall predicted in weather forecasts. Weather forecast data indicates that annual average temperatures have increased by 1°C over the last century in regions where wheat, barley, maize, and soya are cultivated. It can jeopardize domestic development in areas already vulnerable (Godfray et al. 2010; Coumou and Rahmstorf 2012). In this analysis, high temperatures > 30°C have a direct and an indirect effect on crop yields. By losing surface water and with rising demand for ambient water, high temperatures will lead to water scarcity. This helps to close the stomata to avoid dessication (thus to minimize carbon dioxide absorption) and improve root production at the costs of the overground biomass (Lobell et al. 2011; Liu et al. 2016). In addition, high temperatures can lead to direct damage to enzymes and tissues (Lobell 2007; Food and Agriculture Organization of the United Nations 2014), improper flowering (Tian et al. 2012) and damage to stress-oxidation (Ewert et al. 2015). Overall, higher photosynthesis level (Schlenker and Roberts 2009; Long et al. 2006) or lower due to higher integration of carbon (C) and/or breathing level (Brisson et al. 2010; Ray et al. 2012) have been known to exist.

In addition to the above analysis, rising global climate instability endangers cotton production worldwide due to extreme temperatures, drought stress, and erratic precipitation patterns. These stresses worldwide accounted for more than 50% yield reduction in arable crops (Boyer 1982). Due to these abiotic and other biotic stresses, cotton production is significantly affected, which leads to lower yields and less harvesting efficiency. A cotton crop with a limitless growth habit has a complex fruiting pattern that is considered severely susceptible to climatic interactions and various management techniques.

Upland cotton (*Gossypium hirsutum L.*) is one of the most important cash crops that provide fibre to the embroidery industries worldwide. 80% comes from Brazil, China, India, Pakistan, Turkey, the USA, and Uzbekistan according to world body for cotton production's rough estimates. The crop makes an important contribution to many countries' gross national product (GNP). The accumulation in the atmosphere of greenhouse gasses and carbon dioxide could threaten agriculture in the near future. Global warming (GW) has been mentioned, leading to most of the world's agricultural areas to become susceptible to greater environmental fluctuations (Solomon et al. 2007). GW can be attributed to the increase of 0.6 kb C during the period from 1990–2000 (Tebaldi et al. 2007). By 2080, continuous GHG emissions (from frequency, intensity, and period of radiation) in major crop-growing zones around the world would lead to significant changes in annual mean temperature (2,5–4,3 livres).

8.3 Plant Dehydration Tolerance Mechanisms

Plants possess multiple ways of dealing with loss of water. Drought tolerance can be defined as tolerance to inadequate water availability, and drought-tolerant plants are known to use methods to manage water in the cell, e.g., stomatal closure or osmotic adjustment. Dehydration resistance is a cell water depletion resistance. Just a few plants demonstrate dessication resistance, described as drying resistance to air balance. To determine dehydration stresses, the plant's water level must be correctly calculated. Water state is defined differently in different scientific reports as water capacity, relative water content (RWC), or tissue water content. In contrast to a pure water level (0 MPa) at the same temperature, water potential is the chemical potential for water in a cell structure. Many plants withstand 0 to –4 MPa dehydration. The RWC corresponds to and is represented as a proportion of water volume left after tissue dehydration contrasted with water at maximum turgor and is denoted by % RWC. A fraction or a percentage of a specimen of fresh or dry weight is defined as water content, for example, H2O per g DW (Oliver et al. 2010).

Dessication tolerance (DT) is a dynamic characteristic of vascular plant breeding structures (pollen and seeds). DT is also generally widespread in less complex plant species, including bryophytes and lichens, but is uncommon in pteridophytes, angiosperms, or known by their absence in gymnosperms. It is known that there are 275–325 vascular dessication resistant plant species: nine pteridophyte families and seven angiosperm families include drying and resistant sporophytes. Among the dessication resistant plants, mention may be made of the family Pteridaceae and genera Cheilanthe and Selaginella, Ferns, and allied species which seem comparatively rich by variety. The Velloziaceae and Poaceae monocotyledonous families have respectively 200 and 39 dessication-tolerant species. Myrothamnaceae are fully tolerant of premature drying among the angiosperms of a limited family (Oliver et al. 2000).

The initial stage of DT formation in aerial tissues was essential in the invasion of the soil by primitive plants. Phylogenetic studies have indicated DT might look like the DT contained in modern DT mooses in earlier plants. With more diverse morphologies and structures formed and accreted in plants to prevent lack of water, vegetative DT in most plants was lost and retrieved and preserved in seeds and pollen in reproductive organs. Vegetative DT subsequently re-evolved as an extension of the DT seed drying resistance taking place at least eight times separately (Oliver et al. 2000; Farrant and Moore 2011). Various mechanisms were identified for the dessication tolerance. DT is constitutive in less complicated crops, such as mooses, paired with an inductive recovery process after the tissue is rehydrated. Club mosses, e.g., Selaginella lepidophylla, all of which have DT as inducible and constitutive. Ferns have a seasonally controlled resistance mechanism for dessccation, and angiosperms have an inducible mechanism (Oliver et al. 2000).

8.4 Signal Transduction in Drought Stress Response

Dehydration induced by the mechanisms of signal transduction, signal recognition, or osmotic change to separate gene expression, and stress signalling pathways in plants have not been studied at length or in depth. Mechanisms for the signal transduction of drought stress response contributing to a better understanding of yeast and animal models were investigated (Figure 8.2). Two-part mechanisms function to detect the osmotic stresses in bacteria and the yeast where the Arabidopsis cDNA *(ATAHKl)* extracted in the cell had histidine kinase composed of two constituents and a yeast osmosensor Slnl homologue. The *(ATHKl)* has a regular histidine kinase domain and a domain

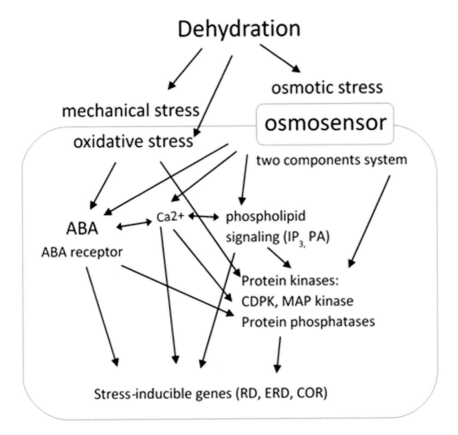

FIGURE 8.2 Signal sensing and drought-stress signal transduction related secondary messengers and factors.

receiver like Slnl (Urao et aI. 1999). In normal growth conditions, transcription of *(ATHKl)* were more abundant in the roots and collected in lower temperatures and higher salinity than other tissues. *(ATHKl)* has been injected into mutant yeasts without two osmosensors *(SLNI)* and *(SH0I)* which enabled a natural *HOG1 (MAPK)* cascade growth and activation of high osmolarity. These results show that an external osmolarity signal *(ATHKl)* may be identified and converted into downstream yeast targets. We have produced Arabidopsis transgenic plants which exaggerate the predominantly negative cDNAs (ATHKl). The transgenics had a break in shooting and rooting production. Any stress-induced genes were upregulated under dysfunctional conditions. We believe *(ATHKl)* in Arabidopsis during drought stress may function in signal perception. However, there might be alternative mechanisms for stress sensing, such as mechanical cytoskeleton sensors and superoxide stress-generated sensors. In the signalling pathways, the cell proliferation and environmental stresses of the yeast and animals sensed by the growth factor are involved in the production of mitogen kinase *(MAPK)*, a protein kinase triggering. The prompt *(ATMPK4)* and *(AtMPK6)* activation of the MAP kinase in Arabidopsis was tested for cold, changes in moisture, injuries, and interactions (Ichimura et aI. 2000). Osmotic stress is often found in alfalfa to activate MPK kinase. These results suggest that some MAP kinase cascades in signalling pathways can function in a response to drought stress.

Many genes, such as calcium-dependent protein kinases and enzymes related to phospholipid metabolism such as PIPS kinase, phosphatase C, and D, are up-regulated by dehydration and cold (Shinozaki and Yamaguchi-Shinozaki 2003; Arshad et al. 2016). Such genes could be used to intensify stress signals and adaptation of plant cells to the conditions of drought stress (Figure 8.2). More details would be given on their gene products' role by transgenic plants that alter their expression and mutants with disorders.

8.5 Salt Stress Tolerance

Soil salinity has a detrimental effect on global farm development and yield. Salt deposition on agricultural land is primarily dependent on irrigation water comprising sodium chloride (NaCl) and seawater trace quantities. The elevated concentration of soil salt limits the plant's capacity to absorb water (dehydration). Na+ and Cl- accumulations in tissues adversely affect development, as photosynthesis and other metabolic processes are impaired. Therefore, osmotic and ionic stresses are components of salinity stress. Plants have developed osmotic stress mitigation mechanisms through the accumulation of osmolytes (ions, solutes, or organic compounding) to reduce the osmotic potential and water losses. Moreover, by excluding na-+ from leaf tissue and mainly in vacuoles, plants can counteract the effects of Na+ ionicity (Munns and Tester 2008).

The first organs to feel the hyperosmotic component are the plant roots while the ionic Na+ is increasing. Plants are suggested to have two different sensory mechanisms to deal with both components of salinity stress since some of NaCl's response to ionicity stress differs from osmolytes response to osmotic stress. The molecular identity of a Na+ sensor or plant osmotic sensors is however unknown (Ali et al. 2018; Deinlein et al. 2014). Pflanze osmotic sensors are probably located in close proximity to Ca2 + channels as [Ca2+] cytes increase in NaCl response in seconds very rapidly. Therefore, a Ca2 + channel can be connected closely with an osmosensor. Furthermore, Ca2 + osmosensory channels act as osmosensors for bacteria and animals. An osmosensor has recently been proposed for OSCA1, a previously unknown plasma membrane protein. OSCA1 is responsible for [Ca2+]cytes of increases in osmotic stress-induced hyperosmolality-gated calcium-permeable channels (Yuan et al. 2014). The mechanosensitive (MS) channels are also candidates to play a role as osmosensors. Osmotic stress disrupts the plasma membrane, which can be sensed on MS channels, but as a result of the high-functional disruption in the family to date, no MS channel has been affirmative of an osmosensor (Haswell and Verslues 2015).

Calcineurine B-like protein (CBL), CBL (CiPKs) and Calcium-dependent protein kinases (CDPKs) have been activated by elevated cytosolic Ca2 + for a wide range of sensor proteins including CA2 +. Ca2 Plus is also enabling RBOHD to produce ROS, which in turn can turn on Ca2 + permeable channels such as annexes to support further Ca2 + mobility (Laohavisit et al. 2013). The network of Ca2 +-and ROS-based signallings is involved in regulating detoxification mechanisms such as the Ca2+-dependent activation of Na+/H+ exchangers, SOS1 (SALT OVERLY SENSITIVE 1) and NHX1 (Na+/H+ exchanger 1), Na+ efflux from the cytosol, regulated xylem loading of Na+, Na-+ exclusion from leaf and osmoprotective proteins, and osmotic induction (Deinlein et al. 2014; Hasegawa 2013). Plants also have a rapid Ca2 +-based stress signalling system that transmits signals to the aerial tissues through the cortex and endodermal cell layers of the roots. These waves of Ca2 + evoke structural molecular responses in target organ and lead to a tolerance for whole plant tension (Choi et al. 2014).

8.6 Biotic and Abiotic Stresses and Their Crosstalk

While most of the literature has examined one stress in isolation, plants are subject to environmental abiotic stress variations under natural circumstances, and often have to combat pathogens' biotic stresses (Fahad and Bano 2012; Fahad et al. 2017; Fahad et al. 2013; Fahad et al. 2014a, 2014b; Fahad et al. 2016a, 2016b, 2016c, 2016d; Fahad et al. 2015a, 2015b; Fahad et al. 2018; Fahad et al. 2019a, 2019b; Hesham and Fahad 2020; Iqra et al. 2020; Akbar et al. 2020; Mahar et al. 2020; Noor et al. 2020; Bayram et al. 2020; Amanullah et al. 2020; Rashid et al. 2020; Arif et al. 2020; Amir et al. 2020; Saman et al. 2020; Muhammad Tahir et al. 2020; Md Jakirand Allah 2020; Farah et al. 2020; Sadam et al. 2020; Unsar et al. 2020; Fazli et al. 2020; Md. Enamul et al. 2020; Gopakumar et al. 2020; Zia-ur-Rehman 2020; Ayman et al. 2020; Al-Wabel et al. 2020a, 2020b; Senol 2020; Amjad et al. 2020; Ibrar et al. 2020; Sajid et al. 2020. Plants need to react to these multiple stress combinations rapidly and efficiently. Therefore it is no wonder that many regulatory structures are part of broader, interconnected networks, which allow environmental responses to be incorporated. The 'crosstalk' takes place in a diverse regulatory network that involves Ca2 + sources, oxidative bursts, kinasis cascades, and an interaction between genes of stress response (Kissoudis et al.

2014). Abiotic stress also influences biotic stress resistance, and vice versa. Two potential pathways may influence the relationship of plants and pathogens through abiotic stress. It may be possible that pathogens and abiotic stress have negative effects that may be additive. Tomatoes that have been pre-irrigated with water, for example, are more seriously impaired by coronary and root rot (*Fusarium oxysporum sp. radicis-lycopersici*) (Triky-Dotan et al. 2005) and extreme vulnerability to bacterial P. syringae pathogens was shown by drought-stressing arabidopsis plants (Mohr and Cahill 2003).

Another choice is to increase the tolerance to abiotic stress or vice versa by plant pathogen. This effect has also been shown with numerous kinds of tension, such as cross-tolerance, cross resistance, or multiple stress resistance. For instance, when exposed to drought and salt stress, the barley and tomato plants expressed tolerance to the fungus barley powdery mildew (*Blumeria graminis*), B-cinerea, and Oidium neolycopersici respectively (Wiese et al. 2004; Achuo et al. 2006). Interestingly enough, drought stress showed tolerance to both necrotrophic fungus B. cinerea and biotrophic O. neolycopersici, whereas, salt stress shows tolerance to O. *neolycopersici* only. It indicated that drought and salt stress poses different mechanisms of pathogen defence pathways (Achuo et al. 2006). Similarly, pathogens can increase abiotic stress resistance. For instance, Arabidopsis infected with the Verticillium longisporum that result in de-novo xylem formation and develop tolerance to drought stress (Reusches et al. 2012).

In addition to abiotic or biotic pressures, certain signalling components play a dual function. To react to necrotrophic fungal pathogen Botrytis cinerea and to salinity stress, e.g., BOS1 is a transcribing factor of R2R3, MYB its interactor BOI is an E3-ligase (Mengiste et al. 2011). It has been proposed that certain LysM-RLKs should play a part between abiotic and biotic stresses in the crossroads. At LYK3 is a significant crosstalk between pathogens and abscisic acid (ABA) signalling pathways (Paparella et al. 2014) apparently. At LYK3 negatively controls defence gene basal expression and resistance to B. cinerea and Pectobacterium carotovorum infection. The inhibition of PHYTOALEXIN-DEFICIENT (PAD3) expressions in plants missing the LYK3 functionality has resulted in decreased physiological responses to ABA and a decrease in ABA (Paparella et al. 2014). LYK3 has previously been active in removing the inherent immunity of plants in reaction to chitin-short-chain [degree of polymerization (dp) < 6] (Liang et al. 2013). Significant research study has reported that the salinity stress resistance of Arabidopsis plants was enhanced by treatment with chitin. Similar findings were observed when fungal chitinases were ectopically expressed in Arabidopsis, as salinity tolerances were enhanced and the tolerance for *B. cinerea* was strengthened (Brotman et al. 2012). Interestingly, ectopic expression of chitinase has not increased tolerance for salinity and necrotrophic fungus in cerk1 mutant plants and, thus, indicated the role of AtCERK1 in cross-talking between fungal and saline stress (Brotman et al. 2012).

While a small range of experiments is performed, many specifically-enabled signalling mechanisms, including various transcriptome, metabolomes, and proteome monitoring plants, have revealed in reaction to numerous stress combinations (Suzuki et al. 2014). These studies found that the plantation's reaction to two or three stresses varies from one stress concurrently (Suzuki et al. 2014; Pandey et al. 2015). Further experiments are also important to explore the bio-chemical and molecular pathways that govern the reaction to natural stress combinations. This knowledge allows one to devise methods for improving plant and seed tolerance.

REFERENCES

Achuo EA, Prinsen E, Hofte M (2006) Influence of drought, salt stress and abscisic acid on the resistance of tomato to Botrytis cinerea and Oidium neolycopersici. *Plant Pathol* 55(2):178–186.

Adnan M, Shah Z, Sharif M, Rahman H (2018a) Liming induces carbon dioxide (CO_2) emission in PSB inoculated alkaline soil supplemented with different phosphorus sources. *Environ Sci Poll Res* 25(10): 9501–9509.

Adnan M, Fahad S, Khan IA, Saeed M, Ihsan MZ, Saud S, Riaz M, Wang D, Wu C (2019) Integration of poultry manure and phosphate solubilizing bacteria improved availability of Ca bound P in calcareous soils. *3 Biotech* 9(10):368

Adnan M, Fahad S, Muhammad Z, Shahen S, Ishaq AM, Subhan D, Zafar-ul-Hye M, Martin LB, Raja MMN, Beena S, Saud S, Imran A, Zhen Y, Martin B, Jiri H, Rahul D (2020) Coupling phosphate-solubilizing bacteria with phosphorus supplements improve maize phosphorus acquisition and growth under lime induced salinity stress. *Plants* 9: 900. doi: 10.3390/plants9070900

Adnan M, Zahir S, Fahad S, Arif M, Mukhtar A, Imtiaz AK, Ishaq AM, Abdul B, Hidayat U, Muhammad A, Inayat-Ur R, Saud S, Muhammad ZI, Yousaf J, Amanullah, Hafiz MH, Wajid N (2018b) Phosphate-solubilizing bacteria nullify the antagonistic effect of soil calcification on bioavailability of phosphorus in alkaline soils. *Sci Rep* 8:4339. doi: 10.1038/s41598-018-22653-7

Ahmad S, Kamran M, Ding R, Meng X, Wang H, Ahmad I, Fahad S, Han Q (2019) Exogenous melatonin confers drought stress by promoting plant growth, photosynthetic capacity and antioxidant defense system of maize seedlings. *PeerJ* 7:e7793 http://doi.org/10.7717/peerj.7793

Akbar H, Timothy JK, Jagadish T, M. Golam M, Apurbo KC, Muhammad F, Rajan B, Fahad S, Hasanuzzaman M (2020) Agricultural land degradation: processes and problems undermining future food security. In: Fahad S, Hasanuzzaman M, Alam M, Ullah H, Saeed M, Khan AK, Adnan M (eds) *Environment, climate, plant and vegetation growth.* Springer Publ Ltd, Springer Nature Switzerland AG. Part of Springer Nature, pp 17–62. doi: 10.1007/978-3-030-49732-3

Akram R, Turan V, Wahid A, Ijaz M, Shahid MA, Kaleem S, Hafeez A, Maqbool MM, Chaudhary HJ, Munis, MFH, Mubeen M, Sadiq N, Murtaza R, Kazmi DH, Ali S, Khan N, Sultana SR, Fahad S, Amin A, Nasim W (2018b) Paddy Land Pollutants and Their Role in Climate Change, in: Hashmi, MZ and Varma, A (Ed.), *Environmental Pollution of Paddy Soils, Soil Biology.* Springer International Publishing Ag, GEWERBESTRASSE 11, CHAM, CH-6330, Switzerland, pp 113–124.

Akram R, Turan V, Hammad HM, Ahmad S, Hussain S, Hasnain A, Maqbool MM, Rehmani MIA, Rasool A, Masood N, Mahmood F, Mubeen M, Sultana SR, Fahad S, Amanet K, Saleem M, Abbas Y, Akhtar HM, Waseem F, Murtaza R, Amin A, Zahoor SA, ul Din MS, Nasim W (2018a) Fate of organic and inorganic pollutants in paddy soils. In: Hashmi MZ, Varma A (eds) *Environmental pollution of paddy soils, soil biology.* Springer International Publishing Ag, GEWERBESTRASSE 11, CHAM, CH-6330, Switzerland, pp 197–214.

Ali S, Xu Y, Jia Q, Ma X, Ahmad I, Adnan M, Gerard R, Ren X, Zhang P, Cai T, Zhang J (2018) Interactive effects of plasticfilm mulching with supplemental irrigation on winter wheat photosynthesis, chlor-ophyllfluorescence and yield under simulated precipitation conditions. *Agric Water Manag* 207:1–14.

Allakhverdiev SI, Kreslavski VD, Klimov VV, Los DA, Carpentier R, Mohanty P (2008) Heat stress: an overview of molecular responses in photosynthesis. *Photosynth Res* 98:541–550.

Al-Wabel MI, Abdelazeem S, Munir A, Khalid E, Adel RAU (2020a) Extent of climate change in Saudi Arabia and its impacts on agriculture: a case study from Qassim region. In: Fahad S, Hasanuzzaman M, Alam M, Ullah H, Saeed M, Khan AK, Adnan M (eds) *Environment, climate, plant and vegetation growth.* Springer Publ Ltd, Springer Nature Switzerland AG. Part of Springer Nature, pp 635–658. doi: 10.1007/978-3-030-49732-3

Al-Wabel MI, Ahmad M, Usman ARA, Akanji M, Rafique MI (2020b) Advances in pyrolytic technologies with improved carbon capture and storage to combat climate change. In: Fahad S, Hasanuzzaman M, Alam M, Ullah H, Saeed M, Khan AK, Adnan M (eds) *Environment, climate, plant and vegetation growth.* Springer Publ Ltd, Springer Nature Switzerland AG. Part of Springer Nature, pp 535–576. doi: 10.1007/978-3-030-49732-3

Amanullah, Shah K, Imran, Hamdan AK, Muhammad A, Abdel RA, Muhammad A, Fahad S, Azizullah S, Brajendra P (2020) Effects of climate change on irrigation water quality. In: Fahad S, Hasanuzzaman M, Alam M, Ullah H, Saeed M, Khan AK, Adnan M (eds) *Environment, climate, plant and vegetation growth.* Springer Publ Ltd, Springer Nature Switzerland AG. Part of Springer Nature, pp 123–132. doi: 10.1007/978-3-030-49732-3

Amir M, Muhammad A, Allah B, Sevgi Ç, Haroon ZK, Muhammad A, Emre A (2020) Bio fortification under climate change: the fight between quality and quantity. In: Fahad S, Hasanuzzaman M, Alam M, Ullah H, Saeed M, Khan AK, Adnan M (eds) *Environment, climate, plant and vegetation growth.* Springer Publ Ltd, Springer Nature Switzerland AG. Part of Springer Nature, pp 173–228. doi: 10.1007/978-3-030-49732-3

Amjad I, Muhammad H, Farooq S, Anwar H (2020) Role of plant bioactives in sustainable agriculture. In: Fahad S, Hasanuzzaman M, Alam M, Ullah H, Saeed M, Khan AK, Adnan M (eds) *Environment, climate, plant and vegetation growth.* Springer Publ Ltd, Springer Nature Switzerland AG. Part of Springer Nature, pp 591–606. doi: 10.1007/978-3-030-49732-3

Arif M, Talha J, Muhammad R, Fahad S, Muhammad A, Amanullah, Kawsar A, Ishaq AM, Bushra K, Fahd R (2020) Biochar; a remedy for climate change. In: Fahad S, Hasanuzzaman M, Alam M, Ullah H, Saeed M, Khan AK, Adnan M (eds) *Environment, climate, plant and vegetation growth.* Springer Publ Ltd,

Springer Nature Switzerland AG. Part of Springer Nature. pp 151–172. doi: 10.1007/978-3-030-4 9732-3

Arshad M, Adnan M, Ahmed S, Khan AK, Ali I, Ali M, Ali A, Khan A, Kamal MA, Gul F, Khan MA (2016) Integrated effect of phosphorus and zinc on wheat crop. *American-Eurasian J Agric Environ Sci* 16(3):455–459.

Aziz K, Daniel KYT, Muhammad ZA, Honghai L, Shahbaz AT, Mir A, Fahad S (2017b) Nitrogen fertility and abiotic stresses management in cotton crop: a review. *Environ Sci Pollut Res* 24:14551–14566. doi: 10.1007/s11356-017-8920-x

Aziz K, Daniel KYT, Fazal M,Muhammad ZA, Farooq S, Fan W, Fahad S, Ruiyang Z (2017a) Nitrogen nutrition in cotton and control strategies for greenhouse gas emissions: a review. *Environ Sci Pollut Res* 24:23471–23487. doi: 10.1007/s11356-017-0131-y

Balla K, Bencze S, Janda T, Veisz O (2009) Analysis of heat stress tolerance in winter wheat. *Acta Agron Hung* 57:437–444.

Baseer M, Adnan M, Fazal M, Fahad S, Muhammad S, Fazli W, Muhammad A, Jr. Amanullah, Depeng W, Saud S, Muhammad N, Muhammad Z, Fazli S, Beena S, Mian AR, Ishaq AM (2019) Substituting urea by organic wastes for improving maize yield in alkaline soil. *J Plant Nutrition*. doi: 10.1080/019041 67.2019.1659344

Bayram AY, Seher Ö, Nazlican A (2020) Climate change forecasting and modeling for the year of 2050. In: Fahad S, Hasanuzzaman M, Alam M, Ullah H,Saeed M, Khan AK, Adnan M (eds) *Environment, climate, plant and vegetation growth*. Springer Publ Ltd, Springer Nature Switzerland AG. Part of Springer Nature, pp 109–122. doi: 10.1007/978-3-030-49732-3

Boyer JS (1982) Plant productivity and environment. *Science* 218(4571):443–448

Brisson N, et al (2010) Why are wheat yields stagnating in Europe? A comprehensive data analysis for France. *Field Crops Res* 119:201–212.

Brotman, Y, et al (2012) The LysM receptor-like kinase LysM RLK1 is required to activate defense and abiotic-stress responses induced by overexpression of fungal chitinases in arabidopsis plants. *Molecular Plant* 5(5):1113–1124.

Choi, WG, et al (2014) Salt stress-induced Ca2+ waves are associated with rapid, long-distance root-to-shoot signaling in plants. *Proc Natl Acad Sci* 111(17):6497–6502.

Coumou D, Rahmstorf S (2012) A decade of weather extremes. *Nat Clim Change* 2:491–496.

Deinlein, U, et al (2014) Plant salt-tolerance mechanisms. *Trends Plant Sci* 19(6):371–379.

Depeng W, Fahad S, Saud S, Muhammad K, Aziz K, Mohammad NK, Hafiz MH, Wajid N (2018) Morphological acclimation to agronomic manipulation in leaf dispersion and orientation to promote "Ideotype" breeding: evidence from 3D visual modeling of "super" rice (Oryza sativa L.). *Plant Physiol Biochem* 135:499–510. doi: 10.1016/j.plaphy.2018.11.010

Djanaguiraman M, Prasad PVV, Al-Khatib K (2011) Ethylene perception inhibitor 1-MCP decreases oxidative damage of leaves through enhanced antioxidant defense mechanisms in soybean plants grown under high temperature stress. *Environ Exp Bot* 71:215–223.

Djanaguiraman M, Sheeba JA, Devi DD, Bangarusamy U (2009) Cotton leaf senescence can be delayed by nitrophenolate spray through enhanced antioxidant defense system. *J Agron Crop Sci* 195:213–224.

EL Sabagh A, Hossain A, Barutçular C, Iqbal MA, Islam MS, Fahad S, Sytar O, Çig F, Meena RS, Erman M (2020) Consequences of Salinity Stress on the Quality of Crops and Its Mitigation Strategies for Sustainable Crop Production: An Outlook of Arid and Semi- arid Regions. In: Fahad S, Hasanuzzaman M, Alam M, Ullah H, Saeed M, Khan AK, Adnan M (eds) *Environment, climate, plant and vegetation growth*. Springer Publ Ltd, Springer Nature Switzerland AG. Part of Springer Nature, pp 503–534. doi: 10.1007/978-3-030-49732-3

Ewert F, et al (2015) Crop modelling for integrated assessment of risk to food production from climate change. *Environ Model Softw* 72:287–303.

Fahad S, Bano A (2012) Effect of salicylic acid on physiological and biochemical characterization of maize grown in saline area. *Pak J Bot* 44:1433–1438.

Fahad S, Hussain S, Saud S, Hassan S, Chauhan BS, Khan F et al (2016a) Responses of rapid viscoanalyzer profile and other rice grain qualities to exogenously applied plant growth regulators under high day and high night temperatures. *PLoS One* 11(7):e0159590. doi: 10.1371/journal.pone.0159590

Fahad S, Nie L, Chen Y, Wu C, Xiong D, Saud S, Hongyan L, Cui K, Huang J (2015b) Crop plant hormones and environmental stress. *Sustain Agric Rev* 15:371–400.

Fahad S, Chen Y, Saud S, Wang K, Xiong D, Chen C, Wu C, Shah F, Nie L, Huang J (2013) Ultraviolet radiation effect on photosynthetic pigments, biochemical attributes, antioxidant enzyme activity and hormonal contents of wheat. *J Food Agri Environ* 11(3&4):1635–1641.

Fahad S, Hussain S, Saud S, Khan F, Hassan S, Jr A, Nasim W, Arif M, Wang F, Huang J (2016d) Exogenously applied plant growth regulators affect heat-stressed rice pollens. *J Agron Crop Sci* 202:139–150.

Fahad S, Rehman A, Shahzad B, Tanveer M, Saud S, Kamran M, Ihtisham M, Khan SU, Turan V, Rahman MHU (2019b). Rice responses and tolerance to metal/metalloid toxicity. In: Hasanuzzaman M, Fujita M, Nahar K, Biswas JK (eds) *Advances in rice research for abiotic stress tolerance*. Woodhead Publ Ltd, Abington Hall Abington, Cambridge CB1 6AH, CAMBS, England, pp 299–312.

Fahad S, Hussain S, Saud S, Hassan S, Ihsan Z, Shah AN, Wu C, Yousaf M, Nasim W, Alharby H, Alghabari F, Huang J (2016b) Exogenously applied plant growth regulators enhance the morpho-physiological growth and yield of rice under high temperature. *Front Plant Sci* 7:1250. doi: 10.3389/fpls.2016.01250

Fahad S, Muhammad ZI, Abdul K, Ihsanullah D, Saud S, Saleh A, Wajid N, Muhammad A, Imtiaz AK, Chao W, Depeng W, Jianliang H (2018) Consequences of high temperature under changing climate optima for rice pollen characteristics-concepts and perspectives, *Archives Agron Soil Sci* 64:1473–1488. doi: 10.1080/03650340.2018.1443213

Fahad S, Adnan M, Hassan S, Saud S, Hussain S, Wu C, Wang D, Hakeem KR, Alharby HF, Turan V, Khan MA, Huang J (2019a) Rice responses and tolerance to high temperature. In: Hasanuzzaman M, Fujita M, Nahar K, Biswas JK (eds) *Advances in rice research for abiotic stress tolerance*. Woodhead Publ Ltd, Abington Hall Abington, Cambridge CB1 6AH, CAMBS, England, pp 201–224.

Fahad S, Hussain S, Saud S, Hassan S, Tanveer M, Ihsan MZ, Shah AN, Ullah A, Nasrullah KF, Ullah S, Alharby HNW, Wu C, Huang J (2016c) A combined application of biochar and phosphorus alleviates heat-induced adversities on physiological, agronomical and quality attributes of rice. *Plant Physiol Biochem* 103:191–198.

Fahad S, Hussain S, Saud S, Tanveer M, Bajwa AA, Hassan S, Shah AN, Ullah A,Wu C, Khan FA, Shah F, Ullah S, Chen Y, Huang J (2015a) A biochar application protects rice pollen from high-temperature stress. *Plant Physiol Biochem* 96:281–287.

Fahad S, Bajwa AA, Nazir U, Anjum SA, Farooq A, Zohaib A, Sadia S, Nasim W, Adkins S, Saud S, Ihsan MZ, Alharby H, Wu C, Wang D, Huang J (2017) Crop production under drought and heat stress: plant responses and management options. *Front Plant Sci* 8:1147. doi: 10.3389/fpls.2017.01147

Fahad S, Hussain S, Bano A, Saud S, Hassan S, Shan D, Khan FA, Khan F, Chen Y, Wu C, Tabassum MA, Chun MX, Afzal M, Jan A, Jan MT, Huang J (2014a) Potential role of phytohormones and plant growth-promoting rhizobacteria in abiotic stresses: consequences for changing environment. *Environ Sci Pollut Res* 22(7):4907–4921. doi: 10.1007/s11356-014-3754-2

Fahad S, Hussain S, Matloob A, Khan FA, Khaliq A, Saud S, Hassan S, Shan D, Khan F, Ullah N, Faiq M, Khan MR, Tareen AK, Khan A, Ullah A, Ullah N, Huang J (2014b) Phytohormones and plant responses to salinity stress: a review. *Plant Growth Regul* 75(2):391–404. doi: 10.1007/s10725-014-0013-y

Farah R, Muhammad R, Muhammad SA, Tahira Y, Muhammad AA, Maryam A, Shafaqat A, Rashid M, Muhammad R, Qaiser H, Afia Z, Muhammad AA, Muhammad A, Fahad S (2020) Alternative and non-conventional soil and crop management strategies for increasing water use efficiency. In: Fahad S, Hasanuzzaman M, Alam M, Ullah H, Saeed M, Khan AK, Adnan M (eds) *Environment, climate, plant and vegetation growth*. Springer Publ Ltd, Springer Nature Switzerland AG. Part of Springer Nature, pp 323–338. doi: 10.1007/978-3-030-49732-3

Farhana G, Ishfaq A, Muhammad A, Dawood J, Fahad S, Xiuling L, Depeng W, Muhammad F, Muhammad F, Syed AS (2020) Use of crop growth model to simulate the impact of climate change on yield of various wheat cultivars under different agro-environmental conditions in Khyber Pakhtunkhwa, Pakistan. *Arabian J Geosci* 13:112. doi: 10.1007/s12517-020-5118-1

Farhat A, Hafiz MH, Wajid I, Aitazaz AF, Hafiz FB, Zahida Z, Fahad S, Wajid F, Artemi C (2020) A review of soil carbon dynamics resulting from agricultural practices. *J Environ Manage* 268 (2020):110319.

Farooq M, Basra SMA, Wahid A, et al (2009) Exogenously applied nitric oxide Enhances the drought tolerance in fine grain aromatic rice (Oryza sativa L.) *J Agron Crop Sci*. 195:254–261.

Farrant JM, Moore JP (2011) Programming desiccation-tolerance: from plants to seeds to resurrection plants. *Curr Opin Plant Biol* 14(3):340–345.

Fazli W, Muhmmad S, Amjad A, Fahad S, Muhammad A, Muhammad N, Ishaq AM, Imtiaz AK, Mukhtar A, Muhammad S, Muhammad I, Rafi U, Haroon I, Muhammad A (2020) Plant-microbes interactions and functions in changing climate. In: Fahad S, Hasanuzzaman M, Alam M, Ullah H, Saeed M, Khan AK, Adnan M (eds) *Environment, climate, plant and vegetation growth*. Springer Publ Ltd, Springer Nature Switzerland AG. Part of Springer Nature, pp 397–420. doi: 10.1007/978-3-030-49732-3

Fisher, MC, et al (2012) Emerging fungal threats to animal, plant and ecosystem health. *Nature* 484(7393):186–194.

Food and Agriculture Organization of the United Nations (2014) Faostat. www.fao.org/faostat/en/. Accessed 16 August 2016

Godfray, HC et al (2010) The future of the global food system. *Philos Trans R Soc Lond B Biol Sci* 365:2769–2777.

Gopakumar L, Bernard NO, Donato V (2020) Soil microarthropods and nutrient cycling. In: Fahad S, Hasanuzzaman M, Alam M, Ullah H, Saeed M, Khan AK, Adnan M (eds) *Environment, climate, plant and vegetation growth*. Springer Publ Ltd, Springer Nature Switzerland AG. Part of Springer Nature, pp 453–472. doi: 10.1007/978-3-030-49732-3

Habib ur R, Ashfaq A, Aftab W, Manzoor H, Fahd R, Wajid I, Md. Aminul I, Vakhtang S, Muhammad A, Asmat U, Abdul W, Syeda RS, Shah S, Shahbaz K, Fahad S, Manzoor H, Saddam H, Wajid N (2017) Application of CSM-CROPGRO-Cotton model for cultivars and optimum planting dates: evaluation in changing semi-arid climate. *Field Crops Res* 238:139–152. doi: 10.1016/j.fcr.2017.07.007

Hafiz MH, Wajid F, Farhat A, Fahad S, Shafqat S, Wajid N, Hafiz FB (2016) Maize plant nitrogen uptake dynamics at limited irrigation water and nitrogen. *Environ Sci Pollut Res* 24(3):2549–2557. doi: 10.1 007/s11356-016-8031-0

Hafiz MH, Muhammad A, Farhat A, Hafiz FB, Saeed AQ, Muhammad M, Fahad S, Muhammad A (2019) Environmental factors affecting the frequency of road traffic accidents: a case study of sub-urban area of Pakistan. *Environ Sci Pollut Res.* doi: 10.1007/s11356-019-04752-8

Hafiz MH, Farhat A, Ashfaq A, Hafiz FB, Wajid F, Carol Jo W, Fahad S, Gerrit H (2020b) Predicting Kernel Growth of Maize under Controlled Water and Nitrogen Applications. *Int J Plant Prod.* doi: 10.1007/ s42106-020-00110-8

Hafiz MH, Farhat A, Shafqat S, Fahad S, Artemi C, Wajid F, Chaves CB, Wajid N, Muhammad M, Hafiz FB (2018) Offsetting land degradation through nitrogen and water management during maize cultivation under arid conditions. *Land Degrad Dev* 1–10. doi: 10.1002/ldr.2933

Hafiz MH, Abdul K, Farhat A, Wajid F, Fahad S, Muhammad A, Ghulam MS, Wajid N, Muhammad M, Hafiz FB (2020a) Comparative effects of organic and inorganic fertilizers on soil organic carbon and wheat productivity under arid region. *Commun Soil Sci Plant Anal.* doi: 10.1080/00103624.2020.1 763385

Hakeem KR, Chandna R, Ahmad A, Iqbal M (2011b) Physiological and molecular analysis of applied nitrogen in rice (Oryza sativa L.) genotypes. *Rice Sci* 19(1):213–222.

Hakeem KR, Ahmad A, Iqbal M, Gucel S, Ozturk M (2011a) Nitrogen efficient rice genotype can reduce nitrate pollution. *Environ Sci Pollut Res* 18:1184–1193.

Hasanuzzaman M, Nahar K, Alam MM, Fujita M (2012) Exogenous nitric oxide alleviates high temperature induced oxidative stress in wheat (Triticum aestivum L.) seedlings by modulating the antioxidant defense and glyoxalase system. *Aust J Crop Sci* 6:1314–1323.

Hasegawa, PM (2013) Sodium (Na+) homeostasis and salt tolerance of plants. *Environ Exp Botany* 92(0):19–31.

Haswell ES, Verslues PE (2015) The ongoing search for the molecular basis of plant osmosensing. *J Gen Physiol* 145(5):389–394.

Hesham FA, Fahad S (2020) Melatonin application enhances biochar efficiency for drought tolerance in maize varieties: modifications in physio-biochemical machinery. *Agron J* 112(4):1–22.

Hoffman GJ, Rawlins SI (1970) Infertility of cotton flowers at both high and low relative humidities. *Crop Sci* 10:721–723.

Hussain MA, Fahad S, Rahat S, Muhammad FJ, Muhammad M, Qasid A, Ali A, Husain A, Nooral A, Babatope SA, Changbao S, Liya G, Ibrar A, Zhanmei J, Juncai H (2020) Multifunctional role of brassinosteroid and its analogues in plants. *Plant Growth Regul.* doi: 10.1007/s10725-020-00647-8

Ibrar K, Aneela R, Khola Z, Urooba N, Sana B, Rabia S, Ishtiaq H, Mujaddad Ur Rehman, Salvatore M (2020) Microbes and environment: global warming reverting the frozen zombies. In: Fahad S,

Hasanuzzaman M, Alam M, Ullah H, Saeed M, Khan AK, Adnan M (eds) *Environment, climate, plant and vegetation growth.* Springer Publ Ltd, Springer Nature Switzerland AG. Part of Springer Nature, pp 607–634. doi: 10.1007/978-3-030-49732-3

Ichimura K, Mizoguchi T, Yoshida R, Yuasa T, Shinozaki K (2000) Various abiotic stresses rapidly activate Arabidopsis MAP kinases ATMPK4 and ATMPK6. *Plant J* 24:655–665.

Illing, N., et al (2005) The signature of seeds in resurrection plants: a molecular and physiological comparison of desiccation tolerance in seeds and vegetative tissues. *Integr Comp Biol* 45(5):771–787.

Ilyas M, Mohammad N, Nadeem K, Ali H, Aamir HK, Kashif H, Fahad S, Aziz K, Abid U, (2020) Drought tolerance strategies in plants: a mechanistic approach. *J Plant Growth Regulation.* doi: 10.1007/s00344-020-10174-5.

Iqra M, Amna B, Shakeel I, Fatima K, Sehrish L, Hamza A, Fahad S (2020) Carbon cycle in response to global warming. In: Fahad S, Hasanuzzaman M, Alam M, Ullah H, Saeed M, Khan AK, Adnan M (eds) *Environment, climate, plant and vegetation growth.* Springer Publ Ltd, Springer Nature Switzerland AG. Part of Springer Nature, pp 1–16. doi: 10.1007/978-3-030-49732-3

Jan M, Muhammad Anwar-ul-Haq, Adnan NS, Muhammad Y, Javaid I, Xiuling L, Depeng W, Fahad S (2019) Modulation in growth, gas exchange, and antioxidant activities of salt-stressed rice (Oryza sativa L.) genotypes by zinc fertilization. *Arabian J Geosci* 12:775. doi: 10.1007/s12517-019-4939-2

Kamarn M, Wenwen C, Irshad A, Xiangping M, Xudong Z, Wennan S, Junzhi C, Shakeel A, Fahad S, Qingfang H, Tiening L (2017) Effect of paclobutrazol, a potential growth regulator on stalk mechanical strength, lignin accumulation and its relation with lodging resistance of maize. *Plant Growth Regul* 84:317–332. doi: 10.1007/s10725-017-0342-8

Kappen L (1981) Ecological significance of resistance to high temperature. In: Lange et al (eds) *Encyclopedia of plant physiology, physiological plant ecology I B, New Series*, Vol 12A. Springer, New York, pp 439–474.

Khan A, Gul F, Jamal Y, Ali J, Adnan M, Khan MA, Naeem I, Bacha Z (2017) Weeds management in maize hybrids with night plowing. *Pak J Weed Sci Res* 22(2):179–191.

Kissoudis, C., et al (2014) Enhancing crop resilience to combined abiotic and biotic stress through the dissection of physiological and molecular crosstalk. *Front Plant Sci* 5:207.

Laohavisit, A., et al (2013) Salinity-induced calcium signaling and root adaptation in Arabidopsis thaliana require the calcium regulatory protein annexin1. *Plant Physiol* 163:253–262.

Liang, Y., et al (2013) Nonlegumes respond to rhizobial Nod factors by suppressing the innate immune response. *Science* 341(6152):1384–1387.

Liu B, et al. (2016) Similar estimates of temperature impacts on global wheat yield by three independent methods. *Nat Clim Change* 6:1130–1136.

Lobell DB (2007) Changes in diurnal temperature range and national cereal yields. *Agric Meteorol* 145:229–238.

Lobell DB, Schlenker W, Costa-Roberts J (2011) Climate trends and global crop production since 1980. *Science* 333:616–620.

Long SP, Ainsworth EA, Leakey AD, Nösberger J, Ort DR (2006) Food for thought: lower-than-expected crop yield stimulation with rising CO2 concentrations. *Science* 312:1918–1921.

Mahar A, Amjad A, Altaf HL, Fazli W, Ronghua L, Muhammad A, Fahad S, Muhammad A, Rafiullah, Imtiaz AK, Zengqiang Z (2020) Promising technologies for Cd-contaminated soils: drawbacks and possibilities. In: Fahad S, Hasanuzzaman M, Alam M, Ullah H, Saeed M, Khan AK, Adnan M (eds) *Environment, climate, plant and vegetation growth.* Springer Publ Ltd, Springer Nature Switzerland AG. Part of Springer Nature, pp 63–92. doi: 10.1007/978-3-030-49732-3

Md Jakir H, Allah B (2020) Development and applications of transplastomic plants; a way towards eco-friendly agriculture. In: Fahad S, Hasanuzzaman M, Alam M, Ullah H, Saeed M, Khan AK, Adnan M (eds) *Environment, climate, plant and vegetation growth.* Springer Publ Ltd, Springer Nature Switzerland AG. Part of Springer Nature, pp 285–322. doi: 10.1007/978-3-030-49732-3

Md. Enamul H, Shoeb AZM, Mallik AH, Fahad S, Kamruzzaman MM, Akib J, Nayyer S, Mehedi A KM, Swati AS, Md Yeamin A, Most SS (2020) Measuring vulnerability to environmental hazards: qualitative to quantitative. In: Fahad S, Hasanuzzaman M, Alam M, Ullah H, Saeed M, Khan AK, Adnan M (eds) *Environment, climate, plant and vegetation growth.* Springer Publ Ltd, Springer Nature Switzerland AG. Part of Springer Nature, pp 421–452. doi: 10.1007/978-3-030-49732-3

Mengiste, T., et al (2011) The BOTRYTIS SUSCEPTIBLE1 gene encodes an R2R3MYB transcription factor protein that is required for biotic and abiotic stress responses in Arabidopsis. *Plant Cell Online* 15(11):2551–2565.

Mohr PG, Cahill DM (2003) Abscisic acid influences the susceptibility of Arabidopsis thaliana to Pseudomonas syringae pv. tomato and Peronospora parasitica. *Funct Plant Biol* 30(4):461.

Mubeen M, Ashfaq A, Hafiz MH, Muhammad A, Hafiz UF, Mazhar S, Muhammad SD, Asad A, Amjed A, Fahad S, Wajid N (2020) Evaluating the climate change impact on water use efficiency of cotton-wheat in semi-arid conditions using DSSAT model. *J Water Climate Change*. doi: 10.2166/wcc.2019.179/622 035/jwc2019179.pdf

Muhammad Tahir ul Qamar, Amna F, Amna B, Barira Z, Xitong Z, Ling-Ling C (2020) Effectiveness of conventional crop improvement strategies vs. omics. In: Fahad S, Hasanuzzaman M, Alam M, Ullah H, Saeed M, Khan AK, Adnan M (eds) *Environment, climate, plant and vegetation growth*. Springer Publ Ltd, Springer Nature Switzerland AG. Part of Springer Nature, pp 253–284. doi: 10.1007/978-3-030-4 9732-3

Muhammad Z, Abdul MK, Abdul MS, Kenneth BM, Muhammad S, Shahen S, Ibadullah J, Fahad S (2019) Performance of Aeluropus lagopoides (mangrove grass) ecotypes, a potential turfgrass, under high saline conditions. *Environ Sci Pollut Res*. doi: 10.1007/s11356-019-04838-3.

Munns R, Tester M (2008) Mechanisms of salinity tolerance. *Ann Rev Plant Biol* 59(1):651–681.

Nahar K, Hasanuzzaman M, Ahamed KU, Hakeem KR, Ozturk M, Fujita M (2015). Plant responses and tolerance to high temperature stress: role of exogenous phytoprotectants. In: *Crop production and global environmental issues*. Springer, Cham, pp 385–435

Noor M, Naveed ur R, Ajmal J, Fahad S, Muhammad A, Fazli W, Saud S, Hassan S (2020) Climate change and costal plant lives. In: Fahad S, Hasanuzzaman M, Alam M, Ullah H, Saeed M, Khan AK, Adnan M (eds) *Environment, climate, plant and vegetation growth*. Springer Publ Ltd, Springer Nature Switzerland AG. Part of Springer Nature, pp 93–108. doi: 10.1007/978-3-030-49732-3

Oliver M, Tuba Z, Mishler B (2000) Evolution of desiccation tolerance in land plants. *Plant Ecol* 151:85–100.

Oliver MJ, Velten J, Mishler BD (2005) Desiccation tolerance in bryophytes: a reflection of the primitive strategy for plant survival in dehydrating habitats? *Integr Comp Biol* 45(5):788–799.

Oliver MJ, Cushman JC, Koster KL (2010) Dehydration tolerance in plants. *Methods Mol Biol* 639:3–24.

Ottman MJ, Kimball BA, White JW, Wall GW (2012) Wheat growth response to increased temperature from varied planting dates and supplemental infrared heating. *Agron J* 104:7–16

Pandey P, Ramegowda V, Senthil-Kumar M (2015) Shared and unique responses of plants to multiple individual stresses and stress combinations: physiological and molecular mechanisms. *Front Plant Sci* 6:723.

Paparella, C., et al (2014) The Arabidopsis LYSIN MOTIF-CONTAINING RECEPTOR-LIKE KINASE3 regulates the cross talk between immunity and abscisic acid responses. *Plant Physiol* 165(1):262–276.

Porter JR, Gawith M (1999) Temperatures and the growth and development of wheat: a review. *Eur J Agron* 10:23–36.

Qamar-uz Z, Zubair A, Muhammad Y, Muhammad ZI, Abdul K, Fahad S, Safder B, Ramzani PMA, Muhammad N (2017) Zinc biofortification in rice: leveraging agriculture to moderate hidden hunger in developing countries. *Arch Agron Soil Sci* 64:147–161. doi: 10.1080/03650340.2017.1338343

Rashid M, Qaiser H, Khalid SK, Al-Wabel MI, Zhang A, Muhammad A, Shahzada SI, Rukhsanda A, Ghulam AS, Shahzada MM, Sarosh A, Muhammad FQ (2020) Prospects of biochar in alkaline soils to mitigate climate change. In: Fahad S, Hasanuzzaman M, Alam M, Ullah H,Saeed M, Khan AK, Adnan M (eds) *Environment, climate, plant and vegetation growth*. Springer Publ Ltd, Springer Nature Switzerland AG. Part of Springer Nature, pp 133–150. doi: 10.1007/978-3-030-49732-3

Ray DK, Ramankutty N, Mueller ND, West PC, Foley JA (2012) Recent patterns of crop yield growth and stagnation. *Nat Commun* 3:1293.

Rehman M, Fahad S, Saleem MH, Hafeez M, Muhammad HR, Liu F, Deng G (2020) Red light optimized physiological traits and enhanced the growth of ramie (Boehmeria nivea L.). *Photosynthetica* 58 (4):922–931

Reusche, M., et al (2012) Verticillium infection triggers VASCULAR-RELATED NAC DOMAIN7-dependent de novo xylem formation and enhances drought tolerance in Arabidopsis. *Plant Cell* 24(9):3823–3837.

Sadam M, Muhammad TQ, Ghulam M, Muhammad SK, Faiz AJ (2020) Role of Biotechnology in climate resilient agriculture. In: Fahad S, Hasanuzzaman M, Alam M, Ullah H, Saeed M, Khan AK, Adnan M (eds) *Environment, climate, plant and vegetation growth.* Springer Publ Ltd, Springer Nature Switzerland AG. Part of Springer Nature, pp 339–366. doi: 10.1007/978-3-030-49732-3

Sajid H, Jie H, Jing H, Shakeel A, Satyabrata N, Sumera A, Awais S, Chunquan Z, Lianfeng Z, Xiaochuang C, Qianyu J, Junhua Z (2020) Rice production under climate change: adaptations and mitigating strategies. In: Fahad S, Hasanuzzaman M, Alam M, Ullah H, Saeed M, Khan AK, Adnan M (eds) *Environment, climate, plant and vegetation growth.* Springer Publ Ltd, Springer Nature Switzerland AG. Part of Springer Nature, pp 659–686. doi: 10.1007/978-3-030-49732-3

Sajjad H, Muhammad M, Ashfaq A, Waseem A, Hafiz MH, Mazhar A, Nasir M, Asad A, Hafiz UF, Syeda RS, Fahad S, Depeng W, Wajid N (2019) Using GIS tools to detect the land use/land cover changes during forty years in Lodhran district of Pakistan. *Environ Sci Pollut Res* 10.1007/s11356-019-06072-3.

Saleem MH, Fahad S, Shahid UK, Mairaj D, Abid U, Ayman ELS, Akbar H, Analía L, Lijun L (2020b) Copper-induced oxidative stress, initiation of antioxidants and phytoremediation potential of flax (Linum usitatissimum L.) seedlings grown under the mixing of two different soils of China. *Environ Sci Poll Res.* doi: 10.1007/s11356-019-07264-7.

Saleem MH, Rehman M, Fahad S, Tung SA, Iqbal N, Hassan A, Ayub A, Wahid MA, Shaukat S, Liu L, Deng G (2020c) Leaf gas exchange, oxidative stress, and physiological attributes of rapeseed (Brassica napus L.) grown under different light-emitting diodes. *Photosynthetica* 58 (3):836–845.

Saleem MH, Fahad S, Adnan M, Mohsin A, Muhammad SR, Muhammad K, Qurban A, Inas AH, Parashuram B, Mubassir A, Reem MH (2020a) Foliar application of gibberellic acid endorsed phytoextraction of copper and alleviates oxidative stress in jute (Corchorus capsularis L.) plant grown in highly copper-contaminated soil of China. *Environ Sci Pollution Res.* doi: 10.1007/s11356-020-09764-3.

Saleh J, Maftoun M (2008) Interactive effect of NaCl levels and Zinc sources and levels on the growth and mineral composition of rice. *J Agric Sci Tech* 10: 325–336.

Saman S, Amna B, Bani A, Muhammad TQ, Rana MA, Muhammad SK (2020) QTL mapping for abiotic stresses in cereals. In: Fahad S, Hasanuzzaman M, Alam M, Ullah H, Saeed M, Khan AK, Adnan M (eds) *Environment, climate, plant and vegetation growth.* Springer Publ Ltd, Springer Nature Switzerland AG. Part of Springer Nature, pp 229–252. doi: 10.1007/978-3-030-49732-3

Saud S, Chen Y, Long B, Fahad S, Sadiq A (2013) The different impact on the growth of cool season turf grass under the various conditions on salinity and drought stress. *Int J Agric Sci Res* 3:77–84

Saud S, Fahad S, Cui G, Chen Y, Anwar S (2020) Determining nitrogen isotopes discrimination under drought stress on enzymatic activities, nitrogen isotope abundance and water contents of Kentucky bluegrass. *Sci Rep* 10:6415. doi: 10.1038/s41598-020-63548-w

Saud S, Chen Y, Fahad S, Hussain S, Na L, Xin L, Alhussien SA (2016) Silicate application increases the photosynthesis and its associated metabolic activities in Kentucky bluegrass under drought stress and post-drought recovery. *Environ Sci Pollut Res* 23(17):17647–17655. doi: 10.1007/s11356-016-6957-x

Saud S, Li X, Chen Y, Zhang L, Fahad S, Hussain S, Sadiq A, Chen Y (2014) Silicon application increases drought tolerance of Kentucky bluegrass by improving plant water relations and morph physiological functions. *SciWorld J* 2014:1–10. doi: 10.1155/2014/368694

Saud S, Fahad S, Yajun C, Ihsan MZ, Hammad HM, Nasim W, Amanullah Jr, Arif M, Alharby H (2017) Effects of nitrogen supply on water stress and recovery mechanisms in Kentucky Bluegrass plants. *Front Plant Sci* 8:983. doi: 10.3389/fpls.2017.00983

Schlenker W, Roberts MJ (2009) Nonlinear temperature effects indicate severe damages to U.S. crop yields under climate change. *Proc Natl Acad Sci USA* 106:15594–15598.

Schroeder JI, et al (2013) Using membrane transporters to improve crops for sustainable food production. *Nature* 497(7447):60–66.

Senol C (2020) The effects of climate change on human behaviors. In: Fahad S, Hasanuzzaman M, Alam M, Ullah H, Saeed M, Khan AK, Adnan M (eds) *Environment, climate, plant and vegetation growth.* Springer Publ Ltd, Springer Nature Switzerland AG. Part of Springer Nature, pp 577–590. doi: https://doi.org/10.1007/978-3-030-49732-3

Shafi MI, Adnan M, Fahad S, Fazli W, Ahsan K, Zhen Y, Subhan D, Zafar-ul-Hye M, Martin B, Rahul D (2020) Application of single superphosphate with humic acid improves the growth, yield and phosphorus uptake of wheat (Triticum aestivum L.). *Calcareous Soil Agron* (10):1224. doi: 10.3390/agronomy10091224.

Shah F, Lixiao N, Kehui C, Tariq S, Wei W, Chang C, Liyang Z, Farhan A, Fahad S, Huang J (2013) Rice grain yield and component responses to near 2°C of warming. *Field Crop Res* 157:98–110

Shinozaki K, Yamaguchi-Shinozaki K (2003) Molecular responses to dehydration and low temperature: difference and cross-talk between two stress signaling pathways. *Cur Opin Plant Biol* 3:217–223.

Silva EN, Ferreira-Silva SL, de Vasconcelos Fontenelea A, Ribeirob RV, Viégasc RA, Silveira JAG (2010). Photosynthetic changes and protective mechanisms against oxidative damage subjected to isolated and combined drought and heat stresses in Jatropha curcas plants. *J Plant Physiol* 167:1157–1164.

Solomon D Qin M Manning Z Chen M Marquis KB Averyt et al (2007) Contribution of working group I contribution to the fourth assessment report of the IPCC. In: *Climate change 2007-the physical science basis*, Cambridge University Press, Cambridge, NY.

Subhan D, Zafar-ul-Hye M, Fahad S, Saud S, Martin B, Tereza H, Rahul D (2020) Drought stress alleviation by ACC deaminase producing achromobacter xylosoxidans and enterobacter cloacae, with and without Timber Waste Biochar in Maize. *Sustain* 12(6286). doi:10.3390/su12156286

Suzuki, N., et al (2014) Abiotic and biotic stress combinations. *New Phytologist* 203(1):32–43.

Taha MA, Malik MNA, Chaudhry FI, Makhdum I (1981) Heat induced sterility in cotton sown during early April in West Punjab. *Exp Agric* 17:189–194.

Tariq M, Ahmad S, Fahad S, Abbas G, Hussain S, Fatima Z, Nasim W, Mubeen M, ur Rehman MH, Khan MA, Adnan M (2018). The impact of climate warming and crop management on phenology of sunflower-based cropping systems in Punjab, Pakistan. *Agri and Forest Met* 15(256):270–282.

Tebaldi C, Hayhoe K, Arblaster JM, Meehl GA (2007) Going to the extremes: an intercomparison of model-simulated historical and future changes in extreme events. *Clim Change* 82:233–234. doi: 10.1007/s105 84-007-9247-2.

Tian Y, et al (2012) Warming impacts on winter wheat phenophase and grain yield under field conditions in Yangtze Delta Plain. *China Field Crops Res* 134:193–199.

Triky-Dotan, S., et al (2005) Development of crown and root rot disease of tomato under irrigation with saline water. *Phytopathology* 95(12):1438–1444.

Unsar NU, Muhammad R, Syed HMB, Asad S, Mirza AQ, Naeem I, Muhammad HR, Fahad S, Shafqat S (2020) Insect pests of cotton crop and management under climate change scenarios. In: Fahad S, Hasanuzzaman M, Alam M, Ullah H, Saeed M, Khan AK, Adnan M (eds) *Environment, climate, plant and vegetation growth*. Springer Publ Ltd, Springer Nature Switzerland AG. Part of Springer Nature, pp 367–396. doi: 10.1007/978-3-030-49732-3

Urao T, Yakubov B, Satoh R, Yamaguchi-Shinozaki K, Shinozaki K (1999) A transmembrane hybrid-type histidine kinase in Arabidopsis functions as an osmosensor. *Plant Cell II* 11:1743–1754.

Wahid F, Fahad S, Subhan D, Adnan M,, Zhen Y, Saud S, Manzer HS, Martin B, Tereza H, Rahul D (2020) Sustainable management with Mycorrhizae and phosphate solubilizing bacteria for enhanced phosphorus uptake in Calcareous soils. *Agriculture* 10(334). doi:10.3390/agriculture10080334

Wajid N, Ashfaq A, Asad A, Muhammad T, Muhammad A, Muhammad S, Khawar J, Ghulam MS, Syeda RS, Hafiz MH, Muhammad IAR, Muhammad ZH, Muhammad HR, Veysel T, Fahad S, Suad S, Aziz K, Shahzad A (2017) Radiation efficiency and nitrogen fertilizer impacts on sunflower crop in contrasting environments of Punjab. *Pakistan Environ Sci Pollut Res* 25:1822–1836. doi: 10.1007/ s11356-017-0592-z

Wiese J, Kranz T, Schubert S (2004) Induction of pathogen resistance in barley by abiotic stress. *Plant Biol (Stuttg)* 6(5):529–536.

Wu C, Tang S, Li G, Wang S, Fahad S, Ding Y (2019) Roles of phytohormone changes in the grain yield of rice plants exposed to heat: a review. *PeerJ* 7:e7792. doi: 10.7717/peerj.7792

Wu C, Kehui C, She T, Ganghua L, Shaohua W, Fahad S, Lixiao N, Jianliang H, Shaobing P, Yanfeng D (2020) Intensified pollination and fertilization ameliorate heat injury in rice (Oryza sativa L.) during the flowering stage. *Field Crops Res* 252:107795

Yang G, Rhodes D, Joly RJ (1996) Effects of high temperature on membrane stability and chlorophyll fluorescence in glycinebetaine-deficient and glycinebetaine-containing maize lines. *Aust J Plant Physiol* 23:437–443.

Yang Z, Zhang Z, Zhang T, Fahad S, Cui K, Nie L, Peng S, Huang J (2017) The effect of season-long temperature increases on rice cultivars grown in the central and southern regions of China. *Front Plant Sci* 8:1908. doi: 10.3389/fpls.2017.01908

Yuan, F., et al (2014) OSCA1 mediates osmotic-stress-evoked Ca2+ increases vital for osmosensing in Arabidopsis. *Nature* 514(7522):367–371.

Zafar-ul-Hye M, Ahzeeb-ul-Hassan MT, Muhammad A, Fahad S, Martin B, Tereza D, Rahul D, Subhan D (2020a) Potential role of compost mixed biochar with rhizobacteria in mitigating lead toxicity in spinach. *Scientific Rep* 10:12159. doi: 10.1038/s41598-020-69183-9

Zafar-ul-Hye M, Muhammad N, Subhan D, Fahad S, Rahul D, Mazhar A, Ashfaq AR, Martin B, Jiˇrí H, Zahid HT, Muhammad N (2020b) Alleviation of Cadmium adverse effects by improving nutrients uptake in bitter gourd through Cadmium tolerant Rhizobacteria. *Environment* 7(54). doi: 10.3390/environments7080054

Zahida Z, Hafiz FB, Zulfiqar AS, Ghulam MS, Fahad S, Muhammad RA, Hafiz MH, Wajid N, Muhammad S (2017) Effect of water management and silicon on germination, growth, phosphorus and arsenic uptake in rice. *Ecotoxicol Environ Saf* 144:11–18.

Zia-ur-Rehman M (2020) Environment, climate change and biodiversity. In: Fahad S, Hasanuzzaman M, Alam M, Ullah H, Saeed M, Khan AK, Adnan M (eds) *Environment, climate, plant and vegetation growth*. Springer Publ Ltd, Springer Nature Switzerland AG. Part of Springer Nature, pp 473–502. doi: 10.1007/978-3-030-49732-3

9

Legumes under Drought Stress: Plant Responses, Adaptive Mechanisms, and Management Strategies in Relation to Nitrogen Fixation

Mohammad Sohidul Islam, Shah Fahad, Akbar Hossain, M Kaium Chowdhury, Muhammad Aamir Iqbal, Anamika Dubey, Ashwani Kumar, Karthika Rajendran, Subhan Danish, Muhammad Habib Ur Rahman, Muhammad Ali Raza, Muhammad Arif, Shah Saud, Mohammad Anwar Hossain, Ejaz Waraich, Zahoor Ahmad, Sajjad Hussain, Arzu Çığ, Murat Erman, Fatih Çığ, and Ayman EL Sabagh

CONTENTS

9.1 Introduction

In modern intensive farming systems, nitrogen (N) is one of the most abundantly supplied nutrients to crop plants whereas its suboptimal availability in nature constitutes the major limiting factor to crop growth and yield (Date 2000; Kumar et al. 2018; Iqbal et al. 2019). The environment contains abundant inert N but this N cannot be used directly by higher plants (Iqbal et al. 2019b). Only some prokaryotic organisms can reduce it to an available form for use by plants. Biological transformations of N (e.g., N_2 fixation) through biological nitrogen fixation (BNF) in legumes contributes about 60% of the biologically fixed N from the atmospheric nitrogen (N_2) (Bano and Iqbal 2016; Iqbal et al.

DOI: 10.1201/9781003160717-9

2018a, 2018b, 2018c). Conversion of N_2 into ammonia (NH_3) by *nitrogenase* enzyme complex is a natural process which is commonly known as BNF. Nitrogen must first be fixed, and then reduced to NH_3 or ammonium ions (NH_4^+). In an endergonic process, N_2 is biologically reduced to NH_3, and this process requires energy equivalent to Ca. 960 KJ mol^{-1} of N-fixed (Sprent and Raven 1985; Ahmad et al. 2017). In the biological system, respiration and electron carriers, usually ferredoxin, provides the required ATP and electrons, respectively, for *nitrogenase* functions. The *nitrogenase* enzyme catalyses the reduction of different substrates like H^+, N_2, and C_2H_2. The main effect of dinitrogen reaction is as follows:

$$N_2 + 16\,MgATP + 8\,e^- + 8\,H^+ \rightarrow 2\,NH_3 + H_2 + 16\,MgADP + 16\,Pi$$

The carbonaceous substances fulfil the energy needs of N_2 fixing bacteria involved in this symbiotic reaction. In legumes, *Rhizobium* invades roots to infect the cortex cells and initiate the formation of root nodules. Inside the root nodules, bacteria differentiate into a fully functional form called *bacteroid* which initiates atmospheric N-fixation (Chanway et al. 2014). Globally, legume-*Rhizobium* symbiosis yields a major chunk of biologically fixed N in prevalent agricultural systems (Graham and Vance 2000). In addition, it is a key process in sustainable land management, where N is the limiting nutrient that affects crop yield and productivity (Al-Falih 2002).

Generally, root nodule bacteria which are known as rhizobia, develop a symbiotic association with different leguminous plants (EL Sabagh et al. 2020e). Symbiotic nitrogen fixation (SNF) involves different host leguminous plants and bacterial micro-symbionts which may be belonging to one of the genera; *Rhizobium, Bradyrhizobium* and *Azorhizobium* (Al-Falih 2002). Bacteria are intimately attached to the wall of root hairs which is essential for the induction of morphological response ensuring symbiosis installation. The symbiotic bacteria after differentiation into bacteroids which have peri-bacteroid membrane can fix N inside the root nodules (Nabizadeh et al. 2011). However, N-fixing ability of legumes depends on their genetic makeup, soil, and environmental conditions such as drought, temperature, salinity, alkalinity, acidity etc. On the other hand, crop plants exhibit a wide range of mechanisms like reduced water loss by promoting diffusive resistance, increased water uptake with prolific and deep root systems, and reduced transpirational loss through smaller and succulent leaves etc. to resist drought stress. The success of the nodulation – fixation in root nodules depends on bacterial survival along with the duration and severity of drought stress (DS). In this chapter, we highlight the impacts of DS on legumes to N accumulation, as well as evolving the tolerance mechanisms, breeding, and management approaches to off-set the adverse effects of DS.

9.2 Effects of Drought Stress in Legumes

How far abiotic stresses as a subject that progressively preoccupies the world is still a matter of debate (EL Sabagh et al. 2020b, 2020c; Adnan et al. 2018a, 2018b; Adnan et al. 2019; Akram et al. 2018a, 2018b; Aziz et al. 2017a, 2017b; Habib et al. 2017; Hafiz et al. 2016; Hafiz et al. 2019; Kamaran et al. 2017; Muhmmad et al. 2019; Sajjad et al. 2019; Saud et al. 2013; Saud et al. 2014; Saud et al. 2017; Saud et al. 2016; Shah et al. 2013; Saud et al. 2020; Qamar et al. 2017; Wajid et al. 2017; Yang et al. 2017; Zahida et al. 2017; Depeng et al. 2018; Hussain et al. 2020; Hafiz et al. 2020a, 2020b; Shafi et al. 2020; Wahid et al. 2020; Subhan et al. 2020; Zafar-ul-Hye et al. 2020a, 2020b; Adnan et al. 2020; Ilyas et al. 2020; Saleem et al. 2020a, 2020b, 2020c; Rehman et al. 2020; Farhat et al. 2020; Wu et al. 2020; Mubeen et al. 2020; Farhana et al. 2020; Jan et al. 2019; Wu et al. 2019; Ahmad et al. 2019; Baseer et al. 2019; Hafiz et al. 2018; Tariq et al. 2018; Fahad and Bano 2012; Fahad et al. 2017; Fahad et al. 2013; Fahad et al. 2014a, 2014b; Fahad et al. 2016a, 2016b, 2016c, 2016d; Fahad et al. 2015a, 2015b; Fahad et al. 2018; Fahad et al. 2019a, 2019b; Hesham and Fahad 2020; Iqra et al. 2020; Akbar et al. 2020; Mahar et al. 2020; Noor et al. 2020; Bayram et al. 2020; Amanullah et al. 2020; Rashid et al. 2020; Arif et al. 2020; Amir et al. 2020; Saman et al. 2020; Muhammad Tahir et al. 2020; Md Jakirand Allah 2020;

Farah et al. 2020; Sadam et al. 2020; Unsar et al. 2020; Fazli et al. 2020; Md. Enamul et al. 2020; Gopakumar et al. 2020; Zia-ur-Rehman 2020; Al-Wabel et al. 2020a, 2020b; Senol 2020; Amjad et al. 2020; Ibrar et al. 2020; Sajid et al. 2020. Drought stress is one of the crucial challenges faced by sustainable farming systems under changing climate (Jaleel et al. 2009). However, in the changing climate, the sustainability of the productivity of grain legumes is susceptible to loss due to the severe actions of abiotic stresses conditions (Hossain et al. 2020a, 2020b). Drought is considered the major factor that unfavourably impacts the production and quality of crops (EL Sabagh et al. 2019a, 2020c, 2020d). It is believed after taking in from the crop growth models that drought is the single most devastating environmental stress, which decreases crop productivity more than any other environmental stresses (Lambers et al. 2008). The ongoing global climate change scenarios also indicate that the severity and frequency of drought stress effects are increasing day by day (Intergovernmental Panel on Climate Change IPCC 2007). For sustainable agriculture, legumes are considered economically important crops in many cropping systems due to their atmospheric N-fixation ability as broadening their adaptability to multiple environments having N deficiency. Legumes exhibit higher sensitivity to drought stress which has been considered as one of the major constraints to high crop yield potential. Drought stress induces several devastating effects on legumes, which have been summarized in the upcoming sections.

9.2.1 Morphological Responses

Rhizobial growth and functioning of symbiosis process are adversely affected by drought stress. Viable strains of *Rhizobium* become unviable under drought-induced osmotic stress as it harms the morphological characteristics of *rhizobia* (Shoushtari and Pepper 1985; Busse and Bottomley 1989; Elboutahiri et al. 2010), root-hair colonization, while hampering the infection process (Zahran 1999). Thus, root infection significantly reduces the nodulation process. In addition, drought stress also restricts the development and function of root nodules (Sauvage et al. 1983; Serraj et al. 1999). As low water potential due to drought stress showed the irregular morphology of mesquite *Rhizobium* (Shoushtari and Pepper 1985) and *R. meliloti* (Busse and Bottomley 1989). Moreover, modification of rhizobial cells due to water stress will ultimately lead to a reduction in infection and nodulation of legumes. A failure report of soybean inoculation in soil with a high indigenous population of *R. japonicum* due to low water content in the soil was tabled by Hunt et al. (1981). About 2% reduction of soil moisture (from 5.5 to 3.5%) in sandy loam soil represented a large drop of water potential from -0.36 to -3.6×10^5 Pa, and had severe effects on root hair infection of *Trifolium subterraneum* (Worrall and Roughley 1976; Eaglesham and Ayanaba 1984). Furthermore, lab experiments with polyethylene glycol (PEG) induced water deficit stress significantly reducing infection, thread formation, and nodulation of *Vicia faba* (Zahran 1986; Zahran and Sprent 1986). In fact, a favourable rhizosphere environment is vital for legume-*Rhizobium* interaction and symbiosis. Development of effective nodules in desert soils highlights that some strains can still tolerate extreme conditions in soils with limited moisture levels (Jenkins et al. 1989; Fuhrmann et al. 1986; Fuhrmann and Wollum 1989).

9.2.2 Physiological Responses

Photosynthesis, especially photosystem-II (PSII), is adversely affected by DS which reduces photosynthesis rate, and the maximum quantum yield of PSII photochemistry (Fv/Fm) indicates an uninterrupted process of photosynthesis in beans (Pastenes et al. 2005). Drought-tolerant species/genotypes did not show a remarkable variation in Fv/Fm when compared to the plants grown under normal conditions such as soybean (Ohashi et al. 2006). Drought stress decreased net CO_2 assimilation rate, stomatal conductance, and transpiration rate, and reduced the growth of crops (Saeidi and Abdoli 2015). The ability to maintain a high carbon gain appears to confer stress tolerance in many crops (Ratnayaka and Kincaid 2005). The imposition of DS sharply declines nodule permeability wing to their inner-cortex cells leading to the synthesis of abscisic acid in nodules (Irekti and Drevon 2003). In addition, *nitrogenase* activity is inhibited and nodule respiration is compensated by triggering PO_2 accumulation in the nodulated-roots (Serraj et al. 1994).

In fact, the drought causes damage to cells through chemical alterations. During hydration, the hydroxyl and peroxyl radicles formation induces lipid peroxidation along with denaturation of proteins and nucleic acids (Casteriano 2014). Reducing sugars covalently react with amino acid residues in a process called non-enzymatic browning or Millard reaction, which deteriorates proteins (Casteriano 2014). Further, it reduces persistence and survival of rhizobia, affects root-hair colonization, and limits root nodulation (Zahran 1999; Mhadhbi et al. 2011). Ultimately, N fixation appears to be very sensitive to drought-stressed conditions (Kirda et al. 1989).

Drought disrupts numerous physiological processes occurring in legumes and reduces the functioning of the BNF process (Sinclair et al. 2007). Consequently, the process of carbohydrates accumulation declines in leaves and leaf senescence is initiated (Kaschuk et al. 2010). It has been reported earlier that water stress reduced O_2 diffusion in legume nodules (Durand et al. 1987; Davey and Simpson 1990), and caused metabolic changes (Purcell and Sinclair 1995; Serraj and Sinclair 1996), particularly the activity of an enzyme sucrose synthase (SS), a key enzyme involved in the hydrolysis of sucrose in nodules, is affected (González et al. 1995; Gordon et al. 1997). However, legumes tend to show a serious decline of water potential under drought along with disrupting nodule formation process (Sinclair et al. 1987; Durand et al. 1987; Streeter 1993). In soybean, drought stress reduced N_2 fixation and posed a serious constraint on N accumulation and yield potential (Sinclair et al. 1987). Although drought stress is known to affect all the steps of nodule formation and functioning, most of the work on the mechanisms of drought effects on N_2 fixation has focused only on *nitrogenase* enzyme activity. Three major hypotheses are considered in the present work to analyze the effect of drought stress on *nitrogenase* activity: O_2 limitation, regulation by N metabolism, and carbon shortage.

The reduced water potential under drought inhibits various enzymes involved in sucrose hydrolysis which leads to C deficiency for bacteroids. Subsequently, *nitrogenase* inhibition gives rise to O_2 accumulation which reduces nodule permeability. Moreover, limitation of O_2 is compensated by the decrease in nodule growth, and the formation of a large number of small nodules facilitating the O_2 entry in nodules by increased contact area with the external medium. Streeter (2003) reported that soybean shoots which were exposed to drought for one-week increased the ureides up to 100%. These accumulated ureides are toxic for *rhizobia* and prohibit the synthesis of various enzymes involved in BNF (Vadez et al. 2000). The ureides accumulation in nodules is caused by delayed transportation to shoots owing to drought stress (Collier and Tegeder 2012; Sinclair and Vadez 2012). Cerezini et al. (2016) concluded that the increase of ureides in petioles is more evident in the susceptible soybean genotypes, and suggested that the N-metabolism is affecting the nitrogen fixation drought-tolerant (NFDT) soybean genotypes under drought-stressed condition. However, it has not yet been identified that ureides build-up is due to persistent synthesis or owing to its delayed exportation to shoots, or due to its overflow from phloem which causes BNF process impairment under drought stress (Vadez et al. 2000). Collier and Tegeder (2012) suggested that delayed exportation of ureides in nodules remains the leading cause of urides toxicity. Ladrera et al. (2007) reported rapid ureides accumulation in root nodules of drought-sensitive soybean genotype, Biloxi, that is more than that in NFDT soybean genotype, Jackson. The NFDT genotypes under drought showed higher concentrations of ureides in the nodules compared to susceptible genotypes (Cerezini et al. 2016). It seems that the NFDT genotypes are intrinsically more effective in maintaining their capacity for BNF, even with higher concentrations of ureides. As a result, shoot N accumulations were higher in NFDT soybean genotypes and they were able to provide N during formation of seeds (King and Purcell 2006; King et al. 2014). Studies have reported that the total chlorophyll and carotenoid contents were also found to decline under water restricted conditions (El-Tayeb 2006a, 2006b; Mouradi et al. 2016).

9.2.3 Biochemical Responses

Numerous biochemical indicators like agro-morphological and physiological traits for drought responses have been employed in rapid screening of genotypes for drought tolerance in many crops (Zarafshar et al. 2014). Drought stress alters the functions and synthesis rate of vital enzymes (nitrate reductase) involved in BNF process (Andrews et al. 2004), which decreases protein levels under water deficit conditions (Zhu and Xiong 2002). Drought-tolerant legumes tolerate water stress by exhibiting osmotic adjustments through the accumulation of some osmotically active solutes, called osmotica (Ford 1984), including N-containing compounds such as proline, and carbohydrates such as sucrose, and mannitol which accumulate in the plant cells and play a key role in osmotic adjustment process (Taiz and Zeiger 1998; Zhang et al. 1999). The osmotica accumulation is one of the characteristic responses shown by crops in response to prolonged water stress. The accumulation of proline (osmoprotectant) is associated with drought tolerance in many crops (Verbruggen and Hermans 2008), including faba bean (Khalafallah et al. 2008), chickpea (Rozrokh et al. 2012), *Glycine max* (Fukutoku and Yamada 1982), *Phaseolus vulgaris* (Kapuya et al. 1985), *Vigna unguiculata* (Lobato et al. 2008), and other legumes (Simon-Sarkadi et al. 2005; Kim and Nam 2013). In drought-stressed plants, positive correlations between the accumulation of proline and drought tolerance exist. Proline influences the biosynthesis of essential amino acids and metabolic pathways (Simon-Sarkadi et al. 2005). Similarly, free amino acids and many other solutes of low-molecular-weight including pinitol (*o*-methylinositol) are synthesized in tropical legumes under drought stress (Labanauskas et al. 1981; Ford 1984). The manipulation in trehalose metabolism has been suggested as one of the promising strategies to enhance resistance against drought stress in *Medicago truncatula* and in other legume crops (Duque et al. 2013). It is found that drought stress decreased the physiological functions in plants, thereby increasing total soluble sugar (TSS), and accelerating leaf senescence. The accumulations of various osmoregulators, as well as osmoprotectors (sugars, amino acids, organic acids, and inorganic ions like K^+) impart drought tolerance in many legumes (Bray 1997).

The increase of TSS in nodules under drought stress could be due to the low activity of soluble sugar (SS), which restricts the energy supply required for N_2 reduction (Ladrera et al. 2007). In soybean, initially, the decline in SS activity leads to the accumulation of TSS in the root nodules because of failure in carbon (C) metabolism and hence reduction in the supply of energy to the bacteroids. Consequently, the BNF process is damaged in the susceptible genotypes (Cerezini et al. 2016). However, the NFDT soybean genotypes demonstrated more stable physiological processes, more nutrient uptake, higher net photosynthetic rates, highest chlorophyll concentration, and better N-metabolism than the susceptible ones, highlighting the NFDT trait. It has also been reported that drought stress modifies the antioxidant metabolism and induces oxidative stress in root nodules. After 7 days of partial drought stress treatment in pea (*Pisum sativum* L.), the ascorbate (AsA) and dedydroascorbate (DHA) contents as well as ascorbate redox state decreased significantly which indicates the importance of reactive oxygen species (ROS) homeostasis in N fixation in root nodules under drought stress (Marino et al. 2007).

9.2.4 Nitrogen Deficiency

Many tropical soils are not suitable for cultivation due to severe deficiency of soil N. Cultivation of legumes could be an important strategy to improve soil N and put lands into cultivation that were lying barren in those regions. However, all N_2-fixing legumes are especially sensitive to drought stress (Zahran 1999) which was shown to reduce the *nitrogenase* activity by decreasing nodule water potential (Ψ_w, MPa) (Pankhurst and Sprent 1975) and nitrogenase-linked respiration in soybean (Durand et al. 1987), metabolic potential in faba bean (Guerin et al. 1990), total root respiration, and specific N fixation in peas (González et al. 1998). It is also revealed that N_2 fixation is more sensitive to drought stress than photosynthesis in soybean (Djekoun and Planchon 1991).

Crops yield potential can falter under drought stress depending on its severity, duration, and environmental factors (Sinclair and Vadez 2012), while tolerant legume varieties could ensure that crops achieve higher yield through effective BNF under these conditions. Chen et al. (2007) reported the BNF process enabled soybean to record higher yield under drought. Sinclair et al. (2007) observed that more N accumulating $F_{3:4}$ lines maintained higher grain yield when compared with the commercial check varieties under drought indicating the expression of improved symbiotic N_2 fixation and drought-tolerance in the former. Cerezini et al. (2016) demonstrated that soybean NFDT genotypes accumulated higher N under wet conditions, and performed better under drought conditions, which supported the early findings of Cerezini et al. (2014). However, a negative correlation between shoot N under wet conditions and drought tolerance in soybean genotypes was observed by King and Purcell (2006), and King et al. (2014). Usually, the NFDT genotypes maintain higher physiological processes and BNF (Cerezini et al. 2016). This ability can be attributed to better osmotic regulation because of higher N concentrations and surge of K^+ in the roots, and also better translocation or metabolization of ureides.

Drought stress seriously exploits the functioning of BNF process and, consequently, the fixed N declines despite the presence of fully-functional root nodules (Kirda et al. 1989). Generally, roots and root nodules, the key sensors of drought, and the root traits always have a high correlation with the success of legume crops during reproductive stage under drought conditions. Coleto et al. (2014) depicted that nodules are severely affected by the drought stress than other plant parts of legumes (common bean). Several mechanisms are reported to inhibit N-fixation under drought stress. Mainly, water stress induces a decrease in O_2 diffusion or permeability in legume nodules (Durand et al. 1987; Davey and Simpson 1990), which limits nodule functioning and hence the BNF process (Minchin 1997; Denison 1998; González et al. 1995; Gordon et al. 1997). Gálvez et al. (2005) also reported that N fixation is regulated by carbon supply to the bacteroid under drought stress. Water stress reduces the carbon flux in nodule and the respiration of bacteroid contributes carbon supply in nodule (Arrese-Igor et al. 1999). In reality, accumulation of sucrose and depletion of malate take place in nodules as a result of the down-regulation of sucrose synthase activity (González et al. 1995, 1998; Ga´lvez et al. 2005). Reduced stomatal conductance under drought stress is also responsible for progressively weakening of CO_2 assimilation rates (Mouradi et al. 2016).

9.3 Improvement of Drought Tolerance in Relation to N Fixation

The frequency, occurrence, and severity of drought stress hampers plant growth and grain yield of legumes (Ghassemi-Golezani and Mardfar 2008; Demirta et al. 2010; Baroowa and Gogoi 2013; Ghassemi-Golezani et al. 2013; Akter et al. 2021). The effect of drought stress depends on the intensity and duration of drought stress, crop developmental stage, and genotypic variability (adaptive strategies of species). Therefore, it is urgently needed to develop novel approaches to enhance drought tolerance of legumes in order to offset the adverse effects of drought stress. The development of drought-tolerant legume varieties with improved water use efficiency (WUE) may lead to enhanced legume yield and productivity in dry areas (Ulemale et al. 2013).

9.3.1 Drought Tolerant Rhizobium-Legume Symbiosis

The symbiotic relationship between *Rhizobium* and legumes is the leading system of N-fixing in dryland agriculture. As plant growth-promoting rhizobacteria (PGPR) encourage phytohormones (gibberellins, auxins, cytokinins, ABA, and ethylene) biosynthesis which tends to alleviate the harmful effects of drought in legumes (Dimkpa et al. 2009; Khan et al. 2019). Inoculation of *Azospirillum* significantly and positively enhanced proton efflux activities in root systems in soybean and cowpea (Bashan and Alcaraz 1992). It is also noted that the growth-enhancing products like proline is promoted with the inoculation of *Bacillus*, *Burkholderia*, and *Arthrobacter* in pepper (Sziderics et al. 2007) and *Rhizobacterium*, *Burkholderia phytofirmans* strain PsJN in grapevine (Barka et al. 2006). Dimkpa et al. (2009) revealed that the inoculation with rhizobacteria (RB) enhanced the root hair and lateral root development, and improved water and nutrient uptake in plants under abiotic stress conditions.

Under drought stress, PGPRs enhance plant growth through either direct or indirect mechanisms (Glick et al. 2007; Nadeem et al. 2010; Niu et al. 2018), such as N fixation, phosphorus solubilization, production of siderophores, organic acids, and plant growth-promoting compounds as well as important enzymes such as ACC (1-aminocyclopropane-1-carboxylic acid) deaminase, glucanase, and chitinase (Glick et al. 2007; Hayat et al. 2010). RBs convert ACC into a-ketobutyrate and ammonia due to ACC deaminase activity, thereby protecting crop plants from harmful concentrations of ethylene under drought stress (Nadeem et al. 2014). RB hydrolyses ACC and hinders ethylene production and improves root growth in plants (Long et al. 2008).

Bacillus subtilis produces indole acetic acid (IAA), siderophore, phytase, organic acid, ACC deaminase, cyanogens, lytic enzymes, oxalate oxidase, and also solubilizes various sources of organic and inorganic phosphates as well as potassium and zinc, even as it stimulates the production of phytohormones involved in metabolism, growth, and development (Kumar et al. 2011). *Bacillus megaterium* influences plant growth and development by producing phytohormones such as auxins, gibberellins, and cytokinins (Ortíz-Castro et al. 2008). PGPRs are known to promote crop growth and achieve more dry matter content in chickpea while enhancing its root growth and uptake of water and minerals (Hossain et al. 2018).

Arbuscular mycorrhizal fungi (AMF) facilitate in various ways to improve the plant growth, yield, and uptake of water and nutrients under drought stress (Augé 2001), and lend a hand to improve soil structure and soil water retention ability through stabilization and formation of soil aggregates (Smith et al. 2010). In legume crops, AMF increased plant growth and phosphate uptake in marginal soil (Gaur and Adholeya 2002). AMF remarkably supported crops to regulate leaf water potential, solute accumulation, and oxidative stress to lessen the detrimental effects of drought stress as well as to enhance stress tolerance by increasing levels of osmoprotectants in soybean (Porcel and Ruiz-Lozano 2004), and mungbean (Habibzadeh et al. 2014), even as it helped to decrease lipid peroxidation and increase antioxidant potential in soybean (Porcel and Ruiz-Lozano 2004), and chickpea (Sohrabi et al. 2012), and to increase N content in soybean (Aliasgharzad et al. 2006). Under DS, application of *Paenibacillus polymyxa* and *Rhizobium tropici* increased nodulation, N assimilation, and growth in common bean (Figueiredo et al. 2008); *Bradyrhizobium japonicum* increased N contents; *Glomus etunicatum* maintained high leaf water potential in soybean (Aliasgharzad et al. 2006); *Glomus mosseae* improved water-use efficiency in green gram (Habibzadeh 2014) and high leaf water potential in soybean (Aliasgharzad et al. 2006); *Glomus intraradiecs* improved the stomatal conductance in cowpea (Stancheva et al. 2017) and common bean (Augé 2004) and maintained root hydraulic conductance in common bean (Aroca et al. 2006); *Paenibacillus polymyxa* and *Rhizobium tropici* increased nodulation, N contents, and plant growth in common bean (Figueiredo et al. 2008); *Azospirillium brasilense* improved root growth in common bean (German et al. 2000); *Pseudomonas spp.* alleviated drought stress in pea (Chanway et al. 1989; Saikia et al. 2018); *Pseudomonas putida* increased the root growth and ACC-diaminase production (Mayak et al. 1999); and *Gigaspora margarita* improved dehydration maintenance (Augé 2004).

9.3.2 Intensity of Defacement

The response of nodulation and N_2 fixation to water stress depends on the growth stage of the crop and the severity of the stress imposed at various growth stages. Water stress at vegetative growth is more detrimental for nodulation and N-fixation than in the reproduction stage (Pena-Cabriales and Castellanos 1993). Nodule P concentrations along with use efficiency in soybean declined linearly with soil moisture contents especially at maturity stage (Franson et al. 1991). It is found that a mild drought reduced nodules numbers in soybean, while severe water stress resulted in a sharp decline of number as well as the size of nodules (Williams and De Mallorca 1984). Soil moisture deficit has a marked effect on nodules initiation and growth and, hence, the N_2 fixation process. Water stress during vegetative growth has a direct effect on the development of nodules than any other stages, and the possibility of recovery is almost impossible. Moreover, nodules development and N-fixation in legumes depend on plant development stage along with severity and duration of a drought whose adverse effects have a multiplier effect on crop when two or more abiotic stresses coincide such as drought and heat stresses.

Developing drought-tolerant varieties continue to remain the most feasible strategy to increase yield and productivity of legumes under water-limited conditions. It is worth noting that some rhizobial species are drought tolerant even at –3.5Mpa (Abolhasani et al. 2010). The underlying mechanisms which enable rhizobia to cope with osmotic stress are the intracellular accumulation of various organic and inorganic solutes. For instance, *R. meliloti* wards off osmotic stress ill-effects through the accumulation of proline, proline betaine, glycine betaine, glutamate, trehalose, and K^+ (Boscari et al. 2002). Some compatible solutes can be used as either N or carbon sources for growth, suggesting that their catabolism may be regulated to prevent degradation during osmotic stress. This drought tolerating potential of rhizospheric bacteria may be exploited to ameliorate the adverse effects of drought through inducing physio-chemical and agro-botanical characteristics (Yang et al. 2009).

Some specific rhizobial strains improve the performance of legume crops under drought stress. Athar et al. (1996a, 1996b) reported that the plant weight and shoot N content were increased in two alfalfa accessions (*Medicago sativa* L. and *Medicago falcata* L.) when inoculated with *R. meliloti* strains of UL136, UL210 and UL222 under water stress condition. The strains UL136 and UL222 were more equal competitors with indigenous strains for nodulating alfalfa under drought stress. In soybean, several indigenous strains of *B. japonicum* were superior inoculants over commercial strains under drought conditions (Hunt et al. 1988). Therefore, the characteristics of indigenous rhizobial populations and the selection of strains for inoculation may be important determinants of N_2 fixation under drought stress conditions.

Many rhizobial species tend to endure severe drought by adopting strategies such as synthesis of sugars, chaperones, stress enzymes, and exopolysaccharides (Hussain et al. 2014). In addition, they synthesize low molecular weight organic compounds such as trehalose, phytohormones, and siderophores along with phosphate solubilizing compounds (Hussain et al. 2014). Under water deficit conditions, aerobic bacteria utilize N oxides as terminal electron acceptors which enable them to survive anoxia.

A number of temperate and tropical legumes, such as *Medicago sativa* (Aparicio-Tejo and Sanchez-Diaz 1982; Abdel-Wahab and Zahran 1983), *Pisum sativum* (Abdel-Wahab and Zahran 1979), *Arachis hypogaea* (Simpson and Daft 1991), *Vicia faba* (Abdel-Wahab and Zahran 1979; Guerin et al. 1990; Zahran and Sprent 1986), *Glycine max* (Devries et al. 1989; Kirda et al. 1989; Randall Weiss et al. 1985), *Vigna* sp. (Pararjasingham and Knievel 1990, Venkateswarlu et al. 1989), *Aeschynomene* (Albrecht et al. 1981), and the shrub legume *Adenocarpus decorticand* (Moro et al. 1992) demonstrate a reduction in N fixation under drought stress. Nodule initiation, growth, and activity are more sensitive to water stress than the root and shoot metabolism (Zahran and Sprent 1986; Albrecht et al. 1994). Sellstedt et al. (1993) found that N derived from N_2 fixation process was decreased by 26% as a result of water deficiency.

Drought-tolerant N_2-fixing legumes can be selected, although the majority of legumes are sensitive to drought stress. Moisture stress had little or no effect on N_2 fixation by some forage crop legumes, e.g., *M. sativa* (Keck et al. 1984), grain legumes, e.g., groundnut (*Arachis hypogaea*) (Venkateswarlu et al. 1989), and some tropical legumes, e.g., *Desmodium intortum* (Ahmed and Quilt 1980). Guar (*Cyamopsis tetragonoloba*) is a drought-tolerant legume and is known to be adapted to the conditions of prevailing water stress in arid regions (Venkateswarlu et al. 1983). Variability for N fixation under drought stress was found among genotypes of *Vigna radiata* (Rai and Prasad 1983) and *Trifolium repens* (Robin et al. 1989). It is also reported that legume species (determinate or indeterminate structures of nodule) may vary in their response to water stress (Swaraj 1987; Venkateswarlu et al. 1990). Indeterminate nodules, which have prolonged meristematic activity appear somewhat resistant to low soil-water deficit condition (Swaraj 1987). Engin and Sprent (1973) reported that the recovery of nodule activity in white clover (*Trifolium repens*) involves both rehydrations of the existing N_2-fixing tissue and renewed growth of the nodule meristem. In contrast, determinate nodules have limited meristematic activity and appear more sensitive to soil dehydration, with low capacity to recover from water deficits.

Diverse rhizobial strains showed a response to water stress, and hence rhizobial strains with different sensitivity to soil moisture and various ecosystems can be selected for N fixation. The sensitivity to moisture stress varies across the variety of rhizobial strains, such as *R. leguminosarum* bv. trifolii

(Fuhrmann et al. 1986), *R. meliloti* (Busse and Bottomley 1989), cowpea rhizobia (Osa-Afina and Alexander 1982), and *B. japonicum* (Mahler and Wollum 1980). Therefore, suitable rhizobial strains should be identified for optimum soil moisture levels and with respect to their specific legume hosts. Optimization of soil moisture for the growth of the host plant is more important for effective legume-rhizobium symbiosis as legumes tend to be more sensitive to moisture deficit conditions compared to rhizobia (Tate 1995; Arif et al. 2015). However, these findings also inferred that rhizobium-legume symbioses improve soil fertility in arid and semiarid regions and this symbiosis can be further exploited to reclaim marginal soils under changing climate.

The genetic variability among legume species may be induced by cultured cells exposed to various physicochemical mutagens employed under controlled conditions. Recently, *in vitro* based tissue culture techniques have been effective in developing stress-tolerant legume cultivars. Tolerant plants against abiotic stresses can be produced by applying substances like NaCl (for salt tolerance), PEG or mannitol (for drought tolerance) etc. In addition, somaclonal variations may be genetically stable in regenerated plants and can be integrated into breeding drought-tolerant legumes through crop improvement programme (Manoj et al. 2011).

Environmental factors (temperature, precipitation etc.) greatly influence the legume nodulation development as well as N-fixation processes. The genetic make-up determines nodulation which also influences host specificity and host response to environmental upheavals (Spaink et al. 1987). The *nodD* gene can be viewed as a regulatory gene which in conjunction with flavonoid-signal molecules activates transcription of nodulation genes (Long 2001). The flavonoid *nod*-gene inducers are legume specific which is influenced by environmental variables such as soil fertility, pH etc. (Schmidt et al. 1994), while repressor proteins also regulate *nod*-gene inducers (Stacey et al. 2002).

Managing macro-nutrients and micro-nutrients availability as per plants requirement may assist in developing drought tolerance as balanced nutrition tend to mitigate adverse effects of drought (Waraich et al. 2011; Hansel et al. 2017; Saleem et al. 2015). For instance, appropriate provision of N boosts agro-botanical traits and protein contents of chickpea (Palta et al. 2005). In addition, P nutrition trigger photosynthesis process by improving stomatal conductance and leaf water potential along with ensuring root growth and membrane stability of soybean under water-limited conditions (Jin et al. 2006). Likewise, the role of K in imparting tolerance against environmental stresses has been recognized. A study suggests that K has the potential to alleviate the deleterious effects of drought on symbiotic N_2 fixation in *V. faba* and *P. vulgaris* (Sangakkara et al. 1996). The availability of K^+ (0.8 or 0.3 mM) boosted the number of nodules which lead to a substantial increase in the amount of fixed N through BNF under a high-water regime. It has been established that the symbiotic systems are more prone to K limitation compared to legume plants. Moreover, K maintains sufficient tissue water potential which multiplies photosynthesis rate and efficiency in legumes during prolonged drought periods (Sangakkara et al. 1999). The provisions of primary nutrients (N, P, and K) through combined organic and chemical fertilizers improve soil organic carbon (SOC) contents, and ultimately improve the water holding capacity of the soil (Hati et al. 2006; Bandyopadhyay et al. 2010). Similarly, zinc (Zn) foliar application increased yield attributes and grain yield of chickpeas under severe water deficiency (Shaban et al. 2012). Thalooth et al. (2006) also noted the effectiveness of Zn in boosting growth and yield of mungbean. The combined application of iron (Fe) and zinc (Zn) has the potential to improve relative water content and grain protein content of legumes (Yadavi et al. 2014). Among micronutrients, boron (B) as foliage application also triggered nodule development in soybean during drought (Bellaloui and Mengistu 2015; Bellaloui et al. 2013). Among trace elements, selenium (Se) applied in smaller concentrations enhanced water uptake by the root system during mild drought (Mali and Aery 2008; Nawaz et al. 2015). Furthermore, foliage applied Se enhances antioxidant enzymes (GPX and SOD) activity and reduces lipid peroxidation in chickpea and soybean under water-deficient conditions (Mohammadi et al. 2011). The foliar-applied Se also remains effective in promoting the vegetative growth of ageing plants and delays the leaf senescence in wilting plants (Grams et al. 2007). It improves relative water content (RWC) in legumes through biosynthesis of proline and glycine betaine (Hattori et al. 2005). Moreover, silicon (Si) improves the structure of tonoplast and plasma membrane under water deficit (Xu et al. 2015). The combined foliar application of Si and K improved dry matter biomass of chickpea during dry periods (Kurdali et al. 2016; Kurdali et al. 2019). In addition to nutrients, organic solutes also impart tolerance in legumes against abiotic stresses. Legumes vary in their potential to accumulate organic solutes intracellularly under water-stressed environments.

9.4 Management Approaches

Drought stress is considered as one the most challenging factors which adversely affect the growth and productivity of the legume crop worldwide (Dubey et al. 2019). Drought stress can cause yield reduction of up to 50% in arid and semiarid regions (Savita et al. 2020). Plants often have different strategies through morphological, biochemical, physiological, and molecular modifications to cope with the negative impact of drought stresses (Dubey et al. 2020). Studies conducted by different researchers have shown that mechanisms of plant tolerance to drought stresses can be achieved through the development of drought-tolerant plant genotypes, seed treatments, genetic modifications, application of plant growth microbes, plant mineral nutrients, and compatible solutes (Figure 9.1) (Hussain et al. 2018; Dubey et al. 2019, 2020; Savita et al. 2020).

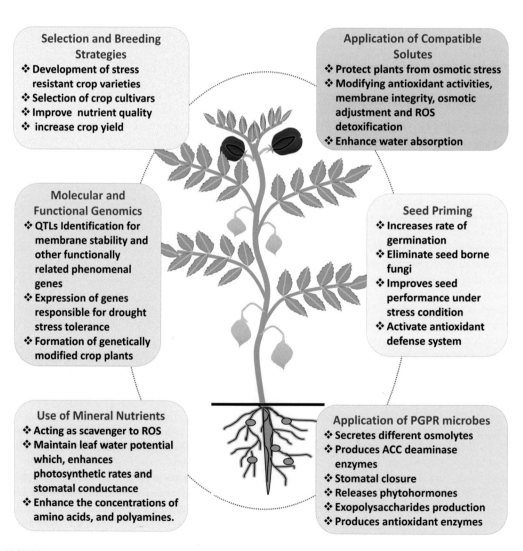

FIGURE 9.1 Different breeding and management strategies for improving drought tolerance in legume plants.

9.4.1 Selection and Breeding Strategies

Crop breeding using conventional and cutting-edge techniques have remained successful in enhancing drought tolerance potential in different legume crops like soybean, chickpea, mung bean, etc. However, these traditional breeding approaches need big investments in labour, land, and money with little success rate. Therefore, integrated use of conventional, molecular marker-aided selection and omic-based approaches can successfully be used for developing drought-tolerant legume genotypes. Traditional breeding has made a noteworthy contribution for soybean enhancement and development of more than 200 improved varieties of chickpea in the last 50 years. Genetic modification of the fatty acid composition of soybean oil has been successful in better meeting the needs of end-users than would have been possible with conventional oil (Fehr 2007). Marker-assisted selection (MAS) can potentially target the desired genes more effectively and accurately. In addition, co-dominant microsatellite-based markers have assisted in genetic mapping of chickpea, which previously remained hindered owing to the varying genome of chickpea cultivars (Dubey et al. 2019). It is expected that the combination of conventional breeding and genomic-assisted breeding will help in the development of climate-resilient crop varieties within a short time.

9.4.2 Molecular and Functional Genomics Approaches

Numerous biochemical and molecular studies have identified different stress-responsive genes and transcription factors responsible for the biosynthesis of the abscisic acid and other stimulations in crop plants. Many fields and laboratory studies have presented that transgenic expression of some of the stress-induced genes with results in augmented drought tolerance. These transgenic approaches are presently the conventional method to bioengineer drought tolerance in different crop plants. The drought tolerance potential of legumes are genetically controlled, and thus many quantitative trait loci involved in maintaining membrane stability and other functional genes have been identified using bioinformatics related tools (Dubey et al. 2019). Judicious use of biotechnological tools holds immense potential in relieving some of the major restraints to the productivity of crops like chickpea, pigeon pea, groundnut, pearl millet, and sorghum (Dubey et al. 2019).

9.4.3 Application of Compatible Solutes

Compatible solutes guard the plants against osmotic stress by setting-off damaging impacts on membranes integrity, biosynthesis of enzymes, and other macromolecules (Ashraf et al. 2011). Compatible solutes include proline, glycine betaine, sugar alcohols, trehalose, soluble sugars, and organic acids (Ahanger et al. 2014). Among different osmolytes, proline has been known to have several roles in plant stress tolerance and regulation. Under severe drought stress conditions, proline accumulates in the cytosol and plays an osmoprotectant role by stabilizing proteins, membranes, and subcellular structures, and shields cellular functions by scavenging on ROS. The positive role of proline in enhancing drought tolerance in pea cultivars has been well testified (Alexieva et al. 2001). Genotypic variation has been reported to be responsible for varying proline accumulation in drought-tolerant and susceptible cultivars of cowpea (Hamidou et al. 2007), chickpea (Mafakheri et al. 2010), and soybean (Masoumi et al. 2011) under varying drought stress levels. However, proline that accumulates in the drought-tolerant cultivars is of higher concentration level than what the susceptible genotypes show. In addition, soluble sugars improve water absorption and reduce osmotic potential under water-deficient conditions, and thus increase stress tolerance in plants. Sugar alcohols and soluble sugars work as signaling molecules, osmoregulatory, and cryoprotectants in plants. Moreover, glycine betaine exogenous application imparts tolerance against abiotic stresses through altering antioxidant activities, osmotic adjustment, boosting membrane integrity, and ROS detoxification (Hossain et al. 2010; Annunziata et al. 2019).

9.4.4 Application of Plant Growth Promoting Microbes

Microbes having growth-promoting potential and ability to survive severe environmental conditions can find their use as biostimulants, especially for grain legumes. The role of these microbes as biostimulants for alleviating the effect of abiotic stress on different legume crop plants has already been recognized (Tiwari et al. 2016; Dubey et al. 2019, 2020; Hashem et al. 2019). These microbes exist in varying ecosystems including in saline, alkaline, and acid soils under arid to tropical climates. Most of these bacteria belong to these genera such as *Rhizobium, Bradyrhizobium, Azospirillum, Azotobacter, Pseudomonas,* and *Bacillus* which have evolved their potential to thrive well under adverse pedoclimatic conditions (Dubey et al. 2020). The study conducted by Porcel and Ruiz-Lozano (2004) reported that AM fungi influence proline accumulation, leaf water potential, and reduced oxidative stress in soybean (*Glycine max*) plants under drought stress condition. The study conducted by Sarma and Saikia (2014) used *Pseudomonas aeruginosa*, and Kumari et al. (2016) used *Pseudomonas simiae* for studying increasing drought stress tolerance in faba bean (*Vicia faba*). The study reported by Tiwari et al. (2016) used *Pseudomonas putida,* Hashem et al. (2019) used AM fungi with biochar, and Kumar et al. (2016) used *Pseudomonas putida* and *Bacillus amyloliquefaciens* for ameliorating the effect of drought stress chickpea (*Cicer arietinum* L.). A similar study conducted by Saikia et al. (2018) used a bacterial consortium including *Bacillus subtilis, Pseudomonas,* and *Ochrobactrum pseudogrignonense* for augmenting the yield of black gram (*Vigna mungo* L.) and pea plant (*Pisum sativum* L.) under drought stress condition.

9.4.5 Seed Priming

Seed priming stimulates metabolic functions related to germination without projection of radicle from the seeds (Hussain et al. 2018). It is an extremely beneficial technique to help crops adapt against stress conditions by modulating physiological and biochemical processes in developing seedlings. Legumes seeds priming with plant hormones can potentially alleviate the drastic effects caused by drought at seedling stages (Hussain et al. 2018). In addition, AsA, salicyclic acid, and H_2O_2 as seed priming agents improved establishment of seedling under heat stress. The priming of bean (*Phaseolus vulgaris*) and tomato (*Lycopersicum esculentum*) seeds with 0.1–0.5 mM SA enhanced drought tolerance potential (Senaratna et al. 2000). The study conducted by Senaratna et al. (2003) used sulfosalicylic acid, methyl salicylic acid, and benzoic acid for testing their efficiency in a monitoring role in persuading multiple stress tolerance in tomato and bean plants. Above indications suggest that seed priming offers an accurate solution to increase the drought stress tolerance in legume plants.

9.4.6 Use of Mineral Nutrients

Mineral nutrients impart tolerance against abiotic stresses in grain legumes. N-fertilization enables plants to combat the adverse effects of abiotic stresses. The availability of N as nitric oxide (NO) has been found to be highly reactive and triggers plants defence systems against stresses through biosynthesis of ROS scavenger (Hussain et al. 2018). Another mineral nutrient phosphorus (P) supports plants to maintain water potential in the leaf which, enhances photosynthetic rates and stomatal conductivity under drought stress condition (Kumar et al. 2015). Plants treated with boron showed enhanced water uptake and stomatal exchange under abiotic stress (Martinez-Ballesta et al. 2008), and increased growth and yield of mungbean (Islam et al. 2017a, 2017b). Waraich et al. (2012) inferred that calcium foliage application in *Vicia faba* improved plant water relations, enhanced chlorophyll contents, and boosted plant biomass via plugging membrane leakage. In addition, foliar-applied silicon increased biomass production of *Vicia faba* under drought stress by improving plant water status (Desoky et al. 2020).

9.5 Conclusion

N is a major limiting factor in crop productivity, however, the continuous use of inorganic-N has become a serious threat to soil and the environment. Therefore, environment-friendly approaches like BNF, organic manures, biofertilizers and similar strategies to bind N to help cultivation of crops should be adopted for healthy and sustainable agriculture. Legume crops are an important source of N in the soil which increases N accumulation through symbiotic N fixation. Legume crops show high sensitivity to drought stress that severely reduces the legume-rhizobium-induced N fixation. Osmotic stress caused by DS changes the morphology of rhizobia, inhibits rhizobium growth, symbiosis and functions, and finally reduces nodulation in legume crops. The water-limited conditions decrease assimilation rate of CO_2, transpiration rate, and stomatal conductance, while it lends a hand in protein denaturation, nucleic acids impairment, and in spurring a significant reduction in plants growth and economic yields of legumes. The biochemical process is also negatively influenced by DS such as where DS inhibits enzymatic activity that leads to the accumulation of TSS due to failure of C metabolism, consequently, and acts as an impediment to the supply of energy to bacteroids and, finally, to weaken the BNF process. Legumes inoculation with plant growth-promoting rhizobacteria has the potential to alleviate the adverse impact of drought. Application of macro-nutrients and micro-nutrients can efficiently mitigate the harmful effects of DS. Therefore, the selection of suitable N-fixing legume crops in the cropping system as well as the application of PGRs and plant nutrients should be done for sustainable agricultural production.

Conflicts of Interest

The authors declare that they have no conflicts of interest to report regarding the present study.

REFERENCES

Abdel-Wahab HH, Zahran HH (1979) Salt tolerance of Rhizobium species in broth culture. *Zeitschrift für allgemeine Mikrobiologie* 19(10):681–685. doi: 10.1002/jobm.19790191002.

Abdel-Wahab HH, Zahran HH (1983) The effects of water stress on N_2 (C_2H_2)-fixation and growth of *Medicago sativa* L. *Acta Agron Acad Sci Hung* 32:114–118.

Abolhasani M, Lakzian A, Tajabadipour A, Haghnia G (2010) The study salt and drought tolerance of Sinorhizobium bacteria to the adaptation to alkaline condition. *Aust J Basic Appl Sci* 4(5):882–886.

Adnan M, Shah Z, Sharif M, Rahman H (2018a). Liming induces carbon dioxide (CO_2) emission in PSB inoculated alkaline soil supplemented with different phosphorus sources. *Environ Sci Poll Res* 25(10):9501–9509.

Adnan M, Zahir S, Fahad S, Arif M, Mukhtar A, Imtiaz AK, Ishaq AM, Abdul B, Hidayat U, Muhammad A, Inayat-Ur R, Saud S, Muhammad ZI, Yousaf J, Amanullah, Hafiz MH, Wajid N (2018b) Phosphate-solubilizing bacteria nullify the antagonistic effect of soil calcification on bioavailability of phosphorus in alkaline soils. *Sci Rep* 8:4339. doi: 10.1038/s41598-018-22653-7.

Adnan M, Fahad S, Khan IA, Saeed M, Ihsan MZ, Saud S, Riaz M, Wang D, Wu C (2019). Integration of poultry manure and phosphate solubilizing bacteria improved availability of Ca bound P in calcareous soils. *3 Biotech* 9(10):368.

Adnan M, Fahad S, Muhammad Z, Shahen S, Ishaq AM, Subhan D, Zafar-ul-Hye M, Martin LB, Raja MMN, Beena S, Saud S, Imran A, Zhen Y, Martin B, Jiri H, Rahul D (2020) Coupling phosphate-solubilizing bacteria with phosphorus supplements improve maize phosphorus acquisition and growth under lime induced salinity stress. *Plants* 9(900). doi: 10.3390/plants9070900.

Ahanger MA, Wani MR, Ahmad P (2014) Drought tolerance: Role of organic osmolytes, growth regulators, and mineral nutrients. In: Ahmad P, Wani MR (eds) *Physiological mechanisms and adaptation strategies in plants under changing environment*, Vol 1. Springer, New York, pp 25–55. doi: 10.1007/978-1-4614-8591-9_2.

Ahmad I, Jadoon SA, Said A, Adnan M, Mohammad F, Munsif F (2017) Response of sunflower varieties to NPK fertilization. *Pure Appl Biol* 6(1):272–277.

Ahmad S, Kamran M, Ding R, Meng X, Wang H, Ahmad I, Fahad S, Han Q (2019) Exogenous melatonin confers drought stress by promoting plant growth, photosynthetic capacity and antioxidant defense system of maize seedlings. *Peer J* 7:e7793. doi: 10.7717/peerj.7793.

Ahmed B, Quilt P (1980) Effect of soil moisture stress on yield, nodulation and nitrogenase activity of *Macroptilium atropurpureum* cv. Sirato and *Desmodium intortum* cv. Greenleaf. *Plant Soil* 57:187–194.

Akbar H, Timothy JK, Jagadish T, Golam M, Apurbo KC, Muhammad F, Rajan B, Fahad S, Hasanuzzaman M (2020) Agricultural land degradation: Processes and problems undermining future food security. In: Fahad S, Hasanuzzaman M, Alam M, Ullah H, Saeed M, Khan AK, Adnan M (eds) *Environment, climate, plant and vegetation growth*. Springer Publ Ltd, Springer Nature Switzerland AG. Part of Springer Nature, pp. 17–62. doi: 10.1007/978-3-030-49732-3.

Akram R, Turan V, Hammad HM, Ahmad S, Hussain S, Hasnain A, Maqbool MM, Rehmani MIA, Rasool A, Masood N, Mahmood F, Mubeen M, Sultana SR, Fahad S, Amanet K, Saleem M, Abbas Y, Akhtar HM, Waseem F, Murtaza R, Amin A, Zahoor SA, ul Din MS, Nasim W (2018a) Fate of organic and inorganic pollutants in paddy soils. In: Hashmi MZ, Varma A (eds) *Environmental pollution of paddy soils, soil biology*. Springer International Publishing AG, GEWERBESTRASSE 11, CHAM, CH-6330, Switzerland, pp. 197–214.

Akram R, Turan V, Wahid A, Ijaz M, Shahid MA, Kaleem S, Hafeez A, Maqbool MM, Chaudhary HJ, Munis MFH, Mubeen M, Sadiq N, Murtaza R, Kazmi DH, Ali S, Khan N, Sultana SR, Fahad S, Amin A, Nasim W (2018b) Paddy land pollutants and their role in climate change. In: Hashmi MZ, Varma A (eds) *Environmental pollution of paddy soils, soil biology*. Springer International Publishing AG, GEWERBESTRASSE 11, CHAM, CH-6330, Switzerland, pp. 113–124.

Akter S, Jahan I, Hossain MA, Hossain MA (2021) Laboratory-and field-phenotyping for drought stress tolerance and diversity study in lentil (*Lens culinaris* Medik.). *Phyton-Int J Exp Botany* (in press).

Albrecht SL, Bennett JM, Quesenberry KH (1981) Growth and nitrogen fixation of Aeschynomene under water stressed conditions. *Plant Soil* 60:309–315.

Albrecht SL, Bennett JM, Boote KJ (1994) Relationship of nitrogenase activity to plant water stress in field grown soybeans. *Field Crop Res* 8:61–71.

Alexieva V, Sergiev I, Mapelli S, Karanov E (2001) The effect of drought and ultraviolet radiation on growth and stress markers in pea and wheat. *Plant Cell Environ* 24(12):1337–1344. doi: 10.1046/j.1365-3040.2001.00778.x.

Al-Falih AMK (2002) Factors affecting the efficiency of symbiotic nitrogen fixation by Rhizobium. *Pak J Biol Sci* 5(11):1277–1293.

Aliasgharzad N, Neyshabouri MR, Salimi G (2006) Effects of arbuscular mycorrhizal fungi and *Bradyrhizobium japonicum* on drought stress of soybean. *Biol Bratislava* 61:324–328.

Al-Wabel MI, Abdelazeem S, Munir A, Khalid E, Adel RAU (2020a) Extent of climate change in Saudi Arabia and its impacts on agriculture: A case study from Qassim Region. In: Fahad S, Hasanuzzaman M, Alam M, Ullah H, Saeed M, Khan AK, Adnan M (eds) *Environment, climate, plant and vegetation growth*. Springer Publ Ltd, Springer Nature Switzerland AG. Part of Springer Nature, pp. 635–658. doi: 10.1007/978-3-030-49732-3.

Al-Wabel MI, Ahmad M, Usman ARA, Akanji M, Rafique MI (2020b) Advances in pyrolytic technologies with improved carbon capture and storage to combat climate change. In: Fahad S, Hasanuzzaman M, Alam M, Ullah H, Saeed M, Khan AK, Adnan M (eds) *Environment, climate, plant and vegetation growth*. Springer Publ Ltd, Springer Nature Switzerland AG. Part of Springer Nature, pp. 535–576. doi: 10.1007/978-3-030-49732-3.

Amanullah S, Imran K, Hamdan AK, Muhammad A, Abdel RA, Muhammad A, Fahad S, Azizullah S, Brajendra P (2020) Effects of climate change on irrigation water quality. In: Fahad S, Hasanuzzaman M, Alam M, Ullah H,Saeed M, Khan AK, Adnan M (eds) *Environment, climate, plant and vegetation growth*. Springer Publ Ltd, Springer Nature Switzerland AG. Part of Springer Nature. pp. 123–132. doi: 10.1007/978-3-030-49732-3.

Amir M, Muhammad A, Allah B, Sevgi Ç, Haroon ZK, Muhammad A, Emre A (2020) Bio fortification under climate change: The fight between quality and quantity. In: Fahad S, Hasanuzzaman M, Alam M, Ullah H, Saeed M, Khan AK, Adnan M (eds) *Environment, climate, plant and vegetation growth*. Springer Publ Ltd, Springer Nature Switzerland AG. Part of Springer Nature, pp. 173–228. doi: 10.1007/978-3-030-49732-3.

Amjad I, Muhammad H, Farooq S, Anwar H (2020) Role of plant bioactives in sustainable agriculture. In: Fahad S, Hasanuzzaman M, Alam M, Ullah H, Saeed M, Khan AK, Adnan M (eds) *Environment, climate, plant and vegetation growth*. Springer Publ Ltd, Springer Nature Switzerland AG. Part of Springer Nature, pp. 591–606. doi: 10.1007/978-3-030-49732-3.

Andrews M, Lea PJ, Raven JA, Lindsey K (2004) Can genetic manipulation of plant nitrogen assimilation enzymes result in increased crop yield and greater N-use efficiency? An assessment. *Ann Appl Biol* 145:25–40.

Annunziata MG, Ciarmiello LF, Woodrow P, Dell'Aversana E, Carillo P (2019) Spatial and temporal profile of glycine betaine accumulation in plants under abiotic stresses. *Front Plant Sci* 10:230.

Aparicio-Tejo P, Sanchez-Diaz M (1982) Nodule and leaf nitrate reductase and nitrogen fixation in *Medicago sativa* L. under water stress. *Plant Physiol* 69:479–482.

Arif M, Shah T, Ilyas M, Ahmad W, Mian AA, Jadoon MA, Adnan M (2015) Effect of organic manures and their levels on weeds density and maize yield. *Pak J Weed Sci Res* 21(4):517–522.

Arif M, Talha J, Muhammad R, Fahad S, Muhammad A, Amanullah, Kawsar A, Ishaq AM, Bushra K, Fahd R (2020) Biochar; a remedy for climate change. In: Fahad S, Hasanuzzaman M, Alam M, Ullah H, Saeed M, Khan AK, Adnan M (eds) *Environment, climate, plant and vegetation growth*. Springer Publ Ltd, Springer Nature Switzerland AG. Part of Springer Nature, pp. 151–172. doi: 10.1007/978-3-030-4 9732-3.

Aroca R, Porcel R, Ruiz-lozano JM, Aroca, R (2006) How does arbuscular mycorrhizal symbiosis regulate root hydraulic properties and plasma membrane aquaporins in *Phaseolus vulgaris* under drought, cold or salinity stresses? *New Phytol* 173:808–816.

Arrese-Igor C, González EM, Gordon AJ, Minchin FR, Gálvez L, Royuela M, Cabrerizo PM, Aparicio-Tejo PM (1999) Sucrose synthase and nodule nitrogen fixation under drought and other environmental stresses. *Symbiosis* 27:189–212.

Ashraf M, Akram NA, Al-Qurainy F, Foolad MR (2011) Drought tolerance. Roles of organic osmolytes, growth regulators, and mineral nutrients. *Adv Agron* 111:249–296. doi: 10.1016/B978-0-12-387689-8.00002-3.

Athar M, Johnson DA (1996a) Influence of drought on competition between selected Rhizobium meliloti and naturalized soil rhizobia in alfalfa. *Plant Soil* 184:231–241.

Athar M, Johnson DA (1996b) Nodulation, biomass production, and nitrogen fixation in alfalfa under drought. *J Plant Nutr* 19:185–199.

Augé RM (2001) Water relations, drought and vesicular-arbuscular mycorrhizal symbiosis. *Mycorrhiza* 11:3–42. doi: 10.1007/s005720100097.

Augé RM (2004) Arbuscular mycorrhizae and soil/plant water relations. *Can J Soil Sci* 84(4):373–381. doi: 10.4141/S04-002.

Aziz K, Daniel KYT, Fazal M, Muhammad ZA, Farooq S, Fan W, Fahad S, Ruiyang Z (2017a) Nitrogen nutrition in cotton and control strategies for greenhouse gas emissions: A review. *Environ Sci Pollut Res* 24:23471–23487. doi: 10.1007/s11356-017-0131-y.

Aziz K, Daniel KYT, Muhammad ZA, Honghai L, Shahbaz AT, Mir A, Fahad S (2017b) Nitrogen fertility and abiotic stresses management in cotton crop: A review. *Environ Sci Pollut Res* 24:14551–14566. doi: 10.1007/s11356-017-8920-x.

Bandyopadhyay KK, Misra AK, Ghosh PK, Hati KM (2010) Effect of integrated use of farmyard manure and chemical fertilizers on soil physical properties and productivity of soybean. *Soil Tillage Res* 110:115–125.

Bano SA, Iqbal SM (2016) Biological nitrogen fixation to improve plant growth and productivity. *Int J Agric Innov Res* 4:596–599.

Barka EA, Nowak J, Cle C (2006) Enhancement of chilling resistance of inoculated grapevine plantlets with a plant growth-promoting rhizobacterium, *Burkholderia phytofirmans* strain PsJN. *Appl Environ Microbiol* 72:7246–7252.

Baroowa B, Gogoi N (2013) Biochemical changes in two *Vigna spp.* during drought and subsequent recovery. *Indian J Plant Physiol* 18:319–325.

Baseer M, Adnan M, Fazal M, Fahad S, Muhammad S, Fazli W, Muhammad A, Jr. Amanullah, Depeng W, Saud S, Muhammad N, Muhammad Z, Fazli S, Beena S, Mian AR, Ishaq AM (2019) Substituting urea by organic wastes for improving maize yield in alkaline soil. *J Plant Nutr*. doi: 10.1080/01904167.201 9.1659344.

Bashan Y, Alcaraz L (1992) Responses of soybean and cowpea root membranes to inoculation with *Azospirillum brasilense*. *Symbiosis* 13:217–228.

Bayram AY, Seher Ö, Nazlican A (2020) Climate change forecasting and modeling for the year of 2050. In: Fahad S, Hasanuzzaman M, Alam M, Ullah H,Saeed M, Khan AK, Adnan M (eds) *Environment, climate, plant and vegetation growth*. Springer Publ Ltd, Springer Nature Switzerland AG. Part of Springer Nature, pp. 109–122. doi: 10.1007/978-3-030-49732-3.

Bellaloui N, Mengistu A (2015) Effects of boron nutrition and water stress on nitrogen fixation, seed δ15N and δ13C dynamics, and seed composition in soybean cultivars differing in maturities. *Sci World J* 2015:407872.

Bellaloui N, Hu Y, Mengistu A, Kassem MA, Abel CA (2013) Effects of foliar boron application on seed composition, cell wall boron, and seed δ15 N and δ13 C isotopes in water-stressed soybean plants. *Front Plant Sci* 4:270.

Boscari A, Mandon K, Dupont L, Poggi MC, Le Rudulier D (2002) Bet S is a major glycine betaine/proline betaine transporter required for early osmotic adjustment in *Sinorhizobium meliloti*. *J Bacteriol* 184:2654–2663.

Bray E (1997) Plant responses to water deficit. *Trends Plant Sci* 2(2):48–54.

Busse MD, Bottomley PJ (1989) Growth and nodulation responses of *Rhizobium meliloti* to water stress induced by permeating and nonpermeating solutes. *Appl Environ Microbiol* 55:2431–2436.

Casteriano AV (2014) *Physiological mechanisms of desiccation tolerance in Rhizobia*. PhD thesis, University Sydney, Australia.

Cerezini P, Pípolo AE, Hungria M, Nogueira MA (2014) Gas exchanges and biological nitrogen fixation in soybean under water restriction. *Ame J Plant Sci* 5:4011–4017.

Cerezini P, Dáfila dos Santos LF, Antonio EP, Mariangela H, Marco AN (2016) Water restriction and physiological traits in soybean genotypes contrasting for nitrogen fixation drought tolerance. *Sci Agric* 74(2):110–117.

Chanway CP, Hynes RK, Nelson LM (1989) Plant growth-promoting rhizobacteria: Effects on growth and nitrogen fixation of lentil (*Lens esculenta* Moench) and pea (*Pisum sativum* L.). *Soil Biol Biochem* 21:511–517.

Chanway CP, Anand R, Yang H (2014) *Nitrogen fixation outside and inside plant tissues: Advances in biology and ecology of nitrogen fixation*. Ohyama T (ed) InTech Open. doi: 10.5772/57532.

Chen P, Sneller CH, Purcell LC, Sinclair TR, King CA, Ishibashi T (2007) Registration of soybean germplasm lines R01-416F and R01-581F for improved yield and nitrogen fixation under drought stress. *J Plant Registra* 1:166–167.

Coleto I, Pineda M, Rodino AP, De Ron AM, Alamillo JM (2014) Comparison of inhibition of N_2 fixation and ureide accumulation under water deficit in four common bean genotypes of contrasting drought tolerance. *Ann Bot* 103:1071–1082.

Collier R, Tegeder M (2012) Soybean ureide transporters play a critical role in nodule development, function and nitrogen export. *The Plant J* 72:355–367.

Date RA (2000) Inoculated legumes in cropping systems of the tropics. *Field Crops Res* 65:123–136.

Davey AG, Simpson RJ (1990) Nitrogen fixation by subterranean clover at varying stages of nodule dehydration. II Efficiency of nitrogenase functioning. *J Exp Bot* 41:1189–1197.

Demirta C, Yazgan S, Candogan BN, Sincik M (2010) Quality and yield response of soybean (*Glycine max* L. Merrill) to drought stress in sub-humid environment. *Afr J Biotechnol* 9:6873–6881.

Denison RF (1998) Decreased oxygen permeability: A universal stress response in legume root nodules. *Botan Acta* 111:191–192.

Depeng W, Fahad S, Saud S, Muhammad K, Aziz K, Mohammad NK, Hafiz MH, Wajid N (2018) Morphological acclimation to agronomic manipulation in leaf dispersion and orientation to promote "Ideotype" breeding: Evidence from 3D visual modeling of "super" rice (*Oryza sativa* L.). *Plant Physiol Biochem*. 135:499–510. doi: 10.1016/j.plaphy.2018.11.010.

Desoky ESM, Mansour E, Yasin MAT, Elsobky EA, Rady M (2020) Improvement of drought tolerance in five different cultivars of Vicia faba with foliar application of ascorbic acid or silicon. *Spanish J Agric Res* 18(2):e0802. doi: 10.5424/sjar/2020182-16122.

Devries JD, Bennett JM, Albrecht SL, Boote KJ (1989) Water relations, nitrogenase activity and root development of three grain legumes in response to soil water deficits. *Field Crop Res* 21:215–226.

Dimkpa C, Weinand T, Asch F (2009) Plant-rhizobacteria interactions alleviate abiotic stress conditions. *Plant Cell Environ* 32:1682–1694.

Djekoun A, Planchon C (1991) Water status effect on dinitrogen fixation and photosynthesis in soybean. *Agron J* 83:316–322.

Dubey A, Malla MA, Khan F et al (2019) Soil microbiome: A key player for conservation of soil health under changing climate. *Biodivers Conserv*. doi: 10.1007/s10531-019-01760-5.

Dubey A, Kumar A, Khan ML (2020) Role of biostimulants for enhancing abiotic stress tolerance in Fabaceae plants: The plant family Fabaceae. pp. 223–236. doi: 10.1007/978-981-15-4752-2_8.

Duque AS, Almeida A, Silva AB, Silva JM, Farinha AP, Santos D, Fevereiro P, Araujo SS (2013) Abiotic stress responses in plants: Unraveling the complexity of genes and networks to survive. In: Vahdati K, Leslie C (eds) *Abiotic stress: Plant responses and applications in agriculture*. InTech Open. doi: 10.5 772/52779.

Durand JL, Sheehy JE, Minchin FR (1987) Nitrogenase activity, photosynthesis and nodule water potential in soybean plants experiencing water deprivation. *J Exp Bot* 38:311–321.

Eaglesham ARJ, Ayanaba A (1984) Tropical stress ecology of rhizobia, root elongation and legume fixation. In: Subra Rao NS (ed) *Currents developments in biological nitrogen fixation*. Edward Arnold, Maryland, USA. pp. 1–35.

EL Sabagh A, Hossain A, Barutçular C, Islam MS, Ratnasekera D, Kumar N, Meena RS, Gharib HS, Saneoka H, Teixeira da Silva JA (2019a) Drought and salinity stress management for higher and sustainable canola (*Brassica napus* L.) production: A critical review. *Aust J Crop Sci* 13(1): 88–96.

EL Sabagh A, Hossain A, Islam MS, Barutçular C, Ratnasekera D, Kumar N, Meena RS, Gharib HS, Saneoka H, Teixeira da Silva JA (2019b) Sustainable soybean production and abiotic stress management in saline environments: A critical review. *Aust J. Crop Sci* 13(2):228–236.

EL Sabagh A et al (2020a) Drought and heat stress in cotton (*Gossypium hirsutum* L.): Consequences and their possible mitigation strategies. In Hasanuzzaman M (ed) *Agronomic Crops*. Springer, Singapore. doi: 10.1007/978-981-15-0025-1_30.

EL Sabagh A, Hossain A, Barutçular C, Islam MS, Ahmad Z, Wasaya A, Meena RS (2020b) Adverse effect of drought on quality of major cereal crops: Implications and their possible mitigation strategies. In *Agronomic crops*. Springer, Singapore, pp. 635–658. doi: 10.1007/978-981-15-0025-1_31.

EL Sabagh AE, Hossain A, Barutçular C, Iqbal MA, Islam MS, Fahad S,… & Erman M (2020c) Consequences of salinity stress on the quality of crops and its mitigation strategies for sustainable crop production: An outlook of arid and semi-arid regions. In: *Environment, climate, plant and vegetation growth*. Springer, Cham, pp. 503–533.

EL Sabagh A, Hossain A, Barutçular C, Iqbal MA, Islam MS, Fahad S, Sytar O, Çig F, Meena RS, Erman M (2020d) Consequences of salinity stress on the quality of crops and its mitigation strategies for sustainable crop production: An outlook of arid and semi-arid regions. In: Fahad S, Hasanuzzaman M, Alam M, Ullah H, Saeed M, Khan AK, Adnan M (eds) *Environment, climate, plant and vegetation growth*. Springer Publ Ltd, Springer Nature Switzerland AG. Part of Springer Nature, pp. 503–534. doi: 10.1007/978-3-030-49732-3.

EL Sabagh A, Hossain A, Islam MS, Fahad S, Ratnasekera D, Meena RS, Wasaya A, Yasir TA, Ikram M, Mubeen M, Fatima M, Nasim W, Çig A, Çig F, Erman M, Hasanuzzaman M (2020e) Nitrogen fixation of legumes under the family Fabaceae: Adverse effect of abiotic stresses and mitigation strategies. In Hasanuzzaman M, Araújo S, Gill S (eds) *The plant family Fabaceae*. Springer, Singapore. pp 75–111.

Elboutahiri N, Thami-Alami I, Udupa SM (2010) *Phenotypic and genetic diversity in Sinorhizobium meliloti and S. medicae from drought and salt-affected regions of Morocco*, Institut National de la RechercheAgronomique (INRA).

El-Tayeb MA (2006a) Differential responses of pigments, lipid per-oxidation, organic solutes, catalase and per-oxidase activity in the leaves of two *Vicia faba* L. cultivars to drought. *Int J Agric Biol* 8(1):126-122. doi: http://www.fspublishers.org 1560-8530/2006/08-1-116-122.

El-Tayeb MA (2006b) Differential response of two *Vicia faba* cultivars to drought: Growth, pigments, lipid peroxidation, organic solutes, catalase and peroxidase activity. *Acta Agron Hungarica* 54:25–37. doi: 10.1556/AAgr.54.2006.1.3.

Engin M, Sprent JI (1973) Effects of water stress on growth and nitrogen fixing activity of Trifolium repens. *New Phytologist* 72:117–126.

Fahad S, Bano A (2012) Effect of salicylic acid on physiological and biochemical characterization of maize grown in saline area. *Pak J Bot* 44:1433–1438.

Fahad S, Chen Y, Saud S,Wang K, Xiong D, Chen C,Wu C, Shah F, Nie L, Huang J (2013) Ultraviolet radiation effect on photosynthetic pigments, biochemical attributes, antioxidant enzyme activity and hormonal contents of wheat. *J Food, Agri Environ* 11(3&4):1635– 1641.

Fahad S, Hussain S, Bano A, Saud S, Hassan S, Shan D, Khan FA, Khan F, Chen Y, Wu C, Tabassum MA, Chun MX, Afzal M, Jan A, Jan MT, Huang J (2014a) Potential role of phytohormones and plant growth-promoting rhizobacteria in abiotic stresses: Consequences for changing environment. *Environ Sci Pollut Res* 22(7):4907– 4921. doi: 10.1007/s11356-014-3754-2.

Fahad S, Hussain S, Matloob A, Khan FA, Khaliq A, Saud S, Hassan S, Shan D, Khan F, Ullah N, Faiq M, Khan MR, Tareen AK, Khan A, Ullah A, Ullah N, Huang J (2014b) Phytohormones and plant responses to salinity stress: A review. *Plant Growth Regul* 75(2):391– 404. doi: 10.1007/s10725-014-0013-y.

Fahad S, Hussain S, Saud S, Tanveer M, Bajwa AA, Hassan S, Shah AN, Ullah A,Wu C, Khan FA, Shah F, Ullah S, Chen Y, Huang J (2015a) A biochar application protects rice pollen from high-temperature stress. *Plant Physiol Biochem* 96:281–287.

Fahad S, Nie L, Chen Y, Wu C, Xiong D, Saud S, Hongyan L, Cui K, Huang J (2015b) Crop plant hormones and environmental stress. *Sustain Agric Rev* 15:371–400.

Fahad S, Hussain S, Saud S, Hassan S, Chauhan BS, Khan F et al (2016a) Responses of rapid viscoanalyzer profile and other rice grain qualities to exogenously applied plant growth regulators under high day and high night temperatures. *PLoS One* 11(7):e0159590. doi: 10.1371/journal.pone.0159590.

Fahad S, Hussain S, Saud S, Hassan S, Ihsan Z, Shah AN,Wu C, Yousaf M, Nasim W, Alharby H, Alghabari F, Huang J (2016b) Exogenously applied plant growth regulators enhance the morphophysiological growth and yield of rice under high temperature. *Front Plant Sci* 7:(1250). doi: 10.3389/fpls.201 6.01250.

Fahad S, Hussain S, Saud S, Hassan S, Tanveer M, Ihsan MZ, Shah AN, Ullah A, Nasrullah KF, Ullah S, Alharby HNW, Wu C, Huang J (2016c) A combined application of biochar and phosphorus alleviates heat-induced adversities on physiological, agronomical and quality attributes of rice. *Plant Physiol Biochem* 103:191–198.

Fahad S, Hussain S, Saud S, Khan F, Hassan S, Jr A, Nasim W, Arif M, Wang F, Huang J (2016d) Exogenously applied plant growth regulators affect heat-stressed rice pollens. *J Agron Crop Sci* 202:139– 150.

Fahad S, Bajwa AA, Nazir U, Anjum SA, Farooq A, Zohaib A, Sadia S, Nasim W, Adkins S, Saud S, Ihsan MZ, Alharby H, Wu C, Wang D, Huang J (2017) Crop production under drought and heat stress: Plant responses and management options. *Front Plant Sci* 8:1147. doi: 10.3389/fpls.2017.01147.

Fahad S, Muhammad ZI, Abdul K, Ihsanullah D, Saud S, Saleh A, Wajid N, Muhammad A, Imtiaz AK, Chao W, Depeng W, Jianliang H (2018) Consequences of high temperature under changing climate optima for rice pollen characteristics-concepts and perspectives. *Arch Agron Soil Sci*. doi: 10.1080/03650340.2 018.1443213.

Fahad S, Adnan M, Hassan S, Saud S, Hussain S, Wu C, Wang D, Hakeem KR, Alharby HF, Turan V, Khan MA, Huang J (2019a) Rice responses and tolerance to high temperature. In: Hasanuzzaman M, Fujita M, Nahar K, Biswas JK (eds) *Advances in rice research for abiotic stress tolerance.* Woodhead Publ Ltd, ABINGTON HALL ABINGTON, CAMBRIDGE CB1 6AH, CAMBS, England, pp. 201–224.

Fahad S, Rehman A, Shahzad B, Tanveer M, Saud S, Kamran M, Ihtisham M, Khan SU, Turan V, Rahman MHU (2019b) Rice responses and tolerance to metal/metalloid toxicity. In Hasanuzzaman M, Fujita M, Nahar K, Biswas JK (eds) *Advances in rice research for abiotic stress tolerance.* Woodhead Publ Ltd, ABINGTON HALL ABINGTON, CAMBRIDGE CB1 6AH, CAMBS, England, pp. 299–312.

Farah R, Muhammad R, Muhammad SA, Tahira Y, Muhammad AA, Maryam A, Shafaqat A, Rashid M, Muhammad R, Qaiser H, Afia Z, Muhammad AA, Muhammad A, Fahad S (2020) Alternative and non-conventional soil and crop management strategies for increasing water use efficiency. In: Fahad S, Hasanuzzaman M, Alam M, Ullah H, Saeed M, Khan AK, Adnan M (eds) *Environment, climate, plant and vegetation growth.* Springer Publ Ltd, Springer Nature Switzerland AG. Part of Springer Nature, pp. 323–338. doi: 10.1007/978-3-030-49732-3.

Farhana G, Ishfaq A, Muhammad A, Dawood J, Fahad S, Xiuling L, Depeng W, Muhammad F, Muhammad F, Syed AS (2020) Use of crop growth model to simulate the impact of climate change on yield of various wheat cultivars under different agro-environmental conditions in Khyber Pakhtunkhwa, Pakistan. *Arabian J Geosci* 13:112. doi: 10.1007/s12517-020-5118-1.

Farhat A, Hafiz MH, Wajid I, Aitazaz AF, Hafiz FB, Zahida Z, Fahad S, Wajid F, Artemi C (2020) A review of soil carbon dynamics resulting from agricultural practices. *J Environ Manage* 268(2020): 110319.

Fazli W, Muhmmad S, Amjad A, Fahad S, Muhammad A, Muhammad N, Ishaq AM, Imtiaz AK, Mukhtar A, Muhammad S, Muhammad I, Rafi U, Haroon I, Muhammad A (2020) Plant-microbes interactions and functions in changing climate. In: Fahad S, Hasanuzzaman M, Alam M, Ullah H, Saeed M, Khan AK, Adnan M (eds) *Environment, climate, plant and vegetation growth.* Springer Publ Ltd, Springer Nature Switzerland AG. Part of Springer Nature, pp. 397–420. doi: 10.1007/978-3-030-49732-3.

Fehr WR (2007) Breeding for modified fatty acid composition in soybean. In: *Crop science. Crop science society of America 677 S. Segoe Rd., Madison, WI 53711, USA.* 47(S3) S72–S87. doi: 10.2135/cropsci2 007.04.0004IPBS .

Figueiredo VB, Burity A, Martı CR, Chanway CP (2008) Alleviation of drought stress in the common bean (*Phaseolus vulgaris* L.) by co-inoculation with Paenibacillus polymyxa and Rhizobium tropici. *Appl Soil Ecol* 40(1):182–188. doi: 10.1016/j.apsoil.2008.04.005.

Ford CW (1984) Accumulation of low molecular weight solutes in water stressed tropical legumes. *Phytochem* 23:1007–1015. doi: 10.1016/S0031-9422(00)82601-1.

Franson RL, Brown MS, Bethlenfalvay GJ (1991) The Glycine-Glomus-Bradyrhizobium symbiosis. XI. Nodule gas exchange and efficiency as a function of soil and root water status in mycorrhizal soybean. *Physiol Planta* 83:476–482. doi: 10.1111/j.1399-3054.1991.tb00123.x.

Fuhrmann J, Davey CB, Wollum AG (1986) Desiccation tolerance in clover rhizobia in sterile soils. *Soil Sci Soc Am J* 50:639–644. doi: 10.2136/sssaj1986.03615995005000030019x.

Fuhrmann J, Wollum AG (1989) Nodulation competition among *Bradyrhizobium japonicum* strains as influenced by rhizosphere bacteria and iron availability. *Biol Fert Soils* 7:108–112. doi: 10.1007/BF002 92567.

Fukutoku Y, Yamada Y (1982) Accumulation of carbohydrates and proline in water stressed soybean (*Glycine max* L.). *Soil Sci Plant Nutr* 28(1):147–151. doi: 10.1080/00380768.1982.10432380.

Gálvez L, González EM, Arrese-Igor C (2005) Evidence for carbon flux shortage and strong carbon/nitrogen interactions in pea nodules at early stages of water stress. *J Exp Bot* 56(419):2551–2561. doi: 10.1093/ jxb/eri249.

Gaur A, Adholeya A (2002) Arbuscular-mycorrhizal inoculation of five tropical fodder crops and inoculums production in marginal soil amended with organic matter. *Biol Fertil Soils* 35:214–218. doi: 10.1007/ s00374-002-0457-5.

German MA, Burdman S, Okon Y (2000) Effects of *Azospirillum brasilense* on root morphology of common bean (*Phaseolus vulgaris* L.) under different water regimes. *Biol Fertil Soils* 32:259–264. doi: 10.1007/ s003740000245.

Ghassemi-Golezani K, Mardfar RA (2008) Effects of limited irrigation on growth and grain yield of common bean. *J Plant Sci* 3:230–235.

Ghassemi-Golezani K, Ghassemi S, Bandehhagh A (2013) Effects of water supply on field performance of chickpea (*Cicer arietinum* L.) cultivars. *Int J Agron Plant Prod* 4:94–97. doi: http://eprints.icrisat.ac.in /id/eprint/9503.

Glick BR, Cheng Z, Czarny J (2007) Promotion of plant growth by ACC deaminase-producing soil bacteria. *Eur J Plant Pathal* 119:329–339.

González EM, Gordon AJ, James CL, Arrese-Igor C (1995) The role of sucrose synthase in the response of soybean nodules to drought. *J Exp Bot* 46:1515–1523. doi:10.1093/jxb/46.10.1515.

González EM, Aparicio-Tejo PM, Gordon AJ, Minchin FR, Royuela M, Arrese-Igor C (1998) Water-deficit effects on carbon and nitrogen metabolism of pea nodules. *J Exp Bot* 49(327):1705–1714. doi: 10.1093/ jexbot/49.327.1705.

Gopakumar L, Bernard NO, Donato V (2020) Soil microarthropods and nutrient cycling. In: Fahad S, Hasanuzzaman M, Alam M, Ullah H, Saeed M, Khan AK, Adnan M (cds) *Environment, climate, plant and vegetation growth.* Springer Publ Ltd, Springer Nature Switzerland AG. Part of Springer Nature, pp. 453–472. doi: 10.1007/978-3-030-49732-3.

Gordon AJ, Minchin FR, Skot L, James CL (1997) Stress-induced declines in soybean N_2 fixation are related to nodule sucrose synthase activity. *Plant Physiol* 114:937–946. doi: 10.1104/pp.114.3.937.

Graham PH, Vance CP (2000) Nitrogen fixation in perspective, an overview of research and extension needs. *Field Crops Res* 65:93–106. doi: 10.1016/S0378-4290(99)00080-5.

Grams TEE, Koziolek C, Lautner S, Matyssek R (2007) Distinct roles of electric and hydraulic signals on the reaction of leaf gas exchange upon re-irrigation in *Zea mays* L. *Plant Cell Environ* 30(1):79–84. doi: 10.1111/j.1365-3040.2006.01607.x.

Guerin V, Trinchant JC, Rigaud J (1990) Nitrogen fixation (C_2H_2 reduction) by broad bean (*Vicia faba* L.) nodules and bacteriods under water restricted conditions. *Plant Physiol* 92(2):595–601. doi: 10.1104/pp.92.3.595.

Habib ur R, Ashfaq A, Aftab W, Manzoor H, Fahd R, Wajid I, Md. Aminul I, Vakhtang S, Muhammad A, Asmat U, Abdul W, Syeda RS, Shah S, Shahbaz K, Fahad S, Manzoor H, Saddam H, Wajid N (2017) Application of CSM-CROPGRO-Cotton model for cultivars and optimum planting dates: Evaluation in changing semi-arid climate. *Field Crops Res*. doi: 10.1016/j.fcr.2017.07.007.

Habibzadeh Y (2014) Response of mung bean plants to arbuscular mycorrhiza and phosphorus in drought stress. *Int J Innov Appl Stud* 6(1):14–20.

Intergovernmental Panel on Climate Change (IPCC) (2007) Climate change: The physical science basis. In: *Contribution of working group I to the Fourth Assessment Report of the Intergovernmental Panel on Climate Change*, Vol. 1009. Cambridge University Press, Cambridge.

Habibzadeh Y, Evazi AR, Abedi M (2014) Alleviation drought stress of mungbean (*Vigna radiata* L.) plants by using arbuscular mycorrhizal fungi. *Int J Agric Sci Nat Resour* 1(1):1–6. doi: http://www.aascit.org/journal/asnr.

Hafiz MH, Wajid F, Farhat A, Fahad S, Shafqat S, Wajid N, Hafiz FB (2016) Maize plant nitrogen uptake dynamics at limited irrigation water and nitrogen. *Environ Sci Pollut Res* 24(3):2549–2557. doi: 10.1007/s11356-016-8031-0.

Hafiz MH, Muhammad A, Farhat A, Hafiz FB, Saeed AQ, Muhammad M, Fahad S, Muhammad A (2019) Environmental factors affecting the frequency of road traffic accidents: A case study of sub-urban area of Pakistan. *Environ Sci Pollut Res*. doi: 10.1007/s11356-019-04752-8.

Hafiz MH, Farhat A, Ashfaq A, Hafiz FB, Wajid F, Carol Jo W, Fahad S, Gerrit H (2020b) Predicting kernel growth of maize under controlled water and nitrogen applications. *Int J Plant Prod* 10.1007/s42106-020-00110-8.

Hafiz MH, Farhat A, Shafqat S, Fahad S, Artemi C, Wajid F, Chaves CB, Wajid N, Muhammad M, Hafiz FB (2018) Offsetting land degradation through nitrogen and water management during maize cultivation under arid conditions. *Land Degrad Dev* 1–10. doi: 10.1002/ldr.2933.

Hafiz MH, Abdul K, Farhat A, Wajid F, Fahad S, Muhammad A, Ghulam MS, Wajid N, Muhammad M, Hafiz FB (2020a) Comparative effects of organic and inorganic fertilizers on soil organic carbon and wheat productivity under arid region. *Commun Soil Sci Plant Anal*. doi: 10.1080/00103624.2020.1763385.

Hamidou F, Zombre G, Braconnier S (2007) Physiological and biochemical responses of cowpea genotypes to water stress under glasshouse and field conditions. *J Agron Crop Sci* 193(4):229–237. doi: 10.1111/j.1439-037X.2007.00253.x.

Hansel FD, Amado TJC, Diaz DAR, Rosso LHM, Nicoloso FT, Schorr M (2017) Phosphorus fertilizer placement and tillage affect soybean root growth and drought tolerance. *Agron J* 109:2936–2944. doi: 10.2134/agronj2017.04.0202.

Hashem A, Kumar A, Al-Dbass AM, et al (2019) Arbuscular mycorrhizal fungi and biochar improves drought tolerance in chickpea. *Saudi J Biol Sci*. doi: 10.1016/j.sjbs.2018.11.005.

Hati KM, Mandal KG, Misra AK, Ghosh PK, Bandyopadhyay KK (2006) Effect of inorganic fertilizer and farmyard manure on soil physical properties, root distribution, and water-use efficiency of soybean in Vertisols of central India. *Bioresour Technol* 97(16):2182–2188. doi: 10.1016/j.biortech.2005.09.033.

Hattori T, Inanaga S, Araki H, An P, Morita S, Luxová M, Lux, A (2005) Application of silicon enhanced drought tolerance in Sorghum bicolor. *Physiol Plant* 123(4):459–466. doi: 10.1111/j.1399-3054.2005.00481.x.

Hayat R, Ali S, Amara U (2010) Soil beneficial bacteria and their role in plant growth promotion: A review. *Ann Microbiol* 60:579–598. doi: 10.1007/s13213-010-0117-1.

Hesham FA, Fahad S (2020) Melatonin application enhances biochar efficiency for drought tolerance in maize varieties: Modifications in physio-biochemical machinery. *Agron J* 112(4):1–22.

Hossain A et al (2020a) Nutrient management for improving abiotic stress tolerance in legumes of the family

Fabaceae. In: Hasanuzzaman M, Araújo S, Gill S (eds) *The plant family Fabaceae*. Springer, Singapore. doi: 10.1007/978-981-15-4752-2_15.

Hossain MA, Hasanuzzaman M, Fujita M (2010) Up-regulation of antioxidant defense and methylglyoxal detoxification system by exogenous glycinebetaine and proline confer tolerance to cadmium stress in mung bean seedlings. *Physiol Molecular Biol Plants* 16(3): 259–272.

Hossain MA, Li ZG, Hoque TS, Burritt DJ, Fujita M, Munne´-Bosch S (2018) Heat or cold priming-induced cross-tolerance to abiotic stresses in plants: Key regulators and possible mechanisms. *Protoplasma* 255(1):399–412. doi: 10.1007/s00709-017-1150-8.

Hossain A, Farooq M, EL Sabagh A, Hasanuzzaman M, Erman M, Islam T (2020b) Morphological, physiobiochemical and molecular adaptability of legumes of Fabaceae to drought stress, with special reference to Medicago Sativa L. In: Hasanuzzaman M, Araújo S, Gill S (eds) *The Plant Family Fabaceae*. Springer, Singapore. doi: 10.1007/978-981-15-4752-2_11.

Hunt PJ, Wollum AG, Matheny TA (1981) Effects of soil water on *Rhizobium japonicum* infection nitrogen accumulation and yield in bragg soybean. *Agric J* 73(3):501–505. doi: 10.2134/agronj1981.00021962 007300030024x.

Hunt PG, Matheny TA, Wollum AG (1988) Yield and N accumulation responses of late-season determinate soybean to irrigation and inoculation with various strains of *Bradyrhizobium japonicum*. *Communi Soil Sci Plant Anal* 19(14):601–1612. doi: 10.1080/00103628809368038.

Hussain MB, Zahir ZA, Asghar HN, Asghar M (2014) Can catalase and exopolysaccharides producing rhizobia ameliorate drought stress in wheat? *Int J Agric Biol* 16:3–13.

Hussain HA, Hussain S, Khaliq A, Ashraf U, Anjum SA, Men S, Wang L (2018) Chilling and drought stresses in crop plants: Implications, cross talk, and potential management opportunities. *Front Plant Sci* 9:393. doi: 10.3389/fpls.2018.00393.

Hussain MA, Fahad S, Rahat S, Muhammad FJ, Muhammad M, Qasid A, Ali A, Husain A, Nooral A, Babatope SA, Changbao S, Liya G, Ibrar A, Zhanmei J, Juncai H (2020) Multifunctional role of brassinosteroid and its analogues in plants. *Plant Growth Regul*. doi: 10.1007/s10725-020-00647-8.

Ibrar K, Aneela R, Khola Z, Urooba N, Sana B, Rabia S, Ishtiaq H, Rehman MU, Salvatore M (2020) Microbes and environment: Global warming reverting the frozen zombies. In: Fahad S, Hasanuzzaman M, Alam M, Ullah H, Saeed M, Khan AK, Adnan M (eds), *Environment, climate, plant and vegetation growth*. Springer Publ Ltd, Springer Nature Switzerland AG. Part of Springer Nature, pp. 607–634. doi: 10.1007/978-3-030-49732-3.

Ilyas M, Mohammad N, Nadeem K, Ali H, Aamir HK, Kashif H, Fahad S, Aziz K, Abid U (2020) Drought tolerance strategies in plants: A mechanistic approach. *J Plant Growth Regulation*. doi: 10.1007/s00344-020-10174-5.

Iqbal MA, Iqbal A, Siddiqui MH, Maqbool Z (2018a) Bio-agronomic evaluation of forage sorghum-legumes binary crops on Haplic Yermosol soil of Pakistan. *Pakistan J Bot* 50(5):1991–1997.

Iqbal MA, Siddiqui MH, Afzal S, Ahmad Z, Maqsood Q, Khan RD (2018c) Forage productivity of cowpea [*Vigna unguiculata* (L.) Walp] cultivars improves by optimization of spatial arrangements. *Revista Mexicana De Ciencias Pecuarias* 9(2):203–219.

Iqbal MA, Iqbal I, Maqbool Z, Ahmad Z, Ali E, Siddiqui MH, Ali S (2018b) Revamping soil quality and correlation studies for yield and yield attributes in sorghum-legumes intercropping systems. *Biosci J* 34(3):1165–1176.

Iqbal MA, Hamid A, Hussain I, Siddiqui MH, Ahmad T, Khaliq A, Ahmad Z (2019a) Competitive indices in cereal and legume mixtures in a South Asian environment. *Agron J* 111(1):242–249.

Iqbal MA, Hamid A, TAhmad T, Hussain I, Ali S, Ali A, Ahmad Z (2019b) Forage sorghum-legumes intercropping: Effect on growth, yields, nutritional quality and economic returns. *Bragantia* 78(1): 82–95.

Iqra M, Amna B, Shakeel I, Fatima K, Sehrish L, Hamza A, Fahad S (2020) Carbon cycle in response to global warming. In: Fahad S, Hasanuzzaman M, Alam M, Ullah H, Saeed M, Khan AK, Adnan M (eds) *Environment, climate, plant and vegetation growth*. Springer Publ Ltd, Springer Nature Switzerland AG. Part of Springer Nature, pp. 1–16. doi: 10.1007/978-3-030-49732-3.

Irekti H, Drevon JJ (2003) Acide abcissique et conductance a` la diffusion de l'oxyge`ne dans les nodosite´s de haricot soumises a` un choc salin. In: Drevon JJ, Sifi B (eds) *Fixation Symbiotique de l'Azote et De´veloppement Durable dans le Bassin Me´diterrane´en, INRA Editions*, Vol 100. les colloques.

Islam MS, EL Sabagh A, Hasan K, Akhter M, Barutçulard C (2017a) Growth and yield response of mungbean

(*Vigna radiata* L.) as influenced by sulphur and boron application. *Sci J Crop Sci* 6(1) 153–160. doi: 10.14196/sjcs.v6i1.2383.

Islam MS, Hasan K, Sarkar NA, EL Sabagh A, RashwanE, Barutçulare C (2017b) Yield and yield contributing characters of mungbean as influenced by zinc and boron. *Sci J Agril Adv* 6(1):391–397 doi: 10.14196/aa.v6i1.2362.

Jaleel CA, Manivanannan P, Wahid AM, Froog HJ, Al-Juburi R, Somasundaram R (2009) Drought stress in plant: A review on morphological characters and pigments composition. *Int J Agric Biol* 11:110–115.

Jan M, Muhammad AH, Adnan NS, Muhammad Y, Javaid I, Xiuling L, Depeng W, Fahad S, (2019) Modulation in growth, gas exchange, and antioxidant activities of salt-stressed rice (Oryza sativa L.) genotypes by zinc fertilization. *Arabian J geosci* 12:775. doi: 10.1007/s12517-019-4939-2.

Jenkins MB, Virginia RA, Jarrel WM (1989) Ecology of fastgrowing and slow-growing mesquite-nodulating rhizobia in Chihuahua and Sonoran desert ecosystems. *Soil Sci Soc Am J* 53:543–549.

Jin J, Wang G, Liu X, Pan X, Herbert SJ, Tang C (2006) Interaction between phosphorus nutrition and drought on grain yield, and assimilation of phosphorus and nitrogen in two soybean cultivars differing in protein concentration in grains. *J Plant Nutr* 29:1433–1449.

Kamarn M, Wenwen C, Irshad A, Xiangping M, Xudong Z, Wennan S, Junzhi C, Shakeel A, Fahad S, Qingfang H, Tiening L (2017) Effect of paclobutrazol, a potential growth regulator on stalk mechanical strength, lignin accumulation and its relation with lodging resistance of maize. *Plant Growth Regul* 84:317–332. doi: 10.1007/s10725-017-0342-8.

Kapuya JA, Barendse GWM, Linskens HF (1985) Water stress tolerance and proline accumulation in *Phaseolus vulgaris*. *Acta Bot Neerl* 34:295–300.

Kaschuk G, Hungria M, Leffelaar PA, Giller KE, Kuyper TW (2010) Differences in photosynthetic behaviour and leaf senescence of soybean [*Glycine max* (L.) Merrill] dependent on N_2 fixation or nitrate supply. *Plant Biol* 12:60–69.

Keck TJ, Wagenet RJ, Campbell WF, Knighton RE (1984) Effects of water and salt stress on growth and acetylene reduction in alfalfa. *Soil Sci Soc Am J* 48:1310–1316.

Khalafallah AA, Tawfik KM, Abd El-Gawad ZA (2008) Tolerance of seven faba bean varieties to drought and salt stresses. *Res J Agric Biol Sci* 4:175–186.

Khan N, Bano A, Babar A (2019) Metabolic and physiological changes induced by plant growth regulators and plant growth promoting rhizobacteria and their impact on drought tolerance in *Cicer arietinum* L. *PLoS ONE* 13:e0213040. doi: 10.1371/journal.pone.0213040.

Kim GB, Nam YW (2013) A novel Δ1-pyrroline-5-carboxylate synthetase gene of *Medicago truncatula* plays a predominant role in stress induced proline accumulation during symbiotic nitrogen fixation. *J Plant Physiol* 170(3):291–302. doi: 10.1016/j.jplph.2012.10.004.

King CA, Purcell LC (2006) Genotypic variation for shoot N concentration and response to water deficits in soybean. *Crop Sci* 46:2396–2402. doi: 10.2135/cropsci2006.03.0165.

King CA, Purcell LC, Bolton A, Specht JE (2014) A possible relationship between shoot N concentration and the sensitivity of N_2 fixation to drought in soybean. *Crop Sci* 54:746–756. doi: 10.2135/cropsci2013.04 .0271.

Kirda C, Danso SKA, Zapata F (1989) Temporal water-stress effects on nodulation, nitrogen accumulation and growth of soybean. *Plant Soil* 120:49–55. doi: 10.1007/BF02370289.

Kumar A, Prakash A, Johri BN (2011) Bacillus as PGPR in crop ecosystem. In *Bacteria in Agrobiology: Crop Ecosystems*. Springer, Berlin, Heidelberg, pp. 37–59. doi: 10.1007/978-3-642-18357-7_2.

Kumar A, Dames JF, Gupta A, Sharma S, Gilbert JA, Ahmad P (2015) Current developments in arbuscular mycorrhizal fungi research and its role in salinity stress alleviation: A biotechnological perspective. *Crit Rev Biotechnol* 35(4):461–474. doi: 10.3109/07388551.2014.899964.

Kumar M, Mishra S, Dixit V, Kumar M, Agarwal L, Chauhan PS, Nautiyal CS. (2016) Synergistic effect of *Pseudomonas putida* and *Bacillus amyloliquefaciens* ameliorates drought stress in chickpea (*Cicer arietinum* L.). *Plant Signal Behav* 11(1): e1071004. doi: 10.1080/15592324.2015.1071004.

Kumar S, Meena RS, Lal R, Yadav GS, Mitran T, Meena BL, Dotaniya ML, EL-Sabagh A (2018) Role of legumes in soil carbon sequestration. In: Meena RS et al. (eds) *Legumes for soil health and sustainable management*. Springer, Singapore, pp. 109–138. doi: 10.1007/978-981-13-0253-4_4.

Kumari S, Vaishnav A, Jain S, Verma A, Choudhary DK (2016) Induced drought tolerance through wild and mutant bacterial strain Pseudomonas simiae in mung bean (*Vigna radiata* L.). *World J Microbiol Biotechnol* 32:1–10. doi: 10.1007/s11274-015-1974-3.

Kurdali F, Al-chammaa M, Mouasess A (2016) Growth and nitrogen fixation in silicon and/or potassium fed chickpeas grown under drought and well watered conditions. *J Stress Physiol Biochem* 9(3): 385–406.

Kurdali F, Al-Chammaa M, Al-Ain F (2019) Growth and N_2 fixation in saline and/or water stressed Sesbania aculeata plants in response to silicon application. *Silicon* 11:781–788. doi: 10.1007/s12633-018-9884-2.

Labanauskas CK, Shouse P, Stolzy LH, Handy MF (1981) Protein and free amino acids in field-grown cowpea seeds as affected by water stress at various stages. *Plant Soil* 63:355–368.

Ladrera R, Marino D, Larrainzar E, González EM, Arrese-Igor C (2007) Reduced carbon availability to bacteroids and elevated ureides in nodules, but not in shoots, are involved in the nitrogen fixation response to early drought in soybean. *Plant Physiol* 145(2):539–546. doi: 10.1104/pp.107.102491.

Lambers H, Chapin FS, Pons TL (2008) *Plant physiological ecology*, 2nd edn. Springer, New York, USA.

Lobato AKS, Oliveira Neto CF, Costa RCL, Santos Filho BG, Cruz FJR, Laughinghouse IV HD (2008) Biochemical and physiological behavior of *Vigna unguiculata* (L.) Walp. under drought stress during the vegetative phase. *Asian J Plant Sci* 7(1):44–49. doi: 10.3923/ajps.2008.44.49.

Long SR (2001) Genes and signals in the Rhizobium-legume symbiosis. *Plant Physiol* 125:69–72. 10.1104/pp.125.1.69.

Long HH, Schmidt DD, Baldwin IT (2008) Native bacterial endophytes promote host growth in a species-specific manner; phytohormone manipulations do not result in common growth responses. *PLoS ONE* 3:e2702. doi: 10.1371/journal.pone.0002702.

Mafakheri A, Siosemardeh A, Bahramnejad B, Struik PC, Sohrabi E (2010) Effect of drought stress on yield, proline and chlorophyll contents in three chickpea cultivars. *Aust J Crop Sci* 4(8):580–585.

Mahar A, Amjad A, Altaf HL, Fazli W, Ronghua L, Muhammad A, Fahad S, Muhammad A, Rafiullah, Imtiaz AK, Zengqiang Z (2020) Promising technologies for Cd-contaminated soils: Drawbacks and possibilities. In: Fahad S, Hasanuzzaman M, Alam M, Ullah H,Saeed M, Khan AK, Adnan M (eds) *Environment, climate, plant and vegetation growth*. Springer Publ Ltd, Springer Nature Switzerland AG. Part of Springer Nature, pp. 63–92. doi: 10.1007/978-3-030-49732-3.

Mahler RL, Wollum II AG (1980) Influence of water potential on the survival of rhizobia in Goldsboro loamy sand. *Soil Sci Am J* 44:988–992.

Mali M, Aery NC (2008) Silicon effects on nodule growth, dry-matter production, and mineral nutrition of cowpea (*Vigna unguiculata* L.). *J Plant Nutr Soil Sci* 171:835–840. doi: 10.1002/jpln.200700362.

Manoj KR, Kalia RK, Singh R, Gangola MP, Dhawan AK (2011) Developing stress tolerant plants through in vitro selection-an overview of the recent progress. *Environ Exp Bot* 71:89–98. doi: 10.1016/j.envexpbot.2010.10.021.

Marino D, Frendo P, Ladrera R, Zabalza A, Puppo A, et al. (2007) Nitrogen fixation control under drought stress. Localized or systemic?. *Plant Physiol* 143:1968–1974.

Martinez-Ballesta MC, Bastias E, Carvajal M (2008) Combined effect of boron and salinity on water transport. *Plant Sig Behav* 3(10):844–845. doi: 10.4161/psb.3.10.5990.

Masoumi H, Darvish F, Daneshian J, Normohammadi G, Habib D (2011) Effects of water deficit stress on seed yield and antioxidants content in soybean (*Glycine max* L.) cultivars. *African J Agril Res* 6(5):1209–1218. http://www.academicjournals.org/AJARDOI:10.5897/AJAR10.821.

Mayak S, Tirosh T, Glick BR (1999) Effect of wild-type and mutant plant growth-promoting rhizobacteria on the rooting of mung bean cuttings. *J Plant Growth Regul* 18(2):49–53. doi: 10.1007/pl00007047.

Md Jakir H, Allah B (2020) Development and applications of transplastomic plants; a way towards eco-friendly agriculture. In: Fahad S, Hasanuzzaman M, Alam M, Ullah H, Saeed M, Khan AK, Adnan M (eds) *Environment, climate, plant and vegetation growth*. Springer Publ Ltd, Springer Nature Switzerland AG. Part of Springer Nature, pp. 285–322. doi: https://doi.org/10.1007/978-3-030-49732-3.

Md. Enamul H, Shoeb AZM, Mallik AH, Fahad S, Kamruzzaman MM, Akib J, Nayyer S, Mehedi AKM, Swati AS, Md Yeamin A, Most SS (2020) Measuring vulnerability to environmental hazards: Qualitative to quantitative. In: Fahad S, Hasanuzzaman M, Alam M, Ullah H, Saeed M, Khan AK, Adnan M (eds) *Environment, climate, plant and vegetation growth*. Springer Publ Ltd, Springer Nature Switzerland AG. Part of Springer Nature, pp. 421–452. doi: 10.1007/978-3-030-49732-3.

Mhadhbi H, Chihaoui S, Mhamdi R, Mnasri B, Jebara M (2011) A highly osmotolerant rhizobial strain confers a better tolerance of nitrogen fixation and enhances protective activities to nodules of *Phseolus vulgaris* under drought stress. *Afr J Biotechnol* 10(22):4555–4563. doi: 10.5897/AJB10.1991.

Minchin FR (1997) Regulation of oxygen diffusion in legume nodules. *Soil Biol Biochem* 29(5-6):881–888. doi: 10.1016/S0038-0717(96)00204-0.

Mohammadi A, Habibi D, Rohami M, Mafakheri S (2011) Effect of drought stress on antioxidant enzymes activity of some chickpea cultivars. *Am-Eurasian J Agric Env Sci* 11:782–785.

Moro MJ, Domingo F, Bermudez-de-Castro F (1992) Acetylene reduction activity (ARA) by the shrub legume *Adenocarpus decorticans* Boiss. in southern Spain (Almeria). *Acta Oecol* 13:325–333.

Mouradi M, Farissi M, Bouizgaren A, Makoudi B, Kabbadj A, Very AA (2016) Effects of water deficit on growth, nodulation and physiological and biochemical processes in *Medicago sativa*-rhizobia symbiotic association. *Arid Land Res Manage* 30(2):193–208. doi: 10.1080/15324982.2015.1073194.

Mubeen M, Ashfaq A, Hafiz MH, Muhammad A, Hafiz UF, Mazhar S, Muhammad Sami ul Din, Asad A, Amjed A, Fahad S, Wajid N (2020) Evaluating the climate change impact on water use efficiency of cotton-wheat in semi-arid conditions using DSSAT model. *J Water Climate Change*. doi/10.2166/wcc.2 019.179/622035/jwc2019179.pdf.

Muhammad TQ, Amna F, Amna B, Barira Z, Xitong Z, Ling-Ling C (2020) Effectiveness of conventional crop improvement strategies vs. omics. In: Fahad S, Hasanuzzaman M, Alam M, Ullah H, Saeed M, Khan AK, Adnan M (eds) *Environment, climate, plant and vegetation growth*. Springer Publ Ltd, Springer Nature Switzerland AG. Part of Springer Nature, pp. 253–284. doi: 10.1007/978-3-030-4 9732-3.

Muhammad Z, Abdul MK, Abdul MS, Kenneth BM, Muhammad S, Shahen S, Ibadullah J, Fahad S (2019) Performance of Aeluropus lagopoides (mangrove grass) ecotypes, a potential turfgrass, under high saline conditions. *Environ Sci Pollut Res*. doi: 10.1007/s11356-019-04838-3.

Nabizadeh E, Jalilnejad N, Armakani M (2011) Effect of salinity on growth and nitrogen fixation of alfalfa (*Medicago sativa*). *World Appl Sci J* 13(8):1895–1900.

Nadeem MS, Ahmad M, Ahmad Z, Javaid A, Ashraf M (2014) The role of mycorrhizae and plant growth promoting rhizobacteria (PGPR) in improving crop productivity under stressful environments. *Biotechnol Adv* 32:429–448.

Nadeem SM, Zahir ZA, Naveed M, Ashraf M, Nadeem SM, Zahir ZA, Naveed M (2010) Microbial ACC-deaminase: prospects and applications for inducing salt tolerance in plants microbial. *Crit Rev Plant Sci* 29:360–393.

Nawaz F, Ashraf MY, Ahmad R, Waraich EA, Shabbir RN, Bukhari MA (2015) Supplemental selenium improves wheat grain yield and quality through alterations in biochemical processes under normal and water deficit conditions. *Food Chem* 175:350–357.

Niu X, Song L, Xiao Y, Ge W, Job D (2018) Drought-tolerant plant growth-promoting rhizobacteria associated with Foxtail Millet in a semi-arid agroecosystem and their potential in alleviating drought stress. *Front Microbiol* 8:2580.

Noor M, Naveed ur R, Ajmal J, Fahad S, Muhammad A, Fazli W, Saud S, Hassan S (2020) Climate change and costal plant lives. In: Fahad S, Hasanuzzaman M, Alam M, Ullah H,Saeed M, Khan AK, Adnan M (eds) *Environment, climate, plant and vegetation growth*. Springer Publ Ltd, Springer Nature Switzerland AG. Part of Springer Nature, pp. 93–108. doi: 10.1007/978-3-030-49732-3.

Ohashi Y, Nakayama N, Saneoka H, Fujita K (2006) Effects of drought stress on photosynthetic gas exchange, chlorophyll fluorescence and stem diameter of soybean plants. *Biol Plant* 50:138–141.

Ortíz-Castro R, Valencia-Cantero E, López-Bucio J (2008) Plant growth promotion by Bacillus megaterium involves cytokinin signaling. *Plant Signal Behav* 3(4):263–265. doi: 10.4161/psb.3.4.5204.

Osa-Afina LO, Alexander M (1982) Difference among cowpea rhizobia in tolerance to high temperature and desiccation in soil. *Appl Environ Microbiol* 43:435–439.

Palta JA, Kumari S, Spring C, Turner NC (2005) Foliar nitrogen applications increase the seed yield and protein content in chickpea (*Cicer arietinum* L.) subject to terminal drought. *Aust J Agril Res* 56:105–112. doi: 10.1071/AR04118.

Pankhurst CE, Sprent JI (1975) Effects of water stress on the respiratory and nitrogen-fixing activity of soybean root nodules. *J Exp Bot* 26(91):287–304.

Pararjasingham S, Knievel DP (1990) Nitrogenase activity of cowpea (*Vigna unguiculata* (L.) Walp.) during and after water stress. *Can J Plant Sci* 70:163–171.

Pastenes C, Pimentel P, Lillo J (2005) Leaf movements and photoinhibition in relation to water stress in field-grown beans. *J Exp Bot* 56(411):425–433. doi: 10.1093/jxb/eri061.

Pena-Cabriales JJ, Castellanos JZ (1993) Effect of water stress on N_2 fixation and grain yield of *Phaseolus vulgaris* L. *Plant Soil* 152:151–155. doi: 10.1007/BF00016345.

Porcel R, Ruiz-Lozano JM (2004) Arbuscular mycorrhizal influence on leaf water potential, solute accumulation, and oxidative stress in soybean plants subjected to drought stress. *J Exp Bot* 55(403): 1743–1750. doi: 10.1093/jxb/erh188.

Purcell LC, Sinclair TR (1995) Nodule gas-exchange and water potential response to rapid imposition of water-deficit. *Plant Cell Environ* 18:179–187. doi: 10.1111/j.1365-3040.1995.tb00351.x.

Qamar-uz Z, Zubair A, Muhammad Y, Muhammad ZI, Abdul K, Fahad S, Safder B, Ramzani PMA, Muhammad N (2017) Zinc biofortification in rice: Leveraging agriculture to moderate hidden hunger in developing countries. *Arch Agron Soil Sci* 64:147–161. doi: 10.1080/03650340.2017.1338343.

Rai R, Prasad V (1983) Studies on compatibility of nitrogen fixation by high temperature-adapted Rhizobium strains and *Vigna radiata* genotype at two levels in calcareous soil. *J Agric Sci* 101:377–381. doi: 10.1 017/S0021859600037692.

Randall Weiss P, Denison RF, Sinclair TR (1985) Response to drought stress of nitrogen fixation (acetylene reduction) rates by field grown soybeans. *Plant Physiol* 78:525–536.

Rashid M, Qaiser H, Khalid SK, Al-Wabel MI, Zhang A, Muhammad A, Shahzada SI, Rukhsanda A, Ghulam AS, Shahzada MM, Sarosh A, Muhammad FQ (2020) Prospects of biochar in alkaline soils to mitigate climate change. In: Fahad S, Hasanuzzaman M, Alam M, Ullah H, Saeed M, Khan AK, Adnan M (eds) *Environment, climate, plant and vegetation growth*. Springer Publ Ltd, Springer Nature Switzerland AG. Part of Springer Nature, pp. 133–150. doi: 10.1007/978-3-030-49732-3.

Ratnayaka HH, Kincaid D (2005) Gas exchange and leaf ultrastructure of Tinnevelly senna, *Cassia angustifolia*, under drought and nitrogen stress. *Crop Sci* 45(3):840–847. doi: 10.2135/cropsci2003.737.

Rehman M, Fahad S, Saleem MH, Hafeez M, Muhammad HR, Liu F, Deng G (2020) Red light optimized physiological traits and enhanced the growth of ramie (Boehmeria nivea L.). *Photosynthetica* 58 (4): 922–931.

Robin C, Shamsun-Noor L, Guckert A (1989) Effect of potassium on the tolerance to PEG-induced water stress of two white clover varieties (*Trifolium repens* L.). *Plant Soil* 120:143–158. doi: 10.1007/BF023 77063.

Rozrokh M, Sabaghpour SH, Armin M, Asgharipour M (2012) The effects of drought stress on some biochemical traits in twenty genotypes of chickpea. *Eur J Exp Biol* 2(6):1980–1987.

Sadam M, Muhammad TQ, Ghulam M, Muhammad SK, Faiz AJ (2020) Role of biotechnology in climate resilient agriculture. In: Fahad S, Hasanuzzaman M, Alam M, Ullah H, Saeed M, Khan AK, Adnan M (eds) *Environment, climate, plant and vegetation growth*. Springer Publ Ltd, Springer Nature Switzerland AG. Part of Springer Nature, pp. 339–366. doi: 10.1007/978-3-030-49732-3.

Saeidi M, Abdoli M (2015) Effect of drought stress during grain filling on yield and its components, gas exchange variables, and some physiological traits of wheat cultivars. *J Agric Sci Technol* 17: 885–898.

Saikia J, Sarma RK, Dhandia R, Yadav A, Bharali R, Gupta VK, Saikia R (2018) Alleviation of drought stress in pulse crops with ACC deaminase producing rhizobacteria isolated from acidic soil of Northeast India. *Sci Rep* 8(1): 3560. doi: 10.1038/s41598-018-25174-5.

Sajid H, Jie H, Jing H, Shakeel A, Satyabrata N, Sumera A, Awais S, Chunquan Z, Lianfeng Z, Xiaochuang C, Qianyu J, Junhua Z (2020) Rice production under climate change: Adaptations and mitigating strategies. In: Fahad S, Hasanuzzaman M, Alam M, Ullah H, Saeed M, Khan AK, Adnan M (eds) *Environment, climate, plant and vegetation growth*. Springer Publ Ltd, Springer Nature Switzerland AG. Part of Springer Nature, pp. 659–686. doi: 10.1007/978-3-030-49732-3.

Sajjad H, Muhammad M, Ashfaq A, Waseem A, Hafiz MH, Mazhar A, Nasir M, Asad A, Hafiz UF, Syeda RS, Fahad S, Depeng W, Wajid N (2019) Using GIS tools to detect the land use/land cover changes during forty years in Lodhran district of Pakistan. *Environ Sci Pollut Res* 10.1007/s11356-019-06072-3.

Saleem MH, Fahad S, Shahid UK, Mairaj D, Abid U, Ayman ELS, Akbar H, Analía L, Lijun L (2020b) Copper-induced oxidative stress, initiation of antioxidants and phytoremediation potential of flax (Linum usitatissimum L.) seedlings grown under the mixing of two different soils of China. *Environ Sci Poll Res*. doi: 10.1007/s11356-019-07264-7.

Saleem MH, Fahad S, Adnan M, Mohsin A, Muhammad SR, Muhammad K, Qurban A, Inas AH, Parashuram B, Mubassir A, Reem MH (2020a) Foliar application of gibberellic acid endorsed phytoextraction of copper and alleviates oxidative stress in jute (Corchorus capsularis L.) plant grown in highly copper-contaminated soil of China. *Environ Sci Pollution Res*. doi: 10.1007/s11356-020-09764-3.

Saleem MH, Rehman M, Fahad S, Tung SA, Iqbal N, Hassan A, Ayub A, Wahid MA, Shaukat S, Liu L, Deng G (2020c) Leaf gas exchange, oxidative stress, and physiological attributes of rapeseed (Brassica napus L.) grown under different light-emitting diodes. *Photosynthetica* 58 (3):836–845.

Saleem N, Adnan M, Khan NA, Zaheer S, Jalal F, Amin M, Khan WM, Arif M, Rahman I, Ibrahim M, Jamal Y, Shah SRA, Junaid K, Ali M (2015) Dual purpose canola: Grazing and grains options. *Pak J Weed Sci Res* 21(2):295–304.

Saman S, Amna B, Bani A, Muhammad TQ, Rana MA, Muhammad SK (2020) QTL mapping for abiotic stresses in cereals. In: Fahad S, Hasanuzzaman M, Alam M, Ullah H, Saeed M, Khan AK, Adnan M (eds) *Environment, climate, plant and vegetation growth*. Springer Publ Ltd, Springer Nature Switzerland AG. Part of Springer Nature, pp. 229–252. doi: 10.1007/978-3-030-49732-3.

Sangakkara UR, Hartwig UA, Noesberger J (1996) Soil moisture and potassium affect the performance of symbiotic nitrogen fixation in faba bean and common bean. *Plant Soil* 184:123–130. doi: 10.1007/BF00029282.

Sangakkara UR, Frehner M, Nosberger J (1999) Effect of soil moisture and potassium fertilizer on shoot water potential, photosynthesis and partitioning of carbon in mungbean and cowpea. *J Agron Crop Sci* 185(3):201–207. doi: 10.1046/j.1439-037x.2000.00422.x.

Sarma RK, Saikia R (2014) Alleviation of drought stress in mung bean by strain *Pseudomonas aeruginosa* GGRJ21. *Plant Soil* 377:111–126. doi: 10.1007/s11104-013-1981-9.

Saud S, Chen Y, Long B, Fahad S, Sadiq A (2013) The different impact on the growth of cool season turf grass under the various conditions on salinity and drought stress. *Int J Agric Sci Res* 3:77–84.

Saud S, Fahad S, Cui G, Chen Y, Anwar S (2020) Determining nitrogen isotopes discrimination under drought stress on enzymatic activities, nitrogen isotope abundance and water contents of Kentucky bluegrass. *Sci Rep* 10:6415.doi: 10.1038/s41598-020-63548-w.

Saud S, Chen Y, Fahad S, Hussain S, Na L, Xin L, Alhussien SA (2016) Silicate application increases the photosynthesis and its associated metabolic activities in Kentucky bluegrass under drought stress and post-drought recovery. *Environ Sci Pollut Res* 23(17):17647– 17655. doi: 10.1007/s11356-016-6957-x.

Saud S, Li X, Chen Y, Zhang L, Fahad S, Hussain S, Sadiq A, Chen Y (2014) Silicon application increases drought tolerance of Kentucky bluegrass by improving plant water relations and morph physiological functions. *SciWorld J* 2014:1–10. doi: 10.1155/2014/368694.

Saud S, Fahad S, Yajun C, Ihsan MZ, Hammad HM, Nasim W, Amanullah Jr, Arif M, Alharby H (2017) Effects of nitrogen supply on water stress and recovery mechanisms in Kentucky Bluegrass plants. *Front Plant Sci* 8:983. doi: 10.3389/fpls.2017.00983.

Sauvage D, Hamelin J, Larher F (1983) Glycine betaine and other structurally related compounds improve the salt tolerance of *Rhizobium meliloti*. *Plant Sci Letters* 31:291–302.

Savita S, Tomer A, Singh SK (2020) Drought Stress Tolerance in Legume Crops. In: Hasanuzzaman M (eds) *Agronomic Crops*. Springer, Singapore. doi: 10.1007/978-981-15-0025-1_9.

Schmidt PE, Broughton WJ, Werner D (1994) Nod factors of *Bradyrhizobium japonicum* and *Rhizobium* sp. NGR 234 induce flavonoid accumulation in soybean root exudates. *Mol Plant Microbe Interact* 7:384–390.

Sellstedt A, Ståhl L, Mattsson U, Jonsson K, Högberg P (1993) Can the 15N dilution technique be used to study N_2 fixation in tropical tree symbioses as affected by water deficit? *J Exp Bot* 44:1749–1755. doi: 10.1093/jxb/44.12.1749.

Senaratna T, Touchell D, Bunn E, Dixon K (2000) Acetyl salicylic acid (Aspirin) and salicylic acid induce multiple stress tolerance in bean and tomato plants. *Plant Growth Regul* 30:157–161. doi: 10.1023/A:1006386800974.

Senaratna T, Mcrritt D, Dixon K, Bunn E, Touchell D, Sivasithamparam K (2003) Benzoic acid may act as the functional group in salicylic acid and derivatives in the induction of multiple stress tolerance in plants. *Plant Growth Regul* 39(1):77–81. doi: 10.1023/A:1021865029762.

Senol C (2020) The effects of climate change on human behaviors. In: Fahad S, Hasanuzzaman M, Alam M, Ullah H, Saeed M, Khan AK, Adnan M (eds) *Environment, climate, plant and vegetation growth*. Springer Publ Ltd, Springer Nature Switzerland AG. Part of Springer Nature, pp. 577–590. doi: 10.1007/978-3-030-49732-3.

Serraj R, Sinclair TR (1996) Inhibition of nitrogenase activity and nodule oxygen permeability by water deficit. *J Exp Bot* 47:1067–1073.

Serraj R, Roy G, Drevon JJ (1994) Salt stress induces a decrease in the oxygen uptake of soybean nodules and their permeability to oxygen diffusion. *Physiol Planta* 91:161–168. doi: 10.1111/j.1399-3054.1994 .tb00414.x.

Serraj R, Sinclair TR, Purcell LC (1999) Symbiotic N_2 fixation response to drought. *J Exp Bot* 50(331): 143–155. doi: 10.1093/jexbot/50.331.143.

Shaban M, Lak M, Hamidvand Y, Nabaty E, Khodaei F (2012) Response of chickpea (*Cicer arietinum* L.) cultivars to integrated application of Zinc nutrient with water stress. *Int J Agric Crop Sci* 4:1074–1082.

Shafi MI, Adnan M, Fahad S, Fazli W, Ahsan K, Zhen Y, Subhan D, Zafar-ul-Hye M, Martin B, Rahul D (2020) Application of single superphosphate with humic acid improves the growth, yield and phosphorus uptake of wheat (Triticum aestivum L.) in Calcareous soil. *Agron* (10):1224. doi: 10.3390/ agronomy10091224.

Shah F, Lixiao N, Kehui C, Tariq S, Wei W, Chang C, Liyang Z, Farhan A, Fahad S, Huang J (2013) Rice grain yield and component responses to near 2°C of warming. *Field Crop Res* 157:98–110.

Shoushtari NH, Pepper IL (1985) Mesquite rhizobia isolated from the Sonoran desert: Competitiveness and survival in soil. *Soil Biol Biochem* 17:803–806.

Simon-Sarkadi L, Kocsy G, V´arhegyi A, Galiba G, de Ronde JA (2005) Genetic manipulation of proline accumulation influences the concentrations of other amino acids in soybean subjected to simultaneous drought and heat stress. *J Agric Food Chem* 53:7512–7517.

Simpson D, Daft MJ (1991) Effects of Glomus clarum and water stress on growth and nitrogen fixation in two genotypes of groundnut. *Afr Ecosyst Environ* 35:47–54.

Sinclair TR, Vadez V (2012) The future of grain legumes in cropping systems. *Crop Pasture Sci* 63:501–512. doi: 10.1071/CP12128.

Sinclair TR, Muchow RC, Bennett JH, Hammmond LC (1987) Relative sensitivity of nitrogen and biomass accumulation to drought in filed-grown soybean. *Agron J* 79(6):986–991. doi: 10.2134/agronj1 987.00021962007900060007x.

Sinclair TR, Purcell LC, King A, Sneller CH, Chen P, Vadez V (2007) Drought tolerance and yield increase of soybean resulting from improved symbiotic N_2 fixation. *Field Crops Res* 101:68–71. doi:10.1016/ j.fcr.2006.09.010.

Smith SE, Facelli E, Pope S, Smith FA (2010) Plant performance in stressful environments: Interpreting new and established knowledge of the roles of arbuscular mycorrhizas. *Plant Soil* 326:3–20. doi: 10.1007/ s11104-009-9981-5.

Sohrabi Y, Heidari G, Weisany W, Golezani KG, Mohammadi K (2012) Some physiological responses of chickpea cultivars to arbuscular mycorrhiza under drought stress. *Russ J Plant Physiol* 59:708–716.

Spaink HP, Wijffelman CA, Pees E, Okker RJH, Lugtenberg BJJ (1987) Rhizobium nodulation gene nodD as a determinant of host specificity. *Nature* 328:337–340. doi: 10.1038/328337a0.

Sprent JI, Raven JA (1985) Evolution of nitrogen fixing symbiosis. *Proc R Soc Edinburgh Sec B Biol Sci* 85(3-4):215–237.

Stacey G, Finan M, O'Brian MR (2002) Signal exchange during the early events of soybean nodulation. In: *Nitrogen fixation global perspectives*. CABI Publishing, Wallingford, UK, pp. 118–122.

Stancheva I, Geneva M, Hristozkova M, Sichanova M, Donkova R, Petkova G, Djonova E (2017) Response of *Vigna unguiculata* grown under different soil moisture regimes to the dual inoculation with nitrogen-fixing bacteria and Arbuscular Mycorrhizal Fungi. *Commun Soil Sci Plant Anal* 48(12):1378–1386. doi: 10.1080/00103624.2017.1358740.

Streeter JG (1993) Translocation-a key factor limiting the efficiency of nitrogen fixation in legume nodules. *Physiol Planta* 87:616–623. doi: 10.1111/j.1399-3054.1993.tb02514.x.

Streeter JG (2003) Effects of drought on nitrogen fixation in soybean root nodules. *Plant Cell Environ* 26:1199–1204. doi: 10.1046/j.1365-3040.2003.01041.x.

Subhan D, Zafar-ul-Hye M, Fahad S, Saud S, Martin B, Tereza H, Rahul D (2020) Drought stress alleviation by ACC deaminase producing *Achromobacter xylosoxidans* and *Enterobacter cloacae*, with and without Timber Waste Biochar in Maize. *Sustain* 12(6286). doi:10.3390/su12156286.

Swaraj K (1987) Environmental stress and symbiotic N_2 fixation in legumes. *Plant Physiol Biochem* 14:117–130.

Sziderics AH, Rasche F, Trognitz F, Sessitsch A (2007) Wilhelm E Bacterial endophytes contribute to abiotic stress adaptation in pepper plants (*Capsicum annuum* L.). *Can J Microbiol* 53:1195–1202. doi: 10.113 9/W07-082.

Taiz L, Zeiger E (1998) *Plant Physiology*. Sinauer Associates, Massachusetts, USA. p 792.

Tariq M, Ahmad S, Fahad S, Abbas G, Hussain S, Fatima Z, Nasim W, Mubeen M, ur Rehman MH, Khan MA, Adnan M. (2018). The impact of climate warming and crop management on phenology of sunflower-based cropping systems in Punjab, Pakistan. *Agric Forest Met* 15(256):270–282.

Tate RL (1995) *Soil microbiology (symbiotic nitrogen fixation)*. John Wiley & Sons, Inc, New York, USA. pp. 307–333.

Thalooth AT, Tawfik MM, Mohamed HM (2006) A Comparative study on the effect of foliar application of zinc, potassium and magnesium on growth, yield and some chemical constituents of mungbean plants grown under water stress conditions. *World J Agric Sci* 2:37–46.

Tiwari S, Lata C, Chauhan PS, Nautiyal CS (2016) Pseudomonas putida attunes morphophysiological, bio-chemical and molecular responses in *Cicer arietinum* L. during drought stress and recovery. *Plant Physiol Biochem* 99:108–117. doi: 10.1016/j.plaphy.2015.11.001.

Ulemale CS, Mate SN, Deshmukh DV (2013) Physiological indices for drought tolerance in chickpea (*Cicer arietinum* L.). *World J Agric Sci* 9:123–131.

Unsar Naeem-U, Muhammad R, Syed HMB, Asad S, Mirza AQ, Naeem I, Muhammad H ur R, Fahad S, Shafqat S (2020) Insect pests of cotton crop and management under climate change scenarios. In: Fahad S, Hasanuzzaman M, Alam M, Ullah H, Saeed M, Khan AK, Adnan M (eds) *Environment, climate, plant and vegetation growth*. Springer Publ Ltd, Springer Nature Switzerland AG. Part of Springer Nature, pp. 367–396. doi: 10.1007/978-3-030-49732-3.

Vadez V, Sinclair TS, Serraj R (2000) Asparagine and ureide accumulation in nodules and shoots as feedback inhibitors of N_2 fixation in soybean. *Physiol Planta* 110:215–223.

Venkateswarlu B, Rao, AV, Lahiri AN (1983) Effect of water stress on nodulation and nitrogenase activity of guar (*Cyamopsis tetragonoloba* (L.) Taub.). *Proc Indian Acad Sci Plant Sci* 92:297–301.

Venkateswarlu B, Maheswari M, Karan NS (1989) Effects of water deficits on N_2 (C_2H_2) fixation in cowpea and groundnut. *Plant Soil* 114:69–74.

Venkateswarlu B, Saharan N, Maheswari M (1990) Nodulation and N_2 fixation in cowpea and groundnut during water stress. *Field Crops Res* 25:223–232.

Verbruggen N, Hermans C (2008) Proline accumulation in plants: A review. *Amino Acids* 35:753–759.

Wahid F, Fahad S, Subhan D, Adnan M, Zhen Y, Saud S, Manzer HS, Martin B, Tereza H, Rahul D (2020) Sustainable management with Mycorrhizae and phosphate solubilizing bacteria for enhanced phos-phorus uptake in calcareous soils. *Agriculture* 10 (334). doi:10.3390/agriculture10080334.

Wajid N, Ashfaq A, Asad A, Muhammad T, Muhammad A,Muhammad S, Khawar J, Ghulam MS, Syeda RS, Hafiz MH, Muhammad IAR, Muhammad ZH, Muhammad Habib ur R, Veysel T, Fahad S, Suad S, Aziz K, Shahzad A (2017) Radiation efficiency and nitrogen fertilizer impacts on sunflower crop in contrasting environments of Punjab. *Pakistan Environ Sci Pollut Res* 25:1822–1836. doi: 10.1007/ s11356-017-0592-z.

Waraich EA, Ahmad R, Halim A, Aziz T (2012) Alleviation of temperature stress by nutrient management in crop plants: A review. *J Soil Sci Plant Nutr* 12(2):221–224. doi: 10.4067/S0718-95162012000200003.

Waraich EA, Ahmad R, Saifullah, Ashraf MY, Ehsanullah (2011) Role of mineral nutrition in alleviation of drought stress in plants. *Aust J Crop Sci* 5(6):764–777.

Williams PM, De Mallorca MS (1984) Effect of osmotically induced leaf moisture stress on nodulation and nitrogenase activity of *Glycine* max. *Plant Soil* 80:267–283. doi: https://www.jstor.org/stable/42 935513.

Worrall VS, Roughley RJ 1976. The effect of moisture stress on infection of *Trifolium subterraneum* L. by *Rhizobium trifolii* Dang. *J Exp Bot* 27:1233–1241.

Wu C, Tang S, Li G, Wang S, Fahad S, Ding Y (2019) Roles of phytohormone changes in the grain yield of rice plants exposed to heat: A review. *PeerJ* 7:e7792. doi: 10.7717/peerj.7792.

Wu C, Kehui C, She T, Ganghua L, Shaohua W, Fahad S, Lixiao N, Jianliang H, Shaobing P, Yanfeng D (2020) Intensified pollination and fertilization ameliorate heat injury in rice (Oryza sativa L.) during the flowering stage. *Field Crops Res* 252: 107795.

Xu CX, Ma YP, Liu YL (2015) Effects of silicon (Si) on growth, quality and ionic homeostasis of aloe under salt stress. *South Afr J Bot* 98:26–36. doi: 10.1016/j.sajb.2015.01.008.

Yadavi A, Movahhedi-dehnavi M, Balouchi H (2014) Effect of micronutrients foliar application on grain qualitative characteristics and some physiological traits of bean (*Phaseolus vulgaris* L.) under drought stress. *Indian J Fundam Appl Life Sci* 4:124–131.

Yang J, Kloepper JW, Ryu CM (2009) Rhizosphere bacteria help plants tolerance abiotic stress. *Trends Plant Sci* 14:1–4.

Yang Z, Zhang Z, Zhang T, Fahad S, Cui K, Nie L, Peng S, Huang J (2017) The effect of season-long temperature increases on rice cultivars grown in the central and southern regions of China. *Front Plant Sci* 8:1908. doi: 10.3389/fpls.2017.01908.

Zafar-ul-Hye M, Muhammad TH, Muhammad A, Fahad S, Martin B, Tereza D, Rahul D, Subhan D (2020a) Potential role of compost mixed biochar with rhizobacteria in mitigating lead toxicity in spinach. *Scientific Rep* 10:12159. doi: 10.1038/s41598-020-69183-9.

Zafar-ul-Hye M, Muhammad N, Subhan D, Fahad S, Rahul D, Mazhar A, Ashfaq AR, Martin B,Jˇrí H, Zahid HT, Muhammad N (2020b) Alleviation of cadmium adverse effects by improving nutrients uptake in bitter gourd through Cadmium tolerant Rhizobacteria. *Environment* 7 (54). doi:10.3390/environments 7080054.

Zahida Z, Hafiz FB, Zulfiqar AS, Ghulam MS, Fahad S, Muhammad RA, Hafiz MH, Wajid N, Muhammad S (2017) Effect of water management and silicon on germination, growth, phosphorus and arsenic uptake in rice. *Ecotoxicol Environ Saf* 144:11–18.

Zahran HH (1986) Effect of sodium chloride and polyethylene glycol on rhizobial root hair infection, root nodule structure and symbiotic nitrogen fixation in *Vicia faba* L plants. PhD thesis, Dundee University, Dundee, Scotland.

Zahran HH (1999) Rhizobium-legume symbiosis and nitrogen fixation under severe conditions and in an arid climate. *Microbiol Mole Biol Rev* 63(4):968–989.

Zahran HH, Sprent JI (1986) Effects of sodium chloride and polyethylene glycol on root hair infection and nodulation of *Vicia faba* L. plants by *Rhizobium leguminosarum*. *Planta* 167:303–309.

Zarafshar M, Akbarinia M, Askari H, Hosseini SM, Rahaie M, Struve D, Striker GG (2014) Morphological, physiological and biochemical responses to soil water deficit in seedlings of three populations of wild pear tree (*Pyrusbois seriana*). *Biotechnol Agron Soc Environ* 18(3): 353–366.

Zhang J, Nguyen HT, Blum A (1999) Genetic analysis of osmotic adjustment in crop plants. *J Exp Bot* 50:291–302.

Zhu JK, Xiong L (2002) Molecularand genetic aspects of plant responses to osmotic stress. *Plant Cell Environ* 25(2):131–139. doi: 10.1046/j.1365-3040.2002.00782.x.

Zia-ur-Rehman M (2020) Environment, climate change and biodiversity. In: Fahad S, Hasanuzzaman M, Alam M, Ullah H, Saeed M, Khan AK, Adnan M (eds) *Environment, climate, plant and vegetation growth*. Springer Publ Ltd, Springer Nature Switzerland AG. Part of Springer Nature, pp. 473–502. doi: 10.1007/978-3-030-49732-3.

10

Auxin's Role in Plant Development in Response to Stress

Dr. Nazish Huma Khan, Dr. Mohammad Nafees, Mr. Fazli Zuljalal, and Ms. Tooba Saeed

CONTENTS

10.1 Introduction

Plants react to various environmental stresses at the molecular level by changing the expression of several genes that are involved in different pathways. Plant hormones, ethylene, abscisic acid, jasmonic acid, and salicylic acid are involved in both biological and non-biological stress responses (Wang et al. 2007). Being sessile species, plants respond to various unfavourable environmental conditions. This has been particularly important in recent decades in the context of current climate variability and forecasts of continued warming and low precipitation. The effect of environmental changes has significantly influenced global agricultural production (Mba et al. 2012: Khedun and Singh 2014; Adnan et al. 2018a, 2018b; Adnan et al. 2019; Akram et al. 2018a, 2018b; Aziz et al. 2017a, 2017b; Habib et al. 2017; Hafiz et al. 2016; Hafiz et al. 2019; Kamaran et al. 2017; Muhmmad et al. 2019; Sajjad et al. 2019; Saud et al. 2013; Saud et al. 2014; Saud et al. 2017; Saud et al. 2016; Shah et al. 2013; Saud et al. 2020; Qamar et al. 2017; Wajid et al. 2017; Yang et al. 2017; Zahida et al. 2017; Depeng et al. 2018; Hussain et al. 2020; Hafiz et al. 2020a, 2020b; Shafi et al. 2020; Wahid et al. 2020; Subhan et al. 2020; Zafar-ul-Hye et al. 2020a, 2020b; Adnan et al. 2020; Ilyas et al. 2020; Saleem et al. 2020a, 2020b, 2020c; Rehman et al. 2020; Farhat et al. 2020; Wu et al. 2020; Mubeen et al. 2020; Farhana et al. 2020; Jan et al. 2019; Wu et al. 2019; Ahmad et al. 2019; Baseer et al. 2019; Hafiz et al. 2018; Tariq et al. 2018).

Various studies reported that considerable losses in crop productivity (more than 50%) are caused by non-biological/abiotic stresses such as drought, increased salinity, and extreme temperature inducing food shortage at global food security level. Statistics reported about 6% of the global land area is affected by salinity, 64% by drought, 13% by flooding, and about 57% by extreme temperatures (Munns and Tester 2008, Cramer et al. 2011, Ismail et al. 2014; Fahad and Bano 2012; Fahad et al. 2017; Fahad et al. 2013; Fahad et al. 2014a,

2014b; Fahad et al. 2016a, 2016b, 2016c, 2016d; Fahad et al. 2015a, 2015b; Fahad et al. 2018; Fahad et al. 2019a, 2019b; Hesham and Fahad 2020; Iqra et al. 2020; Akbar et al. 2020; Mahar et al. 2020; Noor et al. 2020; Bayram et al. 2020; Amanullah et al. 2020; Rashid et al. 2020; Arif et al. 2020; Amir et al. 2020; Saman et al. 2020; Muhammad et al. 2020; Md Jakir and Allah 2020; Farah et al. 2020; Sadam et al. 2020; Unsar et al. 2020; Fazli et al. 2020; Md. et al. 2020; Gopakumar et al. 2020; Zia-ur-Rehman 2020; EL Sabagh et al. 2020; Al-Wabel et al. 2020a, 2020b; Senol 2020; Amjad et al. 2020; Ibrar et al. 2020; Sajid et al. 2020.

The period between 2000–2009 is reported as the warmest decade, with a decrease in net primary production due to large-scale droughts in the Southern Hemisphere (Khedun and Singh 2014). To perceive these external signals, plants have a mechanized system that responds accordingly by changing the expression of various defence-signalling molecules (Derksen et al. 2013). Growth retardation and reduced metabolism are the most common responses of plants to stresses that improve their survival under stress conditions. They coordinate a wide range of processes such as modification of photosynthesis, increased antioxidant activities, secondary metabolites accumulation, changes in gene expression, and similar activities to protect itself (Krasensky and Jonak 2012). To improve the plant's tolerance against unfavourable conditions (abiotic stress), it is important to investigate and understand the mechanisms of plant stress responses and tolerance. In this context, the plant growth hormones known as Phytohormones are important to enhance the plant's ability to adapt to stressful environmental conditions. Such hormones help in mediating a wide range of adaptive responses including growth, development, nutrient allocation, and transition source (Khan et al. 2012; Wilkinson et al. 2012). The literature revealed that among various phytohormones, Auxin plays a central role to overcome the stress conditions and helps in the developmental process of plants (Kohli et al. 2013; Rahman 2013). To participate in adaptive responses against stress environmental conditions, Auxin signalling is proposed (Iglesias et al. 2010). Furthermore, Auxin, ROS (reactive oxygen species), antioxidants, and related proteins are proposed for the formation of redox signalling modules that coordinate plant growth with environmental factors (De Tullio et al 2010; Tognetti et al 2012). While other plant hormones such as cytokinins, brassinosteroids, ethylene, abscisic acid, gibberellins, jasmonic acid, and strigolactones coordinate plant growth with environmental factors either synergistically or antagonistically with Auxin to trigger event cascading, leading to stress reactions (Salopek-Sondi et al. 2017).

10.2 Auxin Isolation and Its Structure

Auxin is a phytohormone that plays a vital role in plant growth and development (Davies 2010). Auxin was the first plant growth hormones discovered, and its name was derived from the Greek word αυξειν (auxein means 'to grow or to increase'). Auxin or indole-acetic-acid (IAA) was discovered 70 years back. But its role in the plant development and growth was discovered later on. By the year 1980s, Auxin research took off with the discovery of a battery of genes, TIR1 Auxin receptor family and various Auxin transporters. These discoveries sparked a wave of research into Auxin signalling and the role of newly discovered genes in plant growth and development by using the model plant species of Arabidopsis Thaliana in the experimental systems (Tivendale et al. 2012; Calderón-Villalobos et al. 2010).

The IPUC name of Auxin is 2-(1H-indole-3-yl) acetic acid. It is a weak organic acid that is composed of a planer indole ring structure grouped to a side chain harbouring a carboxyl group in the terminal (Figure 10.1). The carboxyl group is protonated at low pH that makes the molecule less polar. The less polar form diffuses across cell membranes while the molecule when negatively charged is too polar to diffuse (Rosquete et al. 2012).

Auxin has an important part to play in the entire life span of a plant. This little organic acid influences cell elongation, cell differentiation, cell division, and has a role, not insignificant, in the function of cells and tissues in higher plants (Ljung 2013). Its hand in promoting plant growth was first noted by Charles Darwin and his son Francis in studying the phototropism of coleoptile of canary grass (Phalaris canariensis) and was documented in the remarkable book entitled 'The Power of Movement in Plants' published in 1888. The existence of Auxin in the tip of oat (Avena sativa) that regulates phototropism of

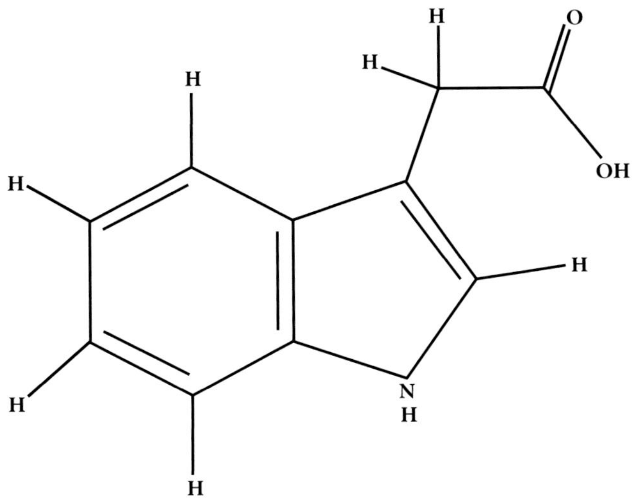

FIGURE 10.1 Chemical structure of Auxin.

coleoptile of oat was unequivocally demonstrated by Frits Went in 1926. Indole-3-acetic acid (IAA) is the primary Auxin in higher plants and was isolated by Thimann in the 1930s (Thimann 1936). IAA and several other chemicals with similar structure and physiological activity in inducing cell elongation of stems were named Auxin in 1954 (Stowe and Thimann 1954).

10.3 Auxin Biosynthesis

Plants produce various phytohormones including Cytokinin, Auxin, Ethylene, Gibberellins, Aabscisic acid, and Brassinosteriod. The production of plant hormones is precisely regulated to manage the process of cell production, plant growth, and development (Mashiguchi et al. 2011). Among these phytohormones, Auxin is responsible for various processes like cell division, root formation, flowering, expansion and differentiation, tropic responses, and senescence. Therefore, among plant hormones, Auxin is considered important as a regulator of plant growth and development (Zhao 2010). Furthermore, in plants, Auxin also responds against environmental stresses such as salinity, drought, and pathogenic attack.

Plants contain various types of Auxins such as indole-3-acetic acid (IAA), indole-3-butyric acid (IBA), indole-3-propionic acid (IPA), 4-choroindole-3-acetic acid (4-CI-IAA), and 2-phenyleacetic acid (PAA). Among these types, IAA also referred as Auxin is considered as the most abundant endogenous type of Auxin that runs majority of Auxin-mediated effects in plants (Korver et al. 2018).

The plant Auxin such as IAA (Indole-3-acetic acid) is synthesized by Trp (Tryptophan)-dependent and Trp-independent pathways. Only the physiological functions and molecular components of the Trp-dependent pathway is well-known that proceeds through 4-metabolic intermediates. Such pathways are known as IPA (indole-3-pyruvic acid), IAOx (indole-3-acetaldoxime), TAM (tryptamine), and IAM (indole-3-acetamide). As the Trp-dependent pathway is responsible for plant developmental process, therefore, the majority of the Trp-dependent pathway is involved in the biosynthesis of Auxin (Won et al. 2011). In Auxin synthesis, YUCCA6 protein is responsible to work as a limiting factor that converts IPA into IAA. It also plays a role in the drought stress signalling rout-way by keeping the ROS equilibrium in plants (Figure 10.2). The IPA pathway is a simple 2-step pathway involved in the conversion of IPA by a family of TAA (tryptophan-aminotransferase of Arabidopsis). The family of TAA consisting of 3-closely linked genes of TAA1, TAR1, TAR2 in Arabidopsis constitute important components in Auxin synthesis (Park et al. 2013a, 2013b).

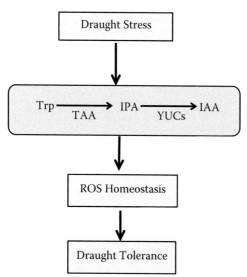

FIGURE 10.2   Shows the function of YUCs protein in drought stress.

10.4 Auxin's Role in Plant Growth

Auxin is a plant hormone that plays an important role in various aspects of plant growth and development. Therefore, it is also known as the plant growth regulator hormone. As a growth regulator, it is involved in many developmental processes of plants such as phototropism, gravitropism, cell division, meristem organization, fruit development, maintenance and development of organ polarity (Naser and Shani 2016). Auxin is also responsible for increasing the plant's tolerance to abiotic stress. To maximise its reach, Auxin is produced in the young growing leaves and then transported to various parts of the plant by both passive diffusion and active Auxin influx and efflux transport systems. In plants, the distribution and transport of Auxin regulate root growth, initiation of lateral root, and Auxin-mediated signalling. The literature revealed that Auxin is a central regulator of root growth (Overvoorde et al. 2010). Therefore, both endogenous and synthetic Auxin has been widely used in global horticulture and agriculture for more than 60 years (Ljung et al. 2005).

10.4.1 Auxin-Mediated Stress Responses in Plants

Being sessile organisms, plants often feel the stresses of both biotic and abiotic. Stresses such as changes in the climatic or weather factors including cold, heat, drought, and salinity are expressed as abiotic stresses. Whereas the small organisms such as bacteria, fungi, and nematodes induce biotic stresses in plants. To handle such stresses, plants have a well-developed mechanism comprising of hormones that will perceive the stress conditions and then regulate the developmental process of plants. Among plant hormones, Auxin's stress responses in plants are many such as many of the Auxin-responsive transcripts induced by drought and cold stress (Jain and Khurana 2009; Khan et al. 2015). The Auxin's transporting proteins in plants are caused by cold stress conditions. Since without Auxin plants can be helpless against stress events, therefore, to enhance the level of endogenous Auxin in Arabidopsis, and to improve the tolerance level against drought, there is the activation of the YUC7 gene. While, OsGH3-1 and OsGH3–8 genes are responsible as effective fighters against fungi, resistance to diseases, and regulating plant development (Lee and Luan 2012; Wang and Fu 2011). In hotter temperatures, there is IAA production in various parts of plants such as cotyledons, hypocotyl, and roots. In such inhospitable conditions, the level of endogenous Auxin is decreased due to the reduction of YUC gene thus leading to male sterility in plants. Alternately, the application of exogenous IAA can rescue the pollen viability in plants under heat stress conditions (Sakata et al. 2010).

    The abiotic stresses such as drought and salinity cause major decrease in crop production. Such stresses are a threat to arable land. Water deficiency (drought) causes crop loss while the soil salinity induces ionic and osmotic stress in plants. Therefore, to improve the levels crop tolerance to such conditions has attracted the attention of researchers as how to improve the yield of edible parts of plant (Morales-Tapia and Cruz-Ramirez 2016). Some of the recent studies suggested that modifications in the root system of plants can give good yield. In this regard, Auxin has been studied to verify its function in the optimal conditions and has proved to be effective agents to improve the crop production. Again, because Auxin plays an important part during abiotic stress-induced changes in the plant root, researchers have studied its various processes involved in root-related changes to determine Auxin-mediated regulation of plant growth under stress conditions, like biosynthesis, Auxin transport, conjugation, perception, and Auxin-signalling. In this regard, the role of polar auxin transport (PAT) through Auxin carrier protein has been well-established. The main mechanism to control the seamless Auxin flow in the root system is PAT. While, ABCB is also another numerous family of Auxin carriers that enhance Auxin flow in the plant root during salt stress conditions (Korver et al. 2018).

10.4.2 Auxin and ROS-Plant Stress Responses

Auxin has been reported to help avert induction of stress-related genes in association with abiotic stress-induced anthocyanin production and regulation of ROS homoeostasis. Therefore, there is a signalling interchange of Auxin to ROS (Gao et al. 2014) affecting plant growth and development by the associations between Auxin and ROS under several environmental stresses. Due to different abiotic stresses, ROS is produced which results in common oxidative stress, and given a higher level of ROS leads to stress conditions in plants that again leads to oxidative damage at molecular and cellular levels, therefore,to prevent oxidative damage, the excessive production of ROS needs to be controlled at the basal level. The antioxidant scavenging systems help to maintain the ROS homoeostasis and also detoxify its toxicity (Gill and Tuteja 2010). The oxidative stress-responsive genes are induced in response to oxidative stress that can be regulated by ARFs. The exogenous Auxins induced some glutaredoxins. Accumulation of Auxin has positive effects on ROS accumulation. To promote the ARF action, ROS has been proposed to regulate the depredation of Auxin/IAA protein (Figure 10.3). Because, Auxin

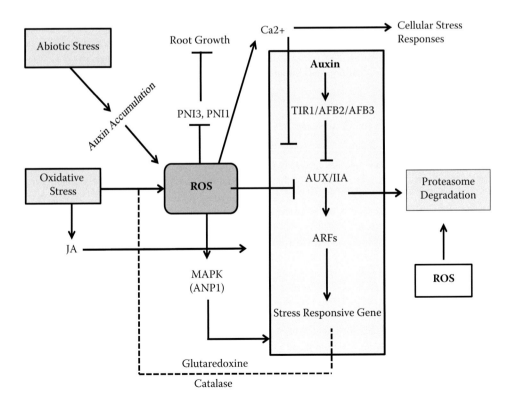

FIGURE 10.3 The crosstalk between oxidative stresses to Auxin.

is responsible for regulating different processes such as cell-cycle progression, cell death, and cell-viability that depend on ROS gesturing.

10.4.3 Auxin's Role in Pathogen Defence of Plants

In the pathogen defence system of plants, Auxin plays both a direct and indirect role. In the cell wall modification, root morphology, and stomatal anatomy Auxin has an indirect roles to play (Kazan and Manners 2009). Plants' defence responses are initiated after perceiving pathogen-associated molecular pattern during pathogen attack. Abolishing of Auxin signalling causes strong resistance inPseudomonas syringe pv and Hyaloperonospora Arabidopsis (Navarro et al. 2006). While induction of salicylic acid facilitated resistance in biotrophic pathogens. Phytohormones (salicylic acid and ABA) and plant stresses (biotic and abiotic stress) suppress the Auxin signalling with the induction of GH3 homologue WES1 to target free Auxin for conjugation (Park et al. 2007). Auxin induces resistance to necrotrophs such as Alternaria brassicicola and Botrytis cinerea. Besides altering Auxin signalling, plant pathogens also modulate Auxin biosynthesis and homoeostasis, such as Ralstonia solanacearum and Xanthomonas species induce Auxin biosynthesis in plants and therefore, increase the vulnerability of host plants. Pathogens induced the availability of Auxin that caused the induction of lateral root and facilitated colonization and infection of the pathogen (Wang and Fu 2011).

10.4.4 Auxin as a Root Forming Hormone

Auxin is the central phytohormone that plays a major part in root growth and development. Therefore it is called a 'root forming hormone' (Overvoorde et al. 2010; Ahmad et al. 2015). It inhibits and promotes and regulates root growth in a dose-dependent way. In dicot Arabidopsis, the development of primary, lateral, and adventitious roots and, in monocot rice, the development of crown and the lateral root is due to Auxin signalling (Yin et al. 2011; Wang et al. 2011). The wet and dry surfaces of plant roots lead to change in the pattern of aerenchyma formation, root growth, and root hairs in plants. To manage these processes as normal, Auxin acts as the central player controlling and regulating their triggering and rate of activity. To maintain the root apical meristem and initiation of the lateral root, a gradient distribution of Auxin is important. For root morphogenesis, the Auxin gradient is dependent on its biosynthesis in tips of the root, transportation, conjugation, signalling, and intracellular compartmentation (Barbez et al. 2012).

10.5 Nutrients as Auxin Transporters in Plants

Plant roots have the potential to adapt to the availability of micro-nutrients and macro-nutrients in plants. Auxin has a prominent role in regulating root responses to the availability of nutrients and water in plants. Some nutrients are responsible for the transportation of high Auxin by the plant root system (Krouk et al. 2010). In Arabidopsis, the NRT1.1 protein acts as a nitrate transporter. Under low nitrate availability, the NRT1.1 functions as an Auxin transporter and helps to inhibit the production and growth of lateral root (Gojon et al. 2011). While miR393/AFB3 is an important N-responsive module that regulates primary and lateral root growth (Vidal et al. 2010).

Like the nitrate transporters, the Arabidopsis TRH1 gene encodes the potassium transporter that works as a putative Auxin transporter. The TRH1 gene altered the gravitropism and development of root hair that decrease Potassium uptakes (K+). The developmental defects of TRH1 in plants can be restored with the application of exogenous Auxin as TRH1 acts as a regulator of Auxin that helps in the gravity perception of roots (Rigas et al. 2001). Moreover, phosphate deficiency induces suppression of primary root growth and improves the lateral root growth in plants. Root responses to Phosphate deficient conditions depend upon the functional Auxin signalling. The pericyclic cells of primary roots with phosphate deficient seedlings were found hypersensitive to Auxin (Pérez-Torres et al. 2008). Under phosphate deficient conditions, the SIZ1 gene inhibits Auxin transport as it enhances the development of the lateral root system (Miura et al. 2011). Iron deficiency in plants is exerted by AUX1 to mediate the transport of Auxin. Auxin plays an important part in plant root adaptation for nutrients availability (Giehl et al. 2012).

10.6 Auxin Transport and Hormonal Crosstalk

Different hormones interact with Auxin to maintain the plant's growth and development. Plant development is regulated by hormonal crosstalk. One of the important targets in crosstalk is Auxin transport (Rahman 2013). Auxin transport modifies the intracellular Auxin gradient that is responsible for the maintenance of hormonal crosstalk and plant development (Figure 10.4). In plants, the growth and development is regulated by a complex web of hormonal interactions where Auxin has been found as a common factor interlacing these interactions. The development of plant shoot and root has been accelerated by both Auxin and Cytokinin (Dello Ioio et al. 2008). While ethylene hormone has also been found close in associations with Auxin. Ethylene and Auxin show synergistic and antagonistic interactions in plants developmental processes like the elongation of shoot and root, the formation of apical hook, development of lateral root, initiation and elongation of root hairs, shoot and root gravitropism and others. The combination of Auxin and Gibberellin (GA) are responsible for elongation of pea stem and Parthenocarpy, inflorescence and root development in Arabidopsis, and development of lateral root in populous (Willige et al. 2011). Brassinolide (BR) interacts with Auxin to maintain the process of cell elongation in shoot and root parts of the pant. While Jasmonic acid (JA) acts synergistically and antagonistically with Auxin to regulate the development of root, flowering, and coleoptile elongation respectively, similarly, the ABA and Auxin regulate the development of lateral root and formation of the embryonic axis in plants (Belin et al. 2009). In plants, the Auxin transport in hormonal crosstalk is the interactions of regulated plant parts development at biosynthesis levels such as Auxin-ethylene, Auxin-GA, Auxin-cytokinin, and Auxin-BR (Figure 10.4).

10.7 Conclusion

Plant species are sessile that are often exposed to both biotic and abiotic stresses. The abiotic stresses include temperature variation (heat, cold), salinity, and drought while the biotic conditions are from infection from bacteria, fungi, insects, and nematodes, and other parasites. To cope with these adverse conditions, a well-developed mechanism has evolved in plants that help to perceive the stresses and support optimal growth. This mechanism is a plant hormone called 'Phytohormones' that help the plants in the adaptation to adverse environmental conditions. Among these hormones, Auxin is considered effective for mediating defence responses. In plant development, the role of Auxin is well-known particularly, its function in root response to abiotic stress. Besides, Auxin play's a crucial role in plant

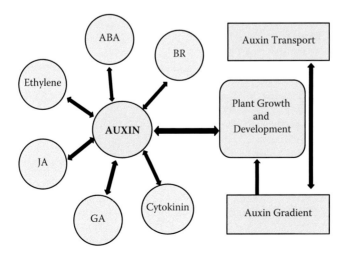

FIGURE 10.4 Hormonal crosstalk for plant growth and development.

tolerance to oxidative stress. Literature revealed that Auxin is an important part of the plant's overall biotic and stress tolerance mechanism. Being a sessile species, plants face both optimal and non-optimal phases during the life cycle. Different biotic and abiotic stresses affect the whole plant developmental processes. But hormones such as jasmonate, salicylic acid, and ethylene are considered effective in regulating plant growth under stress conditions. Under optimum conditions, Auxin regulates all aspects of the plant developmental process making it an indispensable actor in the drama of plant survival in inhospitable environments fulfilling its promise with an allround, stellar performance.

REFERENCES

Adnan M, Fahad S, Khan IA, Saeed M, Ihsan MZ, Saud S, Riaz M, Wang D, Wu C (2019). Integration of poultry manure and phosphate solubilizing bacteria improved availability of Ca bound P in calcareous soils. *3 Biotech. 1* 9(10):368.

Adnan M, Fahad S, Muhammad Z, Shahen S, Ishaq AM, Subhan D, Zafar-ul-Hye M, Martin LB, Raja MMN, Beena S, Saud S, Imran A, Zhen Y, Martin B, Jiri H, Rahul D (2020) Coupling phosphate-solubilizing bacteria with phosphorus supplements improve maize phosphorus acquisition and growth under lime induced salinity stress. *Plants* 9(900). doi: 10.3390/plants9070900

Adnan M, Shah Z, Sharif M, Rahman H (2018a). Liming induces carbon dioxide (CO2) emission in PSB inoculated alkaline soil supplemented with different phosphorus sources. *Environ Sci Poll Res 1* 25(10): 9501–9509.

Adnan M, Zahir S, Fahad S, Arif M, Mukhtar A, Imtiaz AK, Ishaq AM, Abdul B, Hidayat U, Muhammad A, Inayat-Ur R, Saud S, Muhammad ZI, Yousaf J, Amanullah, Hafiz MH, Wajid N (2018b) Phosphate-solubilizing bacteria nullify the antagonistic effect of soil calcification on bioavailability of phosphorus in alkaline soils. *Sci Rep* 8:4339. doi: 10.1038/s41598-018-22653-7

Ahmad K, Adnan M, Khan MA, Hussain Z, Junaid K, Saleem N, Ali M, Basir A (2015) Bioactive neem leaf powder enhances the shelf life of stored mungbean grains and extends protection from pulse beetle. *Pak J Weed Sci Res* 21(1):71–81.

Ahmad S, Kamran M, Ding R, Meng X, Wang H, Ahmad I, Fahad S, Han Q (2019) Exogenous melatonin confers drought stress by promoting plant growth, photosynthetic capacity and antioxidant defense system of maize seedlings. *PeerJ* 7:e7793. doi: 10.7717/peerj.7793

Akbar H, Timothy JK, Jagadish T, M. Golam M, Apurbo KC, Muhammad F, Rajan B, Fahad S, Hasanuzzaman M (2020) Agricultural land degradation: processes and problems undermining future food security. In: Fahad S, Hasanuzzaman M, Alam M, Ullah H, Saeed M, Khan AK, Adnan M (eds) *Environment, climate, plant and vegetation growth.* Springer Publ Ltd, Springer Nature Switzerland AG. Part of Springer Nature, pp 17–62. doi: 10.1007/978-3-030-49732-3

Akram R, Turan V, Hammad HM, Ahmad S, Hussain S, Hasnain A, Maqbool MM, Rehmani MIA, Rasool A, Masood N, Mahmood F, Mubeen M, Sultana SR, Fahad S, Amanet K, Saleem M, Abbas Y, Akhtar HM, Waseem F, Murtaza R, Amin A, Zahoor SA, ul Din MS, Nasim W (2018a) Fate of organic and inorganic pollutants in paddy soils. In: Hashmi MZ, Varma A (eds) *Environmental pollution of paddy soils, soil biology.* Springer International Publishing Ag, GEWERBESTRASSE 11, CHAM, CH-6330, Switzerland, pp 197–214.

Akram R, Turan V, Wahid A, Ijaz M, Shahid MA, Kaleem S, Hafeez A, Maqbool MM, Chaudhary HJ, Munis, MFH, Mubeen M, Sadiq N, Murtaza R, Kazmi DH, Ali S, Khan N, Sultana SR, Fahad S, Amin A, Nasim W (2018b) Paddy land pollutants and their role in climate change. In: Hashmi MZ, Varma A (eds) *Environmental pollution of paddy soils, soil biology.* Springer International Publishing Ag, GEWERBESTRASSE 11, CHAM, CH-6330, Switzerland, pp 113–124.

Al-Wabel MI, Abdelazeem S, Munir A, Khalid E, Adel RAU (2020a) Extent of climate change in Saudi Arabia and its impacts on agriculture: a case study from Qassim region. In: Fahad S, Hasanuzzaman M, Alam M, Ullah H, Saeed M, Khan AK, Adnan M (eds) *Environment, climate, plant and vegetation growth.* Springer Publ Ltd, Springer Nature Switzerland AG. Part of Springer Nature. pp 635–658. doi: 10.1007/978-3-030-49732-3

Al-Wabel MI, Ahmad M, Usman ARA, Akanji M, Rafique MI (2020b) Advances in pyrolytic technologies with improved carbon capture and storage to combat climate change. In: Fahad S, Hasanuzzaman M, Alam M, Ullah H, Saeed M, Khan AK, Adnan M (eds) *Environment, climate, plant and vegetation growth*. Springer Publ Ltd, Springer Nature Switzerland AG. Part of Springer Nature, pp 535–576. doi: 10.1007/978-3-030-49732-3

Amanullah, Shah K, Imran, Hamdan AK, Muhammad A, Abdel RA, Muhammad A, Fahad S, Azizullah S, Brajendra P (2020) Effects of climate change on irrigation water quality. In: Fahad S, Hasanuzzaman M, Alam M, Ullah H, Saeed M, Khan AK, Adnan M (eds) *Environment, climate, plant and vegetation growth*. Springer Publ Ltd, Springer Nature Switzerland AG. Part of Springer Nature, pp 123–132. doi: 10.1007/978-3-030-49732-3

Amir M, Muhammad A, Allah B, Sevgi Ç, Haroon ZK, Muhammad A, Emre A (2020) Bio fortification under climate change: the fight between quality and quantity. In: Fahad S, Hasanuzzaman M, Alam M, Ullah H, Saeed M, Khan AK, Adnan M (eds) *Environment, climate, plant and vegetation growth*. Springer Publ Ltd, Springer Nature Switzerland AG. Part of Springer Nature, pp 173–228. doi: 10.1007/978-3-030-49732-3

Amjad I, Muhammad H, Farooq S, Anwar H (2020) Role of plant bioactives in sustainable agriculture. In: Fahad S, Hasanuzzaman M, Alam M, Ullah H, Saeed M, Khan AK, Adnan M (eds) *Environment, climate, plant and vegetation growth*. Springer Publ Ltd, Springer Nature Switzerland AG. Part of Springer Nature, pp 591–606. doi: 10.1007/978-3-030-49732-3

Arif M, Talha J, Muhammad R, Fahad S, Muhammad A, Amanullah, Kawsar A, Ishaq AM, Bushra K, Fahd R (2020) Biochar; a remedy for climate change. In: Fahad S, Hasanuzzaman M, Alam M, Ullah H, Saeed M, Khan AK, Adnan M (eds) *Environment, climate, plant and vegetation growth*. Springer Publ Ltd, Springer Nature Switzerland AG. Part of Springer Nature, pp 151–172. doi: 10.1007/978-3-030-4 9732-3

Aziz K, Daniel KYT, Muhammad ZA, Honghai L, Shahbaz AT, Mir A, Fahad S (2017b) Nitrogen fertility and abiotic stresses management in cotton crop: a review. *Environ Sci Pollut Res* 24:14551–14566. doi: 10.1007/s11356-017-8920-x

Aziz K, Daniel KYT, Fazal M,Muhammad ZA, Farooq S, FanW, Fahad S, Ruiyang Z (2017a) Nitrogen nutrition in cotton and control strategies for greenhouse gas emissions: a review. *Environ Sci Pollut Res* 24:23471–23487. doi: 10.1007/s11356-017-0131-y

Barbez E, Kubez M, Rolcik J, Beziat C, Pencik A, Wang B, Kleine-Vehn J (2012) A novel putative auxin carrier family regulates intracellular auxin homeostasis in plants. *Nature* 485(7396):119–122.

Baseer M, Adnan M, Fazal M, Fahad S, Muhammad S, Fazli W, Muhammad A, Jr. Amanullah, Depeng W, Saud S, Muhammad N, Muhammad Z, Fazli S, Beena S, Mian AR, Ishaq AM (2019) Substituting urea by organic wastes for improving maize yield in alkaline soil. *J Plant Nutrition*. doi: doi.org/10.1080/01 904167.2019.1659344

Bayram AY, Seher Ö, Nazlican A (2020) Climate change forecasting and modeling for the year of 2050. In: Fahad S, Hasanuzzaman M, Alam M, Ullah H, Saeed M, Khan AK, Adnan M (eds) *Environment, climate, plant and vegetation growth*. Springer Publ Ltd, Springer Nature Switzerland AG. Part of Springer Nature, pp 109–122. doi: 10.1007/978-3-030-49732-3

Belin C, Megies C, Hauserova E, Lopez-Molina L (2009). Abscisic acid represses growth of the Arabidopsis embryonic axis after germination by enhancing auxin signaling. *Plant Cell* 21:2253–2268.

Calderón-Villalobos LI, Tan X, Zheng N, Estelle M (2010) Auxin perception structural insights. *Cold Spring Harb Perspect Biol* 2:a005546.

Cramer GR, Urano K, Delrot S, Pezzotti M, Shinozak K (2011) Effects of abiotic stress on plants: a systems biology perspective. *BMC Plant Biol* 11:163.

Davies PJ (2010) The plant hormones: their nature, occurrence, and functions. In: *Plant hormones*. Springer, Dordrecht, pp 1–15.

De Tullio MC, Jiang K, Feldman LJ (2010) Redox regulation of root apical meristem organization: connecting root development to its environment. *Plant Physiol Biochem* 48:328–336.

Dello Ioio R, Linhares FS, Sabatini S (2008). Emerging role of cytokinin as a regulator of cellular differentiation. *Curr Opin Plant Biol* 11:23–27.

Depeng W, Fahad S, Saud S, Muhammad K, Aziz K, Mohammad NK, Hafiz MH, Wajid N (2018) Morphological acclimation to agronomic manipulation in leaf dispersion and orientation to promote "Ideotype" breeding: Evidence from 3D visual modeling of "super" rice (Oryza sativa L.). *Plant Physiol Biochem* 135:499–510. doi: 10.1016/j.plaphy.2018.11.010

Derksen H, Rampitsch C, Daayf F (2013) Signaling cross-talk in plant disease resistance. *Plant Sci* 207:79–87.

EL Sabagh A, Hossain A, Barutçular C, Iqbal MA, Islam MS, Fahad S, Sytar O, Çig F, Meena RS, Erman M (2020) Consequences of salinity stress on the quality of crops and its mitigation strategies for sustainable crop production: an outlook of arid and semi-arid regions. In: Fahad S, Hasanuzzaman M, Alam M, Ullah H, Saeed M, Khan AK, Adnan M (eds) *Environment, climate, plant and vegetation growth*. Springer Publ Ltd, Springer Nature Switzerland AG. Part of Springer Nature, pp 503–534. doi: 10.1007/978-3-030-49732-3

Fahad S, Adnan M, Hassan S, Saud S, Hussain S, Wu C, Wang D, Hakeem KR, Alharby HF, Turan V, Khan MA, Huang J (2019a). Rice responses and tolerance to high temperature. In: Hasanuzzaman M, Fujita M, Nahar K, Biswas JK (eds) *Advances in rice research for abiotic stress tolerance*. Woodhead Publ Ltd, Abington Hall Abington, Cambridge CB1 6AH, CAMBS, England, pp 201–224.

Fahad S, Bano A (2012) Effect of salicylic acid on physiological and biochemical characterization of maize grown in saline area. *Pak J Bot* 44:1433–1438.

Fahad S, Bajwa AA, Nazir U, Anjum SA, Farooq A, Zohaib A, Sadia S, Nasim W, Adkins S, Saud S, Ihsan MZ, Alharby H,Wu C,Wang D, Huang J (2017) Crop production under drought and heat stress: plant responses and management options. *Front Plant Sci* 8:1147. doi: 10.3389/fpls.2017.01147

Fahad S, Chen Y, Saud S, Wang K, Xiong D, Chen C, Wu C, Shah F, Nie L, Huang J (2013) Ultraviolet radiation effect on photosynthetic pigments, biochemical attributes, antioxidant enzyme activity and hormonal contents of wheat. *J Food, Agri Environ* 11(3&4):1635–1641.

Fahad S, Hussain S, Bano A, Saud S, Hassan S, Shan D, Khan FA, Khan F, Chen Y, Wu C, Tabassum MA, Chun MX, Afzal M, Jan A, Jan MT, Huang J (2014a) Potential role of phytohormones and plant growth-promoting rhizobacteria in abiotic stresses: consequences for changing environment. *Environ Sci Pollut Res* 22(7):4907– 4921. doi: 10.1007/s11356-014-3754-2

Fahad S, Hussain S, Matloob A, Khan FA, Khaliq A, Saud S, Hassan S, Shan D, Khan F, Ullah N, Faiq M, Khan MR, Tareen AK, Khan A, Ullah A, Ullah N, Huang J (2014b) Phytohormones and plant responses to salinity stress: a review. *Plant Growth Regul* 75(2):391– 404. doi: 10.1007/s10725-014-0013-y

Fahad S, Hussain S, Saud S, Hassan S, Chauhan BS, Khan F et al (2016a) Responses of rapid viscoanalyzer profile and other rice grain qualities to exogenously applied plant growth regulators under high day and high night temperatures. *PLoS One* 11(7):e0159590. doi: 10.1371/journal.pone.0159590

Fahad S, Hussain S, Saud S, Hassan S, Ihsan Z, Shah AN,Wu C, Yousaf M, Nasim W, Alharby H, Alghabari F, Huang J (2016b) Exogenously applied plant growth regulators enhance the morphophysiological growth and yield of rice under high temperature. *Front Plant Sci* 7(1250). doi: 10.3389/fpls.2016.01250

Fahad S, Hussain S, Saud S, Hassan S, Tanveer M, Ihsan MZ, Shah AN, Ullah A, Nasrullah KF, Ullah S, AlharbyH NW, Wu C, Huang J (2016c) A combined application of biochar and phosphorus alleviates heat-induced adversities on physiological, agronomical and quality attributes of rice. *Plant Physiol Biochem* 103:191–198

Fahad S, Hussain S, Saud S, Khan F, Hassan S, Jr A, Nasim W, Arif M, Wang F, Huang J (2016d) Exogenously applied plant growth regulators affect heat-stressed rice pollens. *J Agron Crop Sci* 202:139–150.

Fahad S, Hussain S, Saud S, Tanveer M, Bajwa AA, Hassan S, Shah AN, Ullah A,Wu C, Khan FA, Shah F, Ullah S, Chen Y, Huang J (2015a) A biochar application protects rice pollen from high-temperature stress. *Plant Physiol Biochem* 96:281–287.

Fahad S, Muhammad ZI, Abdul K, Ihsanullah D, Saud S, Saleh A, Wajid N, Muhammad A, Imtiaz AK, Chao W, Depeng W, Jianliang H (2018) Consequences of high temperature under changing climate optima for rice pollen characteristics-concepts and perspectives. *Archives Agron Soil Sci.* doi: 10.1080/0365 0340.2018.1443213

Fahad S, Nie L, Chen Y, Wu C, Xiong D, Saud S, Hongyan L, Cui K, Huang J (2015b) Crop plant hormones and environmental stress. *Sustain Agric Rev* 15:371–400.

Fahad S, Rehman A, Shahzad B, Tanveer M, Saud S, Kamran M, Ihtisham M, Khan SU, Turan V, Rahman MHU (2019b). Rice responses and tolerance to metal/metalloid toxicity. In: Hasanuzzaman M, Fujita M, Nahar K, Biswas JK (eds) *Advances in rice research for abiotic stress tolerance*. Woodhead Publ Ltd, Abington Hall Abington, Cambridge CB1 6AH, CAMBS, England, pp 299–312.

Farah R, Muhammad R, Muhammad SA, Tahira Y, Muhammad AA, Maryam A, Shafaqat A, Rashid M, Muhammad R, Qaiser H, Afia Z, Muhammad AA, Muhammad A, Fahad S (2020) Alternative and non-conventional soil and crop management strategies for increasing water use efficiency. In: Fahad S, Hasanuzzaman M, Alam M, Ullah H, Saeed M, Khan AK, Adnan M (eds) *Environment, climate, plant and vegetation growth*. Springer Publ Ltd, Springer Nature Switzerland AG. Part of Springer Nature, pp 323–338. doi: 10.1007/978-3-030-49732-3

Farhana G, Ishfaq A, Muhammad A, Dawood J, Fahad S, Xiuling L, Depeng W, Muhammad F, Muhammad F, Syed AS (2020) Use of crop growth model to simulate the impact of climate change on yield of various wheat cultivars under different agro-environmental conditions in Khyber Pakhtunkhwa, Pakistan. *Arabian J Geosci* 13:112. doi: 10.1007/s12517-020-5118-1

Farhat A, Hafiz MH, Wajid I, Aitazaz AF, Hafiz FB, Zahida Z, Fahad S, Wajid F, Artemi C (2020) A review of soil carbon dynamics resulting from agricultural practices. *J Environ Manage* 268 (2020): 110319.

Fazli W, Muhmmad S, Amjad A, Fahad S, Muhammad A, Muhammad N, Ishaq AM, Imtiaz AK, Mukhtar A, Muhammad S, Muhammad I, Rafi U, Haroon I, Muhammad A (2020) Plant-microbes interactions and Ffnctions in changing climate. In: Fahad S, Hasanuzzaman M, Alam M, Ullah H, Saeed M, Khan AK, Adnan M (eds) *Environment, climate, plant and vegetation growth*. Springer Publ Ltd, Springer Nature Switzerland AG. Part of Springer Nature, pp 397–420. doi: 10.1007/978-3-030-49732-3

Gao X, Yuan HM, Hu YQ, Li J, Lu YT (2014) Mutation of arabidopsis CATALASE2 results in hyponastic leaves by changes of auxin levels. *Plant Cell Environ* 37:175–188.

Giehl RF, Lima JE, von-Wiren N (2012). Localized iron supply triggers lateral root elongation in Arabidopsis by altering the AUX1-mediated auxin distribution. *Plant Cell* 24(1): 33–49.

Gill SS, Tuteja N (2010) Reactive oxygen species and antioxidant machinery in abiotic stress tolerance in crop plants. *Plant Physiol Biochem* 48:909–930.

Gojon A, Krouk G, Perrine-Walker F, Laugier E (2011) Nitrate transceptor in plants. *J Exp Bot* 62: 2299–2308.

Gopakumar L, Bernard NO, Donato V (2020) Soil microarthropods and nutrient cycling. In: Fahad S, Hasanuzzaman M, Alam M, Ullah H, Saeed M, Khan AK, Adnan M (eds) *Environment, climate, plant and vegetation growth*. Springer Publ Ltd, Springer Nature Switzerland AG. Part of Springer Nature, pp 453–472. doi: 10.1007/978-3-030-49732-3

Habib ur R, Ashfaq A, Aftab W, Manzoor H, Fahd R, Wajid I, Md. Aminul I, Vakhtang S, Muhammad A, Asmat U, Abdul W, Syeda RS, Shah S, Shahbaz K, Fahad S, Manzoor H, Saddam H, Wajid N (2017) Application of CSM-CROPGRO-Cotton model for cultivars and optimum planting dates: evaluation in changing semi-arid climate. *Field Crops Res*. doi: 10.1016/j.fcr.2017.07.007

Hafiz MH, Abdul K, Farhat A, Wajid F, Fahad S, Muhammad A, Ghulam MS, Wajid N, Muhammad M, Hafiz FB (2020a) Comparative effects of organic and inorganic fertilizers on soil organic carbon and wheat productivity under arid region. *Commun Soil Sci Plant Anal*. doi: 10.1080/00103624.2020.1 763385

Hafiz MH, Farhat A, Ashfaq A, Hafiz FB, Wajid F, Carol Jo W, Fahad S, Gerrit H (2020b) Predicting kernel growth of maize under controlled water and nitrogen applications. *Int J Plant Prod*. doi: 10.1007/s421 06-020-00110-8

Hafiz MH, Farhat A, Shafqat S, Fahad S, Artemi C, Wajid F, Chaves CB, Wajid N, Muhammad M, Hafiz FB (2018) Offsetting land degradation through nitrogen and water management during maize cultivation under arid conditions. *Land Degrad Dev* 1–10. doi: 10.1002/ldr.2933.

Hafiz MH, Muhammad A, Farhat A, Hafiz FB, Saeed AQ, Muhammad M, Fahad S, Muhammad A (2019) Environmental factors affecting the frequency of road traffic accidents: a case study of sub-urban area of Pakistan. *Environ Sci Pollut Res*. doi: 10.1007/s11356-019-04752-8

Hafiz MH, Wajid F, Farhat A, Fahad S, Shafqat S, Wajid N, Hafiz FB (2016) Maize plant nitrogen uptake dynamics at limited irrigation water and nitrogen. *Environ Sci Pollut Res* 24(3):2549–2557. doi: 10.1007/s11356-016-8031-0

Hesham FA, Fahad S (2020) Melatonin application enhances biochar efficiency for drought tolerance in maize varieties:Modifications in physio-biochemical machinery. *Agron J* 112(4):1–22.

Hussain MA, Fahad S, Rahat S, Muhammad FJ, Muhammad M, Qasid A, Ali A, Husain A, Nooral A, Babatope SA, Changbao S, Liya G, Ibrar A, Zhanmei J, Juncai H (2020) Multifunctional role of brassinosteroid and its analogues in plants. *Plant Growth Regul*. doi: 10.1007/s10725-020-00647-8

Ibrar K, Aneela R, Khola Z, Urooba N, Sana B, Rabia S, Ishtiaq H, Mujaddad UR, Salvatore M (2020) Microbes and environment: global warming reverting the frozen zombies. In: Fahad S, Hasanuzzaman M, Alam M, Ullah H, Saeed M, Khan AK, Adnan M (eds) *Environment, climate, plant and vegetation growth*. Springer Publ Ltd, Springer Nature Switzerland AG. Part of Springer Nature, pp 607–634. doi: 10.1007/978-3-030-49732-3

Iglesias MJ, Terrile MC, Bartoli CG, Ippolito S, Casalongue CA (2010) Auxin signaling participates in the adaptive response against oxidative stress and salinity by interacting with redox metabolism in Arabidopsis. *Plant Mol Biol* 74:215–222.

Ilyas M, Mohammad N, Nadeem K, Ali H, Aamir HK, Kashif H, Fahad S, Aziz K, Abid U (2020) Drought tolerance strategies in plants: a mechanistic approach. *J Plant Growth Regul.* doi: 10.1007/s00344-020-10174-5

Iqra M, Amna B, Shakeel I, Fatima K,Sehrish L, Hamza A, Fahad S (2020) Carbon cycle in response to global warming. In: Fahad S, Hasanuzzaman M, Alam M, Ullah H, Saeed M, Khan AK, Adnan M (eds) *Environment, climate, plant and vegetation growth.* Springer Publ Ltd, Springer Nature Switzerland AG. Part of Springer Nature, pp 1–16. doi: 10.1007/978-3-030-49732-3

Ismail A, Takeda S, Nick P (2014) Life and death under salt stress: same players, different timing. *J Exp Bot* 65:2963–2979.

Jain M, Khurana JP (2009) Transcript profiling reveals diverse roles of auxin-responsive genes during reproductive development and abiotic stress in rice. *Febs J* 276(11):3148–3162.

Jan M, Muhammad AH, Adnan NS, Muhammad Y, Javaid I, Xiuling L, Depeng W, Fahad S, (2019) Modulation in growth, gas exchange, and antioxidant activities of salt-stressed rice (Oryza sativa L.) genotypes by zinc fertilization. *Arabian J Geosci* 12:775. doi: 10.1007/s12517-019-4939-2

Kamarn M, Wenwen C, Irshad A, Xiangping M, Xudong Z, Wennan S, Junzhi C, Shakeel A, Fahad S, Qingfang H, Tiening L (2017) Effect of paclobutrazol, a potential growth regulator on stalk mechanical strength, lignin accumulation and its relation with lodging resistance of maize. *Plant Growth Regul* 84:317–332. doi: 10.1007/s10725-017-0342-8

Kazan K, Manners JM (2009) Linking development to defense: auxin in plant–pathogen interactions. *Trends in Plant Sci* 14(7):373–382.

Khan MA, Basir A, Adnan M, Saleem N, Khan A, Shah SRA, Shah JA, Ali K (2015) Effect of tillage, organic and inorganic nitrogen on maize yield. *Amer Eurasian J Agric Environ Sci* 15(12):2489–2494.

Khan NA, Nazar R, Iqbal N, Anjum NA(2012) *Phytohormones and abiotic stress tolerance in plants.* Springer Verlag, Berlin Heidelberg.

Khedun CP, Singh VP (2014) Climate change, water, and health: a review of regional challenges. *Water Qual Expo Health* 6:7–17.

Kohli A, Sreenivasulu N, Lakshmanan P, Kumar PP (2013) The phytohormone crosstalk paradigm takes center stage in understanding how plants respond to abiotic stresses. *Plant Cell Rep* 32:945–957.

Korver RA, Koevoets IT, Testerink C (2018) Out of shape during stress: a key role for auxin. *Trends Plant Sci* 23(9):783–793. doi: 10.1016/j.tplants.2018.05.011

Krasensky J, Jonak C. 2012. Drought, salt, and temperature stress-induced metabolic rearrangements and regulatory networks. *J. Exp. Bot* 63:1593–1608.

Krouk G, Lacombe B, Bielach A, Perrine-Walker F, Malinska K, Mounier E (2010) Nitrate-regulated auxin transport by NRT1. 1 defines a mechanism for nutrient sensing in plants. *Developmental Cell* 18(6):927–937.

Lee SC, Luan S (2012) ABA signal transduction at the crossroad of biotic and abiotic stress responses. *Plant cell Environ* 35:53–60.

Ljung K (2013) Auxin metabolism and homeostasis during plant development. *Development* 140:943–950. doi:10.1242/dev.086363

Ljung K, Hull AK, Celenza J, Yamada M, Estelle M, Normanly J, Sandberg G (2005) Sites and regulation of auxin biosynthesis in Arabidopsis roots. *Plant Cell* 17(4):1090–1104.

Mahar A, Amjad A, Altaf HL, Fazli W, Ronghua L, Muhammad A, Fahad S, Muhammad A, Rafiullah, Imtiaz AK, Zengqiang Z (2020) Promising technologies for Cd-contaminated soils: drawbacks and possibilities. In: Fahad S, Hasanuzzaman M, Alam M, Ullah H,Saeed M, Khan AK, Adnan M (eds) *Environment, climate, plant and vegetation growth.* Springer Publ Ltd, Springer Nature Switzerland AG. Part of Springer Nature, pp 63–92. doi: 10.1007/978-3-030-49732-3

Mashiguchi K, Tanaka K, Sakai T, Sugawara S, Kawaide H, Natsume M (2011) The main auxin biosynthesis pathway in Arabidopsis. *Proc Natl Acad Sci USA* 108:18512–18517. doi: 10.1073/pnas.1108434108

Mba C, Guimaraes EP, Ghosh K (2012) Re-orienting crop improvement for the changing climatic conditions of the 21st century. *Agric Food Secur* 1:1–17.

Md Jakir H, Allah B (2020) Development and applications of transplastomic plants; a way towards eco-friendly agriculture. In: Fahad S, Hasanuzzaman M, Alam M, Ullah H, Saeed M, Khan AK, Adnan M (eds) *Environment, climate, plant and vegetation growth*. Springer Publ Ltd, Springer Nature Switzerland AG. Part of Springer Nature, pp 285–322. doi: 10.1007/978-3-030-49732-3

Md. EH, Shoeb AZM, Mallik AH, Fahad S, Kamruzzaman MM, Akib J, Nayyer S, Mehedi A KM, Swati AS, Md Yeamin A, Most SS (2020) Measuring vulnerability to environmental hazards: qualitative to quantitative. In: Fahad S, Hasanuzzaman M, Alam M, Ullah H, Saeed M, Khan AK, Adnan M (eds) *Environment, climate, plant and vegetation growth*. Springer Publ Ltd, Springer Nature Switzerland AG. Part of Springer Nature, pp 421–452. doi: 10.1007/978-3-030-49732-3

Miura K, Lee J, Gong Q, Ma S, Jin JB, Yoo CY (2011) SIZ1 regulation of phosphate starvation-induced root architecture remodeling involves the control of auxin accumulation. *Plant Physiol* 155(2):1000–1012.

Morales-Tapia A, Cruz-Ramirez A (2016) Computational modeling of auxin: a foundation for plant engineering. *Front Plant Sci* 7:1881.

Mubeen M, Ashfaq A, Hafiz MH, Muhammad A, Hafiz UF, Mazhar S, Muhammad SD, Asad A, Amjed A, Fahad S, Wajid N (2020) Evaluating the climate change impact on water use efficiency of cotton-wheat in semi-arid conditions using DSSAT model. *J Water Climate Change*. doi/10.2166/wcc.2019.179/622 035/jwc2019179.pdf

Muhammad TQ, Amna F, Amna B, Barira Z, Xitong Z, Ling-Ling C (2020) Effectiveness of conventional crop improvement strategies vs. omics. In: Fahad S, Hasanuzzaman M, Alam M, Ullah H, Saeed M, Khan AK, Adnan M (eds) *Environment, climate, plant and vegetation growth*. Springer Publ Ltd, Springer Nature Switzerland AG. Part of Springer Nature, pp 253–284. doi: 10.1007/978-3-030-4 9732-3

Muhammad Z, Abdul MK, Abdul MS, Kenneth BM, Muhammad S, Shahen S, Ibadullah J, Fahad S (2019) Performance of Aeluropus lagopoides (mangrove grass) ecotypes, a potential turfgrass, under high saline conditions. *Environ Sci Pollut Res*. doi: 10.1007/s11356-019-04838-3

Munns R, Tester M (2008) Mechanisms of salinity tolerance. *Annu Rev Plant Biol* 59:651–681.

Naser V, Shani E (2016) Auxin response under osmotic stress. *Plant Mol Biol* 91:661–672. doi: 10.1007/s111 03-016-0476-5

Navarro L, Dunoyer P, Jay F, Arnold B, Dharmasiri N, Estelle M (2006) A plant miRNA contributes to antibacterial resistance by repressing auxin signaling. *Science* 312:436–439.

Noor M, Naveed ur R, Ajmal J, Fahad S, Muhammad A, Fazli W, Saud S, Hassan S (2020) Climate change and costal plant lives. In: Fahad S, Hasanuzzaman M, Alam M, Ullah H,Saeed M, Khan AK, Adnan M (eds) *Environment, climate, plant and vegetation growth*. Springer Publ Ltd, Springer Nature Switzerland AG. Part of Springer Nature, pp 93–108. doi: 10.1007/978-3-030-49732-3

Overvoorde P, Fukaki H, Beeckman T (2010) Auxin control of root development. *Cold Spring Harb Perspect Biol* 2(6):a001537.

Park HC, Cha JY, Yun DJ (2013a) Roles of YUCCAs in auxin biosynthesis and drought stress responses in plants. *Plant Signal Behav* 8(6):e24495.

Park HC, Joon-Yung C, Dae-Jin Y (2013b) Roles of YUCCAs in auxin biosynthesis and drought stress responses in plants, *Plant Signal Behav* 8:6–e24495. doi: 10.4161/psb.24495

Park JE, Park JY, Kim YS, Staswick PE, Jeon J (2007) GH3-mediated auxin homeostasis links growth regulation with stress adaptation response in Arabidopsis. *J Biol Chem* 282:10036–10046.

Pérez-Torres CA, Lopez-Bucio J, Cruz-Ramirez A, Ibarra-Laclette E, Dharmasiri S, Estelle M, Herrera-Estrella L (2008) Phosphate availability alters lateral root development in Arabidopsis by modulating auxin sensitivity via a mechanism involving the TIR1 auxin receptor. *Plant Cell* 20(12): 3258–3272.

Qamar-uz Z, Zubair A, Muhammad Y, Muhammad ZI, Abdul K, Fahad S, Safder B, Ramzani PMA, Muhammad N (2017) Zinc biofortification in rice: leveraging agriculture to moderate hidden hunger in developing countries. *Arch Agron Soil Sci* 64:147–161. doi: 10.1080/03650340.2017.1338343

Rahman A (2013) Auxin: a regulator of cold stress response. *Physiol Plantarum* 147:28–35.

Rashid M, Qaiser H, Khalid SK, Al-Wabel MI, Zhang A, Muhammad A, Shahzada SI, Rukhsanda A, Ghulam AS, Shahzada MM, Sarosh A, Muhammad FQ (2020) Prospects of biochar in alkaline soils to mitigate climate change. In: Fahad S, Hasanuzzaman M, Alam M, Ullah H,Saeed M, Khan AK, Adnan M (eds) *Environment, climate, plant and vegetation growth*. Springer Publ Ltd, Springer Nature Switzerland AG. Part of Springer Nature, pp 133–150. doi: 10.1007/978-3-030-49732-3

Rehman M, Fahad S, Saleem MH, Hafeez M, Muhammad HR, Liu F, Deng G (2020) Red light optimized physiological traits and enhanced the growth of ramie (Boehmeria nivea L.). *Photosynthetica* 58 (4): 922–931.

Rigas S, Debrosses G, Haralampidis K, Vicente-Agullo F, Feldmann KA, Grabov A (2001) TRH1 encodes a potassium transporter required for tip growth in Arabidopsis root hairs. *Plant Cell* 13(1):139–151.

Rosquete MR, Barbez E, Kleine-Vehn J (2012) Cellular auxin homeostasis: gatekeeping is housekeeping. *Mol Plant* 5:772–786.

Sadam M, Muhammad TQ, Ghulam M, Muhammad SK, Faiz AJ (2020) Role of biotechnology in climate resilient agriculture. In: Fahad S, Hasanuzzaman M, Alam M, Ullah H, Saeed M, Khan AK, Adnan M (eds) *Environment, climate, plant and vegetation growth*. Springer Publ Ltd, Springer Nature Switzerland AG. Part of Springer Nature, pp 339–366. doi: 10.1007/978-3-030-49732-3

Sajid H, Jie H, Jing H, Shakeel A, Satyabrata N, Sumera A, Awais S, Chunquan Z, Lianfeng Z, Xiaochuang C, Qianyu J, Junhua Z (2020) Rice production under climate change: adaptations and mitigating strategies. In: Fahad S, Hasanuzzaman M, Alam M, Ullah H, Saeed M, Khan AK, Adnan M (eds) *Environment, climate, plant and vegetation growth*. Springer Publ Ltd, Springer Nature Switzerland AG. Part of Springer Nature, pp 659–686. doi: 10.1007/978-3-030-49732-3

Sajjad H, Muhammad M, Ashfaq A, Waseem A, Hafiz MH, Mazhar A, Nasir M, Asad A, Hafiz UF, Syeda RS, Fahad S, Depeng W, Wajid N (2019) Using GIS tools to detect the land use/land cover changes during forty years in Lodhran district of Pakistan. *Environ Sci Pollut Res*. doi: 10.1007/s11356-019-06072-3

Sakata T, Takeshi O, Shinya M, Mari T, Yuta T, Nahoko H (2010) Auxins reverse plant male sterility caused by high temperatures. Proceedings of the National Academy of Sciences, 8569–8574.

Saleem MH, Fahad S, Adnan M, Mohsin A, Muhammad SR, Muhammad K, Qurban A, Inas AH, Parashuram B, Mubassir A, Reem MH (2020a) Foliar application of gibberellic acid endorsed phytoextraction of copper and alleviates oxidative stress in jute (Corchorus capsularis L.) plant grown in highly copper-contaminated soil of China. *Environ Sci Pollution Res* 10.1007/s11356-020-09764-3

Saleem MH, Fahad S, Shahid UK, Mairaj D, Abid U, Ayman ELS, Akbar H, Analía L, Lijun L (2020b) Copper-induced oxidative stress, initiation of antioxidants and phytoremediation potential of flax (Linum usitatissimum L.) seedlings grown under the mixing of two different soils of China. *Environ Sci Poll Res* 10.1007/s11356-019-07264-7

Saleem MH, Rehman M, Fahad S, Tung SA, Iqbal N, Hassan A, Ayub A, Wahid MA, Shaukat S, Liu L, Deng G (2020c) Leaf gas exchange, oxidative stress, and physiological attributes of rapeseed (Brassica napus L.) grown under different light-emitting diodes. *Photosynthetica* 58 (3):836–845.

Salopek-Sondi B, Pavlovi I, Smolko A, Same D (2017) *Auxin as a mediator of abiotic stress response. Mechanism of plant hormone signaling under stress*. In: Pandey G (ed), 1st edn, Vol. 1. JohnWiley & Sons, Inc.

Saman S, Amna B, Bani A, Muhammad TQ Rana MA, Muhammad SK (2020) QTL mapping for abiotic stresses in cereals. In: Fahad S, Hasanuzzaman M, Alam M, Ullah H, Saeed M, Khan AK, Adnan M (eds) *Environment, climate, plant and vegetation growth*. Springer Publ Ltd, Springer Nature Switzerland AG. Part of Springer Nature, pp 229–252. doi: 10.1007/978-3-030-49732-3

Saud S, Chen Y, Fahad S, Hussain S, Na L, Xin L, Alhussien SA (2016) Silicate application increases the photosynthesis and its associated metabolic activities in Kentucky bluegrass under drought stress and post-drought recovery. *Environ Sci Pollut Res* 23(17):17647– 17655. 10.1007/s11356-016-6957-x

Saud S, Chen Y, Long B, Fahad S, Sadiq A (2013) The different impact on the growth of cool season turf grass under the various conditions on salinity and drought stress. *Int J Agric Sci Res* 3:77–84.

Saud S, Fahad S, Cui G, Chen Y, Anwar S (2020) Determining nitrogen isotopes discrimination under drought stress on enzymatic activities, nitrogen isotope abundance and water contents of Kentucky bluegrass. *Sci Rep* 10:6415. doi: 10.1038/s41598-020-63548-w

Saud S, Fahad S, Yajun C, Ihsan MZ, Hammad HM, Nasim W, Amanullah Jr, Arif M, Alharby H (2017) Effects of nitrogen supply on water stress and recovery mechanisms in Kentucky Bluegrass plants. *Front Plant Sci* 8:983. doi: 10.3389/fpls.2017.00983

Saud S, Li X, Chen Y, Zhang L, Fahad S, Hussain S, Sadiq A, Chen Y (2014) Silicon application increases drought tolerance of Kentucky bluegrass by improving plant water relations and morph physiological functions. *SciWorld J* 2014:1–10. doi: 10.1155/2014/368694

Senol C (2020) The effects of climate change on human behaviors. In: Fahad S, Hasanuzzaman M, Alam M, Ullah H, Saeed M, Khan AK, Adnan M (eds) *Environment, climate, plant and vegetation growth*. Springer Publ Ltd, Springer Nature Switzerland AG. Part of Springer Nature, pp 577–590. doi: 10.1 007/978-3-030-49732-3

Shafi MI, Adnan M, Fahad S, Fazli W, Ahsan K, Zhen Y, Subhan D, Zafar-ul-Hye M, Martin B, Rahul D (2020) Application of single superphosphate with humic acid improves the growth, yield and phosphorus uptake of wheat (Triticum aestivum L.) in calcareous soil. *Agron* (10):1224. doi:10.3390/agronomy10091224

Shah F, Lixiao N, Kehui C, Tariq S, Wei W, Chang C, Liyang Z, Farhan A, Fahad S, Huang J (2013) Rice grain yield and component responses to near 2°C of warming. *Field Crop Res* 157:98–110.

Stowe BB, Thimann KV (1954) The paper chromatography of indole compounds and some indolecontaining auxins of plant tissues. *Arch Biochem Biophys* 51(2):499–516.

Subhan D, Zafar-ul-Hye M, Fahad S, Saud S, Martin B, Tereza H, Rahul D (2020) Drought stress alleviation by ACC deaminase producing achromobacter xylosoxidans and enterobacter cloacae, with and without timber waste biochar in maize. *Sustain* 12:6286. doi: 10.3390/su12156286

Tariq M, Ahmad S, Fahad S, Abbas G, Hussain S, Fatima Z, Nasim W, Mubeen M, ur Rehman MH, Khan MA, Adnan M (2018). The impact of climate warming and crop management on phenology of sunflower-based cropping systems in Punjab, Pakistan. *Agri Forest Met.* 15(256):270–282.

Thimann KV (1936) Auxins and the growth of roots. *Am J Bot*, 561–569.

Tivendale ND, Davidson SE, Davies NW, Smith JA, Dalmais M, Bendahmane AI, Quittenden LJ, Sutton L, Bala RK, Le Signor C (2012) Biosynthesis of the halogenated auxin, 4-chloroindole-3-acetic acid. *Plant Physiol* 159:1055–1063.

Tognetti VB, Muhlenbock P, van Breusegem F (2012) Stress homeostasis – the redox and auxin perspective. *Plant Cell Env* 35:321–333.

Unsar NU, Muhammad R, Syed HMB, Asad S, Mirza AQ, Naeem I, Muhammad H ur R, Fahad S, Shafqat S (2020) Insect pests of cotton crop and management under climate change scenarios. In: Fahad S, Hasanuzzaman M, Alam M, Ullah H, Saeed M, Khan AK, Adnan M (eds) *Environment, climate, plant and vegetation growth*. Springer Publ Ltd, Springer Nature Switzerland AG. Part of Springer Nature, pp 367–396. doi: 10.1007/978-3-030-49732-3

Vidal EA, Araus V, Lu C, Parry G, Green PJ, Coruzzi GM, Gutierrez RA (2010) Nitrate-responsive miR393/AFB3 regulatory module controls root system architecture in Arabidopsis thaliana. *Proc Natl Acad Sci U S A* 107:4477–4482. doi: 10.1073/pnas.0909571107.

Wahid F, Fahad S, Subhan D, Adnan M, Zhen Y, Saud S, Manzer HS, Martin B, Tereza H, Rahul D (2020) Sustainable management with mycorrhizae and phosphate solubilizing bacteria for enhanced phosphorus uptake in calcareous soils. *Agriculture* 10 (334). doi:10.3390/agriculture10080334

Wajid N, Ashfaq A, Asad A, Muhammad T, Muhammad A, Muhammad S, Khawar J, Ghulam MS, Syeda RS, Hafiz MH, Muhammad IAR, Muhammad ZH, Muhammad Habib ur R, Veysel T, Fahad S, Suad S, Aziz K, Shahzad A (2017) Radiation efficiency and nitrogen fertilizer impacts on sunflower crop in contrasting environments of Punjab. *Pakistan Environ Sci Pollut Res* 25:1822–1836. doi: 10.1007/s11356-017-0592-z

Wang S, Fu J (2011) Insights into auxin signaling in plant–pathogen interactions. *Front Plant Sci* 2:74.

Wang D, Pajerowska-Mukhtar K, Culler AH, Dong X (2007) Salicylic acid inhibits pathogen growth in plants through repression of the auxin signaling pathway. *Curr Biol* 17:1784–1790.

Wang XF, He FF, Ma XX, Mao CZ, Hodgman C, Lu CG, Wu P (2011) OsCAND1 is required for crown root emergence in rice. *Molecular Plant* 4(2):289–299.

Wilkinson S, Kudoyarova GR, Veselov DS, Arkhipova TN, Davies WJ (2012) Plant hormone interactions: innovative targets for crop breeding and management. *J Exp Bot* 63:3499–3509.

Willige BC, Isono E, Richter R, Zourelidou M, Schwechheimer C (2011) Gibberellin regulates PIN-FORMED abundance and is required for auxin transport-dependent growth and development in Arabidopsis thaliana. *Plant Cell* 23:2184–2195.

Won C, Shen X, Mashiguchi K, Zheng Z, Dai X, Cheng Y (2011) Conversion of tryptophan to indole-3-acetic acid by Tryptophan Aminotransferases of Arabidopsis and YUCCAs in Arabidopsis. *Proc Natl Acad Sci USA* 108:18518–18523. doi: 10.1073/pnas.1108436108

Wu C, Tang S, Li G, Wang S, Fahad S, Ding Y (2019) Roles of phytohormone changes in the grain yield of rice plants exposed to heat: a review. *PeerJ* 7:e7792. doi: 10.7717/peerj.7792

Wu C, Kehui C, She T, Ganghua L, Shaohua W, Fahad S, Lixiao N, Jianliang H, Shaobing P, Yanfeng D (2020) Intensified pollination and fertilization ameliorate heat injury in rice (Oryza sativa L.) during the flowering stage. *Field Crops Res* 252:107795.

Yang Z, Zhang Z, Zhang T, Fahad S, Cui K, Nie L, Peng S, Huang J (2017) The effect of season-long temperature increases on rice cultivars grown in the central and southern regions of China. *Front Plant Sci* 8:1908. doi: 10.3389/fpls.2017.01908

Yin C, Wu Q, Zeng H, Xia K, Xu J, Li R (2011) Endogenous auxin is required but supraoptimal for rapid growth of rice (Oryza sativa L.) seminal roots, and auxin inhibition of rice seminal root growth is not caused by ethylene. *J Plant Growth Regul* 30(1):20–29.

Zafar-ul-Hye M, Muhammad N, Subhan D, Fahad S, Rahul D, Mazhar A, Ashfaq AR, Martin B, Jiřrí H, Zahid HT, Muhammad N (2020a) Alleviation of cadmium adverse effects by improving nutrients uptake in bitter gourd through cadmium tolerant Rhizobacteria. *Environment* 7 (54); doi:10.3390/environments7080054

Zafar-ul-Hye M, Muhammad TH, Muhammad A, Fahad S, Martin B, Tereza D, Rahul D, Subhan D (2020b) Potential role of compost mixed biochar with rhizobacteria in mitigating lead toxicity in spinach. *Scientific Rep* 10:12159. doi: 10.1038/s41598-020-69183-9.

Zahida Z, Hafiz FB, Zulfiqar AS, Ghulam MS, Fahad S, Muhammad RA, Hafiz MH,Wajid N, Muhammad S (2017) Effect of water management and silicon on germination, growth, phosphorus and arsenic uptake in rice. *Ecotoxicol Environ Saf* 144:11–18.

Zhao Y (2010) Auxin biosynthesis and its role in plant development. *Annu Rev Plant Biol* 61:49–64. doi: 10.1146/annurev-arplant-042809-112308

Zia-ur-Rehman M (2020) Environment, climate change and biodiversity. In: Fahad S, Hasanuzzaman M, Alam M, Ullah H, Saeed M, Khan AK, Adnan M (eds) *Environment, climate, plant and vegetation growth*. Springer Publ Ltd, Springer Nature Switzerland AG. Part of Springer Nature, pp 473–502. doi: 10.1007/978-3-030-49732-3

11

Existing Water Scarcity Scenario and Climate Change Impact on the Transboundary Water Resources in Pakistan

Muhammad Mohsin Waqas, Muhammad Habib Ur Rahman, Sikandar Ali, and Haroon Rasheed

CONTENTS

11.1 Introduction

Water demand is increasing at the global level, if not the regional level, to ensure food security especially under the era of a changing climate. The advancement of civilization with its paraphernalia of water-hungry technologies and its population needs access to safe and adequate water supplies on a regular and sustainable basis. Life demands water which is essential to economic growth and prosperity, agriculture, and industrialization where the quality of life meets with the quality of the water. In brief, sufficient quantity and adequate quality of the water is required for the proper functioning of the ecosystem.

Worldwide, there are more than 286 rivers that spill across the boundaries of the different countries. The river Nile is a major river that crosses into 11 countries. The river Indus flows across China, India, and Pakistan. While the 600 aquifer shares the transboundary borders worldwide, almost 40% of the world population shared the transboundary water resources. While 90% of the USA population depends upon the shared water between states. The 14 river basins are critical due to the population of 1.4 billion having an economic dependence on the water resources. For the sustainable development and to avoid water riots in the shared basins, there is a dire need for transboundary water cooperation under the existing and the changing climatic condition.

Climate change is the main driver to affect the long-term water resources management polices (Adnan et al. 2018a, 2018b; Adnan et al. 2019; Akram et al. 2018a, 2018b; Aziz et al. 2017a, 2017b; Habib et al. 2017; Hafiz et al. 2016; Hafiz et al. 2019; Kamaran et al. 2017; Muhmmad et al. 2019; Sajjad et al.

2019; Saud et al. 2013; Saud et al. 2014; Saud et al. 2017; Saud et al. 2016; Shah et al. 2013; Saud et al. 2020; Qamar et al. 2017; Wajid et al. 2017; Yang et al. 2017; Zahida et al. 2017; Depeng et al. 2018; Hussain et al. 2020; Hafiz et al. 2020a, 2020b; Shafi et al. 2020; Wahid et al. 2020; Subhan et al. 2020; Zafar-ul-Hye et al. 2020a, 2020b; Adnan et al. 2020; Ilyas et al. 2020; Saleem 2020a, 2020b, 2020c; Rehman et al. 2020; Farhat et al. 2020; Wu et al. 2020; Mubeen et al. 2020; Farhana et al. 2020; Jan et al. 2019; Wu et al. 2019; Ahmad et al. 2019; Baseer et al. 2019; Hafiz et al. 2018; Tariq et al. 2018; Fahad and Bano 2012; Fahad et al. 2017; Fahad et al. 2013; Fahad et al. 2014a, 2014b; Fahad et al. 2016a, 2016b, 2016c, 2016d; Fahad et al. 2015a, 2015b; Fahad et al. 2018; Fahad et al. 2019a, 2019b; Hesham and Fahad 2020; Iqra et al. 2020; Akbar et al. 2020; Mahar et al. 2020; Noor et al. 2020; Bayram et al. 2020; Amanullah et al. 2020; Rashid et al. 2020; Arif et al. 2020; Amir et al. 2020; Saman et al. 2020; Muhammad et al. 2020; Md Jakir and Allah 2020; Farah et al. 2020; Sadam et al. 2020; Unsar et al. 2020; Fazli et al. 2020; Md. Enamul et al. 2020; Gopakumar et al. 2020; Zia-ur-Rehman 2020; EL Sabagh et al. 2020; Al-Wabel et al. 2020a, 2020b; Senol 2020; Amjad et al. 2020; Ibrar et al. 2020; Sajid et al. 2020).

There is no lack of argument in the opposition to the global warming (Fahad and Bano 2012; Fahad et al. 2017; Fahad et al. 2013; Fahad et al. 2014a, 2014b; Fahad et al. 2016a, 2016b, 2016c, 2016d; Fahad et al. 2015a, 2015b; Fahad et al. 2018; Fahad et al. 2019a, 2019b; Hesham and Fahad 2020). The water availability may or may not be reduced but the availability will not be reliable in the short-term. As far as the evidence goes, climate change indicates the increasing warming of the globe and its climate variability. In IPCC 2007 report, the undeniable threat of climate change that impact us include the rising temperature at the regional scale, and the pressure of melting ice on sea water level. The specific systems that are vulnerable to the changing climate are regions of sea ice, coral reefs, mountains, forests, meditation regions, mangroves, low lying coasts, marshes, agriculture in the low land areas, and human health especially under the less adaptive environment. The impact of climate change on the question of human survival will be worse from the aforementioned places with Arctic, Sub-Saharan Africa, Small islands and Asian mega-deltas deserving mention.

The impacts of these hazards will be more on living beings and it requires the imminent adoption of strategies to counter environmental threats and careful mitigation of risks to life to cope the negative impact of the climate change. The adoption of strategies is necessary to address the leap in the surface temperate of earth due to the presence of gases already released in the air from engineering activities and volcanoes, or strategies to counter the reradiation effects of global warming. Further, the vulnerability of earth and its people from the climate change can be reduced by walking the paths of progress, or a civilization, that adheres to the tenets of sustainable development to take away from the vulnerability its Achilles' heel: Loss of limited natural resources vis-à-vis adjustment in unlimited wants. The working group II (WGII-2007) of the IPPC reported that the on the basis of the evidence garnered the hydrological cycle will be impacted in different ways: (i) The increase in the runoff, (ii) the peak discharge in the spring will be earlier in the glaciers and snow fed rivers, (iii) warming will affect the quality of water, and (iv) the thermal structure in the rivers and the lakes and (v) the expectation of the more catastrophic events with a higher frequency of theirs hitting the earth and with an intensity expected to last in the memories of people far into the 21st century.

11.2 Water Scarcity

Water scarcity is defined as the lack of sufficient water to meet the requisite demand. The World Economic Forum 2019 listed water scarcity as one of the biggest global risks for the near future. The water demand will rise with the economic competition for the quality and quantity of the water, increase in the user's disputes, consistent and irreversible depletion of the groundwater resources, and eventually, the negative impact on the environment. Almost one third of the global population lives under severe water scarcity for a month in each passing year. The severe water scarcity is also estimated to affect the half billion people of the world every passing month of the year. Technically there is sufficient amount of water available for meeting the requirement on a global scale but the unequal water distribution pattern makes it scarce in different regions of the world especially under changing climate conditions. The distributed-water-availability patches of the world varied from the very wet to very dry and make the water scarce in the different regions of the world. The impact of climate change such as flood and drought, deforestation, changing pattern of the precipitation, more numerous occurrence of extreme events, increase in pollution, and wasteful and inefficient use of water makes the supply of water scarcer in the different geographical locations of the world.

The climate change is putting pressure on the water resources as the Himalayan glaciers are melting swiftly with an expected 75% glaciers on retreat and soon to disappear in 2035 (Misra 2014; Hadi et al. 2014). While in the case of the African continent, the rainfall has reduced up to 10% (Misra 2014). Global water resources are decreasing due to the swiftly increasing population, increasing demand for industrialization, agriculture and pollution of water bodies, and the impact of climate change. The highest increase in the population is detected in Asia where 60% of the world's population reside. The population of the USA is increasing at the rate of 0.6% per annum/decade as estimated for the year July 2017 to July 2018.

The World Wildlife Funds for the Nature (WWF) reported that global warming will accelerate the rate at which the Himalayan glaciers are melting with an approximate rate of melting hovering in the range of about 10–15 metres annually. Water scarcity is also increasing at the different spatial and temporal scales in the world. Thus, the water scarcity in the South Asia is increasing due to receding Himalayan glaciers with a recession rate higher than that of surrounding regions. The south Asian countries consist of an almost one-fourth of the world's population but the water sharing quota is 4.5% of the total water available (1,945 billion m^3 from the 43,659 billion m^3). The per-capita water availability is less than the average water availability in these countries as the major water use here is for the purpose of agriculture. These countries utilize 95% of the water for agriculture purpose, while the worldwide average use of the water for agriculture is 70%. The proportion of water use for industrial sector is limited because of comparatively few industrial activities in this region (Babel and Wahid 2008).

In case of Pakistan, the water availability dropped from 1,100 m^3/ capita since 1940-41, with the approximate decrease of the water availability to 60% for the reporting period. The causes of water scarcity in case of Indus basin irrigation system includes the variability of the surface water supplies and the impact of the global warming, decreasing the water storage capacity due to sedimentation, cultivation of high delta crops with low water productivity, increasing demands of water for increasing industrialization, low irrigation efficiency, low efficiency municipal water supply system, deterioration of water quality due to dumping of untreated water in the water bodies, over-exploitation of the groundwater reserve, and the quality degradation of the aquifer and sea water intrusion due to lesser supply of the water through the delta into seas (Ashfaq et al. 2009).

11.3 Major Causes of Water Scarcity

The causes of the water scarcity include the increasing trend of the groundwater salinity, unrestricted and overexploitation of the groundwater, low irrigation system efficiency and productivity, inefficient distribution of the irrigation water, increasing pollution of the water bodies, poor system operation and maintenance, sea water intrusion in the coastal areas, and the climate change impacts. The detailed discussion is given from a global perspective with a case study of the Indus basin irrigation system.

11.3.1 Groundwater Over-exploitation and Groundwater Salinity

The impact of the climate change has bearings not only on the recharge into the aquifer or the discharge from the aquifer but also on the quality of the groundwater (Dragoni and Sukhija 2008). Dragoni and Sukhija (2008), further reported that the accumulation of the salt concentration during the study period is more as compared to the wet period. The other impact of climate change on aquifers is the saltwater intrusion from the sea into freshwater aquifers as sea level rises from a rise in sea surface temperature. The intrusion of the saltwater in coastal areas can be triggered when there is excessive groundwater pumping that reduces freshwater storage capacity in aquifer and hastens inadequate and unregulated discharge to flow to the sea from the river's deltas. In fact, the excessive groundwater pumping is directly related to droughts in the coastal areas, even as the increase in the intensity and frequency of the drought takes toll on the coastal areas that are more witness to and demonstrate high salt concentration of the saltwater intrusion in the estuary (Bates et al. 2008), and eventually deteriorate the groundwater supplies and the surface water resources. The rise in the sea level will increase the areas under salinization and eventually will reduce the freshwater availability to humans and the ecosystem (Bates et al. 2008). Further, the abatement in the groundwater recharge under the drought condition provides the opportunity to the attached aquifer water to flow into the nearest aquifer. This is the case of inland aquifer where the decrease in the recharge increase the intrusion of the saline water from the neighbouring saline aquifer (Chen et al. 2004).

The major causes of the increase in the ground water salinity include the mismanagement of the use of irrigation of the crops. The increase in the groundwater salinity directly contributed towards the water scarcity as ground water contributes more the 50% of the agriculture water needs globally. Groundwater is the most flexible water resource that can provide water as and when there is need. The major causes behind the increase in the groundwater salinity in the case of Indus basin irrigation system include the mismanaged usage pattern of irrigation water that led to harmful consequences to take effect, mainly consisting of seepage from the unlined irrigation network, unavailability of the proper drainage system in the flat topographic area, and the use of the poor-quality drainage effluent. The combined effect of the waterlogging and the salinity in Pakistan has affected 23% area under irrigation. In Punjab, 79% area under irrigation is under the fresh aquifer, while in Sindh only 28% is under the fresh aquifer.

Globally the groundwater extraction is at its optimum. The groundwater extraction at the global scale is 982 km^3/year from which 70% is used for agriculture purpose (Margat and Gun 2013). About half of the drinking water requirement is met through the use of the groundwater worldwide (Smith et al. 2016). Globally, the use of groundwater for the purpose of irrigation is about 38% (Siebert et al. 2010). The total groundwater resources in the upper 2 km of the earth's crust are approximately 22.6 million km^3 and 0.1–5 million km^3 water is just 50 year old (Gleeson et al. 2016). The total volume of the groundwater available is that part of the water standing over the continents and accessible to a depth almost equal to 3 m (Gleeson et al. 2016). In case of the Indus basin irrigation system, the groundwater contributes almost 50% of the irrigation water requirement and the extraction is 40% higher at the heads of the irrigation canals networks (Awan et al. 2016).

11.3.2 Surface Water Availability and Low Irrigation System Efficiency

Globally, the irrigation system efficiencies are estimated to come a notch below 50% due to excessive loss of water under the conveyance system or from inefficient methods of water application to the field (Jägermeyr et al. 2015). The water losses from the field in the form of the percolation in the fresh aquifer are not considered as the water losses since it's recovered through the pumping system. While the seepage into the saline aquifer are considered as water losses from the irrigation system.

11.3.3 Rainfall Variability

The impact of the climate change is not insignificant as far as the precipitation variability is concerned. Change in climate can be traced to rainfall occurrence with high intensity and increased frequency. Early setting in of monsoon and its taking full swing is the clear impact of the climate change in the South Asian countries. Generally, the climate of the eastern flange of the Himalayan region is described as the East Asian monsoon and the Indian monsoon is that which brings the bulk of monsoonal precipitation during the June to September. While in case of the Hindu Kush and Karakoram ranges, the precipitation pattern is one of equally distributed precipitation throughout the year.

11.4 Climate Induced Transboundary Water Issues

Climate change is a catalyst to natural catastrophes that will accelerate the flood and drought-related problems causing a water crisis. Disasters under changing climates are not limited by and not shaped by the boundaries that separate countries. Already, almost two billion people is facing water scarcity in one way or the other. Climate change impacts have been projected up to 2030 as affecting almost 54 million people who will have to battle with heavy floods. Many transboundary basins are particularly vulnerable to climate change impacts across the Middle East, Africa, Central Europe and Central Asia, Southern Europe, Asia, and Latin America.

It has been observed that the phenomenon of efficient water sharing at the transboundary level is embattled by the problems of incomplete information about the timing and the quantity of the water sharing among the parties. The timing and the quantity of the water available to the downstream user is based on the historic records and set to happen according to the trends of water sharing phenomenon during the normal period data. Now, the climate change has increased the uncertainty in the timing and the quantity of the water sharing phenomenon between the parties.

The glaciers melting in the Himalayan region will increase the calamity of flooding during the next two to three decades, and then the rivers' amount of water flow will be reduced because of the receding glaciers. More than one billion people will be affected due to insubstantial supply from reduction in the fresh water supply for the central, south, east, and southeast Asia vis-à-vis increase in the water demands here. There are almost 41 transboundary water agreements in the Asian countries and 15 agreements pertaining to the western Asia that are under the risk of climate change and attendant issues rising from decrease in the water available from a changing climate. There is always an uncertainty in the arrival of monsoon in the South Asia which was linked to reduced water availability in the region. In Ganges–Brahmaputra–Meghna basin, a reduction in the freshwater supplies will always result in a reduced freshwater access for Bangladesh which must counter increased saltwater intrusion due to less freshwater flow and increase in the sea level. A low risk agreement is one between China and Mongolia on the Amur river basin due to less impact of the climate change on the monsoon. By 2030, water scarcity problems are going to increase due to the reduction in the precipitation and enhanced evapotranspiration in the New Zealand and the eastern Australia, northland and some eastern regions. Although Australia is not party to any transboundary water sharing agreement, still it will face an acute water shortage in the Murry-Darling Basin from changing climate. In case of European countries, the negative impacts of climate change include increase in flash flood, increased erosion of delta, and the increase in the coastal area flooding. In the southern Europe, the climate change impact will worsen the water scarcity condition. The agreements in the southern Europe are at greater risk from climate change as the vulnerability in the climate is projected to be very high. In case of Latin America, the floods and the droughts will be more frequent. Due to the increase in the phenomenon of aridity under the changing climate, the 1978 Treaty for Amazonian cooperation will be under a great risk of lost habitats and other catastrophies of climate change. While in case of the North America, those agreements will be at the risk of rivers' reduced waterflows given rivers are dependent on timely melting of snow which is slated to be affected because of inadequate deposition of snow cover in winter and early meltdown during spring.

11.5 Transboundary Water Issue and IWT

The upper Indus basin covers the area of 425.000 km^2 covering the Hindu Kush, Karakoram, Himalayas and the Tibetan Plateau (Lutz et al. 2016). The aerial distribution of the UIB in the Pakistan, Afghanistan, India, and China puts it in the list of transboundary rivers basins. Indus basin is the major largest glaciated area on the earth having 22.000 km^2 glaciers surface (53). The diverse climate of the Upper Indus basin is due to the effects of the circulation of monsoon, topography, and the westerlies (Hewitt 2011; Mishra 2015; Maussion et al. 2014; Mölg et al. 2014; Scherler et al. 2011).

The supplies from this transboundary basin is essential for the millions of people who live there and are subject to future variations in the supply of and the demand in water, putting pressure on the basin even as it must deal with the impacts of climate change (Immerzeel and Bierkens 2012). The upper Indus basin is the major water resource to feed the world's largest contiguous irrigation system, and the water storage in the majors dams of Pakistan is more than 50% dependent upon the snow and glacier melting phenomenon from the Upper Indus basin (Immerzeel et al. 2010). Given the pattern of the precipitation is variable and there are concerns of sharp swings in the variability in the streamflow (Reggiani and Rientjes 2014), eventually, the Indus river's waterflow will fluctuate affecting the supply in the downstream. The demand for water here is very high due to high water use for irrigation purpose (Jain et al. 2007). The dependency of the Indus basin on the uncertain mountainous water resources makes it a hot spot of water scarcity under the era of the climate change (Immerzeel and Bierkens 2012). This variability is also propelled by the swift growth of the population (Unied Nations 2015). The major climate change evidence is in the occurrence of extreme events whose numbers at any time interval are more likely to go up in the future than before in this region (Field et al. 2012; Lutz et al. 2016) and the threat is real so far as the phenomenon of floods is concerned with the area consistently witnessing extremely severe floods (Houze et al. 2011). Archer and Hayley (2004) studied the 17 stations for the period of 1961–1990 and found the increasing trends of precipitation at the seasonal and annual time scale. Bocchiola and Diolaiuti (2013) and Zaki et al. (2015) study shows the similar trends for the new time scale of 1980–2009.

There is a hot debate on the impact of the climate change in the upper Indus basin as the glaciers in the Himalayas are losing the mass of ice under the changing climate more rapidly than the other glaciers of the world are losing, but the balance of ice mass vis-à-vis climate change is neutral in case of Karakoram and Pamir mountain ranges (Gardelle et al. 2013; Gardner et al. 2013; Shah et al. 2015; Hewitt 2007; Kääb et al. 2015; Kääb et al. 2012; Quincey et al. 2015; Mushtaq et al. 2015). The longer term analysis of the Terbeela dam inflow shows the declining trends (Reggiani and Rientjes 2014), as it's a major water reservoir fed by the streams of the water resources from the upper Indus basin. The studies in a different location in the upstream reaches of the river also show the declining trends in the runoff (Mukhopadhyay and Khan 2015; Sharif et al. 2013; Rasheed et al. 2015; Tahir et al. 2008).

Mahmood and Jia described the future changes in the water resources at the Mangala dam under the A2 and the B2 scenario of the climate change. In the early century scenario (2020s), and the mid-century scenario (2050s), the projected flow during the spring season was set to increase, while the flow in the autumn and the summer months was going to decrease with different amount of water in the outflow. Overall the mean annual flow was going to increase under both scenarios. The 20–36% increase in the median flow was found with respect to the base line period (1961–1990). The in-adequate flow could be increased for the early century and the late century scenarios by the 1.7–7.75%, and decreased by 12–14% for the midcentury scenario under the both A2 and B2 gas emission scenarios.

Pakistan is crisscrossed by five major rivers sourcing in Pakistan but Pakistan has rights only in three western rivers after signing the Indus Water Treaty (IWT). IWT was signed in 1960 by the President of Pakistan, Field Marshal General Ayub Khan, and the Prime Minister of India, Shri Jawaharlal Nehru, in Karachi, Pakistan. In this treaty, Pakistan was given the rights to unrestricted use of water from the three western rivers, while India had unrestricted water usage rights of the three eastern rivers. The increase in the meltdown of the Himalayan ice, that of Hindu Kush and Karakorum will cause flash floods as both parties have rights to release the water accumulated during floods in the natural stream, that will cause huge damage to the downstream riparian basin of the Indus Basin irrigation system. Similarly, after glacier meltdowns result in receding glaciers, the supply of the water will be reduced in 3 decades, and then the upper riparian basin will try to store more water engendering water stress for the lower riparian people. The water scarcity and the violation of treaty under the drought period due to climate change will trigger new tension between Pakistan and India as both countries have been involved in conflicts on the issue of Kashmir: A United Nations accepted disputed territory of both the countries.

11.6 Conclusion

Climate change will increase water scarcity due to harsh events like occurrence of flash floods and droughts, while the demands for water will increase. It's without doubt that climate change will alter the phenomenon of hydrological cycle of the different river basins. This will create problems between water sharing countries who are parties to transboundary water sharing agreements, so that the impact of the climate change on the transboundary river basin is clear and loud. The water treaties will need a revisit to carve out the distribution pattern for the water sharing parties under the changing climate situation, as the historical trends are no longer the more true representatives for the people on water sharing agreements than world leaders can be with their ability to strategize in the light of new understanding and cooperate and collaborate with the stakeholders in a manner that is circumspective and wise.

REFERENCES

Adnan M, Shah Z, Sharif M, Rahman H (2018a). Liming induces carbon dioxide (CO_2) emission in PSB inoculated alkaline soil supplemented with different phosphorus sources. *Environ Sci Poll Res. 1* 25(10):9501–9509.

Adnan M, Fahad S, Khan IA, Saeed M, Ihsan MZ, Saud S, Riaz M, Wang D, Wu C (2019). Integration of poultry manure and phosphate solubilizing bacteria improved availability of Ca bound P in calcareous soils. *3 Biotech. 1* 9(10):368.

Adnan M, Fahad S, Muhammad Z, Shahen S, Ishaq AM, Subhan D, Zafar-ul-Hye M, Martin LB, Raja MMN, Beena S, Saud S, Imran A, Zhen Y, Martin B, Jiri H, Rahul D (2020) Coupling phosphate-solubilizing bacteria with phosphorus supplements improve maize phosphorus acquisition and growth under lime induced salinity stress. *Plants* 9(900). doi: 10.3390/plants9070900

Adnan M, Zahir S, Fahad S, Arif M, Mukhtar A, Imtiaz AK, Ishaq AM, Abdul B, Hidayat U, Muhammad A, Inayat-Ur R, Saud S, Muhammad ZI, Yousaf J, Amanullah, Hafiz MH, Wajid N (2018b) Phosphate-solubilizing bacteria nullify the antagonistic effect of soil calcification on bioavailability of phosphorus in alkaline soils. *Sci Rep* 8:4339. doi: 10.1038/s41598-018-22653-7

Ahmad S, Kamran M, Ding R, Meng X, Wang H, Ahmad I, Fahad S, Han Q (2019) Exogenous melatonin confers drought stress by promoting plant growth, photosynthetic capacity and antioxidant defense system of maize seedlings. *PeerJ* 7:e7793. doi: 10.7717/peerj.7793

Akbar H, Timothy JK, Jagadish T, M. Golam M, Apurbo KC, Muhammad F, Rajan B, Fahad S, Hasanuzzaman M (2020) Agricultural and degradation: processes and problems undermining future food security. In: Fahad S, Hasanuzzaman M, Alam M, Ullah H, Saeed M, Khan AK, Adnan M (eds) *Environment, climate, plant and vegetation growth.* Springer Publ Ltd, Springer Nature Switzerland AG. Part of Springer Nature, pp 17–62. doi: 10.1007/978-3-030-49732-3

Akram R, Turan V, Hammad HM, Ahmad S, Hussain S, Hasnain A, Maqbool MM, Rehmani MIA, Rasool A, Masood N, Mahmood F, Mubeen M, Sultana SR, Fahad S, Amanet K, Saleem M, Abbas Y, Akhtar HM, Waseem F, Murtaza R, Amin A, Zahoor SA, ul Din MS, Nasim W (2018a) Fate of organic and inorganic pollutants in paddy soils. In: Hashmi MZ, Varma A (eds) *Environmental pollution of paddy soils, soil biology.* Springer International Publishing Ag, GEWERBESTRASSE 11, CHAM, CH-6330, Switzerland, pp 197–214.

Akram R, Turan V, Wahid A, Ijaz M, Shahid MA, Kaleem S, Hafeez A, Maqbool MM, Chaudhary HJ, Munis MFH, Mubeen M, Sadiq N, Murtaza R, Kazmi DH, Ali S, Khan N, Sultana SR, Fahad S, Amin A, Nasim W (2018b) Paddy land pollutants and their role in climate change. In: Hashmi MZ, Varma A (eds) *Environmental pollution of paddy soils, soil biology.* Springer International Publishing Ag, GEWERBESTRASSE 11, CHAM, CH-6330, Switzerland, pp 113–124.

Al-Wabel MI, Ahmad M, Usman ARA, Akanji M, Rafique MI (2020a) Advances in pyrolytic technologies with improved carbon capture and storage to combat climate change. In: Fahad S, Hasanuzzaman M, Alam M, Ullah H, Saeed M, Khan AK, Adnan M (eds) *Environment, climate, plant and vegetation growth.* Springer Publ Ltd, Springer Nature Switzerland AG. Part of Springer Nature, pp 535–576. doi: 10.1007/978-3-030-49732-3

Al-Wabel MI, Abdelazeem S, Munir A, Khalid E, Adel RAU (2020b) Extent of climate change in Saudi Arabia and its impacts on agriculture: a case study from Qassim region. In: Fahad S, Hasanuzzaman M, Alam M, Ullah H, Saeed M, Khan AK, Adnan M (eds) *Environment, climate, plant and vegetation growth.* Springer Publ Ltd, Springer Nature Switzerland AG. Part of Springer Nature, pp 635–658. doi: 10.1007/978-3-030-49732-3

Amanullah SK, Imran, Hamdan AK, Muhammad A, Abdel RA, Muhammad A, Fahad S, Azizullah S, Brajendra P (2020) Effects of climate change on irrigation water quality. In: Fahad S, Hasanuzzaman M, Alam M, Ullah H, Saeed M, Khan AK, Adnan M (eds) *Environment, climate, plant and vegetation growth.* Springer Publ Ltd, Springer Nature Switzerland AG. Part of Springer Nature, pp 123–132. doi: 10.1007/978-3-030-49732-3

Amir M, Muhammad A, Allah B, Sevgi Ç, Haroon ZK, Muhammad A, Emre A (2020) Bio fortification under climate change: the fight between quality and quantity. In: Fahad S, Hasanuzzaman M, Alam M, Ullah H, Saeed M, Khan AK, Adnan M (eds) *Environment, climate, plant and vegetation growth.* Springer Publ Ltd, Springer Nature Switzerland AG. Part of Springer Nature, pp 173–228. doi: 10.1007/978-3-030-49732-3

Amjad I, Muhammad H, Farooq S, Anwar H (2020) Role of plant bioactives in sustainable agriculture. In: Fahad S, Hasanuzzaman M, Alam M, Ullah H, Saeed M, Khan AK, Adnan M (eds) *Environment, climate, plant and vegetation growth.* Springer Publ Ltd, Springer Nature Switzerland AG. Part of Springer Nature, pp 591–606. doi: 10.1007/978-3-030-49732-3

Archer DR, Hayley JF (2004) Spatial and temporal variations in precipitation in the upper Indus Basin, global teleconnections and hydrological implications.

Arif M, Talha J, Muhammad R, Fahad S, Muhammad A, Amanullah, Kawsar A, Ishaq AM, Bushra K, Fahd R (2020) Biochar; a remedy for climate change. In: Fahad S, Hasanuzzaman M, Alam M, Ullah H,Saeed M, Khan AK, Adnan M (eds) *Environment, climate, plant and vegetation growth.* Springer Publ Ltd, Springer Nature Switzerland AG. Part of Springer Nature, pp 151–172. doi: 10.1007/978-3-030-4 9732-3

Ashfaq M, Griffith G, Hussain I (2009) *Economics of water resources in Pakistan.* Pak TM printers, Lahore, Pakistan, p 230.

Awan UK, Anwar A, Ahmad W, Hafeez M (2016) A methodology to estimate equity of canal water and groundwater use at different spatial and temporal scales: a geo-informatics approach. *Environ Earth Sci* 75(5): 1–13.

Aziz K, Daniel KYT, Fazal M, Muhammad ZA, Farooq S, Fan W, Fahad S, Ruiyang Z (2017a) Nitrogen nutrition in cotton and control strategies for greenhouse gas emissions: a review. *Environ Sci Pollut Res* 24:23471–23487. doi: 10.1007/s11356-017-0131-y

Aziz K, Daniel KYT, Muhammad ZA, Honghai L, Shahbaz AT, Mir A, Fahad S (2017b) Nitrogen fertility and abiotic stresses management in cotton crop: a review. *Environ Sci Pollut Res* 24:14551–14566. doi: 10.1007/s11356-017-8920-x

Babel MS, Wahid SM (2008) Freshwater under threat: vulnerability assessment of freshwater resources to environmental change–Ganges-Brahmaputra-Meghna river basin Helmand River Basin Indus River Basin, United Nations Environment Programme (UNEP), Nairobi, Kenya. *United Nations Environment Programme (UNEP, Nairobi), Kenya.* http://www.unep.org/pdf/southasia_report.pdf. Accessed 28 August 2015.

Baseer M, Adnan M, Fazal M, Fahad S, Muhammad S, Fazli W, Muhammad A, Jr. Amanullah, Depeng W, Saud S, Muhammad N, Muhammad Z, Fazli S, Beena S, Mian AR, Ishaq AM (2019) Substituting urea by organic wastes for improving maize yield in alkaline soil. *J Plant Nutrition.* doi: 10.1080/019041 67.2019.1659344

Bates B, Kundzewicz Z, Wu S (2008) *Climate change and water.* Intergovernmental Panel on Climate Change Secretariat.

Bayram AY, Seher Ö, Nazlican A (2020) Climate change forecasting and modeling for the year of 2050. In: Fahad S, Hasanuzzaman M, Alam M, Ullah H,Saeed M, Khan AK, Adnan M (eds) *Environment, climate, plant and vegetation growth.* Springer Publ Ltd, Springer Nature Switzerland AG. Part of Springer Nature, pp 109–122. doi: 10.1007/978-3-030-49732-3

Bocchiola D, Diolaiuti G (2013) Recent (1980–2009) evidence of climate change in the Upper Karakoram, Pakistan *Theor Appl Climatol* 113(3–4):611–641.

Chen Z, Grasby SE, Osadetz KG (2004) Relation between climate variability and groundwater levels in the Upper Carbonate Aquifer, Southern Manitoba, Canada. *J Hydrol* 290(1–2):43–62.

Depeng W, Fahad S, Saud S, Muhammad K, Aziz K, Mohammad NK, Hafiz MH, Wajid N (2018) Morphological acclimation to agronomic manipulation in leaf dispersion and orientation to promote "Ideotype" breeding: Evidence from 3D visual modeling of "super" rice (*Oryza sativa* L.). *Plant Physiol Biochem* 135:499–510. doi: 10.1016/j.plaphy.2018.11.010

Dragoni W, Sukhija BS (2008) Climate change and groundwater: a short review *Geol Soc London Spec Publ* 288(1):1–12.

EL Sabagh A, Hossain A, Barutçular C, Iqbal MA, Sohidul Islam M, Fahad S, Sytar O, Çig F, Meena RS, Murat Erman (2020) Consequences of salinity stress on the quality of crops and its mitigation strategies for sustainable crop production: an outlook of arid and semi-arid regions. In: Fahad S, Hasanuzzaman

M, Alam M, Ullah H, Saeed M, Khan AK, Adnan M (eds) *Environment, climate, plant and vegetation growth*. Springer Publ Ltd, Springer Nature Switzerland AG. Part of Springer Nature, pp 503–534. doi: 10.1007/978-3-030-49732-3

Fahad S, Adnan M, Hassan S, Saud S, Hussain S, Wu C, Wang D, Hakeem KR, Alharby HF, Turan V, Khan MA, Huang J (2019a) Rice responses and tolerance to high temperature. In: Hasanuzzaman M, Fujita M, Nahar K, Biswas JK (eds) *Advances in rice research for abiotic stress tolerance*. Woodhead Publ Ltd, Abington hall Abington, Cambridge CB1 6AH, CAMBS, England, pp 201–224.

Fahad S, Bajwa AA, Nazir U, Anjum SA, Farooq A, Zohaib A, Sadia S, NasimW, Adkins S, Saud S, Ihsan MZ, Alharby H,Wu C, Wang D, Huang J (2017) Crop production under drought and heat stress: plant responses and management options. *Front Plant Sci* 8:1147. doi: 10.3389/fpls.2017.01147

Fahad S, Bano A (2012) Effect of salicylic acid on physiological and biochemical characterization of maize grown in saline area. *Pak J Bot* 44:1433–1438.

Fahad S, Chen Y, Saud S,Wang K, Xiong D, Chen C, Wu C, Shah F, Nie L, Huang J (2013) Ultraviolet radiation effect on photosynthetic pigments, biochemical attributes, antioxidant enzyme activity and hormonal contents of wheat. *J Food Agri Environ* 11(3&4):1635–1641.

Fahad S, Hussain S, Bano A, Saud S, Hassan S, Shan D, Khan FA, Khan F, Chen Y, Wu C, Tabassum MA, Chun MX, Afzal M, Jan A, Jan MT, Huang J (2014a) Potential role of phytohormones and plant growth-promoting rhizobacteria in abiotic stresses: consequences for changing environment. *Environ Sci Pollut Res* 22(7):4907– 4921. doi: 10.1007/s11356-014-3754-2

Fahad S, Hussain S, Matloob A, Khan FA, Khaliq A, Saud S, Hassan S, Shan D, Khan F, Ullah N, Faiq M, Khan MR, Tareen AK, Khan A, Ullah A, Ullah N, Huang J (2014b) Phytohormones and plant responses to salinity stress: a review. *Plant Growth Regul* 75(2):391–404. doi: 10.1007/s10725-014-0013-y

Fahad S, Hussain S, Saud S, Hassan S, Chauhan BS, Khan F et al (2016a) Responses of rapid viscoanalyzer profile and other rice grain qualities to exogenously applied plant growth regulators under high day and high night temperatures. *PLoS One* 11(7):e0159590. doi: 10.1371/journal.pone.0159590

Fahad S, Hussain S, Saud S, Hassan S, Ihsan Z, Shah AN,Wu C, Yousaf M, Nasim W, Alharby H, Alghabari F, Huang J (2016b) Exogenously applied plant growth regulators enhance the morphophysiological growth and yield of rice under high temperature. *Front Plant Sci* 7:1250. doi: 10.3389/fpls.2 016.01250

Fahad S, Hussain S, Saud S, Hassan S, Tanveer M, Ihsan MZ, Shah AN, Ullah A, Nasrullah KF, Ullah S, Alharby HNW, Wu C, Huang J (2016c) A combined application of biochar and phosphorus alleviates heat-induced adversities on physiological, agronomical and quality attributes of rice. *Plant Physiol Biochem* 103:191–198.

Fahad S, Hussain S, Saud S, Khan F, Hassan S, Jr A, Nasim W, Arif M, Wang F, Huang J (2016d) Exogenously applied plant growth regulators affect heat-stressed rice pollens. *J Agron Crop Sci* 202:139– 150.

Fahad S, Hussain S, Saud S, Tanveer M, Bajwa AA, Hassan S, Shah AN, Ullah A, Wu C, Khan FA, Shah F, Ullah S, Chen Y, Huang J (2015a) A biochar application protects rice pollen from high-temperature stress. *Plant Physiol Biochem* 96:281–287.

Fahad S, Muhammad ZI, Abdul K, Ihsanullah D, Saud S, Saleh A, Wajid N, Muhammad A, Imtiaz AK, Chao W, Depeng W, Jianliang H (2018) Consequences of high temperature under changing climate optima for rice pollen characteristics-concepts and perspectives. *Archives Agron Soil Sci*. doi: 10.1080/ 03650340.2018.1443213

Fahad S, Nie L, Chen Y, Wu C, Xiong D, Saud S, Hongyan L, Cui K, Huang J (2015b) Crop plant hormones and environmental stress. *Sustain Agric Rev* 15:371–400.

Fahad S, Rehman A, Shahzad B, Tanveer M, Saud S, Kamran M, Ihtisham M, Khan SU, Turan V, Rahman MHU (2019b) Rice responses and tolerance to metal/metalloid toxicity. In: Hasanuzzaman M, Fujita M, Nahar K, Biswas JK (eds) *Advances in rice research for abiotic stress tolerance*. Woodhead Publ Ltd, Abington hall Abington, Cambridge CB1 6AH, CAMBS, England, pp 299–312.

Farah R, Muhammad R, Muhammad SA, Tahira Y, Muhammad AA, Maryam A, Shafaqat A, Rashid M, Muhammad R, Qaiser H, Afia Z, Muhammad AA, Muhammad A, Fahad S (2020) Alternative and non-conventional soil and crop management strategies for increasing water use efficiency. In: Fahad S, Hasanuzzaman M, Alam M, Ullah H, Saeed M, Khan AK, Adnan M (eds) *Environment, climate, plant and vegetation growth*. Springer Publ Ltd, Springer Nature Switzerland AG. Part of Springer Nature, pp 323–338. doi: 10.1007/978-3-030-49732-3

Farhana G, Ishfaq A, Muhammad A, Dawood J, Fahad S, Xiuling L, Depeng W, Muhammad F, Muhammad F, Syed AS (2020) Use of crop growth model to simulate the impact of climate change on yield of various wheat cultivars under different agro-environmental conditions in Khyber Pakhtunkhwa, Pakistan. *Arabian J Geosci* 13:112. doi: 10.1007/s12517-020-5118-1

Farhat A, Hafiz MH, Wajid I, Aitazaz AF, Hafiz FB, Zahida Z, Fahad S, Wajid F, Artemi C (2020) A review of soil carbon dynamics resulting from agricultural practices. *J Environ Manage* 268(2020):110319.

Fazli W, Muhmmad S, Amjad A, Fahad S, Muhammad A, Muhammad N, Ishaq AM, Imtiaz AK, Mukhtar A, Muhammad S, Muhammad I, Rafi U, Haroon I, Muhammad A (2020) Plant-microbes interactions and functions in changing climate. In: Fahad S, Hasanuzzaman M, Alam M, Ullah H, Saeed M, Khan AK, Adnan M (eds) *Environment, climate, plant and vegetation growth.* Springer Publ Ltd, Springer Nature Switzerland AG. Part of Springer Nature, pp 397–420. doi: 10.1007/978-3-030-49732-3

Field CB, Barros V, Stocker TF, Dahe Q (2012) *Managing the risks of extreme events and disasters to advance climate change adaptation: special report of the Intergovernmental panel on climate change.* Cambridge University Press.

Gardelle J, Berthier E, Arnaud Y, Kaab A (2013) Region-wide glacier mass balances over the Pamir-Karakoram-Himalaya during 1999-2011. *The Cryosphere* 7:1263–1286.

Gardner, Alex S et al. (2013) A reconciled estimate of glacier contributions to sea level rise: 2003 to 2009. *Science* 340(6134): 852–857.

Gleeson, Tom et al. (2016) The global volume and distribution of modern groundwater. *Nature Geosci* 9(2):161.

Gopakumar L, Bernard NO, Donato V (2020) Soil microarthropods and nutrient cycling. In: Fahad S, Hasanuzzaman M, Alam M, Ullah H, Saeed M, Khan AK, Adnan M (eds) *Environment, climate, plant and vegetation growth.* Springer Publ Ltd, Springer Nature Switzerland AG. Part of Springer Nature, pp 453–472. doi: 10.1007/978-3-030-49732-3

Habib ur R, Ashfaq A, Aftab W, Manzoor H, Fahd R, Wajid I, Md. Aminul I, Vakhtang S, Muhammad A, Asmat U, Abdul W, Syeda RS, Shah S, Shahbaz K, Fahad S, Manzoor H, Saddam H, Wajid N (2017) Application of CSM-CROPGRO-Cotton model for cultivars and optimum planting dates: Evaluation in changing semi-arid climate. *Field Crops Res.* doi: 10.1016/j.fcr.2017.07.007

Hadi F, Rahman AU, Ibrar M, Dastagir G, Arif M, Naveed K, Adnan M (2014) Weed diversity with special reference to their ethnomedicinal uses in wheat and maize at Rech valley, Hindokush Range, Chitral, Pakistan. *Pak J Weed Sci Res* 20(3):335–346.

Hafiz MH, Abdul K, Farhat A, Wajid F, Fahad S, Muhammad A, Ghulam MS, Wajid N, Muhammad M, Hafiz FB (2020a) Comparative effects of organic and inorganic fertilizers on soil organic carbon and wheat productivity under arid region. *Commun Soil Sci Plant Anal.* doi: 10.1080/00103624.2020.1 763385

Hafiz MH, Farhat A, Ashfaq A, Hafiz FB, Wajid F, Carol Jo W, Fahad S, Gerrit H (2020b) Predicting kernel growth of maize under controlled water and nitrogen applications. *Int J Plant Prod.* doi: 10.1007/s421 06-020-00110-8

Hafiz MH, Farhat A, Shafqat S, Fahad S, Artemi C, Wajid F, Chaves CB, Wajid N, Muhammad M, Hafiz FB (2018) Offsetting land degradation through nitrogen and water management during maize cultivation under arid conditions. *Land Degrad Dev* 1–10. doi: 10.1002/ldr.2933.

Hafiz MH, Muhammad A, Farhat A, Hafiz FB, Saeed AQ, Muhammad M, Fahad S, Muhammad A (2019) Environmental factors affecting the frequency of road traffic accidents: a case study of sub-urban area of Pakistan. *Environ Sci Pollut Res.* doi: 10.1007/s11356-019-04752-8

Hafiz MH, Wajid F, Farhat A, Fahad S, Shafqat S, Wajid N, Hafiz FB (2016) Maize plant nitrogen uptake dynamics at limited irrigation water and nitrogen. *Environ Sci Pollut Res* 24(3):2549–2557. doi: 10.1007/s11356-016-8031-0

Hesham FA, Fahad S (2020) Melatonin application enhances biochar efficiency for drought tolerance in maize varieties: Modifications in physio-biochemical machinery. *Agron J* 112(4):1–22.

Hewitt K (2007) Tributary glacier surges: an exceptional concentration at Panmah Glacier, Karakoram Himalaya. *J Glaciol* 53(181):181–188.

Hewitt K (2011) Glacier change, concentration, and elevation effects in the Karakoram Himalaya, Upper Indus Basin. *Mountain Res Dev* 31(3):188–201.

Houze Jr, RA et al. (2011) Anomalous atmospheric events leading to the Summer 2010 floods in Pakistan. *Bull Am Meteorol Soc* 92(3):291–298.

Hussain MA, Fahad S, Rahat S, Muhammad FJ, Muhammad M, Qasid A, Ali A, Husain A, Nooral A, Babatope SA, Changbao S, Liya G, Ibrar A, Zhanmei J, Juncai H (2020) Multifunctional role of brassinosteroid and its analogues in plants. *Plant Growth Regul.* doi: 10.1007/s10725-020-00647-8

Ibrar K, Aneela R, Khola Z, Urooba N, Sana B, Rabia S, Ishtiaq H, Rehman MUr, Salvatore M (2020) Microbes and environment: global warming reverting the frozen zombies. In: Fahad S, Hasanuzzaman M, Alam M, Ullah H, Saeed M, Khan AK, Adnan M (eds) *Environment, climate, plant and vegetation growth*. Springer Publ Ltd, Springer Nature Switzerland AG. Part of Springer Nature, pp 607–634. doi: 10.1007/978-3-030-49732-3

Ilyas M, Mohammad N, Nadeem K, Ali H, Aamir HK, Kashif H, Fahad S, Aziz K, Abid U (2020) Drought tolerance strategies in plants: a mechanistic approach. *J Plant Growth Regul.* doi: 10.1007/s00344-02 0-10174-5

Immerzeel WW, Bierkens MFP (2012) Asia's water balance. *Nat Geosci* 5 (12): 841–842.

Immerzeel WW, Ludovicus PHVB, Bierkens MFP (2010) Climate change will affect the Asian Water Towers. *Science* 328(5984):1382–1385.

Iqra M, Amna B, Shakeel I, Fatima K,Sehrish L, Hamza A, Fahad S (2020) Carbon cycle in response to global warming. In: Fahad S, Hasanuzzaman M, Alam M, Ullah H,Saeed M, Khan AK, Adnan M (eds) *Environment, climate, plant and vegetation growth*. Springer Publ Ltd, Springer Nature Switzerland AG. Part of Springer Nature, pp 1–16. doi: 10.1007/978-3-030-49732-3

Jägermeyr J et al. (2015) Water savings potentials of irrigation systems: global simulation of processes and linkages. *Hydrol Earth Syst Sci* 19(7):3073–3091.

Jain SK, Pushpendra KA, Singh VP (2007) Indus Basin. In: *Hydrology and water resources of India*. Springer, pp 473–511.

Jan M, Muhammad AH, Adnan NS, Muhammad Y, Javaid I, Xiuling L, Depeng W, Fahad S, (2019) Modulation in growth, gas exchange, and antioxidant activities of salt-stressed rice (Oryza sativa L.) genotypes by zinc fertilization. *Arabian J Geosci* 12:775. doi: 10.1007/s12517-019-4939-2

Kääb, A et al. (2012) Contrasting patterns of early twenty-first-century glacier mass change in the Himalayas. *Nature* 488(7412):495.

Kääb A, Treichler D, Nuth C, Berthier E (2015) Brief communication: contending estimates of 2003–2008 glacier mass balance over the Pamir–Karakoram–Himalaya. *Cryosphere* 9(2):557–564.

Kamarn M, Wenwen C, Irshad A, Xiangping M, Xudong Z, Wennan S, Junzhi C, Shakeel A, Fahad S, Qingfang H, Tiening L (2017) Effect of paclobutrazol, a potential growth regulator on stalk mechanical strength, lignin accumulation and its relation with lodging resistance of maize. *Plant Growth Regul* 84:317–332. doi: 10.1007/s10725-017-0342-8

Lutz, Arthur F et al. (2016) Selecting representative climate models for climate change impact studies: an advanced envelope-based selection approach. *Int J Climatol* 36(12):3988–4005.

Mahar A, Amjad A, Altaf HL, Fazli W, Ronghua L, Muhammad A, Fahad S, Muhammad A, Rafiullah, Imtiaz AK, Zengqiang Z (2020) Promising technologies for Cd-contaminated soils: drawbacks and possibilities. In: Fahad S, Hasanuzzaman M, Alam M, Ullah H,Saeed M, Khan AK, Adnan M (eds) *Environment, climate, plant and vegetation growth*. Springer Publ Ltd, Springer Nature Switzerland AG. Part of Springer Nature, pp 63–92. doi: 10.1007/978-3-030-49732-3

Margat J, Gun JV (2013) *Groundwater around the world: a geographic synopsis*. CRC Press.

Maussion F et al. (2014) Precipitation seasonality and variability over the Tibetan Plateau as resolved by the high Asia reanalysis. *J Climate* 27(5):1910–1927.

Md Enamul H, Shoeb AZM, Mallik AH, Fahad S, Kamruzzaman MM, Akib J, Nayyer S, Mehedi A KM, Swati AS, Md Yeamin A, Most SS (2020) Measuring vulnerability to environmental hazards: qualitative to quantitative. In: Fahad S, Hasanuzzaman M, Alam M, Ullah H, Saeed M, Khan AK, Adnan M (eds) *Environment, climate, plant and vegetation growth*. Springer Publ Ltd, Springer Nature Switzerland AG. Part of Springer Nature, pp 421–452. doi: 10.1007/978-3-030-49732-3

Md Jakir H, Allah B (2020) Development and applications of transplastomic plants; a way towards eco-friendly agriculture. In: Fahad S, Hasanuzzaman M, Alam M, Ullah H, Saeed M, Khan AK, Adnan M (eds) *Environment, climate, plant and vegetation growth*. Springer Publ Ltd, Springer Nature Switzerland AG. Part of Springer Nature, pp 285–322. doi: 10.1007/978-3-030-49732-3

Mishra V (2015) Climatic uncertainty in Himalayan water towers. *J Geophys Res Atmos* 120(7):2689–2705.

Misra AK (2014) Climate change and challenges of water and food security. *Int J Sustain Built Environ* 3(1):153–165.

Mölg T, Maussion F, Scherer D (2014) Mid-latitude Westerlies as a driver of glacier variability in Monsoonal high Asia. *Nature Climate Change* 4(1):68.

Mubeen M, Ashfaq A, Hafiz MH, Muhammad A, Hafiz UF, Mazhar S, Muhammad SD, Asad A, Amjed A, Fahad S, Wajid N (2020) Evaluating the climate change impact on water use efficiency of cotton-wheat in semi-arid conditions using DSSAT model. *J Water Climate Change*. doi/10.2166/wcc.2019.179/622 035/jwc2019179.pdf

Muhammad TQ, Amna F, Amna B, Barira Z, Xitong Z, Ling-Ling C (2020) Effectiveness of conventional crop improvement strategies vs. omics. In: Fahad S, Hasanuzzaman M, Alam M, Ullah H, Saeed M, Khan AK, Adnan M (eds) *Environment, climate, plant and vegetation growth*. Springer Publ Ltd, Springer Nature Switzerland AG. Part of Springer Nature, pp 253–284. doi: 10.1007/978-3-030-4 9732-3

Muhammad Z, Abdul MK, Abdul MS, Kenneth BM, Muhammad S, Shahen S, Ibadullah J, Fahad S (2019) Performance of Aeluropus lagopoides (mangrove grass) ecotypes, a potential turfgrass, under high saline conditions. *Environ Sci Pollut Res*. doi: 10.1007/s11356-019-04838-3

Mukhopadhyay B, Khan A (2015) Boltzmann–Shannon entropy and river flow stability within Upper Indus Basin in a changing climate. *Int J River Basin Manag* 13(1):87–95.

Mushtaq S, Ahmed N, Khan IA, Shah B, Khan A, Ali M, Rashid MT, Adnan M, Junaid K, Zaki AB, Ahmed S (2015) Study on the efficacy of ladybird beetle as a biological control agent against aphids (Chaitophorus spp.). *J Ento Zool Stud* 3(6):117–119.

Noor M, Naveed ur R, Ajmal J, Fahad S, Muhammad A, Fazli W, Saud S, Hassan S (2020) Climate change and costal plant lives. In: Fahad S, Hasanuzzaman M, Alam M, Ullah H,Saeed M, Khan AK, Adnan M (eds) *Environment, climate, plant and vegetation growth*. Springer Publ Ltd, Springer Nature Switzerland AG. Part of Springer Nature, pp 93–108. doi: 10.1007/978-3-030-49732-3

Qamar-uz Z, Zubair A, Muhammad Y, Muhammad ZI, Abdul K, Fahad S, Safder B, Ramzani PMA, Muhammad N (2017) Zinc biofortification in rice: leveraging agriculture to moderate hidden hunger in developing countries. *Arch Agron Soil Sci* 64:147–161. doi: 10.1080/03650340.2017.1338343

Quincey DJ, Glasser NF, Cook SJ, Luckman A (2015) Heterogeneity in Karakoram Glacier Surges. *J Geophys Res Earth Surf* 120(7):1288–1300.

Rasheed MT, Inayatullah M, Shah B, Ahmed N, Khan A, Ali M, Ahmed S, Junaid K, Adnan M, Huma Z (2015) Relative abundance of insect pollinators on two cultivars of sunflower in Islamabad. *J Ento Zool Stud* 3(6):164–165.

Rashid M, Qaiser H, Khalid SK, Al-Wabel MI , Zhang A, Muhammad A, Shahzada SI, Rukhsanda A, Ghulam AS, Shahzada MM, Sarosh A, Muhammad FQ (2020) Prospects of biochar in alkaline soils to mitigate climate change. In: Fahad S, Hasanuzzaman M, Alam M, Ullah H,Saeed M, Khan AK, Adnan M (eds) *Environment, climate, plant and vegetation growth*. Springer Publ Ltd, Springer Nature Switzerland AG. Part of Springer Nature, pp 133–150. doi: 10.1007/978-3-030-4 9732-3

Reggiani P, Rientjes THM (2014) A reflection on the long-term water balance of the Upper Indus Basin *Hydrol Res* 46(3):446–462.

Rehman M, Fahad S, Saleem MH, Hafeez M, Muhammad HR, Liu F, Deng G (2020) Red light optimized physiological traits and enhanced the growth of ramie (Boehmeria nivea L.). *Photosynthetica* 58 (4):922–931.

Sadam M, Muhammad TQ, Ghulam M, Muhammad SK, Faiz AJ (2020) Role of biotechnology in climate resilient agriculture. In: Fahad S, Hasanuzzaman M, Alam M, Ullah H, Saeed M, Khan AK, Adnan M (eds) *Environment, climate, plant and vegetation growth*. Springer Publ Ltd, Springer Nature Switzerland AG. Part of Springer Nature, pp 339–366. doi: 10.1007/978-3-030-49732-3

Sajid H, Jie H, Jing H, Shakeel A, Satyabrata N, Sumera A, Awais S, Chunquan Z, Lianfeng Z, Xiaochuang C, Qianyu J, Junhua Z (2020) Rice production under climate change: adaptations and mitigating strategies. In: Fahad S, Hasanuzzaman M, Alam M, Ullah H, Saeed M, Khan AK, Adnan M (eds) *Environment, climate, plant and vegetation growth*. Springer Publ Ltd, Springer Nature Switzerland AG. Part of Springer Nature, pp 659–686. doi: 10.1007/978-3-030-49732-3

Sajjad H, Muhammad M, Ashfaq A, Waseem A, Hafiz MH, Mazhar A, Nasir M, Asad A, Hafiz UF, Syeda RS, Fahad S, Depeng W, Wajid N (2019) Using GIS tools to detect the land use/land cover changes during forty years in Lodhran district of Pakistan. *Environ Sci Pollut Res* 10.1007/s11356-019-06072-3

Saleem MH, Fahad S, Adnan M, Mohsin A, Muhammad SR, Muhammad K, Qurban A, Inas AH, Parashuram B, Mubassir A, Reem MH (2020a) Foliar application of gibberellic acid endorsed phytoextraction of copper and alleviates oxidative stress in jute (Corchorus capsularis L.) plant grown in highly copper-contaminated soil of China. *Environ Sci Pollution Res.* doi: 10.1007/s11356-020-09764-3

Saleem MH, Fahad S, Shahid UK, Mairaj D, Abid U, Ayman ELS, Akbar H, Analía L, Lijun L (2020b) Copper-induced oxidative stress, initiation of antioxidants and phytoremediation potential of flax (Linum usitatissimum L.) seedlings grown under the mixing of two different soils of China. *Environ Sci Poll Res.* doi: 10.1007/s11356-019-07264-7

Saleem MH, Rehman M, Fahad S, Tung SA, Iqbal N, Hassan A, Ayub A, Wahid MA, Shaukat S, Liu L, Deng G (2020c) Leaf gas exchange, oxidative stress, and physiological attributes of rapeseed (Brassica napus L.) grown under different light-emitting diodes. *Photosynthetica* 58 (3): 836–845.

Saman S, Amna B, Bani A, Muhammad TQ, Rana MA, Muhammad SK (2020) QTL mapping for abiotic stresses in cereals. In: Fahad S, Hasanuzzaman M, Alam M, Ullah H, Saeed M, Khan AK, Adnan M (eds) *Environment, climate, plant and vegetation growth.* Springer Publ Ltd, Springer Nature Switzerland AG. Part of Springer Nature, pp 229–252. doi: 10.1007/978-3-030-49732-3

Saud S, Chen Y, Fahad S, Hussain S, Na L, Xin L, Alhussien SA (2016) Silicate application increases the photosynthesis and its associated metabolic activities in Kentucky bluegrass under drought stress and post-drought recovery. *Environ Sci Pollut Res* 23(17):17647–17655. doi: 10.1007/s11356-016-6957-x

Saud S, Chen Y, Long B, Fahad S, Sadiq A (2013) The different impact on the growth of cool season turf grass under the various conditions on salinity and drought stress. *Int J Agric Sci Res* 3:77–84.

Saud S, Fahad S, Cui G, Chen Y, Anwar S (2020) Determining nitrogen isotopes discrimination under drought stress on enzymatic activities, nitrogen isotope abundance and water contents of Kentucky bluegrass. *Sci Rep* 10:6415.doi: 10.1038/s41598-020-63548-w

Saud S, Fahad S, Yajun C, Ihsan MZ, Hammad HM, Nasim W, Amanullah Jr, Arif M, Alharby H (2017) Effects of nitrogen supply on water stress and recovery mechanisms in Kentucky Bluegrass plants. *Front Plant Sci* 8:983. doi: 10.3389/fpls.2017.00983

Saud S, Li X, Chen Y, Zhang L, Fahad S, Hussain S, Sadiq A, Chen Y (2014) Silicon application increases drought tolerance of Kentucky bluegrass by improving plant water relations and morph physiological functions. *SciWorld J* 2014:1–10. doi: 10.1155/2014/368694

Scherler D, Bookhagen B, Strecker MR (2011) Spatially variable response of Himalayan Glaciers to climate change affected by Debris cover. *Nature Geosci* 4(3):156.

Senol C (2020) The effects of climate change on human behaviors. In: Fahad S, Hasanuzzaman M, Alam M, Ullah H, Saeed M, Khan AK, Adnan M (eds) *Environment, climate, plant and vegetation growth.* Springer Publ Ltd, Springer Nature Switzerland AG. Part of Springer Nature, pp 577–590. doi: 10.1 007/978-3-030-49732-3

Shafi MI, Adnan M, Fahad S, Fazli W, Ahsan K, Zhen Y, Subhan D, Zafar-ul-Hye M, Martin B, Rahul D (2020) Application of single superphosphate with humic acid improves the growth, yield and phosphorus uptake of wheat (Triticum aestivum L.) in calcareous soil. *Agron* (10):1224. doi:10.3390/agronomy10091224

Shah F, Lixiao N, Kehui C, Tariq S, Wei W, Chang C, Liyang Z, Farhan A, Fahad S, Huang J (2013) Rice grain yield and component responses to near 2°C of warming. *Field Crop Res* 157:98–110.

Shah SAR, Khan SA, Junaid K, Sattar S, Zaman M, Saleem N, Adnan M (2015) Screening of mustard genotypes for antixenosis and multiplication against mustard aphid, Lipaphiserysimi(Kalt) (Aphididae: Homoptera). *J Ento Zool Studies* 3(6):84–87.

Sharif M, Archer DR, Fowler HJ, Forsythe N (2013) Trends in timing and magnitude of flow in the Upper Indus Basin. *Hydrol Earth Syst Sci* 17(4): 1503–1516.

Siebert S et al. (2010) Groundwater use for irrigation–a global inventory. *Hydrol Earth Syst Sci* 14(10):1863–1880.

Smith M, Cross K, Paden M, Laban P (2016) *Spring: managing groundwater sustainability.* IUCN Global Water Programme.

Subhan D, Zafar-ul-Hye M, Fahad S, Saud S, Martin B, Tereza H, Rahul D (2020) Drought stress alleviation by ACC deaminase producing *Achromobacter xylosoxidans* and *Enterobacter cloacae*, with and without timber waste biochar in maize. *Sustain* 12(6286). doi: 10.3390/su12156286

Tahir, M. et al. (2008) Comparative yield performance of different maize (Zea Mays L.) hybrids under local conditions of Faisalabad-Pakistan. *Pakistan J Life Social Sci* 6(2):118–120. http://www.pjlss.edu.pk/pdf_files/2008_2/11_tahir118-120.pdf.

Tariq M, Ahmad S, Fahad S, Abbas G, Hussain S, Fatima Z, Nasim W, Mubeen M, ur Rehman MH, Khan MA, Adnan M (2018). The impact of climate warming and crop management on phenology of sunflower-based cropping systems in Punjab, Pakistan. *Agri and Forest Met.* 15(256):270–282.

Unied Nations. (2015) *World population prospects. The 2015 revision.* United Nations Population Division, New York.

Unsar NU, Muhammad R, Syed HMB, Asad S, Mirza AQ, Naeem I, Muhammad H ur R, Fahad S, Shafqat S (2020) Insect pests of cotton crop and management under climate change scenarios. In: Fahad S, Hasanuzzaman M, Alam M, Ullah H, Saeed M, Khan AK, Adnan M (eds) *Environment, climate, plant and vegetation growth.* Springer Publ Ltd, Springer Nature Switzerland AG. Part of Springer Nature, pp 367–396. doi: 10.1007/978-3-030-49732-3

Wahid F, Fahad S, Subhan D, Adnan M, Zhen Y, Saud S, Manzer HS, Martin B, Tereza H, Rahul D (2020) Sustainable management with mycorrhizae and phosphate solubilizing bacteria for enhanced phosphorus uptake in calcareous soils. *Agriculture* 10 (334). doi: 10.3390/agriculture10080334

Wajid N, Ashfaq A, Asad A, Muhammad T, Muhammad A, Muhammad S, Khawar J, Ghulam MS, Syeda RS, Hafiz MH, Muhammad IAR, Muhammad ZH, Muhammad HR, Veysel T, Fahad S, Suad S, Aziz K, Shahzad A (2017) Radiation efficiency and nitrogen fertilizer impacts on sunflower crop in contrasting environments of Punjab. *Pakistan Environ Sci Pollut Res* 25:1822–1836. doi: 10.1007/s11356-017-0592-z

Wu C, Kehui C, She T, Ganghua L, Shaohua W, Fahad S, Lixiao N, Jianliang H, Shaobing P, Yanfeng D (2020) Intensified pollination and fertilization ameliorate heat injury in rice (Oryza sativa L.) during the flowering stage. *Field Crops Res* 252:107795.

Wu C, Tang S, Li G, Wang S, Fahad S, Ding Y (2019) Roles of phytohormone changes in the grain yield of rice plants exposed to heat: a review. *PeerJ* 7:e7792. doi: 10.7717/peerj.7792

Yang Z, Zhang Z, Zhang T, Fahad S, Cui K, Nie L, Peng S, Huang J (2017) The effect of season-long temperature increases on rice cultivars grown in the central and southern regions of China. *Front Plant Sci* 8:1908. doi: 10.3389/fpls.2017.01908

Zafar-ul-Hye M, Muhammad N, Subhan D, Fahad S, Rahul D, Mazhar A, Ashfaq AR, Martin B, Jiˇrí H, Zahid HT, Muhammad N (2020a) Alleviation of cadmium adverse effects by improving nutrients uptake in bitter gourd through cadmium tolerant Rhizobacteria. *Environment* 7(54). doi: 10.3390/environments7080054

Zafar-ul-Hye M, Muhammad TH, Muhammad A, Fahad S, Martin B, Tereza D, Rahul D, Subhan D (2020b) Potential role of compost mixed biochar with rhizobacteria in mitigating lead toxicity in spinach. *Scientific Rep* 10:12159. doi: 10.1038/s41598-020-69183-9.

Zahida Z, Hafiz FB, Zulfiqar AS, Ghulam MS, Fahad S, Muhammad RA, Hafiz MH, Wajid N, Muhammad S (2017) Effect of water management and silicon on germination, growth, phosphorus and arsenic uptake in rice. *Ecotoxicol Environ Saf* 144:11–18.

Zaki AB, Ahmad N, Khan IA, Shah B, Khan A, Rasheed MT, Adnan M, Junaid K, Huma Z, Ahmed S (2015) Adulticidal efficacy of Azadirachtaindica (neem tree), Sesamumindicum(til) and Pinussabinaena(pine tree) extracts against Aedesaegyptiunder laboratory conditions. *J Ento Zool Stud* 3(6):112–116.

Zia-ur-Rehman M (2020) Environment, climate change and biodiversity. In: Fahad S, Hasanuzzaman M, Alam M, Ullah H, Saeed M, Khan AK, Adnan M (eds) *Environment, climate, plant and vegetation growth.* Springer Publ Ltd, Springer Nature Switzerland AG. Part of Springer Nature, pp 473–502. doi: 10.1007/978-3-030-49732-3

12

Nutrient Deficiency Stress and Relation with Plant Growth and Development

Saghir Abbas, Amna, Muhammad Tariq Javed, Qasim Ali, Muhammad Azeem, and Shafaqat Ali

CONTENTS

DOI: 10.1201/9781003160717-12

239

12.1 Introduction

Plants need an adequate supply of nutrients for physiological functions to run as usual and growth to flourish. Nutrients are absorbed through roots and leaves (Singh et al. 2013; Bouain 2019) but most plants absorb various macro-nutrients and micro-nutrients via roots (Weisany et al. 2013). The nutrients have various functions such as to strengthen the structural components of the cell membrane and cell walls, participate as important components of organic metabolites, implement the role of co-factor in enzymatic reactions as well as to function as a component of the plant signalling system. Deficiency symptoms of these nutrients not fulfilling plant requirements will appear at the plant level (Table 12.1).

The macro-nutrients include nitrogen, phosphorus, potassium, calcium, magnesium, and sulphur, and these are required in more quantities for plant growth than are micro-nutrients like iron, boron, copper, manganese, zinc, molybdenum, nickel, and chlorine (Singh et al. 2015). These 14 macronutrients and micronutrients, along with carbon, oxygen, and hydrogen, are essential for plant physiological functions to happen without disturbance (Singh et al. 2015; Grusak Michael et al. 2016). Some trace minerals such as sodium, selenium, cobalt, silicon, and aluminium are also beneficial, if not essential, in some plants adapted to grow in a particular environment (Singh et al. 2013; Bloom 2015).

Mineral nutrients are present in low concentrations in the soil, and their availability is affected by different environmental factors, including soil pH, nutrient bio-physio-chemical processes, biogeo-chemical cycles, precipitation, climate, and soil organic matter (Maathuis and Diatloff 2013; Kalaji et al. 2018). Plants can absorb nutrients actively by releasing energy or passively without energy loss (Karthika et al. 2018), where plants' nutrient sufficiency range is described as important for normal growth and functioning of plants and nutrient deficiency or toxicity is said to occur when availability of a nutrient is either insufficient vis-à-vis need or surpasses the sufficiency range for normal plant physiological functioning to take place. Nutrient deficiency drastically affects plant overall health, and the characteristic symptoms can become visible at the level of the whole plant (Karthika et al. 2018). Therefore, a proper understanding of nutrients' roles in plants is vital for sustainable crop cultivation and healthy crop growth visible in: nutrient uptake, status of nutrition in soil, assimilation, nutrient deficiency symptoms, and their role in plant development.

12.2 Nutrient Classification

Nutrients are categorized into macro-nutrients and micro-nutrients on the basis of their functions in plant development and growth, and quantity needed by healthy plant tissues.

TABLE 12.1

Functions and Deficiency Symptoms of Macro- and Micro and Beneficial Nutrients in Plants

Element & Symbol	Class	Form Acquired	Functions	Deficiency Symptoms
Nitrogen (N)	Macronutrient	NH_4^+, NO_3^-	Photosynthesis, Metabolites, structural	Chlorosis, wrinkled cereal grains
Potassium (K)	Macronutrient	K^+	Metabolism, CO_2 assimilation	Chlorosis, loss of apical dominance,
Phosphorus (P)	Macronutrient	$H_2PO_4^-$	Metabolites, structural signalling	Chlorosis, necrosis, premature leaf fall
Calcium (Ca)	Macronutrient	Ca^{2+}	Structural signalling	Deformed chlorotic appearance, stunted growth
Magnesium (Mg)	Macronutrient	Mg^{2+}	Metabolism, photosynthesis	Interveinal chlorosis, necrosis in older leaves
Sulphur (S)	Macronutrient	SO_4^{2-}	Osmotic, metabolites, structural	Leaf defoliation, stunted growth
Iron (Fe)	Micronutrient	Fe^{2+}, Fe^{3+}	Redox, photosynthesis	Interveinal chlorosis, necrotic tips
Manganese (Mn)	Micronutrient	Mn^{2+}	Redox, cofactor	Impaired growth
Zinc (Zn)	Micronutrient	Zn^{2+}	Redox, structural,	Stunted root development
Cupper (Cu)	Micronutrient	Cu^+, Cu^{2+}	Electron transport chain, redox	Necrotic spots
Boron (B)	Micronutrient	$B(OH)_3$	Structural,	Deformed leaves, deformed capitulum
Nickel (N)	Micronutrient	Ni^{2+}	Urease	Chlorosis, necrosis
Molybdenum (Mo)	Micronutrient	MoO_4^2	Redox, N_2 fixation	Yellow leaf area
Chlorine (Cl)	Micronutrient	Cl^-	Electrochemical, enzyme component	Leaf curling, wilting
Sodium (Na)	beneficial	Na^+	Electrochemical,	—
Selenium	beneficial	SeO_4^{2-}, SeO_3^{2-}	Redox	—
Cobalt	beneficial	Co^{2+}	Symbiosis	—
Silicon	beneficial	$Si(OH)_4$	Structural	—
Aluminium	beneficial	Al^{3+}	Structural	—

12.2.1 Macronutrient

Nutrients which plants require in large quantities to live a healthy life and consume in large quantities ranging from 0.2–4.0% dry matter weight basis in plant tissues are termed as macronutrients. Main macronutrients include C, H, O, N, P, K, Ca, Mg and S. The primary macronutrients needed in larger quantities include N, P, and K, whereas Ca, Mg, and S are required in moderate amounts and termed as secondary nutrients.

12.2.2 Micronutrients

The nutrients plants require in lesser amounts vis-à-vis that required by macro-nutrients, and which plants consume with an uptake ranging from 5–200 ppm or lesser than 0.02% of dry weight are characterized as micronutrients. These are either cationic (Fe, Mn, Zn, Cu, Ni,) or anionic (B, Mo, Cl).

12.2.3 Beneficial Elements

There are some elements that stimulate plant growth in some plants and are termed as beneficial elements: Sodium (Na), silicon (Si), cobalt (Co), selenium (Se), and aluminium (Al).

12.3 Nutrient Functions, Uptake, Transport, and Deficiency in relation to Plant Development

All plants must absorb mineral nutrients to ensure their normal growth and development. The role of mineral nutrients in plant development, mineral uptake, transport, and the impact of deficiency of certain elements in relation with plant development are discussed here.

12.3.1 Nitrogen

Nitrogen (N) is a valuable part of various plant components, and in combination with C, H, and O, it is the building block of proteins, amino acids, nucleic acids, phospholipids, and several secondary metabolites (Figure 12.1). Again, nitrogen is part of the ring structures of the purine and pyrimidine bases of nucleotide structures, the basis of nucleic acids, like DNA and RNA involved in the transfer of genetic code to the offspring (Virginia-Pérez et al. 2015; Pandey 2018). As an essential component of vitamins that help in metabolism, and of coenzymes that play a vital role in turgor generation as biochemistry analysis shows, nitrogen is the structural part of different alkaloids synthesized from the amino acids with names as diverse as tyrosine, arginine, tryptophan and lysine (Kováčik and Klejdus 2014). Plants with nitrate availability show more tolerance against photodamage. Given it can provide defence against oxidative stress, nitrogen improves defence system against stress by enhancing antioxidant systems and improving quantity of dry matter as a whole in plant tissues in crops (Kausar et al. 2017).

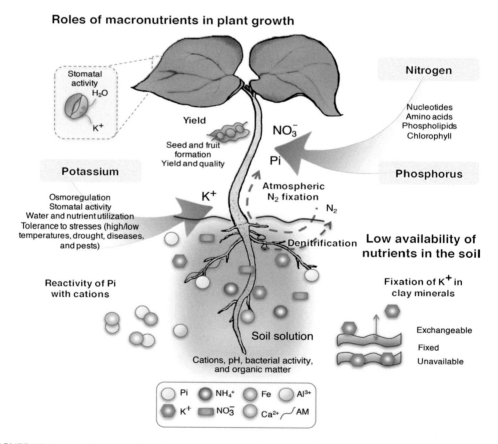

FIGURE 12.1 Role of macronutrient phosphorus, nitrogen, and potassium in plant and their bioavailability in soil. Plant uptake macronutrient in the form of phosphorus (Pi), nitrogen (NO_3^-), and K^+ which are essential structural components and plays different role in cellular processes. Their availability depends on several factors such pH, organic matter, and cations in the soil solution.

12.3.1.1 Nitrogen Uptake and Assimilation

Nitrogen is the most abundant gas present (80%) in atmosphere. However, its stable, atomic form (N_2) is not accessible to plants directly (Mahmud et al. 2020). After carbon, it's nitrogen that is required by plants in perceptively large amounts with a major source of N found to exist for plant roots to take up as ammonium (NH_4) and nitrate (NO_3). To become available to plants in this form, freely-living microorganisms and symbiotic ones fix atmospheric nitrogen into NH_4 or the nitrifying bacteria convert atmospheric nitrogen into NO_3. For roots to absorb nitrate and ammonium, they are assisted by the transport system located on the plasma membrane of the epidermal and cortical roots cells of the plant (Rentsch et al. 2007). Various physiological transport systems exist which mediate nitrate or ammonium uptake and possess different affinities: A high affinity transport system (HATS) works at low concentrations of outer nitrate or ammonium; whereas, at high concentrations of the same, a low affinity transport system (LATS) participates in uptake permitting great influxes of substrate at high substrate accessibility (Karthika et al. 2018). The genes involved in nitrate or ammonium transporters are members of NRT, AMT, and CLC families. Nitrogen translocation in plants occurs when nitrate and amino acids travel across xylem in roots to upper plant parts, where high affinity H^+-coupled symporters as members of NRT family process NO_3 uptake and MT transporters mediate NH_4 uptake in plants (Karthika et al. 2018; Tegeder and Masclaux-Daubresse 2018). In *Arabidopsis,* six transporters are found to exist: NRT1.1 and NRT1.2 operate under high nitrate resource while NRT2.1, NRT2.2, NRT2.4, and NRT2.5 work under nitrate starvation (Li et al. 2008; Kiba et al. 2012). Excess ammonium is noxious to plant cells and its uptake and assimilation are highly regulated, where high affinity ammonium transporters (AMTs) and low affinity uptake systems (aquaporins or cation channels) regulate ammonium transport in plants (Yuan et al. 2007; Ferreira et al. 2015).

12.3.1.2 Nitrogen Deficiency

Nitrogen is a mobile element, and during N deficiency, plants exhibit chlorosis in older leaves followed by necrosis symptoms which causes yellow leaves and leaf to fall off. Triggering slow and stunted growth in plants due to inhibition in cell division and dormancy in lateral buds (Kimura et al. 2016; Kathpalia and Bhatla 2018), nitrogen deficiency in some plants such as maize and tomato show symptoms of purple appearances and chlorosis symptoms on stems as well as on petioles and leaves (Lin et al. 2011). Plants exhibiting N-deficiency indicate smaller leaves and premature shedding of older leaves associated with adverse effects on the root (Costa et al. 2017; Karthika et al. 2018). Nitrogen starvation could increase flavonoids such as rutin, and if ferulic acid forms in excess amounts in plant parts, it can harm plant growth. In cereal crops, nitrogen deficiency causes reduction in amino acids Gln, proteins, and photosynthetic pigments which trigger decrease in tillering and in numbers of ears per unit area as well as small grain size due to decrease in carbohydrate levels in the grains. It was reported that N-deficiency causes excess anthocyanin production in plants that is usually associated with symptoms of purple leaves. Triggering 34%, or even 36%, decrease in plant chlorophyll and carotenoid contents in rice plants (Huang et al. 2004), nitrogen limitation is also associated with reduced enzyme activities during energy metabolism including during photosynthesis and respiration (Lin et al. 2011). The reduction in Rubisco activity in electron transport chain under nitrogen starvation has been reported by Evans and Terashima (1987), while deficiency of nitrogen can lead to excess electron flow possibly causing an increase in oxygen photoreduction in the chloroplast vis-à-vis the Mehler reaction following augmented levels of ROS production. All enzymatic and non-enzymatic antioxidants revealed the presence of primitive antioxidant systems in spinach, maize, barley, and *Arabidopsis* plants (Kandlbinder et al. 2004; Finkemeier et al. 2003; Tewari et al. 2006; Tewari et al. 2007). An increased level of abscisic acid and H_2O_2 in roots was reported in plants deficient in nitrogen (Shin et al. 2005; Zhao et al. 2007). The activities of ascorbate peroxidase, glutathione reductase, peroxidase, and catalase were recorded in rice plants (Lin et al. 2011). Whereas some studies have reported that nitrogen deficiency causes reduction in quantum yield of PSII and electron transport, others say it diminishes chloroplast pigments and production of several enzymes involved in Calvin cycle especially Rubisco. Moreover, N starvation could drastically affect Rubisco-related CO_2 assimilation and photorespiration in cassava plants (Cruz et al. 2003).

12.3.2 Phosphorus

Phosphorus (P), one of the major mineral nutrients that plants cannot do without, is the key component of many biological processes such as energy metabolism and synthesis of nucleic acids, cell membrane, phospholipids, and amino acids (Figure 12.1). Having a key role to play in the energy transmission of plant processes, phosphorus through pyrophosphate bonds in ATP, offers metabolic energy to a multitude of plant processes, like plant photosynthesis and respiration, via addition of phosphate group to different sugars (Taiz and Zeiger 2002). Involved in enzyme regulation and signal transduction through protein phosphorylation and dephosphorylation reactions, phosphorus is necessary for nucleoproteins and phosphorus-comprising compounds formation in young meristematic cells. In plants, phosphorus is available as *phosphate ester* which helps DNA and RNA carry out biosynthesis; and ester is an important structural constituent of phospholipids constituting the cell membranes. It also plays a key role in membrane biochemistry and is a key component of ATP, AMP, ADP, and pyrophosphate which are essential components of energy metabolism. Inositol triphosphate (IP₃) is a secondary messenger involved in signal transduction, whereas in many plants, phosphorus and nitrogen regulate plant growth and maturation (Zhang et al. 2014; Kanno et al. 2016). Playing an important role in seed formation and to enhance root diameter of plants to receive more water and nutrients, phosphorus increases plant tolerance against drought or other abiotic stresses and improves water use efficiency and yield efficacy (Shen et al. 2011). Rhizobium and nitrogen fixing legume's activity have been known to break into a sprint with sufficient phosphorus availability (Karthika et al. 2018).

12.3.2.1 Phosphorus Uptake and Transport in Plants

Most plants require phosphorus as an inorganic phosphate (Pi), and though phosphorus is present in enormous amounts in soil still high chemical fixation rate and low diffusion make it one of minimum available nutrients for plant growth (Wang et al. 2018). Plants acquire phosphorus as primary or secondary food as a monovalent $H_2PO_4^-$ and a divalent HPO_4^{2-} present in low concentrations in the soil solution. Most cations form insoluble salts with phosphorus like in a soil rich in iron and aluminium, phosphorus can react with these minerals to form insoluble salts and, thus, become unavailable to plants (Fang et al. 2009; Wang et al. 2018). Phosphorus is readily absorbed by plants as the monovalent $H_2PO_4^-$ and least absorbed as the divalent HPO_4^{2-}. Either roots can take Pi from soil or plant organs rely on various transport systems like $H_2PO_4^-$ symporters PHT family or other phosphate transporters (Młodzińska and Zboińska 2016) to receive its daily ration of phosphorus. Various studies reported proteins with phosphate transport include PHT family and it involves Pi uptake from soil; on the other hand, SPX-EXS subfamily PHO1 and homologous genes are responsible for Pi translocation from root to shoots (Rausch and Bucher 2002; Wang et al. 2018).

12.3.2.2 Phosphorus Deficiency

Phosphorus is an essential nutrient forming a structural component of nucleic acids, plays a role in signal transduction, and takes part in various biological functions in plant development. Being the primary source of nutrition for plant development, any lack of phosphorus will cause plants to respond with symptoms of P deficiency, particularly when P concentration falls below 0.1%. Affecting plant growth and development, with first symptoms affecting older leaves (Takizawa et al. 2008), phosphorus deficiency causes anthocyanin pigment production to occur with high concentration in plants which gives purple colour to plant leaves and triggers necrotic spots on the leaves. With symptoms detected in dark greenish purple coloured and malformed leaves (Ramaekers et al. 2010), P deficiency triggers 4-fold decrease in leaf tissue when Pi reduces to 20 to 40 in chloroplast; furthermore, reduction in ATP synthase causes decreased growth (Muneer and Jeong 2015). It's also observed that the NADP⁺ levels working as substrate for NADPH synthesis is significantly decreased in P deficient plants. Causing decreased ATP synthase which limits the export of proton to the chloroplast stroma, and thus acidifying the thylakoid lumen (Goltsev et al. 2016; Karlsson et al. 2015), P deficiency in plants causes leaf expansion to be disrupted with reduced leaf area and slowed down overall growth of plants. Scarcity in phosphorus causes disruption in shoots and flower development in fruit trees, and is also a reason for reduced tillering in cereal crops (Karthika et al. 2018), while other symptoms include premature drying and shedding of leaves, and debilitated growth of fronds in coconut palms, with drastic P deficiency possibly causing leaf etiolation with leaves turning yellow. Young plants require more P as compared to mature plants, and a low P environment, for young plants, causes reduced cell division and cell enlargement so that plant development is stunted with dark green leaves as well as reduced leaf surface area (Malhotra et al. 2018). Phosphorus deficiency causes upward curling from base to tip of leaves and tilting of leaves associated with delayed maturity in crops under P limitation (Carstensen et al. 2018). The reduced leaf area triggers decline in photosynthetic efficiency in white clover plants (Schulze et al. 2006); and, in targeted metabolic profiles, phosphorus deficiency has been known to result in amino acids disturbing the nitrogen metabolites and to repress nitrogen fixation in common bean plants. *Arabidopsis* plants deprived of requisite phosphorus have shown altered Ca and Fe availability and increased ROS production, with hampered root development (Matthus et al. 2019).

12.3.3 Potassium

Potassium (K) is the third most essential macronutrient and a richly absorbed cation that performs a major role in plant growth, metabolism, and development (Figure 12.1) (Prajapati and Modi 2012; Karthika et al. 2018). As a crucial nutrient for meristematic development and in carrying out functions such as gas exchange and water regulation in plants, protein synthesis, enzyme activations, photosynthesis, and carbohydrate transport (Pandey and Mahiwal 2020), potassium also neutralizes different organic anions and compounds inside the plants to help stabilize tissue pH between 7 and 8 which is considered +optimal for most enzymes. With major effects on nucleic acid, protein, and vitamin metabolism that facilitate carbon movements and CO_2 assimilation activities (Wang and Wu 2017; Pandey and Mahiwal 2020), this mineral facilitates cell extension, plant development via management of turgor, and gas exchange through stomatal aperture control (Koch et al. 2019). As the element loading into phloem helping long distance nutrient transport via pressure driven flow, it maintains electrochemical gradient through cell membrane, and contributes to the translocation solutes (Karthika et al. 2018; Pandey 2018; Liu and Martinoia 2020). A potassium constituent uptake by cells of 50% by equal weight and in quantities more significant than other nutrients, except nitrogen, is suitable for plant growth (Pandey, 2018). It constitutes 80 mM to 200 mM concentration in cytoplasm and varies in subcellular compartments. The plants' young and active parts, especially buds and young root tips and leaves, contain rich amounts of potassium (Koch et al. 2019), and it is mainly used in its catalytic and regulatory roles while having a say in activation of 60 enzymes involved in photosynthesis and respiration, such as starch synthase and pyruvate kinase. Potassium is significant in maintaining crop growth improving disease resistance and enhancing shelf life of vegetable and fruits (Lei 2019; Koch et al. 2019).

12.3.3.1 Potassium Uptake and Transport in plants

Potassium is absorbed by plant roots in cationic form (K^+) where root epidermal and cortical cells are responsible for K uptake where the concentration of potassium varies from 1 to 5% in healthy plant tissues (Pandey and Mahiwal 2020). Membranes identified as transport proteins facilitate K^+ transport inside the plant cell as the lipid bilayer cell membrane is impervious to ions (Becker et al. 2004; Chao et al. 2016) with potassium uptake following biphasic kinetics, and high affinity potassium transport facilitated by H^+/K^+ symporters (active) as well as low affinity K^+ transporters, mediated by channels (Pandey and Mahiwal 2020). The translocation of K from soil to plant tissues involves different channels and transporters. Potassium is a mobile element and can travel from absorption site to the place it's needed, where the three main transporter families which contribute to K transport are TRK/HKT family, HAK, KT, and KUP family, and (CPA) cation/proton antiporter family. It is reported that in *Arabidopsis thaliana* plant, there are 15 genes encoding potassium channels and three transporter families (Rawat et al. 2016; Pandey and Mahiwal 2020).

12.3.3.2 Potassium Deficiency

Scavenged from older plant parts and travelling to where it is needed, potassium (K) is a mobile element with potassium deficiency symptoms observed in older leaves, having multiplier effects in the form of:

Decreased nitrogen availability affecting protein synthesis it is associated with (Figure 12.2) (Chérel and Gaillard 2019); loss of amino acid, amides, and nitrate buildup being the precursor of decreased protein synthesis where visual symptoms are not immediately perceptive for following up, and, with the visible symptoms of K$^+$ deficiency signing in as chlorosis in older leaves followed by leaf curling and wrinkle advances; and plants appearing with shorter and weak internodes. In soybeans, leaf tissues are damaged, leaf edges become ruptured and ragged with delayed maturity and less uniform beans. Under potassium deficiency, cotton leaves turn yellowish green around margins and veins followed by leaf etiolation where leaves turn brownish and can die prematurely with leaf shedding (Ashley et al. 2006; Karthika et al. 2018). It has been observed that under K deficiency, in apple plants, yellowish green curl along entire leaf and ragged, poorly formed edges, and the shedding of fruits early are most visible symptoms (Yadav and Sidhu 2016). Necrotic spots occur on the older leaf tips and subsequently on younger leaves related to chlorophyll degradation which impedes carbon flow and disrupts the water status of plants. Potassium-deficient plants develop poor root system, and stalks become weak followed by lodging of crops such as corn and legumes. Fruits and vegetables deteriorate early when shipped and have a short shelf life when plucked from potassium deficient plants (Jaiswal et al. 2016). While poor root development, damaged nodal tissues, chaffy ears, and stalk lodging are common symptoms in corn plants (Karthika et al. 2018; Koch et al. 2019), in areca palms, K deficiency causes shrunk trunk diameter and canopy size while in K$^+$ starved coconut palm, growth with reduced trunk and shorter leaf with decreased number of inflorescences and nuts per branch are the most common symptoms (Prajapati and Modi 2012).

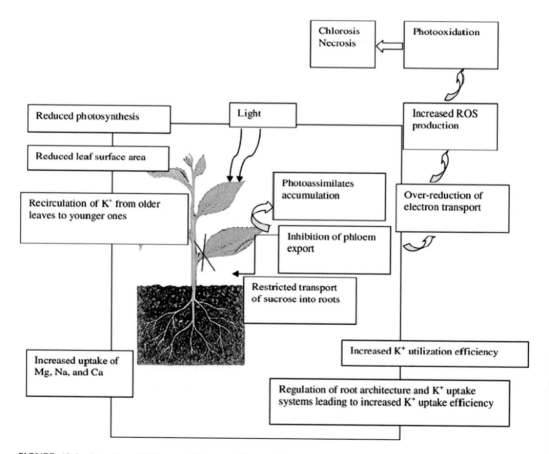

FIGURE 12.2 Potassium deficiency effects on different plant processes. Potassium deficiency triggers decrease in photosynthetic activity and causes oxidative stress leading to severe harmful effects on plant growth while plant develop various defence strategies to enhance potassium uptake and utilization (Hafsi et al. 2014).

12.3.4 Calcium

Calcium (Ca) available to plants in high concentration in the soil system is an essential nutrient for plant growth and development. Ca required by plants works as a secondary messenger of shock during stress events and stabilizes normal functioning of cell membranes, where concentration of free Ca is maintained in the cytoplasm at sub-micromolar levels to facilitate intracellular signalling (Karthika et al. 2018). Calcium plays a vital role in meristematic growth and, in plants, Ca levels range between 0.1% and 0.2%. Acting against hormonal and environment stress in plants, Ca forms calcium complex with calmodulin protein known as calmodulin calcium complex which regulates several key metabolic processes (Demidchik and Shabala 2018). While calcium pectate maintains cell elongation, as well as cell division, and plays a key role in sustaining the structure of chromosome (Karthika et al. 2018; Pandey 2018), high levels of Ca supply are required for symbiotic leguminous plants prone to infection by *Rhizobia* bacteria. Calcium is also involved in enzyme activation such as activation of arginine kinase, phospholipase, amylase, and ATPase. With a large ion range, calcium helps during ion dehydration by subsequently binding to different anionic substances (Makavitskaya et al. 2018). With a strong influence on cell wall strength and cell pH, calcium can help plant cells stick together due to its cation effect on cell medium. Calcium has the ability to influence the synthesis of cell wall polysaccharides including 1,3, β glucan that acts like a glue in cell walls. Calcium is reported to have high binding capacity which makes it an important signalling molecule in the cytosol (Karthika et al. 2018; Huang et al. 2017). Besides, calcium plays an important role in cytosolic signal transduction pathway responsible in activating proteins, kinase, and phospholipases which regulate cellular response to different biotic and abiotic stresses. Protecting root cells when there is imbalance of nutrient in soil, calcium helps to maintain cell membrane integrity (Karthika et al. 2018). Mitochondrial enzymes such as glutamate dehydrogenase are also activated by calcium (Gong et al. 2016).

12.3.4.1 Calcium Uptake and Transport

Calcium is taken up by plant roots as divalent cations (Ca^{2+}) by calcium selective channels and then must be diffused through the plant in the absence of pumps or channels through lipid bilayers (Kathpalia and Bhatla 2018; Pandey 2018). In plants, apoplastic and symplastic pathways are available for calcium movement with the need for calcium to be in the micromolar range to guarantee the cell to generate Ca^{2+} signal (Thor 2019). Calcium enters root xylem at the apical region of the root tip where root endodermis is differentiated and, then, distributed within the plant in its free form or by making a complex with organic acid (Kathpalia and Bhatla 2018; Karthika et al. 2018). The process of Ca uptake requires the help of members of CAX H^+/Ca^{2+} antiporter family ATP driven P-ATPase (Gao et al. 2019). Higher concentration of Ca is toxic to plant growth, so, its movement through the phloem is slow replenishing older leaves containing more Ca after supply to younger leaves is done (Karthika et al. 2018). Higher concentration of Ca (0.1–1.0 mM) is required to maintain the functional and structural integrity of plasma membrane (McAinsh and Pittman 2009).

12.3.4.2 Calcium Deficiency

Calcium is the fifth most abundant element in the environment and constitute around 3.5% of the earth's crust, with plants requiring a stable supply of 1–10 mM for its everyday normal functioning (De Freitas et al. 2016). Calcium deficiency triggers stunted growth of middle lamella in cell expansion, and membrane structure can collapse and molecular weights of solutes show solute leakages (Tian et al. 2016; Karthika et al. 2018). In flowering plants with Ca-low condition, pollen tube growth and direction of pollen tubes are disrupted, and younger leaves exhibit deformed, small, and chlorotic appearance, with leaf etiolation causing leaves to change from flourish to cup-shaped appearances; even as terminal buds deteriorate with collapse of petioles, and the stem is nourished by a weakened root system leading to brownish small branches (Liu et al. 2019). The most common Ca deficiency symptoms are nasty pit in apples, black heart in celery, and blossom end rot in tomatoes (Huang et al. 2017; Karthika et al. 2018). In Ca starved tomato plants, deficiency symptoms usually appear during fruiting where the Ca deficiency condition was named as blossom end rot (BER) (Hermans et al. 2013). Water soaked tissues in the blossom's expanse that later become brown with necrotic spots developing in plants at later stages of Ca deficiency are also common. In vegetables, Ca deficiency causes low transpiration rate in young leaves characterized by stunted growth and brownish spots associated with necrosis due to phenol oxidation of younger leaf tips and meristematic cell in cauliflower plants. In artichoke plants, lack of calcium is known as tip burn, which is characterized by discoloration and immature flower buds (Taiz and Zeiger 2002; Pandey 2018; Karthika et al. 2018).

12.3.5 Magnesium

Magnesium (Mg) in green leafy plants is one mineral plants cannot do without because of its role in chlorophyll production and its functions as an activator of many plant enzymes (Hermans et al. 2013). As a mobile element, magnesium performs diverse functions in plants; and with the central magnesium atom of the chlorophyll molecule with the nitrogen atoms of four pyrrole rings, chlorophyll has a porphyrin like composition (Gransee and Führs 2013; Hermans et al. 2013). Small amounts of Mg^{2+} is essential for different functions to run in cytoplasm, and for functions to be carried out such as counterions for vacuolic organic acid anions and inorganic anions in the middle lamella of plant cell walls (Karthika et al. 2018). Magnesium is essential for pH stabilization and maintenance of ionic balance in plants, while helping sustain the structural integrity of cellular ribosome and attaching ribosomal aggregates to tRNA, a method important for biosynthesis of proteins (Liang et al. 2009; Guo et al. 2016). Magnesium activates not only several key enzymes along with K but also involved in phosphate or carboxyl group enzyme catalase transfer and, in some cases, it freely functions as a dissociable cofactor (Deng et al. 2006; Karthika et al. 2018). With a role in ATP synthesis in plants, Mg is an essential component in all nucleoside triphosphates forming magnesium complexes. Important for the enzyme RNA polymerase to function, magnesium is essential for RNA synthesis. The enzyme RuBP carboxylase and fructose 1,6 diphosphate involve Mg as does higher pH levels for optimum activity (Hermans et al. 2004; Marschner 2012). Magnesium is an important nutrient given 75% of leaf Mg is involved in protein synthesis, and 15–20% of total Mg is linked with chlorophyll pigments (Kathpalia and Bhatla 2018). Counted among the basic components of polyribosome for protein synthesis, Mg levels determine gene transcription and translation in plants (Guo et al. 2016).

12.3.5.1 Magnesium Uptake and Transport

The Mg uptake from soil solution is lower than K though it is present in higher concentrations in soil, with translocation of Mg facilitated by potassium towards fruits and storage tissues. It has been reported that Mg being a mobile element can be transferred from older to younger leaves as need (Mengutay et al. 2013). The Mg uptake is facilitated by MGT family which is similar to bacterial CorA mg transporters. In recent years, Mg homeostasis has been explained with the discovery of *Arabidopsis* Mg^{2+} transporters (Chen et al. 2009; Guo et al. 2016). Recently, an AtMRS2/AtMGT gene family which encodes Mg transport proteins has been identified (Li et al. 2008; Chen et al. 2009; Mao et al. 2014). Ten Cor A family AtMGT genes clustered into high affinity genes (AtMGT1 and AtMGT10), low affinity genes (AtMGT3, AtMGT7, AtMGT9), and dual affinity genes (AtMGT5) was considered as the primary Mg transporter, and is responsible for uptaking, circulation, and homeostasis in plants (Luo et al. 2006; Gilliham and Conn 2010; Conn et al. 2011).

12.3.5.2 Magnesium Deficiency

Magnesium deficient leaves accumulate sugars in source leaves in high quantities caused by restricted loading of phloem which may trigger feedback and disruption of Rubisco carboxylase activity and change to oxygenase activity (Tewari et al. 2006; Guo et al. 2016). Magnesium-starved plants show chlorosis associated with high sugar buildup, whereas severe Mg deficiency can cause reduction in photosynthetic electron transport in PSI and PSII recorded in pine and beat plants (Kathpalia and Bhatla 2018). Even with the possibility of magnesium deficiency triggering ROS production which causes oxidative damage to plants, there are reports of activation of antioxidant defence system in different plants under Mg deficiency stress (Gruber et al. 2013). Oxidative damage caused by Mg deficiency increases malondialdehyde (MDA) and H_2O_2 synthesis (Pandey 2018). Further, magnesium is the central atom in the chlorophyll molecule and its deficiency causes premature leaf abscission in plants associated with interveinal chlorosis and with leaf margins turning reddish purple (Guo et al. 2016). In different plants such as wheat, distinct mottling associated with yellowish green patches are recorded, and in alfalfa plants leaf curls with reddish underside has been reported (Karthika et al. 2018; Pandey 2018). Leaves of Mg-starved sugar beet and potato plants turn stiff and brittle, while it is an established fact that depolymerization of ribosome leads to premature ageing of plants with Mg deficiency symptoms appearing first in older leaves and later extending to younger leaves resulting in drying of tissue and death. In Areca palms and cocoa plants, the symptoms of interveinal chlorosis, necrosis of older leaves, and leaf margins which curve upwards after prolonged Mg deficiency were recorded (Hermans et al. 2005; Guo et al. 2016). Photosynthetic CO_2 fixation impairment can trigger non-utilized accumulation of electrons in the chloroplast which causes ROS generation and photooxidative damage of chloroplast membrane lipids (Guo et al. 2016). Magnesium deficiency inhibits N reductase, glutamate synthase, urease, and glutamic oxaloacetic protease transaminase enzymes associated with reduction in nitrogen metabolism (Yang et al. 2012).

12.3.6 Sulphur

Sulphur (S), an important macronutrient, has for its most important source sulphate (SO_4^{-2}) and plant roots are responsible for its uptake as divalent Sulphate anions and, then, the nutrients' transportation to xylem and phloem (Kathpalia and Bhatla 2018). Aerial parts of plants can also absorb SO_2 from air; however, gaseous S needs to be transformed into Sulphate, and then reduced to sulphide before being incorporated into cysteine and amino acid which is important for metabolism. Essential for protein synthesis and with a role in chlorophyll synthesis, sulphur levels in plants fluctuate between 0.1% and 0.4%. Its need is most important for oilseed crops as they require more S concentration in the tissues than cereal crops do. Sulphur is an essential constituent of amino acids such as cysteine. Cysteine and methionine are also present in different coenzymes and vitamins needed for metabolism (Huang et al. 2019). Required for nodulation, sulphur is needed in flowering and for maintaining the quality of oil seed plants. An important component of tripeptide glutathione compounds involved in ROS detoxification, and the production of phytochelatins and metallothionein involved in metal detoxification (Karthika et al. 2018), sulphur provides root growth, seed production, and also facilitates plant resistance against cold. It is also present as an iron Sulphur protein complex such as ferredoxin in photosynthesis electron transport chain. Sulphur is a component of thiocyanates and isothiocyanates responsible for pungent flavour in cabbage, mustard, and turnips.

12.3.6.1 Sulphur Uptake and Transport

Sulphur is taken up by plant roots against electrochemical gradient generated at plasma membrane due to proton gradient caused by membrane H^+-ATPase. The plants' sulphate uptake is a high affinity transport with expression of transporter genes, but low affinity sulphate transporters do cell-to-cell distribution across plasma membrane and storage of mobilization from vacuoles across cell tonoplast. Sulphite, molybdate, selenite, and chromate compete with sulphate for sulphate transport proteins binding (Huang et al. 2019). Electrochemical gradient favours sulphate transfer, and sulphate transfer across tonoplast follows sulphate-specific channels. Different transporters, that are present in plants roots with variable affinities for sulphate in *Arabidopsis* 14 genes, have been identified which facilitate sulphate transfer (Takahashi 2019). High affinity transporters SULTR1.1 and SULTR1.2 are present in root epidermis, while the four low affinity transporters SULTR1.3, SULTR 2.1, SULTR3.5, and SULTR2.2 have been identified in the vascular system (Sacchi and Nocito 2019; Takahashi 2019).

12.3.6.2 Sulphur Deficiency

Sulphur is an immobile element and its deficiency symptoms include leaf yellowing, anthocyanin accumulation, and stunted growth, with symptoms first appearing in younger leaves (Samborska et al. 2019), and triggering leaf defoliation and reduced nodulation associated with stunted root growth in tomato plants. While sulphur starvation in plants causes reduced stomatal degradation, root hydraulic conductivity, and improper photosynthesis, thus reducing leaf area and leaf numbers (Veliz et al. 2017), other symptoms as reported in wheat plants with sulphur deficiency are decrease in chlorophyll and protein levels of leaves (Sutar et al. 2017). S-containing amino acid methionine, under S deficient conditions, is disrupted from functioning which causes reduction in protein synthesis followed by decreased chlorophyll concentrations. Causing enhanced amide content in plants, sulphur deficiency leads to proportions of amide content to be enhanced in relation to soluble N fractions (Karthika et al. 2018; Pandey 2018). In sulphur deficient plants, increased ROS production enhances the glutathione levels which generate hydrogen peroxidase H_2O_2 in barley leaves (Veliz et al. 2017; Qadeem et al. 2015), with sulphur deficient plants showing symptoms of small stems, that in the lateral stage causes the entire plant to become pale green (Karthika et al. 2018). Sulphur deficiency occurs in plants at less than 0.1 to 0.2% concentrations and deficiency symptoms are characterized by chlorosis. Decrease in plants' leaf numbers and leaf size in sulphur-deficient older palms have been reported, while in cocoa plants symptoms are reported as leaves turning pale yellow and older leaves having yellow blotches (Taiz and Zeiger 2002; Saha et al. 2016).

12.3.7 Iron

Iron (Fe) is an important micronutrient for plant growth required for multiple cellular processes, including chlorophyll biosynthesis, photosynthesis, and respiration (Figure 12.3) (Connorton et al. 2017). An essential component of different cofactors that participate in electron transfer or radical mediated modification, Fe assists in chemical transition such as hydroxylation. Iron cofactors participate in oxygen transport and regulate protein stabilization (Karthika et al. 2018). Leaf chloroplast contains iron sulphur (FeS) proteins i.e., ferredoxins, as do photosystem I and a range of different metabolic enzymes. Iron enzymes are essential

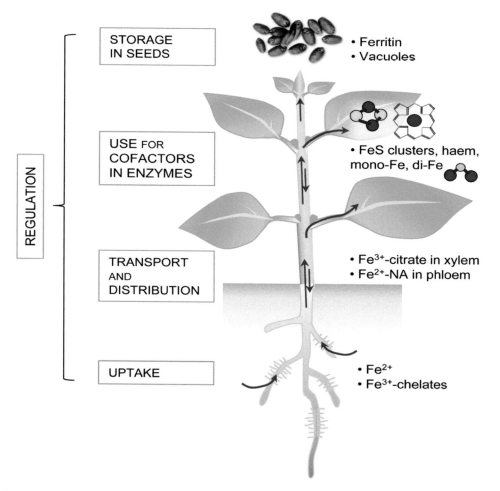

FIGURE 12.3 An overview of Fe homeostasis in plants facilitated by high affinity iron uptake systems; transport, distribution and storage mechanisms (Connorton et al. 2017).

components in respiratory complex comprising of Fe-S clusters, a combination of Fe-S and haem copper in the mitochondria. Involved in synthesis of gibberellic acid, jasmonic acid, and ethylene as well as osmolytes associated with scavenging of ROS under oxidative stress (Pandey 2018), iron as a component of cytochromes plays a role in electron transfer in redox reactions. Three groups of iron-comprising proteins such as ferredoxin work as electron carriers in electron transfer reactions (Gao et al. 2020), while iron plays a key role in nucleic acid metabolism, and chlorophyll synthesis, and is loaded in seeds vacuoles or in ferritin. Important for harvesting light energy, Fe is also known for transfer of electrons yielding from water splitting via photosystem II and I (Kroh and Pilon 2020).

12.3.7.1 Iron Uptake and Transport

Iron constitutes nearly 5% of earth's crust and is present as amorphous Fe $(OH)_3$ precipitate in soil which is a source of iron for plants. Availability depending on redox potential and soil pH (Figure 12.3), iron is absorbed by plant roots as Fe^{2+} via specific channels on the plasma membrane, given Fe^{2+} is closely related to Fe^{3+} reduction (Connorton et al. 2017). Iron (Fe^{3+}) siderophores are transported by diffusion or mass flow to the root spaces which moves to plasmalemma bound Fe^{3+} reductase (Inoue et al. 2009; Connorton et al. 2017). The IRTs of ZIP family transporters (AtIRT1, AtIRT2; OsIRT1, OsIRT2) are associated with high affinity Fe uptake by roots. Second ABC transporters (ArBCB25, AtATM3) a mitochondrial transporter associated in Fe-S biogenesis clusters in different plants (Eide et al. 1996).

12.3.7.2 Iron Deficiency

Iron plays an important role in photosynthetic reactions. However, because iron is an important micronutrient for plant growth and development, its deficiency decreases chlorophyll levels and hampers chlorophyll synthesis characterized by interveinal chlorosis in young leaves and followed by entire leaf turning whitish yellow and which later leads to necrosis (McCauley et al. 2009). Iron deficiency is common in members of Rosaceae family with symptoms of leaves turning irregularly yellow and showing deformed growth (Karthika et al. 2018). Iron deficiency is relatively high in calcareous soils, and when concentrations are lower than 50 mg/kg, plants are considered as Fe deficient. In coconut plants, uniform chlorosis is associated with Fe deficiency where from top of the crown to the base the plant turns dark yellow or pale green (McCauley et al. 2009). Under severe Fe deficiency, newly germinated leaves show necrotic tips and, in later stages, can turn completely yellow (Taiz and Zeiger 2002; Karthika et al. 2018). Causing photodamage because electron transport requires Fe sulphur containing proteins, iron deficiency causes activities of enzymes involved in ROS scavenging to decrease.

12.3.8 Boron

Boron is an important micronutrient, and although required in low amounts, it serves essential metabolic processes in plants (Shah et al. 2017; Wu et al. 2018). Belonging to subgroup III metalloids, it has intermediate characteristics of both metals and non-metals, and its availability in irrigation water and soil is important for agriculture purpose (Archana and Verma 2017). Forming borate esters with apiose RG-II, boron, in the form of this complex, plays a key role in cell wall function and structure, and controls cell wall permeability (Shireen et al. 2018). Boric acids perform complexes with polyols and diols such as uranic acids and sugar alcohols which are important for binding of cell wall polymers. Being a component of haem proteins, which contain Fe porphyrin complexes (i.e., cytochrome, catalase) and iron sulphur proteins (ferredoxins, superoxide dismutase, and aconitase), boron is involved in chlorophyll synthesis and photosynthetic processes (Landi et al. 2019; Lewis 2020). With an important role to play in pollen tube elongation and nutrient use efficiency, boron has a secondary use in its role in cell division, sugar transport, and enzyme synthesis (Brdar-Jokanović 2020; Moghadam et al. 2020).

12.3.8.1 Boron Uptake and Transfer

Boron (B) is present in several forms in soil but the commonest form is as the undissociated boric acid (H_3BO_3). Depending on the availability, plants uptake of boron happens in three different molecular mechanisms: (i) passive diffusion through lipid bilayer (ii) facilitated transport by intrinsic proteins MIP, and (iii) energy supported high affinity transport facilitated by BOR transporters (Karthika et al. 2018; Moghadam et al. 2020). In *Arabidopsis*, a boric acid channel (ArNIO5;1) that belongs to nodulin 24 intrinsic protein (NIP) which is the subfamily of MIPs family has been identified. The AtNIP5;1 expression in B deficient plants is upregulated which suggests its importance as required B uptake under boron starved environments (Archana and Verma 2017; Moghadam et al. 2020).

12.3.8.2 Boron Deficiency

Boron deficiency is the most widespread phenomenon among all the deficiency phenomenons involving micronutrient shortfall in plants, and said to take effect from the tropical regions to the temperate zones (Shah et al. 2017). Boron deficiency symptoms are known as corky core, sickness of tobacco, internal cork, water core of turnip, yellows of alfalfa, heart rot of sugar beet, and hollow stem of broccoli and cauliflower (Moghadam et al. 2020). Appearing in terminal shoots in the form of deformed, small leaves, boron deficiency shows symptoms like growth of apical meristem having a stunted look and necrotic spots with axillary branches eventually causing necrosis (Princi et al. 2016; Archana and Verma 2017). Symptoms in the sunflower petiole and stem appear as longitudinal splits and deformed capitulum, with poor seed set results and weak root system (Archana and Verma 2017). Boron deficient onion plants show brittle leaves and blue green colour associated with necrosis, while leaves of cauliflower and cabbage show bluish appearance (Wu et al. 2018). In higher plants, deficiency symptoms show up in stunted root elongation in the lateral roots and, in leaves, malformed growths, while in the case of severe boron deficiency, the whole plant growth appears stunted (Lewis 2020; Landi et al. 2019). In boron-starved plants, nitrogen fixation and nitrate assimilation are adversely affected (Karthika et al. 2018). Citrus fruits, like pears, grow lumpy with uneven thickness of the peel under boron deficiency. Under drought stress, a boron deficient plant appears to have aggravated symptoms, and with internal tissues of roots appearing to take on a darkened tint and referred to as black heart (Shah et al. 2017; Gracia et al. 2020).

12.3.9 Copper

Copper (Cu), an important micronutrient in plant growth, is considered optimal for plant tissues when it is found in the range 5–30 mg/kg. An essential component of the electron transport chain, it is involved in lignin biosynthesis (Kumar et al. 2020). As the basic component of cytochrome oxidase enzymes and plastocyanin, copper facilitates root metabolism and utilization of proteins. Copper is involved in cell wall lignification, ethylene perception, and carbohydrate metabolism (Karthika et al. 2018), and is one of the components of proteins grouped in three types which includes blue proteins, non-blue protein, and multicopper protein. Copper imparts resistance to plants against different stresses, enhances fertility in male flowers, and functions as oxidizing and reducing agents in biochemical reactions (Pandey 2018).

12.3.9.1 Copper Uptake and Transport

Plants uptake copper as the divalent cupric Cu^{2+} ion in aerated soil or will absorb copper as monovalent Cu^+ ion from wet soil with pH below 7.0. Copper concentration in plant tissues ranges between 1–5 µg g^{-1} of dry weight and 5–20 µg g^{-1} in leaves (Bernal et al. 2012; Kumar et al. 2020). A wide range of copper transporter proteins (COPT), P-type ATPases, and YSL transporters are involved in Cu homeostasis (Yruela 2009). In *Arabidopsis*, copper uptake is regulated by expression of ZIP family transporters through ZIP2 and ZIP2 (Kavitha et al. 2015; Kumar et al. 2020).

12.3.9.2 Copper Deficiency

Copper deficiency in many crops has been reported to cause delayed flowering and reduced number of flowers at a time, delayed hormone signalling, and decrease in photosynthesis reactions. Plastocyanin activity, which is involved in cyclic electron transport coupled to ATPs, decreases photosynthetic activity and disrupts PSII/PSI rate in Cu deficient plants (Kumar et al. 2020; Arshad et al. 2016). Copper-starved plants show lower chloroplast activity and reduced chlorophyll levels, and CO_2 assimilation while necrosis spots starting at the leaf tip and extending down along the margins towards the leaf base (Karthika et al. 2018). Copper deficiency delays senescence and the most common deficiency symptoms include necrosis, white tips, and reclamation disease (Taiz and Zeiger 2002).

12.3.10 Manganese

Manganese (Mn) was described as the essential nutrient for plants in 1922 and is involved in different metabolic processes (Li et al. 2016a, 2016b; Arafat et al. 2016). With an essential role in oxygen evolving photosynthetic apparatus, manganese catalyzes water splitting process in photosystem II (PSII). While carrying out activation of more than 30 enzymes, manganese is counted among the nutrients found in different enzymes such as Mn superoxide dismutase (SOD) (Corpas et al. 2017). Manganese functions as cofactor for many enzymes such as glutamine synthase, arginase, and pyruvate carboxylase (Köllner et al. 2008), and is required in small quantities in plants and is the essential element metalloenzymes cluster of the oxygen evolving complex (OEC) in photosystem II. Interchangeable with different divalent cations i.e., cobalt, copper, Mn is involved in ROS detoxification in plants (Bloom 2019). Oxalate oxidase is an Mn dependent enzyme which catalyzes oxalate oxidation to CO_2 coupled with O_2 reduction to H_2O_2, and facilitates plant defence against pathogens. Being the essential cofactor of enzymes, Mn is involved in isoprenoid biosynthesis and participates with Mg in terpene synthase (Basu et al. 2015; Alejandro et al. 2017). Manganese is considered to play an important role in deposition of waxes in leaves and essential in protein glycosylation and pectin biosynthesis in plants (Venkidasamy et al. 2019).

12.3.10.1 Manganese Uptake and Transport

Manganese (Mn) uptake has been considered to be facilitated by plasma membrane Ca^{2+} transport channels which are permeable to Mn^{2+}. In *Arabidopsis* plants, Mn^{2+} absorption is mediated by AtNRAMP1 transporter localized in plasma membrane of root epidermis and root cortex (Edmond et al. 2009; Alejandro et al. 2020). In addition, Mn^{2+} is facilitated by ZIP transporter AtRT1 which is considered as Fe uptake transporter (Gao et al. 2018). Furthermore, two ZIP transporters facilitate root-to-shoot Mn translocation in *Arabidopsis*, which are: AtZIP1 localized to the tonoplast involved in Mn^{2+} remobilization from vacuoles to cytosol in roots cells; whereas, AtZIP1 localized to plasma membrane mediated by Mn^{2+} uptake to root stele in plants. In barley plants, HvIRT1 plasma membrane transporter is considered to facilitate Mn uptake and transport (Fu et al. 2017; Gao et al. 2018).

12.3.10.2 Manganese Deficiency

Manganese deficiency in plants often occurs with significant, clear symptoms and Mg concentration below 10–20 mg/kg dry weight is considered lethal in plant survival. Manganese-starved plants show impaired growth and reduced biomass caused by lower chloroplast levels and photosynthetic efficiency rate (Hebbern et al. 2009; Schmidt et al. 2016), while there is decreased tolerance to drought and low temperatures, as well as the number of Mn-complexes in photosystem II, and triggering of disintegration and destabilization of PSII complexes which lowers photosynthetic efficiency in plants (Alejandro et al. 2017; Alejandro et al. 2020). Photosystem disruption initiates chlorophyll degradation and leads to interveinal leaf chlorosis development under Mn deficiency (Broadley 2012; Li et al. 2017), with reports that Mn deficient plants show necrotic spots as a consequence of enhanced free oxygen radicals, decreased MnSOD activity, and damaged chloroplast (Broadley 2012; Alejandro et al. 2020). Interveinal chlorosis in different plants such as cherry, peas, apple, onion, and raspberry with small leaf vein has been observed in Mn deficient plants (Shao et al., 2017; Karthika et al. 2018; Alejandro et al. 2020).

12.3.11 Zinc

Zinc (Zn) is the essential plant micronutrient which is involved in many different physiological and metabolic processes, as well as acts as cofactor in more than 300 proteins, RNA polymerase, and DNA polymerase (Gupta et al. 2016). Involved in the activation of enzymes for protein synthesis, carbohydrate metabolism, auxin regulation, and maintaining the functional integrity of cellular membrane (Karthika et al. 2018), zinc is involved in regulation of gene expression needed for combatting environmental stresses in plants, in tryptophan biosynthesis which is the auxin precursor molecule (Rudani et al. 2018), and is essential for maintaining cellular membrane integrity to hasten the structural adjustment of macromolecules and ion transfer system. Zinc involvement with phospholipids and sulfhydryl clusters of membrane proteins promotes maintenance of membrane (Kathpalia and Bhatla 2018). The Zinc-Cu-SOD plays a role in protecting the proteins and lipids of membrane from oxidation by detoxification of superoxide radicals. This nutrient is required for seed development and promotes cytochrome C synthesis (Karthika et al. 2018).

12.3.11.1 Zinc Uptake and Transport

Zinc is taken up by plant roots as divalent cation Zn^{2+} through mass flow and diffusion. However, organic-ligand-Zn complexes' uptake by plants roots is also known to occur. Important families of zinc transporter proteins involved in zinc transport are ZIP permease family ZRT-IRT, the HMA heavy metal ATPases, and the MTP metal tolerance proteins family. The ZIP family facilitates Zn entry into cytosol while HMA participates in Zn efflux to the apoplast, and the MTP family facilitates Zn sequestration in cellular compartments such as vacuole (Gupta et al. 2016; Rudani et al. 2018; Karthika et al. 2018).

12.3.11.2 Zinc Deficiency

Zinc deficiency triggers decrease in protein synthesis and inhibits root development (Karthika et al. 2018). Showing symptoms of disturbed water and nutrient uptake from soil resulting in stunted growth and yield, zinc deficient plants display interveinal chlorosis and inhibited growth in maize, beans, and sorghum leaves (Rudani et al. 2018). Zinc-starved leaves show malformed bushy roseate appearance and delayed maturity leading to crop reduction, with common deficiency symptoms being white maize bud, little leaf of cotton, khaira disease in rice, and mottled leaf or crown choking in areca nut (Cabot et al. 2019). The deficiency of this nutrient causes decreased shoot growth photosynthesis, which leads to decreased biomass accumulation in wheat plants (Kaznina and Titov 2017).

12.3.12 Molybdenum

Molybdenum (Mo) is absorbed by plant roots as molybdate anion (MoO_4^{2-}) soluble form present in soil. The common component in over 30 enzymes which facilitate oxidation reduction reactions, Mo is essential for symbiotic nitrogen fixation in legume and non-leguminous plants (Padhi and Mishra 2019), and it acts as the cofactor of nitrate reductase assimilation and ureide metabolism in plants. With a significant part played in plant reproductive structure growth and seed yield (Manuel et al. 2018), molybdenum is involved in pollen formation and Mo-containing enzyme aldehyde oxidase which is engaged in the final step of catalytic reactions. This nutrient is a component of sulphur oxidase which helps plants combat damaging effects of sulphite after acid rain (Karthika et al. 2018).

12.3.12.1 Molybdenum Uptake and Transport

Molybdenum is present in soil for plants to uptake, and given the total Mo level in the soil depends on soil pH concentration of adsorbing oxides, it is important that there is an interface with organic compound found in soil colloids (Manuel et al. 2018). Molybdenum is absorbed and stored in vacuoles as MoO_4^{2-} under neutral to mild alkaline conditions. Plant proteins facilitating Mo transport belong to high affinity MOT1, and MOT2 family, specifically for MoO_4^{2-} transport in higher plants (Manuel et al. 2018).

12.3.12.2 Molybdenum Deficiency

Molybdenum is a mobile element and deficiency symptoms appear on the whole plant in the form of stunted growth with pale green or yellow area on the leaf edges. In Mo deficient plants, necrosis around leaf edges and veins due to plant's inability to assimilate nitrate and to convert it to proteins is common. The deficiency symptom of 'whiptail' in cauliflower occurs when leaf tissue fails to develop surrounding mid leaf vein. Due to insufficient N supply in legumes, growth stunting, and yellowing have been seen in Mo deficient plants (Padhi and Mishra 2019), while in broccoli plants, twisting of young leaves which die later is reported. Furthermore, interveinal necrosis associated with enhanced nitrate concentration showing lack of nitrate reductase activity has been reported in plants. Deficiency symptoms include flaccid, grey tinting and mottling of plant's leaves at seedling stage (Karthika et al. 2018; Kathpalia and Bhatla 2018).

12.3.13 Nickel

Nickle (Ni) is a micronutrient which was discovered in year 1987 and is required in a low quantities amounting to <0.5 mg/kg and is taken up as Ni^{2+}. Nickel exists in three oxidation states such as I, II, III in plants and is the key component in particular enzymes involved in N metabolism and biological nitrogen fixation, such as urease. Important is N metabolism, nickel also acts as cofactor in urease facilitating substrate hydrolysis to CO_2 and ammonia. Nickel is required in the growth of legume root nodules and hydrogenase enzyme activation (Liu et al. 2011). Beneficial effects of Ni have been reported like increasing the growth of wheat, oats, and tomato crops (Karthika et al. 2018).

12.3.13.1 Nickle Uptake and Transport

Ni comprises about 3% of the earth's crust and is available to plants from soil with alkalinity above pH 6.7. Studies have shown that AtIRT1 iron regulated transporter 1 member of ZIP family is possibly involved in Ni transport in *Arabidopsis* plants (Nishida et al. 2011).

12.3.13.2 Nickle Deficiency

Nickel deficiency has not been studied to the extent needed which affects understanding how urease hydrolysis to carbon dioxide and ammonia happens. In nickel deficient plants, symptoms can include decrease in dry matter weight, amino acid content, and nitrate accumulation. Necrosis of leaf tips, and leaf necrosis and chlorosis are major symptoms of Ni deficiency (De Macedo et al. 2016). Delay in leaf expansion, rosette bronzing, and chlorosis are common symptoms of Ni deficiency (Liu et al. 2011).

12.3.14 Chlorine

Chlorine (Cl) is a halogen element which occurs in plants as free anion Cl^- bound to chlorinated organic compound at exchange site. A structural component of Mn oxygen evolving complex associated with charge accumulation and water oxidation by PSI (Ali et al. 2020), chlorine plays an important role in maintaining turgor and osmoregulation. Furthermore, the osmoregulatory role of chlorine involves turgor driven cell growth and stomatal functions. Chlorine exists in high concentration in shoot and root apices, and functions in turgor-induced growth of cells (Shrivastav et al. 2020). Involved in regulating activities of certain enzymes such as asparagine synthase, chlorine is known to be involved in glutamate dependent asparagine synthesis. This nutrient is important for its role in inducing plant resistance against stem rot, sheath blight in rice, root rot in barley, stalk rot in corn, common root rot in wheat, downy mildew in millet, coconut grey leaf spot, and brown centre in potatoes (Karthika et al. 2018; Ali et al. 2020).

12.3.14.1 Chlorine Uptake and Transport

Chlorine is absorbed by plants as Cl^- ions and concentration ranges from 100–500 mg kg of dry matter in healthy plants. It is a mobile element and its uptake is mediated by Cl^-/H^+ cotransport through plasmalemma.

12.3.14.2 Chlorine Deficiency

Plants are considered chlorine deficient when chlorine concentrations are less than 100mg/kg in plants. Deficiency symptoms of chlorine consist of leave curling, leaf chlorosis, decreased leaf surface area, restricted root branching, and plant wilting (Karthika et al. 2018). Other Cl deficiency symptoms involve leaf wilting near the margins as transpiration is disrupted with chlorine deficiency and followed by plants becoming chlorotic. Chlorine deficiency symptoms are cultivar-specific and causes wilting, swelling, or root tips and leaf change colour to bronze (Kathpalia and Bhatla 2018).

12.4 Conclusion

The chapter emphasizes the physio-biochemical roles of mineral nutrients associated with plant growth and development. The manifestations include macro-nutrient roles and micro-nutrient roles, uptake, and transport. The availability of nutrients fluctuates as plant faces various biotic and abiotic stresses resulting in nutrient deficiency. Nutrient deficiencies cause different physiological disorders identified in the plant displaying deficiency symptoms which causes disrupted growth and yield losses in plants. Present review would help to understand nutrient homeostasis in plants, and to identify deficiency symptoms in relation to plant development which will be beneficial for yield improvement. Molecular approaches and defence mechanisms against nutritional deficiency in plants languishing under different stress conditions should be studied in the future to help understand how to cope with low nutrient availability and leading to improved crop productivity and yield to meet the global food demands.

REFERENCES

Alejandro S, Höller S, Meier B, Peiter E (2020) Manganese in plants: from acquisition to subcellular allocation. *Front Plant Sci* 11:300.

Alejandro S, Cailliatte R, Alcon C, Dirick L, Domergue F, Correia D, Curie C (2017) Intracellular distribution of manganese by the trans-Golgi network transporter NRAMP2 is critical for photosynthesis and cellular redox homeostasis. *Plant Cell* 29(12), 3068–3084.

Ali A, Bhat BA, Rather GA, Malla BA, Ganie SA (2020) Proteomic studies of micronutrient deficiency and toxicity. In: *Plant micronutrients*. Springer, Cham, pp 257–284.

Arafat Y, Shafi M, Khan MA, Adnan M, Basir A, Rahman IU, Arshad M, Khan A, Saleem N, Romman M, Rahman Z, Shah JA (2016) Yield response of Wheat cultivars to zinc application rates and methods. *Pure Appl Biol* 5:1260–1270.

Archana NP, Verma P (2017) Boron deficiency and toxicity and their tolerance in plants: a review. *J Global Biosci* 6:4958–4965.

Arshad M, Adnan M, Ali A, Khan AK, Khan F, Khan A, Kamal MA, Alam M, Ullah H, Saleem A, Hussain A, Shahwar D (2016) Integrated effect of phosphorous and zinc on wheat quality and soil properties. *Adv Environ Biol* 10(2):40–45.

Ashley MK, Grant M, Grabov A (2006) Plant responses to potassium deficiencies: a role for potassium transport proteins. *J Exp Bot* 57(2): 425–436.

Basu D, Tian L, Wang W, Bobbs S, Herock H, Travers A, Showalter A M (2015) A small multigene hydroxyproline-O-galactosyltransferase family functions in arabinogalactan-protein glycosylation, growth and development in Arabidopsis. *BMC Plant Biol* 15(1): 295.

Becker D, Geiger D, Dunkel M, Roller A, Bertl A, Latz A, Schmidt D (2004) AtTPK4, an Arabidopsis tandem-pore K+ channel, poised to control the pollen membrane voltage in a pH-and Ca2+-dependent manner. *PNAS* 101(44):15621–15626.

Bernal M, Casero D, Singh V, Wilson GT, Grande A, Yang H, Merchant SS (2012) Transcriptome sequencing identifies SPL7-regulated copper acquisition genes FRO4/FRO5 and the copper dependence of iron homeostasis in Arabidopsis. *Plant Cell* 24(2):738–761.

Bloom AJ (2015) The increasing importance of distinguishing among plant nitrogen sources. *Curr Opinion Plant Biol* 25:10–16.

Bloom AJ (2019) Metal regulation of metabolism. *Curr Opin Chem Biol* 49:33–38.

Bouain N, Krouk G, Lacombe B, Rouached H (2019) Getting to the root of plant mineral nutrition: combinatorial nutrient stresses reveal emergent properties. *Trends Plant Sci* 24(6):542–552.

Brdar-Jokanović M (2020) Boron toxicity and deficiency in agricultural plants. *Int J Mol Sci* 21(4):1424.

Broadley MR (2012). *Marschner's mineral nutrition of higher plants*. Elsevier/Academic Press.

Cabot C, Martos S, Llugany M, Gallego B, Tolrà R, Poschenrieder C (2019) A role for zinc in plant defense against pathogens and herbivores. *Front Plant Sci* 10:1171.

Carstensen A, Herdean A, Schmidt SB, Sharma A, Spetea C, Pribil M, Husted S (2018) The impacts of phosphorus deficiency on the photosynthetic electron transport chain. *Plant Physiol* 177(1):271–284.

Chao Q, Gao ZF, Wang YF, Li Z, Huang XH, Wang YC, Wang BC (2016) The proteome and phosphoproteome of maize pollen uncovers fertility candidate proteins. *Plant Mol Biol* 91(3):287–304.

Chen J, Li LG, Liu, ZH, Yuan, YJ, Guo LL, Mao, DD, Li DP (2009) Magnesium transporter AtMGT9 is essential for pollen development in Arabidopsis. *Cell Res* 19(7):887–898.

Chérel I, Gaillard I (2019) The complex fine-tuning of K+ fluxes in plants in relation to osmotic and ionic abiotic stresses. *Int J Mol Sci* 20(3):715.

Conn SJ, Conn V, Tyerman SD, Kaiser BN, Leigh RA, Gilliham, M (2011) Magnesium transporters, MGT2/MRS2-1 and MGT3/MRS2-5, are important for magnesium partitioning within Arabidopsis thaliana mesophyll vacuoles. *New Phytologist* 190(3):583–594.

Connorton JM, Balk J, Rodríguez-Celma J (2017) Iron homeostasis in plants–a brief overview. *Metallomics* 9(7):813–823.

Corpas FJ, Barroso JB, Palma JM, Rodriguez-Ruiz M (2017) Plant peroxisomes: a nitro-oxidative cocktail. *Redox Biol* 11:535–542.

Costa LC, Carmona VM, Cecílio Filho AB, Nascimento CS, Nascimento CS (2017). Symptoms of deficiencies macronutrients in watermelon. *Comun Sci* 8(1):80.

Cruz JL, Mosquim PR, Pelacani CR, Araújo WL, DaMatta FM (2003). Photosynthesis impairment in cassava leaves in response to nitrogen deficiency. *Plant Soil* 257(2):417–423.

De Freitas, ST, Amarante CD, Mitcham EJ (2016). Calcium deficiency disorders in plants. *Postharvest ripening physiology of crops*. CRC Press, Boca Raton, pp 477–502.

De Macedo FG, Bresolin JD, Santos EF, Furlan F, Lopes da Silva, WT, Polacco JC, Lavres J (2016) Nickel availability in soil as influenced by liming and its role in soybean nitrogen metabolism. *Front Plant Sci* 7:1358.

Demidchik V, Shabala S (2018) Mechanisms of cytosolic calcium elevation in plants: the role of ion channels, calcium extrusion systems and NADPH oxidase-mediated 'ROS-Ca2+ Hub'. *Funct Plant Biol* 45(2):9–27.

Deng W, Luo K, Li D, Zheng X, Wei X, Smith W, Thammina C, Lu L, Li Y, Pei Y (2006) Overexpression of an Arabidopsis magnesium transport gene, AtMGT1, in Nicotiana benthamiana confers Al tolerance. *J Exp Bot* 57(15):4235–4243.

Edmond C, Shigaki T, Ewert S, Nelson MD, Connorton JM, Chalova V, Pittman JK (2009) Comparative analysis of CAX2-like cation transporters indicates functional and regulatory diversity. *Biochem J* 418(1):145–154.

Eide D, Broderius M, Fett J, Guerinot ML (1996) A novel iron-regulated metal transporter from plants identified by functional expression in yeast. *PNAS* 93(11):5624–5628.

Evans JR, Terashima I (1987) Effects of nitrogen nutrition on electron transport components and photosynthesis in spinach. *Funct Plant Biol* 14(1):59–68.

Fang Z, Shao C, Meng Y, Wu P, Chen M (2009). Phosphate signaling in Arabidopsis and Oryza sativa. *Plant Sci* 176(2):170–180.

Ferreira LM, de Souza, VM, Tavares OCH, Zonta E, Santa-Catarina C, de Souza SR, Santos LA (2015) OsAMT1. 3 expression alters rice ammonium uptake kinetics and root morphology. *Plant Biotechnol Rep* 9(4):221–229.

Finkemeier I, Kluge C, Metwally A, Georgi M, Grotjohann N, Dietz KJ (2003) Alterations in Cd-induced gene expression under nitrogen deficiency in Hordeum vulgare. *Plant Cell Environ* 26(6):821–833.

Fu XZ, Zhou X, Xing F, Ling LL, Chun CP, Cao L, Peng LZ (2017) Genome-wide identification, cloning and functional analysis of the zinc/iron-regulated transporter-like protein (ZIP) gene family in trifoliate orange (Poncirus trifoliata L. Raf.). *Front Plant Sci* 8:588.

Gao F, Robe K, Dubos C (2020) Further insights into the role of bHLH121 in the regulation of iron homeostasis in Arabidopsis thaliana. *Plant Signal Behav*, 1795582.

Gao Q, Xiong T, Li X, Chen W, Zhu X (2019) Calcium and calcium sensors in fruit development and ripening. *Sci Hortic* 253:412–421.

Gao H, Xie W, Yang C, Xu J, Li J, Wang H, Huang CF (2018) NRAMP2, a trans-Golgi network-localized manganese transporter, is required for Arabidopsis root growth under manganese deficiency. *New Phytologist* 217(1):179–193.

Gilliham M, Conn S (2010) Magnesium transporters, MGT2/ MRS2-1 and MGT3/ MRS2-5, are important for magnesium partitioning within Arabidopsis thaliana mesophyll vacuoles. *New Phytologist* 190: 583–594.

Goltsev VN, Kalaji HM, Paunov, M., Bąba, W., Horaczek, T., Mojski, J.,… & Allakhverdiev, SI (2016). Variable chlorophyll fluorescence and its use for assessing physiological condition of plant photosynthetic apparatus. *Russian J Plant Physiol* 63(6):869–893.

Gong X, Liu Y, Huang D, Zeng G, Liu S, Tang H, Tan X (2016) Effects of exogenous calcium and spermidine on cadmium stress moderation and metal accumulation in Boehmeria nivea (L.) Gaudich. *Environ Sci Pollut Res* 23(9):8699–8708.

Gransee A, Führs H (2013) Magnesium mobility in soils as a challenge for soil and plant analysis, magnesium fertilization and root uptake under adverse growth conditions. *Plant Soil* 368(1-2):5–21.

Gruber BD, Giehl RF, Friedel S, von Wirén N (2013) Plasticity of the Arabidopsis root system under nutrient deficiencies. *Plant Physiol* 163(1):161–179.

Grusak Michael A, Broadley MR, White PJ (December 2016) Plant macro- and micronutrient minerals. In: *eLS*. John Wiley & Sons, Ltd: Chichester. doi: 10.1002/9780470015902.a0001306.pub2

Guo W, Nazim, H, Liang Z, Yang D (2016) Magnesium deficiency in plants: an urgent problem. *Crop J* 4(2):83–91.

Gupta N, Ram H, Kuma B (2016) Mechanism of Zinc absorption in plants: uptake, transport, translocation and accumulation. *Rev Enviro Sci Bio* 15(1):89–109.

Hafsi C, Debez A, Abdelly, C (2014) Potassium deficiency in plants: effects and signaling cascades. *Acta Physiol Plant* 36(5):1055–1070.

Hebbern CA, Laursen KH, Ladegaard AH, Schmidt SB, Pedas P, Bruhn D, Husted S (2009). Latent manganese deficiency increases transpiration in barley (Hordeum vulgare). *Physiol Plant* 135(3):307–316.

Hermans C, Johnson GN, Strasser RJ, Verbruggen N (2004) Physiological characterisation of magnesium deficiency in sugar beet: acclimation to low magnesium differentially affects photosystems I and II. *Planta* 220(2):344–355.

Hermans C, Conn SJ, Chen J, Xiao Q, Verbruggen N (2013) An update on magnesium homeostasis mechanisms in plants. *Metallomics* 5(9):1170–1183.

Hermans C, Bourgis F, Faucher M, Strasser RJ, Delrot S, Verbruggen N (2005) Magnesium deficiency in sugar beets alters sugar partitioning and phloem loading in young mature leaves. *Planta* 220(4):541–549.

Huang ZA, Jiang DA, Yang Y, Sun JW, Jin SH (2004) Effects of nitrogen deficiency on gas exchange, chlorophyll fluorescence, and antioxidant enzymes in leaves of rice plants. *Photosynthetica* 42(3):357–364.

Huang XY, Li M, Luo R, Zhao F, Salt DE (2019) Epigenetic regulation of sulfur homeostasis in plants. *J Exp Bot* 70(16):4171–4182.

Huang D, Gong X, Liu Y, Zeng G, Lai C, Bashir H, Wan J (2017) Effects of calcium at toxic concentrations of cadmium in plants. *Planta* 245(5):863–873.

Inoue H, Kobayashi T, Nozoye T, Takahashi M, Kakei Y, Suzuki K, Nishizawa NK (2009) Rice OsYSL15 is an iron-regulated iron (III)-deoxymugineic acid transporter expressed in the roots and is essential for iron uptake in early growth of the seedlings. *J Biol* 284(6):3470–3479.

Jaiswal DK, Verma JP, Prakash S, Meena VS, Meena RS (2016) Potassium as an important plant nutrient in sustainable agriculture: a state of the art. In: *Potassium solubilizing microorganisms for sustainable agriculture*. Springer, New Delhi, pp 21–29.

Kalaji HM, Bąba W, Gediga K, Goltsev V, Samborska IA, Cetner MD, Dankov K (2018) Chlorophyll fluorescence as a tool for nutrient status identification in rapeseed plants. *Photosynth Res* 3: 329–343.

Kandlbinder A, Finkemeier I, Wormuth D, Hanitzsch M, Dietz KJ (2004) The antioxidant status of photosynthesizing leaves under nutrient deficiency: redox regulation, gene expression and antioxidant activity in Arabidopsis thaliana. *Physiol Plant* 120(1):63–73.

Kanno S, Arrighi JF, Chiarenza S, Bayle V, Berthomé R, Péret B, Thibaud C (2016) A novel role for the root cap in phosphate uptake and homeostasis. *Elife* 5:e14577.

Karlsson PM, Herdean A, Adolfsson L, Beebo A, Nziengui H, Irigoyen S, Aronsson H (2015) The Arabidopsis thylakoid transporter PHT 4; 1 influences phosphate availability for ATP synthesis and plant growth. *Plant J* 84(1):99–110.

Karthika KS, Rashmi I, Parvathi MS (2018) Biological functions, uptake and transport of essential nutrients in relation to plant growth. In: *Plant nutrients and abiotic stress tolerance*. Springer, Singapore, pp 1–49.

Kathpalia R, Bhatla SC (2018) Plant mineral nutrition. In: *Plant physiology, development and metabolism*. Springer, Singapore, pp 37–81.

Kausar S, Faizan S, Haneef I (2017) Nitrogen level affects growth and reactive oxygen scavenging of fenugreek irrigated with wastewater. *Trop Plant Res* 4:210–224.

Kavitha PG, Kuruvilla S, Mathew MK (2015) Functional characterization of a transition metal ion transporter, OsZIP6 from rice (Oryza sativa L.). *Plant Physiol Bioch* 97:165–174.

Kaznina NM, Titov AF (2017) Effect of zinc deficiency and excess on the growth and photosynthesis of winter wheat. *J Stress Physiol Biochem* 13(4).

Kiba T, Feria-Bourrellier AB, Lafouge F, Lezhneva L, Boutet-Mercey S, Orsel M, Krapp A (2012) The Arabidopsis nitrate transporter NRT2. 4 plays a double role in roots and shoots of nitrogen-starved plants. *Plant Cell* 24(1):245–258.

Kimura E, Bell J, Trostle C, Neely C, Drake D (2016) Potential causes of yellowing during the tillering stage of wheat in Texas. *Texas A&M AgriLife Ext Serv* 4:1–5.

Koch M, Busse M, Naumann M, Jákli B, Smit I, Cakmak I, Pawelzik E (2019) Differential effects of varied potassium and magnesium nutrition on production and partitioning of photoassimilates in potato plants. *Physiol Planta* 166(4):921–935.

Köllner TG, Held M, Lenk C, Hiltpold I, Turlings TC, Gershenzon J, Degenhardt J (2008) A maize (E)-β-caryophyllene synthase implicated in indirect defense responses against herbivores is not expressed in most American maize varieties. *Plant Cell* 20(2):482–494.

Kováčik J, Klejdus B (2014) Induction of phenolic metabolites and physiological changes in chamomile plants in relation to nitrogen nutrition. *Food Chem* 142:334–341.

Kroh GE, Pilon M (2020) Regulation of Iron Homeostasis and Use in Chloroplasts. *Int J Mol Sci* 21(9):3395.

Kumar V, Pandita S, Sidhu GPS, Sharma A, Khanna K, Kaur P, Setia R (2020) Copper bioavailability, uptake, toxicity and tolerance in plants: a comprehensive review. *Chemosphere* 127810.

Landi M, Margaritopoulou T, Papadakis IE, Araniti F (2019) Boron toxicity in higher plants: an update. *Planta*, 1–22.

Lei Z (2019) Green plant potassium nutrition and green interior design. *Arch Latinoam Nutr* 69(6).

Lewis DH (2020) The status of boron as an essential element for vascular plants. *New Phytologist* 226(5): 1231-1231.

Li L, Xu X, Chen C, Shen Z (2016a) Genome-wide characterization and expression analysis of the germin-like protein family in rice and Arabidopsis. *Int J Mol Sci* 17(10):1622.

Li C, Wang P, Menzies NW, Lombi E, Kopittke PM (2017) Effects of changes in leaf properties mediated by methyl jasmonate (MeJA) on foliar absorption of Zn, Mn and Fe. *Ann Bot* 120(3):405–415.

Li C, Tang Z, Wei J, Qu H, Xie Y, Xu G (2016b) The OsAMT1. 1 gene functions in ammonium uptake and ammonium–potassium homeostasis over low and high ammonium concentration ranges. *JGG* 43(11): 639–649.

Li LG, Sokolov LN, Yang YH, Li DP, Ting J, Pandy GK, Luan S (2008) A mitochondrial magnesium transporter functions in Arabidopsis pollen development. *Mol Plant* 1(4):675–685.

Liang C, Xiao W, Hao H, Xiaoqing L, Chao L, Lei Z, Fashui H (2009) Effects of Mg2+ on spectral characteristics and photosynthetic functions of spinach photosystem II. *Spectrochim Acta Part A Mol Biomol Spectrosc* 72(2):343–347.

Lin YL, Chao YY, Huang WD, Kao CH (2011) Effect of nitrogen deficiency on antioxidant status and Cd toxicity in rice seedlings. *Plant Growth Regul* 64(3):263–273.

Liu G, Martinoia E (2020) How to survive on low potassium. *Nature Plants* 6(4):332–333.

Liu G, Simonne EH, Li Y (2011) Nickel nutrition in plants. In: *IFAS Extension*. University of Florida, p 6.

Liu Y, Riaz M, Yan L, Zeng Y, Cuncang J (2019) Boron and calcium deficiency disturbing the growth of trifoliate rootstock seedlings (Poncirus trifoliate L.) by changing root architecture and cell walE (2013) Roles and functions of plant mineral nutrients. In: *Pla6868iMineral nutrients*. Humana Press, Totowa, NJ, pp 1–21.

Maathuis FJ, Diatloff E (2013) Roles and functions of plant mineral nutrients. . In: *Plant mineral nutrients*. Humana Press, Totowa, NJ, pp 1–21.

Mahmud K, Makaju S, Ibrahim R, Missaoui A (2020). Current progress in nitrogen fixing plants and microbiome research. *Plants* 9(1):97.

Makavitskaya M, Svistunenko D, Navaselsky I, Hryvusevich P, Mackievic V, Rabadanova C, Voitsekhovskaja O (2018) Novel roles of ascorbate in plants: induction of cytosolic Ca2+ signals and efflux from cells via anion channels. *J Exp Bot* 69(14):3477–3489.

Malhotra H, Sharma S, Pandey R (2018) Phosphorus nutrition: plant growth in response to deficiency and excess. In: *Plant nutrients and abiotic stress tolerance*. Springer, Singapore, pp 171–190.

Manuel TJ, Alejandro CA, Angel L, Aurora G, Emilio F (2018) Roles of molybdenum in plants and improvement of its acquisition and use efficiency. In: *Plant micronutrient use efficiency*. Academic Press, pp 137–159.

Mao D, Chen J, Tian L, Liu Z, Yang L, Tang R, Chen L (2014) Arabidopsis transporter MGT6 mediates magnesium uptake and is required for growth under magnesium limitation. *Plant Cell* 26(5):2234–2248.

Matthus E, Wilkins KA, Swarbreck SM, Doddrell NH, Doccula FG, Costa A, Davies JM (2019). Phosphate starvation alters abiotic-stress-induced cytosolic free calcium increases in roots. *Plant Physiol* 179(4):1754–1767.

McAinsh MR, Pittman JK (2009) Shaping the calcium signature. *New Phytologist* 181(2):275–294.

McCauley A, Jones C, Jacobsen J (2009) Plant nutrient functions and deficiency and toxicity symptoms. *Nutr Manag Module* 9:1–16.

Mengutay M, Ceylan, Y, Kutman UB, Cakmak I (2013) Adequate magnesium nutrition mitigates adverse effects of heat stress on maize and wheat. *Plant Soil* 368(1-2):57–72.

Młodzińska E, Zboińska M (2016) Phosphate uptake and allocation–a closer look at Arabidopsis thaliana L. and Oryza sativa L. *Front Plant Sci* 7:1198.

Moghadam NK, Lajayer BA, Ghorbanpour M (2020) Boron tolerance in plants: physiological roles and transport mechanisms. *Met Plants Adv Future Prosp* 301–314.

Muneer S, Jeong BR (2015) Proteomic analysis provides new insights in phosphorus homeostasis subjected to pi (inorganic phosphate) starvation in tomato plants (Solanum lycopersicum L.). *PLoS One* 10(7): e0134103.

Nishida S, Tsuzuki C, Kato A, Aisu A, Yoshida J, Mizuno T (2011) AtIRT1, the primary iron uptake transporter in the root, mediates excess nickel accumulation in Arabidopsis thaliana. *Plant Cell Physiol* 52(8):1433–1442.

Padhi PP, Mishra AP (2019) The role of molybdenum in crop production. *J Pharmacogn Phytochem* 8(5):1400–1403.

Pandey N (2018) Role of plant nutrients in plant growth and physiology. In: *Plant nutrients and abiotic stress tolerance*. Springer, Singapore, pp 51–93.

Pandey GK, Mahiwal S (2020) Potassium uptake and transport system in plant. In: *Role of potassium in plants*. Springer, Cham, pp 19–28.

Prajapati K, Modi HA (2012) The importance of potassium in plant growth-a review. *Indian J Plant Physiol* 1(02):177–186.

Princi MP, Lupini A, Araniti F, Longo C, Mauceri A, Sunseri F, Abenavol MR (2016) Boron toxicity and tolerance in plants: recent advances and future perspectives. In: *Plant metal interaction*. Elsevier, pp 115–147.

Qadeem W, Sattar S, Adnan M, Zaman M, Ali I, Shah SRA, Junaid K, Said F, Rahman IU, Saleem N (2015) Efficacy of botanical and microbial extracts against Angoumois grain moth *Sitotrogacerealella*(Olivier) (Lepidoptera: Gelechiidae) under laboratory conditions. *J Ento Zool Studies* 3(5):451–454.

Ramaekers, L, Remans, R, Rao, IM, Blair, MW, Vanderleyden, J (2010). Strategies for improving phosphorus acquisition efficiency of crop plants. *Field Crops Res* 117(2-3):169–176.

Rausch C, Bucher M (2002) Molecular mechanisms of phosphate transport in plants. *Planta* 216(1):23–37.

Rawat J, Sanwal P, Saxena J (2016) Potassium and its role in sustainable agriculture. In *Potassium solubilizing microorganisms for sustainable agriculture*. Springer, New Delhi, pp 235–253.

Rentsch D, Schmidt S, Tegeder M (2007) Transporters for uptake and allocation of organic nitrogen compounds in plants. *FEBS Letters* 581(12):2281–2289.

Rudani L, Vishal P, Kalavati P (2018) The importance of zinc in plant growth-a review. *IRJNAS* 5(2):38–48.

Sacchi GA, Nocito FF (2019) Plant sulfate transporters in the low phytic acid network: some educated guesses. *Plants* 8(12):616.

Saha S, Samad R, Rashid P, Karmoker JL (2016) Effects of sulphur deficiency on growth, sugars, proline and chlorophyll content in mungbean (Vigna radiata L. var. BARI MUNG-6). *Bangladesh J Bot* 45: 405–410.

Samborska IA, Kalaji HM, Sieczko L, Borucki W, Mazur R, Kouzmanova M, Goltsev V (2019). Can just one-second measurement of chlorophyll a fluorescence be used to predict sulphur deficiency in radish (Raphanus sativus L. sativus) plants?. *Curr Plant Biol* 19:100096.

Schmidt SB, Jensen PE, Husted S (2016) Manganese deficiency in plants: the impact on photosystem II. *Trend Plant Sci* 21(7):622–632.

Schulze J, Temple G, Temple SJ, Beschow H, Vance CP (2006) Nitrogen fixation by white lupin under phosphorus deficiency. *Ann Bot* 98(4):731–740.

Shah A, Wu X, Ullah A, Fahad S, Muhammad R, Yan L, Jiang C (2017) Deficiency and toxicity of boron: alterations in growth, oxidative damage and uptake by citrange orange plants. *Ecotoxicol Environ Saf* 145:575–582.

Shao JF, Yamaji N, Shen RF, Ma JF (2017) The key to Mn homeostasis in plants: regulation of Mn transporters. *Tre Plant Sci* 22(3):215–224.

Shen, J, Yuan L, Zhang J, Li H, Bai Z, Chen X, Zhang F (2011) Phosphorus dynamics: from soil to plant. *Plant Physiol* 156(3):997–1005.

Shin R, Berg RH, Schachtman DP (2005) Reactive oxygen species and root hairs in Arabidopsis root response to nitrogen, phosphorus and potassium deficiency. *Plant Cell Physiol* 46(8):1350–1357.

Shireen F, Nawaz MA, Chen C, Zhang Q, Zheng Z, Sohail H, Bie Z (2018) Boron: functions and approaches to enhance its availability in plants for sustainable agriculture. *Int J Mol Sci* 19(7):1856.

Shrivastav P, Prasad M, Singh TB, Yadav A, Goyal D, Ali A, Dantu PK (2020) Role of nutrients in plant growth and development. In: *Contaminants in agriculture*. Springer, Cham, pp 43–59.

Singh, J, Singh M, Jain, A, Bhardwaj S, Singh A, Singh, DK, Dubey SK (2013) An introduction of plant nutrients and foliar fertilization: a review. *Precision farming: a new approach*, New Delhi: Daya Publishing Company, pp. 252–320.

Singh M, Sidhu HS, Humphreys E, Thind HS, Jat ML, Blackwell J, Singh V (2015) Nitrogen management for zero till wheat with surface retention of rice residues in north-west India. *Field Crops Res* 184: 183–191.

Sutar RK, Pujar AM, Kumar BA, Hebsur NS (2017) Sulphur nutrition in maize-a critical review. *Int J Pure App Biosci* 5(6):1582–1596.

Taiz L, Zeiger E (2002) Mineral nutrition In: *Plant physiology*, 3rd edn. Sinauer Publishers

Takahashi H (2019) Sulfate transport systems in plants: functional diversity and molecular mechanisms underlying regulatory coordination. *J Exp Bot* 70(16):4075–4087.

Takizawa K, Kanazawa A, Kramer DM (2008) Depletion of stromal Pi induces high 'energy-dependent' antenna exciton quenching (qE) by decreasing proton conductivity at CFO-CF1 ATP synthase. *Plant Cell Environ* 31(2):235–243.

Tegeder M, Masclaux-Daubresse C (2018) Source and sink mechanisms of nitrogen transport and use. *New Phytologist* 217(1):35–53.

Tewari RK, Kumar P, Sharma PN (2006) Magnesium deficiency induced oxidative stress and antioxidant responses in mulberry plants. *Sci Hortic* 108(1):7–14.

Tewari RK, Kumar P, Sharma PN (2007) Oxidative stress and antioxidant responses in young leaves of mulberry plants grown under nitrogen, phosphorus or potassium deficiency. *J Integ Plant Biol* 49(3): 313–322.

Thor K (2019) Calcium—nutrient and messenger. *Front Plant Sci* 10:440.

Tian S, Xie R, Wang H, Hu Y, Ge J, Liao X, Lu L(2016) Calcium deficiency triggers phloem remobilization of cadmium in a hyperaccumulating species. *Plant Physio* 172(4):2300–2313.

Veliz CG, Roberts IN, Criado MV, Caputo C (2017). Sulphur deficiency inhibits nitrogen assimilation and recycling in barley plants. *Biologia Plantarum* 61(4):675–684.

Venkidasamy B, Selvaraj D, Ramalingam S (2019) Genome-wide analysis of purple acid phosphatase (PAP) family proteins in Jatropha curcas L. *Int J Biol Macromol* 123:648–656.

Virginia-Pérez V, López-Laredo AR, Sepúlveda-Jiménez G, Zamilpa A, Trejo-Tapia G (2015) Nitrogen deficiency stimulates biosynthesis of bioactive phenylethanoid glycosides in the medicinal plant Castilleja tenuiflora Benth. *Acta Physiol Plant* 37(5):93.

Wang Y, Wu WH (2017) Regulation of potassium transport and signaling in plants. *Curr opin Plant Biol* 39:123–128.

Wang F, Deng M, Xu J, Zhu X, Mao C (2018) Molecular mechanisms of phosphate transport and signaling in higher plants. *Semin Cell Dev Biol* 74: 114–122. Academic Press.

Weisany W, Raei Y, Allahverdipoor KH (2013) Role of some of mineral nutrients in biological nitrogen fixation. *BEPLS* 2(4):77–84.

Wu X, Lu X, Riaz M, Yan L, Jiang C (2018) Boron deficiency and toxicity altered the subcellular structure and cell wall composition architecture in two citrus rootstocks. *Sci Hortic* 238:147–154.

Yadav BK, Sidhu AS (2016) Dynamics of potassium and their bioavailability for plant nutrition. In: *Potassium solubilizing microorganisms for sustainable agriculture*. Springer, New Delhi, pp 187–201.

Yang GH, Yang LT, Jiang HX, Li, Y, Wang, P, Chen LS (2012) Physiological impacts of magnesium-deficiency in Citrus seedlings: photosynthesis, antioxidant system and carbohydrates. *Trees* 26(4): 1237–1250.

Yruela I (2009) Copper in plants: acquisition, transport and interactions. *Funct Plant Biol* 36(5):409–430.

Yuan L, Loqué D, Kojima, S, Rauch S, Ishiyama K, Inoue E, von Wirén N (2007) The organization of high-affinity ammonium uptake in Arabidopsis roots depends on the spatial arrangement and biochemical properties of AMT1-type transporters. *Plant Cell* 19(8):2636–2652.

Zhang Z, Liao H, Lucas WJ (2014) Molecular mechanisms underlying phosphate sensing, signaling, and adaptation in plants. *J Integr plant Biol* 56(3):192–220.

Zhao DY, Tian QY, Li LH, Zhang WH (2007) Nitric oxide is involved in nitrate-induced inhibition of root elongation in Zea mays. *Ann Bot* 100(3):497–503.

13

Agricultural Practices Can Reduce Soil Greenhouse Gas Emissions: Challenges and Future Perspectives

Misbah Naz, Kashif Akhtar, Aziz Khan, and Ghulam Shah Nizamani

CONTENTS

13.1 Introduction

The increase in GHG emissions from soil to atmosphere is an important factor in global climate change, which is mainly caused by overuse of fertilizer and inappropriate farming practices. Carbon dioxide (CO_2), nitrous oxide (N_2O), and methane (CH_4) are the important greenhouse gases contributing to global warming (Wang et al. 2006), where the practice of agriculture is traced to the release of and increase in atmospheric CO_2 and N_2O (Davidson et al. 2008; Monreal 2010; Mosier and Kroeze 2000; Nykanen et al. 1995). Generally, the release of soil C and N to the atmosphere depends on soil temperature and moisture content, soil organic carbon (SOC), microbial biomass, soil aeration, vegetation type, and management of field (Galford et al. 2010). Soil health and its capability in production mainly depend on soil organic matter (Lin et al. 2012). N fertilizers consume large amounts of energy, be it after application or during production, and injudicious application can cause environmental damage, i.e., nitrate in ground water and the production of nitrous oxide, a greenhouse gas (GHG) released to the atmosphere, which is a global concern (Khan et al. 2017).

Fertilizer management practices are responsible for the release of soil C and N to the atmosphere. However, currently, to control or decrease fertilizer overuse, the combined use of manures or crop straw application with

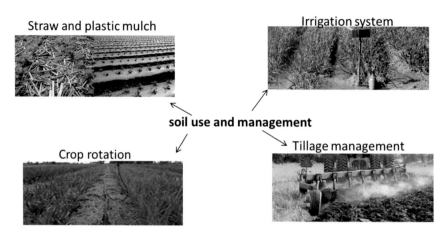

FIGURE 13.1   Soil management practices decrease GHGs emission and improve soil fertility.

fertilizer can efficiently utilize the inorganic fertilizer, and thus improve soil health and agricultural production (Figure 13.1) (Akhtar et al. 2019c). Therefore, we should pay more attention to the proactive use of manures, crop straw, and cultivation practices with benefits detected onto soil C and N cycles, especially in agro-ecosystems, where gas from soils are released through the gas phase. Previous results differed on the effect of fertilizer strategies on soil CO_2 and N_2O emissions. Compared with inorganic fertilizer applications, manure applications significantly stimulate soil CO_2 and N_2O production (Ding et al. 2013). However, there were no significant differences in results for the use of inorganic fertilizers and against manures causing N_2O emissions in sand (Bouwman et al. 2002; Wang and Luo 2018). Therefore, the proper use of manure can reduce soil CO_2 and N_2O emissions (Cayuela et al. 2010). In addition, wheat straw with high C/N ratios did not increase soil N_2O emissions (Cai et al. 2001; Wang and Luo 2018). However, in favourable environments for denitrification, the application of straw with high C/N to pig manure and clay soil significantly increased soil N_2O emissions (Yang et al. 2003). While organic and inorganic fertilizers have varied effects on soil CO_2 and N_2O emissions of different soil types, investigations into decades of fertilization experiments have been particularly rare and thus hamper our ability to assess the impact of continuous fertilization of soil in CO_2 and N_2O emissions. Therefore, long-term experimentation is needed to attain consistent information on soil C and N turnover, and soil CO_2 and N_2O emissions (Xin et al. 2012).

Summer maize and winter wheat are the main crop rotations followed in the semi-arid region of China (Akhtar et al. 2018). Though there are higher soil CO_2 emissions in summer compared with that in winter (Wang et al. 2019), it was reported that atmospheric N_2O concentrations increased by about 0.26% yr^{-1} and reached atmospheric concentrations of 319 ppm in 2005 (IPCC 2007). Though, according to estimates, the soil accounts for about 65% of the total N_2O emissions in the atmosphere (IPCC 2001), a cause for concern to soil experts and environmentalists as N_2O is a greenhouse gas with a global warming potential of 298 over a 100-year period (IPCC 2007) and the largest contributor to stratospheric ozone depletion (Ravishankara et al. 2009), carbon dioxide (CO_2) is the dominant greenhouse gas in the earth's atmosphere (IPCC 2013). Since the Industrial Revolution, CO_2 concentrations have changed significantly, from 280 ppm in 2011 to 391 ppm, with about 20% of the total amount of CO_2 being released from the soil into the atmosphere (Figure 13.2) (Lal 2003). Although, the sources and sinks of N_2O and CO_2 in different environments are increasing, the projections for release of GHGs in time and space remain challenging due to the frequent nature of hot spots (Cakir and Stenstrom 2005; Groffman et al. 2009; Shah et al. 2017). Given the cumulative N_2O flux of 51.1% was observed in Ireland's heavily grazing grasslands (Scanlon and Kiely 2003), it is possible that this large variation in N_2O emission is caused by complex environmental variables such as soil temperature and soil moisture, as well as nutrient availability, which control the nitrification and de-nitrification processes that cause N_2O emissions (Farquharson and Baldock 2008). On the other hand, taking into view that bio-CO_2 variability shows wide fluctuations on a series of time scales, complicating the identification of CO_2 inconsistencies (Risk et al. 2013), it seems the variability of CO_2 is related to a variety of factors, such as soil moisture, soil temperature, soil organic matter content, and matrix

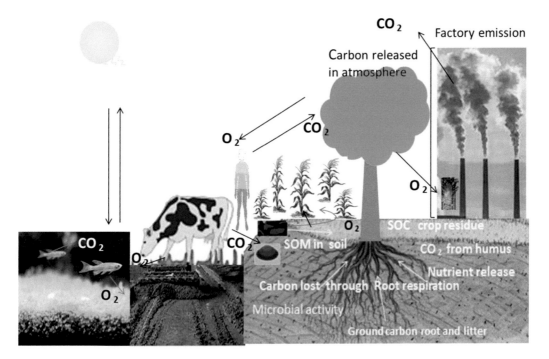

FIGURE 13.2 Carbon dioxide in the atmosphere plays an integral role. Movement of CO_2 from the atmosphere through natural processes such as photosynthesis and carbonate deposition, and the formation of limestone, adds acid to the atmosphere by other natural processes, such as acid dissolution of carbonate deposits. Plants release O_2 and humans and animals release CO_2, and balanced with the weathering of carbon and rock in the oceans, so that life flourished in the ocean during the era of Eocene. Life, ocean chemistry, and atmospheric gases take millions of years to adapt to these levels.

(Bahn et al. 2010; Wang et al. 2013). Therefore, to understand soil CO_2 and N_2O fluxes, it is critical to reliably quantify these gas fluxes for developing future alleviation strategies (Erşahin et al. 2017). From the perspective that the stability of soil C is dependent on biological techniques, especially in temperate soils, from the perspective of carbon accumulation, it is high in semi-arid areas. Global agriculture plays a crucial role in food security, poverty reduction, and sustainable development (Sitthisuntikul et al. 2018), and soil CO_2 fluxes need to be assessed in the backdrop of different environmental conditions and with respect to different management techniques. Different rotations have a different potential for promoting and supporting carbon sequestration (Kong et al. 2019). First, the undisturbed effect of carbon sequestration on soil colour or surface residues reduces soil temperature. However, reducing mechanical farming is sufficient to have a positive impact on crop productivity but, in a hot and dry environment, reducing arable land appears to be the most effective strategy for carbon accumulation and increasing of crop yield. With several management decisions being proposed to reduce GHG emissions to the atmosphere and increase crop productivity, it is a priority to understand the mechanism for GHG emissions to hasten the process of understanding how crop husbandry can suppress GHG emissions. Thus, the purpose of this chapter is to prioritize and synthesize the reported information on efficient soil management practices in relation to crop production and reduction in GHG emissions.

13.2 Accumulation of Soil Organic Carbon and Greenhouse Gases

A warm and humid climate points to high atmospheric CO_2, and high atmospheric CO_2 means increased greenhouse gas emissions, but the mitigation potential of this cause-effect relation has not yet been fully exploited (Farquharson and Baldock 2008). Today's climate change is mainly driven by three greenhouse gases (GHG): carbon dioxide (CO_2), nitrous oxide (N_2O), and methane (CH_2). While, prior to the industrial age, the world's vegetation, soil, and animal communities maintained a vital equilibrium balance between

CO_2, N_2O, and CH_2 emissions (Farquharson and Baldock 2009), the modern industrial civilization has undermined this balance, leading to a sharp rise in atmospheric concentrations of all three greenhouse gases since 1850, leading to global climate change in the late 20th century. Given the agricultural activities affect the climate through direct greenhouse gas emissions and also affect soil and plant biomass components in the global carbon (C) cycle (Pan et al. 2017), soil fertility and productivity depend on soil organic matter. But, the world's agricultural soil organic matter has been lost due to the loss of natural vegetation and, with intensive agricultural farming the current practice, the decomposition of organic matter is accelerated. Therefore, farming-based production systems should be changed in order to achieve future production intensification in a sustainable manner, and it's here that conservation agriculture assumes importance because its practice avoids or minimizes soil disturbances by preserving soil covering through crop diversification (Jones et al. 2006). This study seeks to reduce existing uncertainties about the impact of soil management practices on soil carbon and to promote decision-making on future farming patterns for scientists and policy makers. In addition, cover crops can be grown under trees to protect soils from erosion and to promote soil carbon sequestration (Aguilera et al. 2013).

Though organic and inorganic forms of carbon are present in soil, still the soil organic matter is rich in lignin and lower in carbohydrate, oxygen, and hydrogen levels of organic matter, as mineralization releases oxygen and degrades polysaccharides, resulting in increased concentrations of recalcitrant (or stable) compounds (Aguilera et al. 2013). High variability of SOC varies between the same sampling locations, which requires subsequent sampling to be repeated over time at the same location to eliminate any spatial variability factors (Pan et al. 2017). Given the amount of SOC comes from the decomposition of fresh plant and animal organic matter, the stable pool is also known as recalcitrant SOM, that consists of particles less than 0.053 mm (53 s) in size (Ren et al. 2018). SOM is incorporated into the polymer, and is further protected from decomposition (Aguilera et al. 2013), with soil emission land types and climate zones being the important factors involved in soil health and GHG emissions.

13.3 Farm Practices Improve Soil Functionality and Reduce GHG Emission

13.3.1 Reduced Tillage Improves Soil Health and Reduces Greenhouse Gas Emissions

Tillage is a mechanical and an operational action and, when practiced, can change the natural characteristics of the soil. The purpose of agriculture is to provide an appropriate environment for crops to grow (Rusu et al. 2009). While rotary tillage is a potential crop production activity that greatly reduced crop yields (Wright et al. 2008), agricultural tillage practices inhibit weeds, control soil erosion, and maintain enough soil moisture (Canton et al. 2011). However, minimum tillage has the potential to decrease GHGs emissions without compromising crop yield (Zhang et al. 2015), while the amount of cultivation had no effect on soil C content (Canton et al. 2011). Deep tillage produces more carbon dioxide emissions than undisturbed soil does (no tillage), because tillage makes soil environment better for microbial growth and the acceleration of microbes help in the decomposition of plant and animal residues (Wang et al. 2006). Tillage breaks down soil aggregates, helps mix soil and organic particles, improves the ability to penetrate and retain water, and therefore increases CO_2 production, while reducing CO_2 emissions from tillage is reported to require reducing soil ploughing methods and keeping soil organic C unexposed (Wright et al. 2008). Given tillage increases the ore-forming rate and atmospheric flux of C, it is recommended to reduce the intensity of farming in order to reduce the C loss of soil (Canton et al. 2011). Manipulation of soil environmental conditions, i.e., tillage, irrigation, fertilizer, and manure application, also has an impact on the production and emission of carbon dioxide (Wang et al. 2006), so that the use of C-fixation measures in soil that greatly reduce the increase in atmospheric carbon dioxide are valuable as countermeasures. Greenhouse gas emissions are reported to be reduced by decreasing the use of fossil fuels in field preparation and by increasing soil carbon sequestration. However, the more uniform the distribution of carbon in traditional cultivated soils, the more does reduction of agriculture lead to soil organic carbon stratification (Wang et al. 2006). Under reduced tillage farming conditions, crop residues were accumulated on the soil surface which can lead to the loss of carbon into the atmosphere after decomposition (Wang et al. 2019). Soil under zero farming increases potential N_2O emissions, but this is offset by a significant reduction in potential CO_2 and CH_4 emissions, which are closely related to the geometry of the porous structure of the soil.

13.3.2 Crop Rotation in Relation to Soil Health and Greenhouse Gas Emissions

Rotations with cover crops can reduce greenhouse gas emissions and soil nitrates (MacKenzie et al. 1997), whereas just rice cultivation features high water consumption and high methane emissions (Goglio et al. 2013). Compared with continuous corn, the average yield benefit of corn-soybean rotation was more than 20%, and nitrous oxide emissions decreased by about 35% (Davidson et al. 2008). Nitrous oxide is an extremely deleterious greenhouse gas with the potential for global warming in the atmosphere, and is nearly 300 times more potent than carbon dioxide (Adviento-Borbe et al. 2010; Goglio et al. 2013). It is a by-product of the de-nitrification process, during which bacteria in the soil break down nitrates into inert nitrogen. In this chapter, we discuss whether covering crops have a significant impact on observations for sustainable agriculture. Most crop rotations incorporating cover crops do not affect the yield of cash crops.

Recently, farmers have been seen to follow corn and soybean rotations to avoid continuous corn production. Scientists at the University of Illinois have provided further evidence that rotating crops can increase yields and reduce greenhouse gas emissions compared to continuous corn or soybean (Adviento-Borbe et al. 2010; Ilyas et al. 2016). It has also been noted that nitrous oxide levels are high at the beginning of the season and low at the end of the season (St Luce et al. 2016). A typical farmer would expect these results. For soybeans that are not fertilized, rotation does not affect nitrous oxide emissions compared to continuous soybeans. However, 7% of soybean production is increased by practicing of crop rotation. In this chapter, we discuss soil C change and the Birch effect, two well-known phenomena through which soil organic C loss may be turned around, while the actual mechanism is still unclear. We believe that both phenomena should be evaluated as having a combined impact of high-level destabilizing process. Rotating crops increase yields and reduce greenhouse gas emissions compared to single-grown corn or soybeans (Kong et al. 2019).

13.3.3 Fertilizer Management

The application of synthetic (N) fertilizers (e.g., urea) and the inclusion of crop residues in the soil can lead to CO_2 and N_2O emissions (Kong et al. 2019; Sebilo et al. 2013). When South Africa's field crop program needed to focus more on sustainable improvements in soil fertility, it was advocated the optimal application of synthetic nitrogen fertilizer and crop residues (Aguilera et al. 2013). GHG emission from agricultural practices are unknown compared to other sectors (Niu et al. 2014; Rajaniemi 2002), though we know flooded rice systems emit both methane (CH_4) and nitrous oxide (N_2O), and the increased emissions of CH_4 in rice systems compared with other crops may lead to hike in global warming potential that will require strategies to reduce greenhouse gas emissions, particularly CH_4 (Niu et al. 2014).

13.3.4 Coating of N Fertilizer

In southern China, for rice grown areas, low 20–40% of apparent N recovery (ANR) were noted (Yang et al. 2013). However, ANR would be improved by using nitrogen fertilizer in a split method (Buser et al. 2002), but labour cost and labour availability usually limits their application. Therefore, to improve crop yield and reduce nutrient losses, it is now necessary to provide new types of nitrogen fertilizer, which provide slow release of N, and increase N use efficiency. Some new types of N fertilizers are, urea formaldehyde (UF), also known as slow released N fertilizer, as it contains compounds which are low water soluble, and are slowly available on enzymatic hydrolysis (Fate and efficiency of ^{15}N-labelled slow- and controlled-release fertilizers). Yang et al. (2013) reported that using coating technique, such as coating urea prills with different coating materials, such as polyethylene, sulphur, and other inorganic materials, is also known as controlled release N fertilizer. According to crop demand, coating is designed and, therefore, the overuse of fertilizer is decreased with the release of N fertilizer to the atmosphere curbed. Furthermore, inhibitors like nitrification inhibitors (UNI) or urease are used for the increased stability of nitrogen fertilizer in the soil (Chien et al. 2009). Recently in China, controlled release fertilizers are being used for the enhancement of N fertilizer efficiency and to reduce the nutrient losses (GHGs). However, UNI showed better result for the increase in yield and decrease in GHG emissions in agricultural fields.

13.3.5 Effect of Fertilizer Placement on GHG Emissions

In order to develop effective greenhouse gas emission mitigation strategies, endeavours should be directed at quantifying the impact of GWP calculations of CH_4 and N_2O emissions, studying options for integrated mitigation practices, such as the deep placement of ammonium sulphate, and determining the economic viability of such fertilizer management practices. In order to reduce the CO_2 emissions of farmland, the recommended N fertilization rate can be used for continuous cultivation without tillage (Kong et al. 2019). The main source of carbon dioxide in agricultural soil is the biodegradation of soil organic matter under aerobic conditions and the destruction of soil and vegetation carbon pools through ploughing (Zheng et al. 2007). Carbon is relatively stable in the soil, and though the obduracy of C to degradation in the soil is less clear, yet this analysis focuses primarily on processes and practices for increasing soil C storage. The fertilizer industry recognizes the potential reduction of greenhouse gas emissions with the improvement in agricultural production efficiency. For nutrient management, greenhouse gas emission accounting is included to better reflect sustainable impacts on agricultural practices (Sebilo et al. 2013). Recently, efforts have been developed at most of the country's sub-levels, all of which relate to bottom-up methods of estimating emissions. We need to stop overusing of fertilizers that increase greenhouse gas emissions.

13.3.6 Straw Mulch with Less Nitrogen Fertilizer Improves Soil Properties and Reduces Greenhouse Gas Emissions

Straw burning is a serious problem in China. To fix this problem, covering soil surface by mulch material is considered an ideal management technique for protecting soil moisture, regulating soil temperature, and improving soil quality with the increase in crop yield (Akhtar et al. 2019a, 2019b). It provides a protective barrier around plants and reduces water loss in the form of a layer of mulch used as an insulation barrier to keep the soil warm in winter and cool in summer. A covering of 2.5–5.0 cm around established plants, mulch can hinder the growth of weeds (Moreno et al. 2016), where organic mulch can add nutrients to the soil. Organic mulch can be made of a variety of materials (Moreno et al. 2016), including bark, wood chips, grass shears, pine needles, broken leaves, sawdust, straw, compost, and other organic materials which can all be used to form a layer of material that is broken down to add nutrients to the soil. The use of straw mulch can significantly reduce soil CO_2 emissions with the increase in yield of soybean-wheat cropping systems (Wang et al. 2019).

13.3.7 Effect of Water–Nutrient Management on Greenhouse Gas Emissions

Irrigation may reduce or increase greenhouse gas emissions from the planting system by positive or negative water-to-nutrients relations (Rietz and Haynes 2003). In order to estimate the net impact of irrigation on greenhouse gas emissions, changes in crop yields must be considered (Wallach et al. 2005). Given the intensity of greenhouse gas emissions from irrigation and fertilizer depends to a large extent on water and nutrient use efficiency, therefore, improving water efficiency (technology and management) is an effective way to reduce emissions (Rietz and Haynes 2003). Strengthening the integrated management of water resources utilization, balancing the development and utilization of water resources, avoiding over-consumption of groundwater, and actively improving water efficiency are conducive to reducing greenhouse gas emissions, putting less pressure on water resources, and promoting sustainable agricultural production (Rietz and Haynes 2003). Identifying greenhouse gas emissions in different sectors is a key guide for identifying options in emission reduction. Given the importance of irrigation to agriculture and the multiple pressures on water resources, it is necessary to clarify greenhouse gas emissions from each process and irrigation practices in order to identify emission reductions and the rational development of irrigated agriculture (Playan and Mateos 2006). Based on official statistics and market surveys, this analysis estimates four irrigation processes (groundwater pumping, surface pumping, water delivery, and irrigation equipment production) and five of the most popular irrigation processes for greenhouse gas emissions from China's irrigation technologies (traditional irrigation, irrigation, micro-irrigation, low-pressure pipe irrigation, and canal lining irrigation) (Rietz and Haynes 2003). The main reason for the large proportion of groundwater pumping in the total greenhouse gas emissions of agricultural irrigation is that the pumping volume is large and the pumping head is high (Playan and Mateos 2006). From the 1950s to the beginning of this century, the water table in some parts of China has fallen sharply due to overexploitation of groundwater (Foster et al. 2004), and according to the statistics of 2015 and 2010, the 663 monitoring stations in China showed that the water table continued to decline in most parts of China, with some areas dropping by more than 10 m during this period (Royer et al. 2001). Continued use of groundwater for irrigation will further reduce the water table and increase greenhouse gas emissions from irrigation leading to the thought that proper use of water-to-nutrient application is needed to improve water and nutrients use efficiency, reduce food loss, and increase food production with the decrease in GHG emissions.

The moisture content is a strong prediction of soil C mineralization. Globally, pH is the main variable that controls SOC retention on various minerals. pH is ultimately controlled by soil moisture (Hermle et al. 2008). The overall pattern of pH on organic mineral associations is one of direct proportionality so that as there is increase in soil moisture, there is increase in the proportion of SOC in soil. There is evidence that changes in the availability of other nutrients may be important; nitrogen, in the form of ammonium, is specifically reported to increase in re-humidifying dry soil (Plante et al.). The increase in SOC damage may be compensated by increased carbon inputs from faster-growing plants (Hermle et al. 2008; Khan et al. 2016). Waste layer is the main source of CO_2 and also affects soil-atmosphere exchange between N_2O and CH_4. It is unclear how much of the soil greenhouse gas (GHG) emissions come from waste (Hermle et al. 2008). Forest soil plays an important role in controlling global greenhouse gas budgets as they are primarily sources of carbon dioxide, methane (CH_4) and nitrous oxide (N_2O) (IPCC 2013). Soil microbiome has a strong influence on soil greenhouse gas flux (Hartmann et al. 2014), and usually adapted to plant waste in a given environment (Groffman et al. 2009), although plant waste has the largest input into forest soil C and nutrition (Food and Agriculture Organization of the United Nations 2012). However, there is a lack of knowledge on the significant impact of waste layers on the greenhouse gas flux of forest soils. Atmospheric CO_2 is the main driver of global warming, with CH_4 and N_2O being potently harmful greenhouse gases with 100-year global warming potentials of 28 and 265, respectively (IPCC 2013). The physical properties of soil, or total C and N, are less important in controlling greenhouse gas emissions from farming systems. More CO_2 flux is determined by microbial biomass than soil physical properties (Groffman et al. 2009; Hartmann et al. 2014; Afzal et al. 2017). The activity of microbial C and N affected by farming and irrigation vary, resulting in different levels and combinations of field control of greenhouse gas emissions, and plant residues are being used as soil cover to reduce erosion, maintain soil moisture, and improve soil quality (coverage), and may reduce soil emission rates (Davidson et al. 2008). Management practices such as irrigation, farming systems, and nitrogen fertilization can alter crop residue C inputs, nutrient dynamics, soil temperature, and water content and, thereby, affect soil surface CO_2 emissions (Kong et al. 2019).

13.4 Increased Greenhouse Gas Emissions by Climatic Conditions

Global warming potential (GWP) depends on the efficiency of the molecule as a greenhouse gas and its atmospheric life (van Groenigen et al. 2013). Therefore, if the gas has a high (positive) radiation forcing, but short life, it will have a large GWP in a 20-year scale, and small in 100-year scale. Conversely, if the molecule's atmospheric life is longer than that of CO_2, then from the perspective of the same time scales, the molecule's GWP value will increase from one time to the next (Desjardins et al. 2005). CO_2 is defined as 1 GWP for all time periods. Carbon dioxide is released from the soil through soil respiration and the concerned soil attributes are soil texture, temperature, moisture, pH, available C (epidermal and non-mucus components of soil organic matter) and soil N content that affect soil carbon dioxide production and emissions (Desjardins et al. 2005; van Groenigen et al. 2013). Temperature has a significant effect on CO_2 evolution in soil; because as temperature increases the CO_2 emission increases, as possible global warming will increase CO_2 evolution in soil, accelerating soil carbon consumption and soil fertility

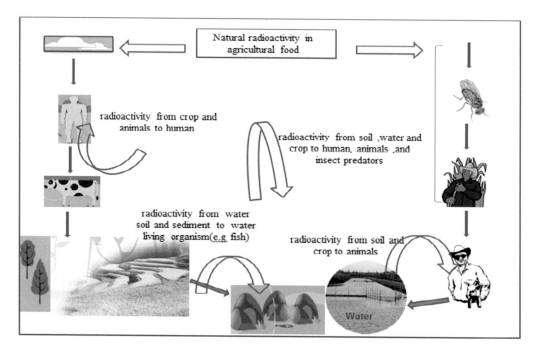

FIGURE 13.3 Changes in the earth's radiation balance, which cause temperatures to rise or fall within a decade, are known as climate forcing. Positive radiation forcing means that the earth receives more energy from the sun than it radiates into space. The net gain of this energy will lead to warming (Desjardins et al. 2005; Lal 2003). On the contrary, negative radiation forcing means that the earth loses more energy in space than it receives from sun. Systems in thermal equilibrium have zero radiation forcing.

(Potma Goncalves et al. 2018; Wang et al. 2019). The changes in soil organic content and land use management have important effects on the respiratory rate of soil (Figure 13.1). Global climate change will accelerate the instability of SOC in several ways. Land-use changes involve the transformation to agriculture and can release the SOC that is actually conserved by farming (Figure 13.3).

Plastic waste emits carbon dioxide when it degrades. In 2018, the study claimed that some of the most common plastics in the environment release greenhouse gases methane and ethylene when exposed to sunlight, and that the extent of their impact can affect the earth's climate (Banger et al. 2012). In 2019, a new report entitled 'Plastics and Climate' will be published according to which plastics will contribute the equivalent of 850 million tons of carbon dioxide (CO_2) to the atmosphere in 2019. At current trends, annual emissions will increase to 1.34 billion tons by 2030, while by 2050, plastics could emit 56 billion tons of greenhouse gases, equivalent to 14% of the planet's remaining carbon budget (Banger et al. 2012). The report says that only reduced consumption solutions can solve the problem, while other solutions such as biodegradable plastics, marine cleansing, and the use of renewable energy in the plastics industry can do harm, and in some cases, may even worsen the problem of GHG. Some technologies eliminate greenhouse gas emissions from the atmosphere (Desjardins et al. 2005), and the most extensive analyses tell us that these technologies concern geological formations that remove CO_2 from the atmosphere, such as bioenergy, carbon capture and storage, and carbon dioxide air capture, or soil, such as biochar (Banger et al. 2012). The IPCC says many long-term climate scenario models require large-scale man-made negative emissions to avoid serious climate change.

13.5 Conclusions and Future Perspectives

Over practice of agricultural activities, such as overuse of chemical fertilizers, deep ploughing, and over irrigation significantly increased greenhouse gas emissions and polluted soil and air environment, which challenges were faced and fixed by adopting different forward-looking technologies like, straw mulch with less N fertilizer, coating of N fertilizer, Water-N management, and reduced tillage practices. Researchers (agronomist, ecologist, and soil scientist) need to pay more attention to the above discussed issues while putting in use the analysis discussed here to form future useful perspectives to improve soil and air environment for better crop production.

Acknowledgements

M.N. and G.S.N. conceived the concept and provided outlines. M.N. drafted the manuscript. M.N. made the graphic models. K.A. and A.K. revised the manuscript and fixed language issues.

Conflicts of Interest

The authors declare no conflict of interest.

REFERENCES

Adviento-Borbe MAA, Kaye JP, Bruns MA, McDaniel MD, McCoy M, Harkcom S (2010) Soil greenhouse gas and ammonia emissions in long-term maize-based cropping systems. *Soil Sci Soc Am J* 74: 1623–1634. doi:10.2136/sssaj2009.0446

Afzal F, Adnan M, Rehman IU, Noor M, Khan A, Shah JA, Khan MA, Roman M, Wahid F, Nawaz S, Perveez R (2017) Growth response of olive cultivars to air layering. *Pure Appl Biol* 6(4): 1403–1409.

Aguilera E, Lassaletta L, Gattinger A, Gimeno BS (2013) Managing soil carbon for climate change mitigation and adaptation in Mediterranean cropping systems: A meta-analysis. *Agric Ecosyst Environ* 168:25–36. doi:10.1016/j.agee.2013.02.003

Akhtar K, Wang W, Khan A, Ren G, Afridi MZ, Feng Y, Yang G (2019a) Wheat straw mulching offset soil moisture deficient for improving physiological and growth performance of summer sown soybean. *Agric Water Manag* 211:16–25. doi:10.1016/j.agwat.2018.09.031

Akhtar K et al (2019b) Straw mulching with fertilizer nitrogen: An approach for improving crop yield, soil nutrients and enzyme activities. *Soil Use Manage* 35:526–535. doi:10.1111/sum.12478

Akhtar K, Wang W, Ren G, Khan A, Feng Y, Yang G (2018) Changes in soil enzymes, soil properties, and maize crop productivity under wheat straw mulching in Guanzhong, China. *Soil Tillage Res* 182: 94–102. doi:10.1016/j.still.2018.05.007

Akhtar K, Wang W, Ren G, Khan A, Feng Y, Yang G, Wang H (2019c) Integrated use of straw mulch with nitrogen fertilizer improves soil functionality and soybean production. *Environ Int* 132. doi:10.1016/j.envint.2019.105092

Bahn M et al (2010) Soil respiration at mean annual temperature predicts annual total across vegetation types and biomes. *Biogeosciences* 7:2147–2157. doi:10.5194/bg-7-2147-2010

Banger K, Tian H, Lu C (2012) Do nitrogen fertilizers stimulate or inhibit methane emissions from rice fields? *Glob Change Biol* 18:3259–3267.

Bouwman AF, Boumans LJM, Batjes NH (2002) Modeling global annual N_2O and NO emissions from fertilized fields. *Glob Biogeochem Cycles* 16. doi:10.1029/2001gb001812

Buser HR, Muller MD, Poiger T, Balmer ME (2002) Environmental behavior of the chiral acetamide pesticide metalaxyl: Enantioselective degradation and chiral stability in soil. *Environ Sci Technol* 36:221–226. doi:10.1021/es010134s

Cai Z, Laughlin RJ, Stevens RJ (2001) Nitrous oxide and dinitrogen emissions from soil under different water regimes and straw amendment. *Chemosphere* 42:113–121. doi:10.1016/s0045-6535(00)00116-8

Cakir FY, Stenstrom MK (2005) Greenhouse gas production: A comparison between aerobic and anaerobic wastewater treatment technology. *Water Res* 39:4197–4203.

Canton Y, Sole-Benet A, de Vente J, Boix-Fayos C, Calvo-Cases A, Asensio C, Puigdefabregas J (2011) A review of runoff generation and soil erosion across scales in semiarid south-eastern Spain. *J Arid Environ* 75:1254–1261. doi:10.1016/j.jaridenv.2011.03.004

Cayuela ML, Oenema O, Kuikman PJ, Bakker RR, Groenigen JWV (2010) Bioenergy by-products as soil amendments? Implications for carbon sequestration and greenhouse gas emissions. *Glob Change Biol Bioenergy* 2:201–213.

Chien SH, Prochnow LI, Cantarella H (2009) Recent developments of fertilizer production and use to improve nutrient efficiency and minimize environmental impacts. In: Sparks DL (ed) *Advances in agronomy*, vol 102. pp 267–322. doi:10.1016/s0065-2113(09)01008-6

Davidson EA, Nepstad DC, Ishida FY, Brando PM (2008) Effects of an experimental drought and recovery on soil emissions of carbon dioxide, methane, nitrous oxide, and nitric oxide in a moist tropical forest. *Glob Change Biol* 14.

Desjardins RL, Smith W, Grant B, Campbell C, Riznek R (2005) Management strategies to sequester carbon in agricultural soils and to mitigate greenhouse gas emissions. *Clim Change* 70:283–297.

Ding W, Luo J, Li J, Yu H, Fan J, Liu D (2013) Effect of long-term compost and inorganic fertilizer application on background N_2O and fertilizer-induced N_2O emissions from an intensively cultivated soil. *Sci Total Environ* 465:115–124. doi:10.1016/j.scitotenv.2012.11.020

Erşahin S, Kapur S, Aydın G, Akça E, Bilgili BC (2017) *Terrestrial ecosystem carbon dynamics as influenced by land use and climate*. Springer International Publishing.

Farquharson R, Baldock J (2008) Concepts in modelling N_2O emissions from land use. *Plant Soil* 309:147–167. doi:10.1007/s11104-007-9485-0

Farquharson R, Baldock J (2009) Concepts in modelling N_2O emissions from land use. *Plant Soil* 325:353–353. doi:10.1007/s11104-009-0155-2

Food and Agriculture Organization of the United Nations (2012) Sustainable diets and biodiversity. Directions and solutions for policy, research and action. In: *Sustainable diets and biodiversity. Directions and solutions for policy, research and action.*

Foster S, Garduno H, Evans R, Olson D, Tian Y, Zhang WZ, Han ZS (2004) Quaternary aquifer of the North China Plain – assessing and achieving groundwater resource sustainability. *Hydrogeol J* 12: 81–93. doi:10.1007/s10040-003-0300-6

Galford GL, Melillo JM, Kicklighter DW, Cronin TW, Cerri CEP, Mustard JF, Cerri CC (2010) Greenhouse gas emissions from alternative futures of deforestation and agricultural management in the southern Amazon. *Proc Natl Acad Sci USA* 107:19649–19654. doi:10.1073/pnas.1000780107

Goglio P, Colnenne-David C, Laville P, Doré T, Gabrielle B (2013) 29% N_2O emission reduction from a modelled low-greenhouse gas cropping system during 2009–2011. *Environ Chem Lett* 11:143–149.

Groffman PM et al (2009) Challenges to incorporating spatially and temporally explicit phenomena (hotspots and hot moments) in denitrification models. *Biogeochemistry* 93:49–77. doi:10.1007/s10533-008-9277-5

Hartmann M et al (2014) Resistance and resilience of the forest soil microbiome to logging-associated compaction. *ISME J* 8:226–244. doi:10.1038/ismej.2013.141

Hermle S, Anken T, Leifeld J, Weisskopf P (2008) The effect of the tillage system on soil organic carbon content under moist, cold-temperate conditions. *Soil Tillage Res* 98:94–105. doi:10.1016/j.still.2007.10.010

Ilyas H, Alam SS, Khan I, Shah B, Naeem A, Khan N, Ullah W, Adnan M, Shah SRA, Junaid K, Ahmed N, Iqbal M (2016) Medicinal plants rhizosphere exploration for the presence of potential biocontrol fungi. *J Entomol Zool Stud* 4(3):108–113.

IPCC (Intergovernmental Panel on Climate Change) (2001) Climate change 2001-the scientific basis

IPCC (Intergovernmental Panel on Climate Change) (2007) Climate change 2007-the scientific basis

IPCC (Intergovernmental Panel on Climate Change) (2013) Climate change 2013-the scientific basis

Jones SK, Rees RM, Kosmas D, Ball BC, Skiba UM (2006) Carbon sequestration in a temperate grassland; management and climatic controls. *Soil Use Manage* 22:132–142.

Khan A et al (2017) Nitrogen nutrition in cotton and control strategies for greenhouse gas emissions: A review. *Environ Sci Pollut Res* 24:23471–23487. doi:10.1007/s11356-017-0131-y

Khan IA, Shah B, Khan A, Zaman M, Din MMU, Shah SRA, Junaid K, Adnan M, Ahmad N, Akbar R, Fayaz W, Rahman IU (2016) Screening of different maize cultivars against maize shootfly and red pumpkin beetle at Peshawar. *J Entomol Zool Stud* 4(1):324–327.

Kong D et al (2019) Soil respiration from fields under three crop rotation treatments and three straw retention treatments. *PloS One* 14:e0219253–e0219253. doi:10.1371/journal.pone.0219253

Lal R (2003) Global potential of soil carbon sequestration to mitigate the greenhouse effect. *Crit Rev Plant Sci* 22:151–184.

Lin Y, Munroe P, Joseph S, Henderson R, Ziolkowski A (2012) Water extractable organic carbon in untreated and chemical treated biochars. *Chemosphere* 87:151–157. doi:10.1016/j.chemosphere.2011.12.007

MacKenzie AF, Fan MX, Cadrin F (1997) Nitrous oxide emission as affected by tillage, corn-soybean-alfalfa rotations and nitrogen fertilization. *Can J Soil Sci* 77:145–152. doi:10.4141/s96-104

Monreal CM (2010) Greenhouse gas (CO_2 and N_2O) emissions from soils: A review. *Chilean Journal of Agricultural Research* 70:485–497.

Moreno MM, Cirujeda A, Aibar J, Moreno C (2016) Soil thermal and productive responses of biodegradable mulch materials in a processing tomato (*Lycopersicon esculentum* Mill.) crop. *Soil Res* 54:207–215. doi:10.1071/sr15065

Mosier A, Kroeze C (2000) Potential impact on the global atmospheric N_2O budget of the increased nitrogen input required to meet future global food demands. *Chemosphere Glob Change Sci* 2:465–473.

Niu K, Choler P, de Bello F, Mirotchnick N, Du G, Sun S (2014) Fertilization decreases species diversity but increases functional diversity: A three-year experiment in a Tibetan alpine meadow. *Agric Ecosyst Environ* 182:106–112. doi:10.1016/j.agee.2013.07.015

Nykanen H, Alm J, Lang K, Silvola J, Martikainen PJ (1995) Emissions of CH_4, N_2O and CO_2 from a virgin fen and a fen drained for grassland in Finland. *J Biogeogr* 22:351–357. doi:10.2307/2845930

Pan WL, Port LE, Xiao Y, Bary AI, Cogger CG (2017) Soil carbon and nitrogen fraction accumulation with long-term biosolids applications. *Soil Sci Soc Am J* 81:1381–1388. doi:10.2136/sssaj2017.03.0075

Plante AF, Conant RT, Stewart CE, Paustian K, Six J. Impact of soil texture on the distribution of soil organic matter in physical and chemical fractions. *Soil Sci Soc Am J* 70:287.

Playan E, Mateos L (2006) Modernization and optimization of irrigation systems to increase water productivity. *Agric Water Manag* 80:100–116. doi:10.1016/j.agwat.2005.07.007

Potma Goncalves DR et al (2018) Soil carbon inventory to quantify the impact of land use change to mitigate greenhouse gas emissions and ecosystem services. *Environ Pollut* 243:940–952. doi:10.1016/j.envpol.2018.07.068

Rajaniemi TK (2002) Why does fertilization reduce plant species diversity? Testing three competition-based hypotheses. *J Ecol* 90:316–324.

Ravishankara AR, Daniel JS, Portmann RW (2009) Nitrous oxide (N_2O): The dominant ozone-depleting substance emitted in the 21st century. *Science* 326:123–125. doi:10.1126/science.1176985

Ren C et al (2018) Differential soil microbial community responses to the linkage of soil organic carbon fractions with respiration across land-use changes. *Forest Ecol Manag* 409:170–178. doi:10.1016/j.foreco.2017.11.011

Rietz DN, Haynes RJ (2003) Effects of irrigation-induced salinity and sodicity on soil microbial activity. *Soil Biol Biochem* 35:845–854. doi:10.1016/s0038-0717(03)00125-1

Risk D, McArthur G, Nickerson N, Phillips C, Hart C, Egan J, Lavoie M (2013) Bulk and isotopic characterization of biogenic CO_2 sources and variability in the Weyburn injection area. *Int J Greenh Gas Control* 16:S263–S275. doi:10.1016/j.ijggc.2013.02.024

Royer DL, Berner RA, Beerling DJ (2001) Phanerozoic atmospheric CO_2 change: Evaluating geochemical and paleobiological approaches. *Earth Sci Rev* 54:349–392. doi:10.1016/s0012-8252(00)00042-8

Rusu T et al (2009) Implications of minimum tillage systems on sustainability of agricultural production and soil conservation. *J Food Agric Environ* 7:335–338.

Scanlon TM, Kiely G (2003) Ecosystem-scale measurements of nitrous oxide fluxes for an intensely grazed, fertilized grassland. *Geophys Res Lett* 30. doi:10.1029/2003gl017454

Sebilo M, Mayer B, Nicolardot B, Pinay G, Mariotti A (2013) Long-term fate of nitrate fertilizer in agricultural soils. *Proc Natl Acad Sci USA* 110:18185–18189.

Shah N, Tanzeela, Khan A, Khisroon M, Adnan M, Jawad SM (2017) Seroprevalence and risk factors of toxoplasmosis in pregnant women of district Swabi. *Pure Appl Biol* 6(4):1306–1313.

Sitthisuntikul K, Yossuck P, Limnirankul B (2018) How does organic agriculture contribute to food security of small land holders?: A case study in the North of Thailand. *Cogent Food Agric* 4. doi:10.1080/23311932.2018.1429698

St Luce M et al (2016) Preceding crops and nitrogen fertilization influence soil nitrogen cycling in no-till canola and wheat cropping systems. *Field Crops Res* 191:20–32. doi:10.1016/j.fcr.2016.02.014

van Groenigen KJ, van Kessel C, Hungate BA (2013) Increased greenhouse-gas intensity of rice production under future atmospheric conditions. *Nat Clim Change* 3:288–291. doi:10.1038/nclimate1712

Wallach R, Ben-Arie O, Graber ER (2005) Soil water repellency induced by long-term irrigation with treated sewage effluent. *J Environ Qual* 34:1910–1920. doi:10.2134/jeq2005.0073

Wang C-K et al (2013) Prediction of soil organic matter content under moist conditions using VIS-NIR diffuse reflectance spectroscopy. *Soil Sci* 178:189–193. doi:10.1097/SS.0b013e3182986735

Wang W, Akhtar K, Ren G, Yang G, Feng Y, Yuan L (2019) Impact of straw management on seasonal soil carbon dioxide emissions, soil water content, and temperature in a semi-arid region of China. *Sci Total Environ* 652:471–482. doi:10.1016/j.scitotenv.2018.10.207

Wang XB, Cai DX, Hoogmoed WB, Oenema O, Perdok UD (2006) Potential effect of conservation tillage on sustainable land use: A review of global long-term studies. *Pedosphere* 16:587–595. doi:10.1016/s1002-0160(06)60092-1

Wang Xg, Luo Y (2018) Crop residue incorporation and nitrogen fertilizer effects on greenhouse gas emissions from a subtropical rice system in southwest China. *J Mt Sci* 15:1972–1986. doi:10.1007/s11629-017-4810-4

Wright AL, Hons FM, Lemon RG, McFarland ML, Nichols RL (2008) Microbial activity and soil C sequestration for reduced and conventional tillage cotton. *Appl Soil Ecol* 38:168–173. doi:10.1016/j.apsoil.2007.10.006

Xin Z-J et al (2012) Effect of stubble heights and treatment duration time on the performance of water dropwort floating treatment wetlands (FTWS). *Ecol Chem Eng S* 19:315–330. doi:10.2478/v10216-011-0023-x

Yang XM, Drury CF, Reynolds WD, Tan CS, McKenney DJ (2003) Interactive effects of composts and liquid pig manure with added nitrate on soil carbon dioxide and nitrous oxide emissions from soil under aerobic and anaerobic conditions. *Can J Soil Sci* 83:343–352. doi:10.4141/s02-067

Yang Y, Zhang M, Zheng L, Cheng D, Liu M, Geng Y, Chen J (2013) Controlled-release urea for rice production and its environmental implications. *J Plant Nutr* 36:781–794. doi:10.1080/01904167.2012.756892

Zhang L et al (2015) Integrative effects of soil tillage and straw management on crop yields and greenhouse gas emissions in a rice–wheat cropping system. *Eur J Agron* 63:47–54.

Zheng J, Zhang X, Li L, Zhang P, Pan G (2007) Effect of long-term fertilization on C mineralization and production of CH_4 and CO_2 under anaerobic incubation from bulk samples and particle size fractions of a typical paddy soil. *Agric Ecosyst Environ* 120:129–138.

14

Plant-Microbial Interactions Confer Tolerance to Abiotic Stress in Plants

Mehtab Muhammad Aslam, Eyalira J. Okal, Muhammad Waseem, Bello H. Jakada, Witness J. Nyimbo, and Joseph K. Karanja

CONTENTS

14.1 Introduction

Abiotic stress is defined as the negative impact of non-living components on living organisms in a particular environmental setting (Ben-Ari and Lavi 2012). Abiotic stresses such as drought, excessive water, extreme temperatures, salinity, and mineral toxicity negatively impact growth, development, yield, and seed quality of crops and other plants (Chequer et al. 2013; Adnan et al. 2018a, 2018b, 2019; Akram et al. 2018a, 2018b; Aziz et al. 2017a, 2017b; Habib et al. 2017; Hafiz et al. 2016, 2019; Kamran et al. 2017; Muhammad et al. 2019; Sajjad et al. 2019; Saud et al. 2013, 2014, 2016, 2017, 2020; Shah et al. 2013; Qamar et al. 2017; Wajid et al. 2017; Yang et al. 2017; Zahida et al. 2017; Depeng 2018; Abd Allah 2010; Hussain et al. 2020; Hafiz 2020; Abd Allah 2010; Shafi 2020; Wahid et al. 2020; Subhan et al. 2020; Zafar-ul-Hye et al. 2020a, 2020b; Adnan et al. 2020; Ilyas et al. 2020; Saleem et al. 2020a, 2020b, 2020c; Rehman 2020; Farhat et al. 2020; Wu et al. 2020; Mubeen et al. 2020; Farhana et al. 2020; Jan et al. 2019; Wu et al. 2019; Ahmad et al. 2019; Baseer et al. 2019; Hafiz et al. 2018; Tariq et al. 2018). These stresses cause several physiological, biochemical, molecular traits of plants to be affected (Kumar and Verma 2018; Fahad and Bano 2012; Fahad et al. 2013, 2014a, 2014b, 2015a, 2015b, 2016a, 2016b, 2016c, 2016d, 2017, 2018, 2019a, 2019b; Hesham and Fahad 2020; Iqra et al. 2020; Akbar et al. 2020; Mahar et al. 2020; Noor et al. 2020; Bayram et al. 2020; Amanullah et al. 2020;

Rashid et al. 2020; Arif et al. 2020; Amir et al. 2020; Saman et al. 2020; Muhammad et al. 2020; Md Jakirand and Allah 2020; Farah et al. 2020; Sadam et al. 2020; Unsar et al. 2020; Fazli et al. 2020; Md Enamul et al. 2020; Gopakumar et al. 2020; Zia-ur-Rehman 2020; Ayman et al. 2020; Mohammad et al. 2020a, 2020b; Senol 2020; Amjad et al. 2020; Ibrar et al. 2020; Sajid et al. 2020). For example, salinity stress causes impaired plant growth and development, crop failure, and cytotoxicity due to excessive uptake of Na^+ and Cl^- ions and causing nutritional imbalance (Rolly et al. 2020). Given abiotic stress can cause decrease in the leaf area and diminutive plant height, with water flow disturbance in higher plants from xylem to the neighbouring elongating cells suppressing cell elongation (Torabi et al. 2016), still the plant's test of endurance to externalities depends on their aptitude to respond to these stresses (Hartman and Tringe 2019). Even though plants developed their own adaptation mechanisms to lessen most of the stresses innate to nature, yet they depend on their ability to associate with microbes to survive and protect themselves against stresses (Hassani et al. 2018). Plant-microbe interaction refers to the communication between plants and microbes which leads to either beneficial or negative impacts to one or both symbiotic partners (Sander et al. 2015), and which can be either profitable, neutral, or hostile directly influencing the plant growth, its health, and development (Imam et al. 2016). It is reported that interactions can influence below-ground-above-ground plants biomass development, thereby, playing a significant role in sustaining plants (Igiehon and Babalola 2018).

Plant-microbe interactions at biochemical, physiological, and molecular levels largely direct plant responses toward combatting abiotic stresses (Meena et al. 2017), and happens when stresses like drought, heavy metal contamination, and soil salinity can occur during which important minerals, such as potassium, phosphorus, and nitrogen present in the soil but unavailable for plant uptake, are made available to plants by microbes using different mechanisms (Chrouqi et al. 2017). Mechanisms used by microbes include the production of cytokinin, indole acetic acid, ACC deaminase, Abscisic acid, trehalose, volatile organic compounds, and exopolysaccharides (Forni et al. 2017). Microbes like plant growth-promoting (PGP) play a role in direct and indirect antagonism of plant pathogens through antibiotic production, fixation, and solubilization of growth-limiting nutrients, activation of plant immune system responses, and through production of plant growth hormones (Hartman and Tringe 2019). The plant-microbial interaction to alleviate stress has recently received considerable attention from researchers all over the world (Arora et al. 2017), given the use of these microorganisms in stressed soils can alleviate the effect of stresses on crop plants, hence creating potential and promising approach in sustainable agriculture (Paul and Lade 2014).

Plant-microbe interactions towards fighting abiotic stress is an important aspect of sustainable agriculture and can help understand the benefits and the pathogenic effects of microbes on crop development (Rodriguez et al. 2019). Recent studies have shown that several plant species require microbial associations for stress tolerance and survival (De Zelicourt et al. 2013), while there is plenty of research on plant-microbial interactions that have focused on three categories of plant-microbe interactions: the early symbiosis between plants and arbuscular mycorrhizae, nitrogen fixation by rhizobia within the roots nodules of legume plant (Farrar et al. 2014). As the climate changes, it is more important to understand how abiotic stresses can alter plant-microbe interaction at molecular and cellular level (Hawkins and Crawford 2018), so as to make best use of the potential in plant-microbe interactions to provide sustainable solutions in raising agricultural crop production and ecosystem management (Lareen et al. 2016).

14.2 How Plants Sense Microbial Signals?

There exists a lot of interest in comprehending the phenomenon of interactions between plants and their associated microorganisms because such knowledge is important in research and development of various agricultural applications. There is a plethora of communication happening between plants and rhizobium microbes during various stages of plant development, whereby signal molecules from both partners exhibit important roles executed (Loh et al. 2002). While, on one hand, plants produce organic compounds such as organic acids, sugars, and vitamins which can be utilized by microorganisms as nutrients or signals, on the other hand, plant microbes secrete phytohormones and volatile compounds which act either directly or indirectly to trigger diverse growth responses in plants (Tyagi et al. 2018). For instance, rhizospheric microorganisms are known to produce substances that control plant growth and development. Or, bacterial signals such as N-acyl-L-homoserine lactone (AHLs) can be detected by plant cells to modulate gene expression, metabolism, and plant growth. Studies have shown AHLs ability to be recognized by plants, change expression of genes in roots and shoots, and regulate cell growth and defence responses in plants (Ortíz-Castro et al. 2009). Again, various plant growth promoting rhizobacteria (PGPR) have been indicated to produce acetoin and 2,3-butanediol which enhance communication between plant-bacteria by triggering plant growth (Yi et al. 2016).

Majority of important molecules produced by plant microbes can be detected by plant cells and will induce an array of biochemical responses that may be either beneficial or harmful to the plant. Since, just like animals, plants are able to recognize microbe-associated molecules and respond by launching appropriate immune responses that can protect the plant, understanding the molecular mechanisms which underlie the plant-microbe interactions and the signalling crosstalk that determine plant responses are important in developing disease-resistant crops. Besides phytohormone recognition receptors, the plants sense infectious microbes by detecting conserved signatures which are referred to as microbe-associated molecular patterns (MAMPs), where MAMPs have been revealed in studies to being detected via pattern recognition receptors (PRR), and mounts pattern triggered immunity (PTI), which are important plant immune sensory complexes (Zhang and Zhou 2010). In addition, plants have evolved to possess resistance (R) proteins which are cytoplasmic immune receptors that often detect the activities of effector proteins in the plant cell. Zhang and Zhou reported various R-proteins associated to receptor like cytoplasmic kinases (RLCKs) including BIK1 and PBLs as components of PTI signalling pathway that allow *Arabidopsis thaliana* to detect *Pseudomonas syringae* effector proteins AvrPphB and trigger effector-triggered immunity (ETI) (Zhang and Zhou 2010).

Studies have shown that plant bacteria and fungi are able to produce auxins and cytokinins which are rapidly recognized by plant cells and thereby induce proliferation of cells in the shoot, sometimes causing tumorous growths similar to *Agrobacterium tumefaciens* or *Ustilago maydis* infection (Boivin et al. 2016). It is a fact that the phytohormones can also induce plants to overproduce lateral roots and root hairs resulting in the increase of nutrient and water intakes, where studies have revealed the respective plant hormone receptors in plants to have detected phytohormones produced by the Rhizobacteria which stimulates plant responses that may confer beneficial effects (Ghosh et al. 2019). Henty-Ridilla et al. (2013) reported that actin cytoskeleton in plants provides a dynamic platform for sensing and responding to a wide range of both pathogenic and non-pathogenic microbes, where the actin cytoskeleton is an important signal target in plants, and it was shown to dramatically change in response to various abiotic and biotic stimuli (Henty-Ridilla et al. 2013; Kalachova et al. 2019; Qian and Xiang 2019). Future endeavours may need to provide more understanding on molecular mechanisms that underlie actin cytoskeleton recognition of various stimuli in the surroundings, including microbial signals, with the still unplumbed need for further research to evaluate how plants are able to differentiate signals from different microbes and generate appropriate responses.

14.3 Plant and Microbial Interactions

Interaction is a fundamental part of the living ecosystem (Meena et al. 2017), but plant performance might be inhibited from negative impacts of interactions, like resource competition and parasitism (Ulrich et al. 2019). The microbe and plant interactions involve all ecological facets of their existence on earth, like physiochemical changes, metabolite exchange, metabolite conversion, signalling, chemotaxis, and genetic exchange which lead to genotypic selection (Braga et al. 2016). Being non-locomotive living matter, plants have coevolved with microbes and obtain different mechanisms that amend the outcome of their association (Ho et al. 2017), with a single interaction often occurring in many different ways and in many different types (Zengler 2009). Interactions can be the one involving microbes inhabiting plant tissue (endophytic) and another in which microbes inhabit the surface of the plant (epiphytic), not to leave out the plant's nearby environment and rhizospheric soil (Schirawski and Perlin 2018), so that these animates have developed varieties of interactions, including mutualism where both species benefit from the interaction, or where positive effects are exerted to affect only one partner (commensalism) or a parasitic interaction where one partner suffers (Moënne-Loccoz et al. 2015). Understanding different types of interactions helps us to be aware of natural phenomena that affect plant life, and could lead to applications resulting in sustainable resources, less impact on the environment, clean-up of pollution, and influence on atmospheric gases on a global scale (Wu et al. 2009), while understanding that the interaction is likely to be strongest at the root surface where roots release substrates that are available to microbial colonizers, and in which microbial metabolites are accessible for plant uptake (Pascale et al. 2020). Plant roots release root exudates which leads to stimulation of the microbial density and activity in the rhizosphere.

One of the most studied plant interactions involving fungi occurs between plants and arbuscular mycorrhizal fungi. Given fungi are important decomposers and recyclers of organic materials, it is also true that they positively or negatively interact with plant roots in the rhizosphere or with above ground plant components (Odelade and Babalola 2019), though most of plant-fungal interactions promote plant growth and development in the plant, improving plant foraging, acquisition of soil resources, and stress tolerance. In turn, the plants deliver carbohydrates to the fungus (Zeilinger et al. 2016). Also, plants are reported to be inhabited by diverse bacterial communities which live in and on the plant's tissues, and it has been demonstrated that plants only recruit a selection of the bacteria present in their immediate surroundings like *Actinobacteria*, *Bacteroidetes*, *Firmicutes*, and *Proteobacteria* which are found mostly in roots (Santhanam et al. 2014).

Communication of plant and microbe is one of the important parts of the interaction (Schulz-Bohm et al. 2017), and is facilitated by chemical signalling within the rhizosphere resulting in priming of defence, or induced resistance in the plant host (Mhlongo et al. 2018). Evidence suggests that chemical exchanges between plants and microbes are the drivers of rhizosphere productivity, and there are specific signals which mediate these interactions (O'Banion et al. 2020), with plants and microorganisms communication occurring during different stages of plant growth where molecular signalling from the two associates plays an important role in enhancing interaction (Vaishnav et al. 2019). For example, if fungal and bacterial species are able to detect the plant host and initiate their colonization strategies in the rhizosphere by producing plant growth-regulating substances such as auxins or cytokinins, then, plants can recognize microbe-derived compounds and adjust their defence and growth responses according to the type of microorganism encountered (Ortíz-Castro et al. 2009).

14.4 Plant Traits Affected by Abiotic Stress

Plant survival, reproduction, growth, and distribution are greatly affected by abiotic stress. This type of stress is the main agent responsible for decrease in crop yield worldwide, given it affects the most important agronomic traits of plants: for example, in rice and wheat, a 27.5% and 25.4% decrease in yield was reported, respectively, due to abiotic stress. Other agronomic traits such as plant height, grain size, fruiting, 1000-grain weight (KGW) are equally affected too (Zhang et al. 2018).

Being sessile, plants endure abiotic stress by adjusting physiological processes and are said to employ some machinery to achieve this: by escape, avoidance, or tolerance. Escape is the successful reproduction before they encounter stress, while avoidance is more reliant on the delay in the initiation of dehydration in plant tissues; however, tolerance is the alteration in physiological and biochemical processes at the molecular level. Over the decades, researchers have reported studies conducted on these types of machinery known as traits (Guo et al. 2017). These traits affected by abiotic stress are majorly physiological traits, morphological traits, and agronomic traits (Figure 14.1).

14.4.1 Physiological Traits

Traits affected by change in physiological activities as a result of exposure to environmental stress are regarded as physiological traits, and involve processes like photosynthesis, respiration, plant nutrition, plant hormone functions, tropisms, photoperiodism, photomorphogenesis, circadian rhythms, seed germination, dormancy, stomata function and transpiration, and plant water relations (Table 14.1). Changes in physiological traits help the plant to maintain tissue water potential (Guo et al. 2017).

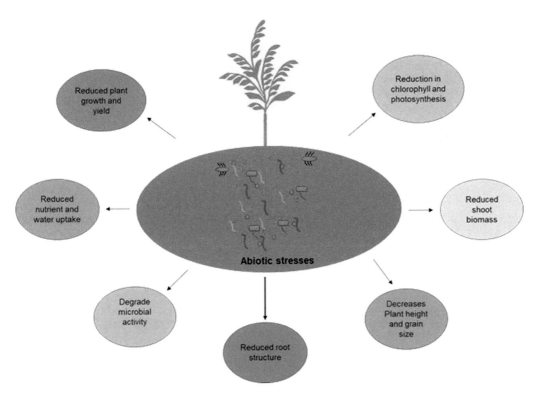

FIGURE 14.1 Physiological and morphological plant traits are greatly affected by abiotic stresses. Drought, salinity, and heat stress significantly alter plant metabolism due to diminished water and nutrient uptake, decreased microbial activity, and reduction in chlorophyll and photosynthetic activity leading to reduced plant growth and yield. Heavy metal contamination and water clogging on the other hand was reported to reduce root structure, decrease plant nutrient uptake, and alter root microbial activity which result in lower crop biomass and yield.

14.4.2 Canopy Temperature

Canopy temperature has been used for long as an indicator of water stress in plants (Guo et al. 2013; Zhang et al. 2019), and is widely used as a parameter to measure water stress or drought stress in plants, using the infrared thermometer (IRT) (Deery et al. 2016). There is a complex relationship between canopy temperature, air temperature, and rate of transpiration under stress conditions because plants with low canopy temperature under drought stress maintain better water status and take up more soil moisture. Being inter-related with other physiological traits such as stomatal conductance and leaf water potential, which in turn affect the rate of photosynthesis (Guo et al. 2013), high levels of stomatal conductance is known to decrease the canopy temperature which dramatically affects the rate of photosynthesis and plant yield (Deery et al. 2016). Contrary to this, canopy temperature was shown to negatively correlate with relative water content, biomass, and yield; and, therefore, the higher the stomatal conductance, the lower the canopy temperature (Casari et al. 2019).

14.4.3 Stay-Green (Plant Chlorophyll Retention)

Stay-green is a physiological trait that is affected by or related to heat stress (Pinto et al. 2016), and defined as the ability of plants to stay in the active state of photosynthesis or to keep their leaves green during stress conditions. This phenomenon is of two types, functional and non-functional stay-green trait: in functional stay-green, the photosynthetic capacity is maintained, and the onset of senescence is delayed; however, in non-functional or cosmetic stay-green, senescence occurs at regular rates with a decline in photosynthetic activity, but the leaf colour is maintained as a result of the failure of the

TABLE 14.1

Abiotic Stress has a Remarkable Impact on Physiological and Morphological Traits of Plants. In General, Common Abiotic Stresses such as Drought, High Temperatures, Salinity, Heavy Metal Contamination, Water Clogging, and Wind are Known to Reduce Plant's Chlorophyll Content, Water Retention Ability, Decreased Root System, Hampered Growth, and Production

Physiological Traits	Reference	Morphological Traits	Reference
Canopy temperature	(Casari et al. 2019; Deery et al. 2016; Guo et al. 2013; Zhang et al. 2019)	Root system and architecture	(Abd Allah et al. 2010; Kim et al. 2020; Pandey et al. 2017)
Stay-green (plant chlorophyll retention)	(de Souza Luche et al. 2015; Guo et al. 2016; Kassahun et al. 2010; Munné-Bosch and Alegre 2004; Pinto et al. 2016; Sklensky and Davies 2011; Thomas and Ougham 2014; Waters et al. 2009)	Cuticular wax	(Hameed et al. 2002; Xue et al. 2017)
Chlorophyll content	(Lopes and Reynolds 2012)	Vegetative and reproductive biomass	(Bustos et al. 2013; Nowicka et al. 2018; Shou et al. 2004; Zhang et al. 2018)
Relative water content	(Füzy et al. 2019; Zhang et al. 2018)	Plant height	(Ahmadikhah and Marufinia 2016; Zhang et al. 2018)
Soluble sugar content	(Zhang et al. 2018)	Leaf mass per area	(Zhang et al. 2018)
Leaf colour, wilting symptoms, relative water content of leaves, membrane stability index of leaves, and stomatal conductance	(Füzy et al. 2019)	Leaf thickness	(Zhang et al. 2018)
Leaf and bud temperature, leaf conductance, and stomatal and cuticular transpiration	(Guo et al. 2013)	Panicle length	(Zhang et al. 2018)

chlorophyll degradation pathway (Pinto et al. 2016; Thomas and Ougham 2014). The association between stay-green and biotic and abiotic stress tolerance, as well as improved yield production, is generally used as a strategy for plants to escape abiotic stress (de Souza Luche et al. 2015; Kassahun et al. 2010). However, this leads to a tradeoff between improved survival and significant yield loss (Guo et al. 2016; Munné-Bosch and Alegre 2004; Pinto et al. 2016). Thus, use of nutrient status in the leaf can provide useful information regarding acceleration or delay in leaf senescence. It is reported that in *Spinacia oleracea*, senescence is a result of the excessive nutrients reallocation from the leaves to flowers (Sklensky and Davies 2011). Additionally, a myriad of transcription factors was reported to facilitate senescence which plays the role with a flourish of the stay-green trait during stress conditions. For example, in *Triticum aestivum* L. NAC-type transcription factor (TF) was shown to delay senescence (Waters et al. 2009), while WRKY TFs also contribute significantly to the stay-green trait pathway via cytokinin and ethylene pathway (Thomas and Ougham 2014).

14.4.4 Chlorophyll Content

Chlorophyll content is an important trait related to abiotic stress said to be directly related to stay green trait and water soluble carbohydrates. It is also a measure of rate of photosynthesis under stress conditions. The authors reported that stay-green trait and chlorophyll content are important for proper light perception and utilization, and they also contribute to increased yield in wheat, with studies establishing the stay-green trait to be targeted for improvement of cultivars, which suggests that chlorophyll content also may be a potential trait for yield- and stress-resistant cultivars (Lopes and Reynolds 2012). This trait is associated with many physiological traits: for example, there is a positive relationship between rate of photosynthesis stomatal conductance, biomass production, and chlorophyll content. This association between physiological and biochemical traits suggests that increase in chlorophyll content affects photosynthetic efficiency, enhances stomatal conductance, and increases the rate of biomass accumulation.

14.5 Morphological Traits

These traits are visible in plants as morphological changes to biotic or abiotic stress. Environmental stress affects morphological traits of plants including plant height, fresh and dry shoot and root weights, general plant architecture (Figure 14.1) (Anglin et al. 2018).

14.5.1 Root System and Architecture

Root system and architecture is majorly affected by several biotic as well as abiotic factors, allowing the plant to sense and respond to the stress accordingly. The density and length of both primary and lateral roots play significant role in stress response, where studies revealed longer root length and increase in root diameter instilling drought stress tolerance in rice (Abd Allah et al. 2010; Kim et al. 2020; Pandey et al. 2017), because long root improves plant growth during drought stress conditions effectively by helping the plant to access moisture at deeper soil depths (Pandey et al. 2017). Similarly, maize with long roots and few lateral roots revealed improved plant water status, increased stomatal conductance and leaf photosynthesis, and increased overall growth when compared to plants with shorter root length and more lateral roots (Pandey et al. 2017).

14.5.2 Cuticular Wax

Cuticular wax is found on the surfaces of terrestrial plants distributed in the shoots, radicles, fruits, flowers, and leaves, and is made up of alkanes, alkenes, branched alkanes, primary and secondary alcohols, aldehydes ketones, wax esters, and unsaturated fatty alcohols, as well as cyclic compounds such as terpenoids, and metabolites such as flavonoids and sterols (Xue et al. 2017). Important for plant protection against pathogen attack and abiotic stress, its function under abiotic stress conditions has been well documented, mainly, the layer is hydrophobic in nature and concentration of its components has been indicated to increase during drought stress, so that drought conditions haul up the synthesis of the cuticular wax (Kosma et al. 2009). Thick cuticular wax layer is therefore associated with enhanced tolerance against abiotic stress including drought and heat. For instance, (Hameed et al. 2002) reported *Triticum aestivum* to develop a thicker cuticular layer during drought stress.

14.5.3 Vegetative and Reproductive Biomass

Basically, high vegetative and reproductive biomass leads to an increased crop yield, where a buildup in vegetative biomass generally correlates with increased photosynthetic activity stomatal conductance, leaf chlorophyll content of a plant (Bustos et al. 2013): for example, in maize, several studies reported drought stress to significantly affect the above ground plant biomass resulting in poor yields (Casari et al. 2019).

There are several studies on the molecular mechanisms that enhance biomass and increase abiotic stress resistance in plants: for example, overexpression of a tobacco MAPK kinase kinase (MAPKKK) in maize resulted in enhanced drought tolerance and enhanced rate of photosynthesis, which led to increase in biomass (Nowicka et al. 2018; Shou et al. 2004). Additionally, overexpression of pineapple CBL1 in *Arabidopsis* enhanced tolerance to salinity and osmotic stresses with increase in the biomass production (Aslam et al. 2019). Similarly, overexpression of LOS5/ABA3 gene which encodes a crucial enzyme of ABA biosynthesis leads to improved grain yield during abiotic stress (Xiao et al. 2009).

14.5.4 Plant Height

Plant height is severely affected by abiotic stress, with diminutive height and reduced chlorophyll content under drought stress in plants (Ahmadikhah and Marufinia 2016). Similarly, a study on *Vigna radiata* revealed a significant reduction in plant height under drought stress (Saima et al. 2018).

14.5.5 Grain Size

As one of the factors determining grain weight, grain size is specified by grain length, grain width, length-to-width ratio, and grain thickness. On the other hand, 1000-grain weight (KGW) is the most reliable trait in assessing grain weight.

14.6 Microbial Contribution towards Abiotic Stress Resilient Crop Production

Microorganisms inhabit a wide range of ecological conditions and they possess strong physiological as well as biochemical abilities to alleviate abiotic stress. Plant-microbe interaction is an important part of the animate ecosystem whereby the two mutually thrive in a symbiotic relationship (Meena et al. 2017). Many studies have outlined plant microbiome to exhibit integral functions that enable plants to alleviate the burden of environmental stresses, while the interaction between plants and rhizospheric microbes was reported as having transpired with plant and microbe eliciting different kinds of systemic responses that modulate biochemical and molecular pathways related to abiotic resilience in plants. Rhizobacteria were demonstrated to induce drought resilience in plants through secretion of phytohormones, useful proteins and enzymes, epoxypolysaccharides, and antioxidants which mediate plant responses (Desbrosses and Stougaard 2011; Rosier et al. 2018). Studies showed that *Piriformospora indica* has the ability to tolerate both salt and drought stress in barley and cabbage, respectively, by producing antioxidants (Ngwene et al. 2016).

On a general note, microbes induce stress resilience responses in plants which promote plant growth and development through nitrogen fixation and the production of nutrients, organic compounds, and hormones. Several works on abiotic stress have mentioned genera *Bacillus, Azospirillum, Pseudomonas, Azotobacter, Enterobacter, Klebsiella, Rhizobium, Bradyrhizobium, Cyanobacteria,* and *Trichorderma* as resilient microbes that can promote plant growth under drought, heavy metal contamination, and high saline conditions (Meena et al. 2017). Evidence from recent studies has indicated that apart from PGPRs helping plants to alleviate environmental stress, they also improve production of crops such as maize, barley, soybean, and rice (Suarez et al. 2015). Samaniego-Gámez et al. (2016) reported that *Bacillus* spp. has the ability to improve photosynthetic metabolism in pepper plants under drought stress (Samaniego-Gámez et al. 2016). Furthermore, *Pseudomonas* spp. and *Bacillus pumilus* were shown to promote root colonization and boost the growth of rice plants in saline soils by producing exopolysaccharides and antioxidants which protect cells against stressful conditions (Israr et al. 2016). In future, studies should provide more information on molecular mechanisms in plant-microbe interactions in improving abiotic stress tolerance in plants.

REFERENCES

Abd Allah A, Badawy SA, Zayed B, El-Gohary AA (2010) The role of root system traits in the drought tolerance of rice (*Oryza sativa* L.). *Int J Plant Prod* 1:621–631.

Adnan M, Fahad S, Khan IA, Saeed M, Ihsan MZ, Saud S, Riaz M, Wang D, Wu C (2019) Integration of poultry manure and phosphate solubilizing bacteria improved availability of Ca bound P in calcareous soils. *3 Biotech* 9(10):368.

Adnan M, Fahad S, Muhammad Z, Shahen S, Ishaq AM, Subhan D, Zafar-ul-Hye M, Martin LB, Raja MMN,

Beena S, Saud S, Imran A, Zhen Y, Martin B, Jiri H, Rahul D (2020) Coupling phosphate-solubilizing bacteria with phosphorus supplements improve maize phosphorus acquisition and growth under lime induced salinity stress. *Plants* 9(900). doi:10.3390/plants9070900

Adnan M, Shah Z, Sharif M, Rahman H (2018b) Liming induces carbon dioxide (CO_2) emission in PSB inoculated alkaline soil supplemented with different phosphorus sources. *Environ Sci Poll Res* 25(10): 9501–9509.

Adnan M, Zahir S, Fahad S, Arif M, Mukhtar A, Imtiaz AK, Ishaq AM, Abdul B, Hidayat U, Muhammad A, Inayat-Ur R, Saud S, Muhammad ZI, Yousaf J, Amanullah, Hafiz MH, Wajid N (2018a) Phosphate-solubilizing bacteria nullify the antagonistic effect of soil calcification on bioavailability of phosphorus in alkaline soils. *Sci Rep* 8:4339. doi:10.1038/s41598-018-22653-7

Ahmad S, Kamran M, Ding R, Meng X, Wang H, Ahmad I, Fahad S, Han Q (2019) Exogenous melatonin confers drought stress by promoting plant growth, photosynthetic capacity and antioxidant defense system of maize seedlings. *PeerJ* 7:e7793. doi:10.7717/peerj.7793

Ahmadikhah A, Marufinia A (2016) Effect of reduced plant height on drought tolerance in rice. *J Biotech* 6:221.

Akbar H, Timothy JK, Jagadish T, Golam M, Apurbo KC, Muhammad F, Rajan B, Fahad S, Hasanuzzaman M (2020) Agricultural land degradation: processes and problems undermining future food security. In: Fahad S, Hasanuzzaman M, Alam M, Ullah H, Saeed M, Khan AK, Adnan M (eds) *Environment, climate, plant and vegetation growth*. Springer Publ Ltd, Springer Nature Switzerland AG. Part of Springer Nature, pp 17–62. doi:10.1007/978-3-030-49732-3

Akram R, Turan V, Hammad HM, Ahmad S, Hussain S, Hasnain A, Maqbool MM, Rehmani MIA, Rasool A, Masood N, Mahmood F, Mubeen M, Sultana SR, Fahad S, Amanet K, Saleem M, Abbas Y, Akhtar HM, Waseem F, Murtaza R, Amin A, Zahoor SA, ul Din MS, Nasim W (2018a) Fate of organic and inorganic pollutants in paddy soils. In: Hashmi MZ, Varma A (eds) *Environmental pollution of paddy soils, soil biology*. Springer International Publishing Ag, Switzerland, pp 197–214.

Akram R, Turan V, Wahid A, Ijaz M, Shahid MA, Kaleem S, Hafeez A, Maqbool MM, Chaudhary HJ, Munis, MFH, Mubeen M, Sadiq N, Murtaza R, Kazmi DH, Ali S, Khan N, Sultana SR, Fahad S, Amin A, Nasim W (2018b) Paddy land pollutants and their role in climate change. In: Hashmi MZ, Varma A (eds) *Environmental pollution of paddy soils, soil biology*. Springer International Publishing Ag, Switzerland, pp 113–124.

Amanullah, Khalid S, Imran, Hamdan AK, Muhammad A, Abdel RA, Muhammad A, Fahad S, Azizullah S, Brajendra P (2020) Effects of climate change on irrigation water quality. In: Fahad S, Hasanuzzaman M, Alam M, Ullah H, Saeed M, Khan AK, Adnan M (eds) *Environment, climate, plant and vegetation growth*. Springer Publ Ltd, Springer Nature Switzerland AG. Part of Springer Nature, pp 123–132. doi:10.1007/978-3-030-49732-3

Amir M, Muhammad A, Allah B, Sevgi Ç, Haroon ZK, Muhammad A, Emre A (2020) Biofortification under climate change: the fight between quality and quantity. In: Fahad S, Hasanuzzaman M, Alam M, Ullah H, Saeed M, Khan AK, Adnan M (eds) *Environment, climate, plant and vegetation growth*. Springer Publ Ltd, Springer Nature Switzerland AG. Part of Springer Nature, pp 173–228. doi:10.1007/978-3-030-49732-3

Amjad I, Muhammad H, Farooq S, Anwar H (2020) Role of plant bioactives in sustainable agriculture. In: Fahad S, Hasanuzzaman M, Alam M, Ullah H, Saeed M, Khan AK, Adnan M (eds) *Environment, climate, plant and vegetation growth*. Springer Publ Ltd, Springer Nature Switzerland AG. Part of Springer Nature, pp 591–606. doi:10.1007/978-3-030-49732-3

Anglin NL, Amri A, Kehel Z, Ellis D (2018) A case of need: linking traits to Genebank accessions. *J Biopreserv Biobank* 16:337–349.

Arif M, Talha J, Muhammad R, Fahad S, Muhammad A, Amanullah, Kawsar A, Ishaq AM, Bushra K, Fahd R (2020) Biochar; a remedy for climate change. In: Fahad S, Hasanuzzaman M, Alam M, Ullah H, Saeed M, Khan AK, Adnan M (eds) *Environment, climate, plant and vegetation growth*. Springer Publ Ltd, Springer Nature Switzerland AG. Part of Springer Nature, pp 151–172. doi:10.1007/978-3-030-4 9732-3

Arora S, Singh AK, Singh YP (2017) *Bioremediation of salt affected soils: an Indian perspective*. Springer.

Aziz K, Daniel KYT, Fazal M, Muhammad ZA, Farooq S, Wei F, Fahad S, Ruiyang Z (2017a) Nitrogen nutrition in cotton and control strategies for greenhouse gas emissions: a review. *Environ Sci Pollut Res* 24:23471–23487. doi:10.1007/s11356-017-0131-y

Aziz K, Daniel KYT, Muhammad ZA, Honghai L, Shahbaz AT, Mir A, Fahad S (2017b) Nitrogen fertility and abiotic stresses management in cotton crop: a review. *Environ Sci Pollut Res* 24:14551–14566. doi:10.1007/s11356-017-8920-x

Baseer M, Adnan M, Fazal M, Fahad S, Muhammad S, Fazli W, Muhammad A, Jr. Amanullah, Depeng W, Saud S, Muhammad N, Muhammad Z, Fazli S, Beena S, Mian AR, Ishaq AM (2019) Substituting urea by organic wastes for improving maize yield in alkaline soil. *J Plant Nutrition.* doi:10.1080/01904167.2 019.1659344

Bayram AY, Seher Ö, Nazlican A (2020) Climate change forecasting and modeling for the year of 2050. In: Fahad S, Hasanuzzaman M, Alam M, Ullah H, Saeed M, Khan AK, Adnan M (eds) *Environment, climate, plant and vegetation growth.* Springer Publ Ltd, Springer Nature Switzerland AG. Part of Springer Nature, pp 109–122. doi:10.1007/978-3-030-49732-3

Ben-Ari G, Lavi U (2012) Marker-assisted selection in plant breeding. In: *Plant biotechnology and agriculture.* Elsevier, pp 163–184.

Boivin S, Fonouni-Farde C, Frugier F (2016) How auxin and cytokinin phytohormones modulate root microbe interactions. *Front Plant Sci* 7:1240.

Braga RM, Dourado MN, Araújo WL (2016) Microbial interactions: ecology in a molecular perspective. *Braz J Microbiol* 47:86–98.

Bustos DV, Hasan AK, Reynolds MP, Calderini DF (2013) Combining high grain number and weight through a DH-population to improve grain yield potential of wheat in high-yielding environments. *Field Crops Res* 145:106–115.

Casari RA et al (2019) Using thermography to confirm genotypic variation for drought response in maize. *Int J Mol Sci* 20:2273.

Chequer FD, de Oliveira GAR, Ferraz EA, Cardoso JC, Zanoni MB, de Oliveira DP (2013) Textile dyes: dyeing process and environmental impact. In: Gunay M (ed) *Eco-friendly textile dyeing and finishing.* InTech, Chapter 6, pp 151–176.

Chrouqi L, Ouahmane L, Jadrane I, Koussa T, Al Feddy MN, Sciences C (2017) Effects of plant growth promoting rhizobacteria (PGPRs) product IAA on the growth of two Moroccan wheat varieties (*Triticum durum* Desf.) *Res J Pharm Biol Chem Sci Res* 8:2296–2302.

Deery DM et al (2016) Methodology for high-throughput field phenotyping of canopy temperature using airborne thermography. *Front Plant Sci* 7:1808.

Depeng W, Fahad S, Saud S, Muhammad K, Aziz K, Mohammad NK, Hafiz MH, Wajid N (2018) Morphological acclimation to agronomic manipulation in leaf dispersion and orientation to promote "Ideotype" breeding: evidence from 3D visual modeling of "super" rice (*Oryza sativa* L.). *Plant Physiol Biochem* 135:499–510. doi:10.1016/j.plaphy.2018.11.010

Desbrosses GJ, Stougaard J (2011) Root nodulation: a paradigm for how plant-microbe symbiosis influences host developmental pathways. *J Cell Host Microbe* 10:348–358

de Souza Luche H et al (2015) Stay-green effects on adaptability and stability in wheat *Afr J Agric Res* 10:1142–1149.

De Zelicourt A, Al-Yousif M, Hirt H (2013) Rhizosphere microbes as essential partners for plant stress tolerance. *Mol Plant* 6:242–245.

Fahad S, Adnan M, Hassan S, Saud S, Hussain S, Wu C, Wang D, Hakeem KR, Alharby HF, Turan V, Khan MA, Huang J (2019a) Rice responses and tolerance to high temperature. In: Hasanuzzaman M, Fujita M, Nahar K, Biswas JK (eds) *Advances in rice research for abiotic stress tolerance.* Woodhead Publ Ltd, UK, pp 201–224.

Fahad S, Bajwa AA, Nazir U, Anjum SA, Farooq A, Zohaib A, Sadia S, Nasim W, Adkins S, Saud S, Ihsan MZ, Alharby H, Wu C, Wang D, Huang J (2017) Crop production under drought and heat stress: plant responses and management options. *Front Plant Sci* 8:1147. doi:10.3389/fpls.2017.01147

Fahad S, Bano A (2012) Effect of salicylic acid on physiological and biochemical characterization of maize grown in saline area. *Pak J Bot* 44:1433–1438.

Fahad S, Chen Y, Saud S, Wang K, Xiong D, Chen C, Wu C, Shah F, Nie L, Huang J (2013) Ultraviolet radiation effect on photosynthetic pigments, biochemical attributes, antioxidant enzyme activity and hormonal contents of wheat. *J Food Agric Environ* 11(3&4):1635–1641.

Fahad S, Hussain S, Bano A, Saud S, Hassan S, Shan D, Khan FA, Khan F, Chen Y, Wu C, Tabassum MA, Chun MX, Afzal M, Jan A, Jan MT, Huang J (2014a) Potential role of phytohormones and plant growth-promoting rhizobacteria in abiotic stresses: consequences for changing environment. *Environ Sci Pollut Res* 22(7):4907– 4921. doi:10.1007/s11356-014-3754-2

Fahad S, Hussain S, Matloob A, Khan FA, Khaliq A, Saud S, Hassan S, Shan D, Khan F, Ullah N, Faiq M, Khan MR, Tareen AK, Khan A, Ullah A, Ullah N, Huang J (2014b) Phytohormones and plant responses to salinity stress: a review. *Plant Growth Regul* 75(2):391–404. doi:10.1007/s10725-014-0013-y

Fahad S, Hussain S, Saud S, Hassan S, Chauhan BS, Khan F et al (2016a) Responses of rapid viscoanalyzer profile and other rice grain qualities to exogenously applied plant growth regulators under high day and high night temperatures. *PLoS One* 11(7):e0159590. doi:10.1371/journal.pone.0159590

Fahad S, Hussain S, Saud S, Khan F, Hassan S, Amanullah, Nasim W, Arif M, Wang F, Huang J (2016b) Exogenously applied plant growth regulators affect heat-stressed rice pollens. *J Agron Crop Sci* 202:139–150.

Fahad S, Hussain S, Saud S, Hassan S, Ihsan Z, Shah AN, Wu C, Yousaf M, Nasim W, Alharby H, Alghabari F, Huang J (2016c) Exogenously applied plant growth regulators enhance the morpho-physiological growth and yield of rice under high temperature. *Front Plant Sci* 7:1250. doi:10.3389/fpls.2016.01250

Fahad S, Hussain S, Saud S, Hassan S, Tanveer M, Ihsan MZ, Shah AN, Ullah A, Nasrullah KF, Ullah S, Alharby H, Nasim W, Wu C, Huang J (2016d) A combined application of biochar and phosphorus alleviates heat-induced adversities on physiological, agronomical and quality attributes of rice. *Plant Physiol Biochem* 103:191–198

Fahad S, Hussain S, Saud S, Tanveer M, Bajwa AA, Hassan S, Shah AN, Ullah A, Wu C, Khan FA, Shah F, Ullah S, Chen Y, Huang J (2015a) A biochar application protects rice pollen from high-temperature stress. *Plant Physiol Biochem* 96:281–287.

Fahad S, Muhammad ZI, Abdul K, Ihsanullah D, Saud S, Saleh A, Wajid N, Muhammad A, Imtiaz AK, Chao W, Depeng W, Jianliang H (2018) Consequences of high temperature under changing climate optima for rice pollen characteristics-concepts and perspectives. *Archives Agron Soil Sci*. doi:10.1080/03650340.2018.1443213

Fahad S, Nie L, Chen Y, Wu C, Xiong D, Saud S, Hongyan L, Cui K, Huang J (2015b) Crop plant hormones and environmental stress. *Sustain Agric Rev* 15:371–400.

Fahad S, Rehman A, Shahzad B, Tanveer M, Saud S, Kamran M, Ihtisham M, Khan SU, Turan V, Rahman MHU (2019b) Rice responses and tolerance to metal/metalloid toxicity. In: Hasanuzzaman M, Fujita M, Nahar K, Biswas JK (eds) *Advances in rice research for abiotic stress tolerance*. Woodhead Publ Ltd, UK, pp 299–312.

Farah R, Muhammad R, Muhammad SA, Tahira Y, Muhammad AA, Maryam A, Shafaqat A, Rashid M, Muhammad R, Qaiser H, Afia Z, Muhammad AA, Muhammad A, Fahad S (2020) Alternative and non-conventional soil and crop management strategies for increasing water use efficiency. In: Fahad S, Hasanuzzaman M, Alam M, Ullah H, Saeed M, Khan AK, Adnan M (eds) *Environment, climate, plant and vegetation growth*. Springer Publ Ltd, Springer Nature Switzerland AG. Part of Springer Nature, pp 323–338. doi:10.1007/978-3-030-49732-3

Farhana G, Ishfaq A, Muhammad A, Dawood J, Fahad S, Xiuling L, Depeng W, Muhammad F, Muhammad F, Syed AS (2020) Use of crop growth model to simulate the impact of climate change on yield of various wheat cultivars under different agro-environmental conditions in Khyber Pakhtunkhwa, Pakistan. *Arabian J Geosci* 13:112. doi:10.1007/s12517-020-5118-1

Farhat A, Hafiz MH, Wajid I, Aitazaz AF, Hafiz FB, Zahida Z, Fahad S, Wajid F, Artemi C (2020) A review of soil carbon dynamics resulting from agricultural practices. *J Environ Manage* 268 (2020):110319.

Farrar K, Bryant D, Cope-Selby N (2014) Understanding and engineering beneficial plant–microbe interactions: plant growth promotion in energy crops. *Plant Biotechnol J* 12:1193–1206.

Fazli W, Muhmmad S, Amjad A, Fahad S, Muhammad A, Muhammad N, Ishaq AM, Imtiaz AK, Mukhtar A, Muhammad S, Muhammad I, Rafi U, Haroon I, Muhammad A (2020) Plant-microbes interactions and functions in changing climate. In: Fahad S, Hasanuzzaman M, Alam M, Ullah H, Saeed M, Khan AK, Adnan M (eds) *Environment, climate, plant and vegetation growth*. Springer Publ Ltd, Springer Nature Switzerland AG. Part of Springer Nature, pp 397–420. doi:10.1007/978-3-030-49732-3

Forni C, Duca D, Glick BR (2017) Mechanisms of plant response to salt and drought stress and their alteration by rhizobacteria. *Plant Soil* 410:335–356.

Füzy A, et al (2019) Selection of plant physiological parameters to detect stress effects in pot experiments using principal component analysis. *J Acta Physiol Plant* 41:56.

Ghosh D, Gupta A, Mohapatra S (2019) Dynamics of endogenous hormone regulation in plants by phyto-hormone secreting rhizobacteria under water-stress. *Symbiosis* 77:265–278.

Gopakumar L, Bernard NO, Donato V (2020) Soil microarthropods and nutrient cycling. In: Fahad S, Hasanuzzaman M, Alam M, Ullah H, Saeed M, Khan AK, Adnan M (eds) *Environment, climate, plant and vegetation growth*. Springer Publ Ltd, Springer Nature Switzerland AG. Part of Springer Nature, pp 453–472. doi:10.1007/978-3-030-49732-3

Guo C, Ma L, Yuan S, Wang R (2017) Morphological, physiological and anatomical traits of plant functional types in temperate grasslands along a large-scale aridity gradient in northeastern China. *Int J Sci Rep* 7:1–10

Guo M, Liu JH, Ma X, Luo DX, Gong ZH, Lu MH (2016) The plant heat stress transcription factors (HSFs): structure, regulation, and function in response to abiotic stresses. *Front Plant Sci* 7:114.

Guo YM, Chen S, Nelson MN, Cowling W, Turner NC (2013) Delayed water loss and temperature rise in floral buds compared with leaves of *Brassica rapa* subjected to a transient water stress during re-productive development. *Funct Plant Biol* 40:690–699.

Habib R, Ashfaq A, Aftab W, Manzoor H, Fahd R, Wajid I, Md Aminul I, Vakhtang S, Muhammad A, Asmat U, Abdul W, Syeda RS, Shah S, Shahbaz K, Fahad S, Manzoor H, Saddam H, Wajid N (2017) Application of CSM-CROPGRO-Cotton model for cultivars and optimum planting dates: evaluation in changing semi-arid climate. *Field Crops Res.* doi:10.1016/j.fcr.2017.07.007

Hafiz MH, Abdul K, Farhat A, Wajid F, Fahad S, Muhammad A, Ghulam MS, Wajid N, Muhammad M, Hafiz FB (2020b) Comparative effects of organic and inorganic fertilizers on soil organic carbon and wheat productivity under arid region. *Commun Soil Sci Plant Anal.* doi:10.1080/00103624.2020.1763385

Hafiz MH, Farhat A, Ashfaq A, Hafiz FB, Wajid F, Carol Jo W, Fahad S, Gerrit H (2020a) Predicting kernel growth of maize under controlled water and nitrogen applications. *Int J Plant Prod.* doi:10.1007/s42106-020-00110-8

Hafiz MH, Farhat A, Shafqat S, Fahad S, Artemi C, Wajid F, Chaves CB, Wajid N, Muhammad M, Hafiz FB (2018) Offsetting land degradation through nitrogen and water management during maize cultivation under arid conditions. *Land Degrad Dev*, 1–10. doi:10.1002/ldr.2933.

Hafiz MH, Muhammad A, Farhat A, Hafiz FB, Saeed AQ, Muhammad M, Fahad S, Muhammad A (2019) Environmental factors affecting the frequency of road traffic accidents: a case study of sub-urban area of Pakistan. *Environ Sci Pollut Res.* doi:10.1007/s11356-019-04752-8

Hafiz MH, Wajid F, Farhat A, Fahad S, Shafqat S, Wajid N, Hafiz FB (2016) Maize plant nitrogen uptake dynamics at limited irrigation water and nitrogen. *Environ Sci Pollut Res* 24(3):2549–2557. doi:10.1007/s11356-016-8031-0

Hameed M, Mansoor U, Ashraf M, Rao A (2002) Variation in leaf anatomy in wheat germplasm from varying drought-hit habitats. *J Int J Agric Biol* 4:12–16.

Hartman K, Tringe SG (2019) Interactions between plants and soil shaping the root microbiome under abiotic stress. *Biochem J* 476:2705–2724.

Hassani MA, Durán P, Hacquard SJM (2018) Microbial interactions within the plant holobiont. *Microbiome* 6:58.

Hawkins AP, Crawford KM (2018) Interactions between plants and soil microbes may alter the relative im-portance of intraspecific and interspecific plant competition in a changing climate. *AoB Plants* 10:p039.

Henty-Ridilla JL, Shimono M, Li J, Chang JH, Day B, Staiger CJ (2013) The plant actin cytoskeleton responds to signals from microbe-associated molecular patterns. *PLoS Pathog* 9:e1003290.

Hesham FA, Fahad S (2020) Melatonin application enhances biochar efficiency for drought tolerance in maize varieties: modifications in physio-biochemical machinery. *Agron J* 112(4):1–22.

Ho YN, Mathew DC, Huang CC (2017) Plant-microbe ecology: interactions of plants and symbiotic microbial communities. *Plant Ecol*, 93–119.

Hussain MA, Fahad S, Rahat S, Muhammad FJ, Muhammad M, Qasid A, Ali A, Husain A, Nooral A, Babatope SA, Changbao S, Liya G, Ibrar A, Zhanmei J, Juncai H (2020) Multifunctional role of brassinosteroid and its analogues in plants. *Plant Growth Regul.* doi:10.1007/s10725-020-00647-8

Ibrar K, Aneela R, Khola Z, Urooba N, Sana B, Rabia S, Ishtiaq H, Mujaddad R, Salvatore M (2020) Microbes and environment: global warming reverting the frozen zombies. In: Fahad S, Hasanuzzaman M, Alam M, Ullah H, Saeed M, Khan AK, Adnan M (eds) *Environment, climate, plant and vegetation growth*. Springer Publ Ltd, Springer Nature Switzerland AG. Part of Springer Nature, pp 607–634. doi:10.1007/978-3-030-49732-3

Igiehon NO, Babalola OO (2018) Below-ground-above-ground plant-microbial interactions: focusing on soybean, rhizobacteria and mycorrhizal fungi. *Open Microbiol J* 12:261.

Ilyas M, Mohammad N, Nadeem K, Ali H, Aamir HK, Kashif H, Fahad S, Aziz K, Abid U (2020) Drought tolerance strategies in plants: a mechanistic approach. *J Plant Growth Regul*. doi:10.1007/s00344-020-10174-5

Imam J, Singh PK, Shukla P (2016) Plant microbe interactions in post genomic era: perspectives and applications. *Front Microbiol* 7:1488.

Iqra M, Amna B, Shakeel I, Fatima K, Sehrish L, Hamza A, Fahad S (2020) Carbon cycle in response to global warming. In: Fahad S, Hasanuzzaman M, Alam M, Ullah H, Saeed M, Khan AK, Adnan M (eds) *Environment, climate, plant and vegetation growth*. Springer Publ Ltd, Springer Nature Switzerland AG. Part of Springer Nature, pp 1–16. doi:10.1007/978-3-030-49732-3

Israr D, Mustafa G, Khan KS, Shahzad M, Ahmad N, Masood S (2016) Interactive effects of phosphorus and *Pseudomonas putida* on chickpea (*Cicer arietinum* L.) growth, nutrient uptake, antioxidant enzymes and organic acids exudation. *Plant Physiol Biochem* 108:304–312.

Jan M, Muhammad AH, Adnan NS, Muhammad Y, Javaid I, Xiuling L, Depeng W, Fahad S (2019) Modulation in growth, gas exchange, and antioxidant activities of salt-stressed rice (*Oryza sativa* L.) genotypes by zinc fertilization. *Arabian J Geosci* 12:775. doi:10.1007/s12517-019-4939-2

Kalachova T et al (2019) Interplay between phosphoinositides and actin cytoskeleton in the regulation of immunity related responses in *Arabidopsis thaliana* seedlings. *Environ Exp Bot* 167:103867.

Kamran M, Wenwen C, Irshad A, Xiangping M, Xudong Z, Wennan S, Junzhi C, Shakeel A, Fahad S, Qingfang H, Tiening L (2017) Effect of paclobutrazol, a potential growth regulator on stalk mechanical strength, lignin accumulation and its relation with lodging resistance of maize. *Plant Growth Regul* 84:317–332. doi:10.1007/s10725-017-0342-8

Kassahun B, Bidinger F, Hash C, Kuruvinashetti M (2010) Stay-green expression in early generation sorghum [*Sorghum bicolor* (L.) Moench] QTL introgression lines. *J Euphytica* 172:351–362.

Kim Y, Chung YS, Lee E, Tripathi P, Heo S, Kim KH (2020) Root response to drought stress in rice (*Oryza sativa* L.). *Int J Mol Sci* 21:1513.

Kosma DK, Bourdenx B, Bernard A, Parsons EP, Lü S, Joubès J, Jenks MA (2009) The impact of water deficiency on leaf cuticle lipids of Arabidopsis. *J Plant Physiol* 151:1918–1929.

Kumar A, Verma JP (2018) Does plant—microbe interaction confer stress tolerance in plants: a review?. *Microbiol Res* 207:41–52.

Lareen A, Burton F, Schäfer P (2016) Plant root-microbe communication in shaping root microbiomes. *J Plant Mol Biol* 90:575–587.

Loh J, Pierson EA, Pierson III LS, Stacey G, Chatterjee A (2002) Quorum sensing in plant-associated bacteria. *Curr Opin Plant Biol* 5:285–290.

Lopes MS, Reynolds MP (2012) Stay-green in spring wheat can be determined by spectral reflectance measurements (normalized difference vegetation index) independently from phenology. *J Exp Bot* 63:3789–3798.

Mahar A, Amjad A, Altaf HL, Fazli W, Ronghua L, Muhammad A, Fahad S, Muhammad A, Rafiullah, Imtiaz AK, Zengqiang Z (2020) Promising technologies for Cd-contaminated soils: drawbacks and possibilities. In: Fahad S, Hasanuzzaman M, Alam M, Ullah H, Saeed M, Khan AK, Adnan M (eds) *Environment, climate, plant and vegetation growth*. Springer Publ Ltd, Springer Nature Switzerland AG. Part of Springer Nature, pp 63–92. doi:10.1007/978-3-030-49732-3

Md Enamul H, Shoeb AZM, Mallik AH, Fahad S, Kamruzzaman MM, Akib J, Nayyer S, KM Mehedi A, Swati AS, Md Yeamin A, Most SS (2020) Measuring vulnerability to environmental hazards: qualitative to quantitative. In: Fahad S, Hasanuzzaman M, Alam M, Ullah H, Saeed M, Khan AK, Adnan M (eds) *Environment, climate, plant and vegetation growth*. Springer Publ Ltd, Springer Nature Switzerland AG. Part of Springer Nature, pp 421–452. doi:10.1007/978-3-030-49732-3

Md Jakir H, Allah B (2020) Development and applications of transplastomic plants; a way towards eco-friendly agriculture. In: Fahad S, Hasanuzzaman M, Alam M, Ullah H, Saeed M, Khan AK, Adnan M (eds) *Environment, climate, plant and vegetation growth*. Springer Publ Ltd, Springer Nature Switzerland AG. Part of Springer Nature, pp 285–322. doi:10.1007/978-3-030-49732-3

Meena KK et al (2017) Abiotic stress responses and microbe-mediated mitigation in plants: the omics strategies. *Front Plant Sci* 8:172.

Mhlongo MI, Piater LA, Madala NE, Labuschagne N, Dubery IA (2018) The chemistry of plant–microbe interactions in the rhizosphere and the potential for metabolomics to reveal signaling related to defense priming and induced systemic resistance. *Front Plant Sci* 9:112.

Moënne-Loccoz Y, Mavingui P, Combes C, Normand P, Steinberg C (2015) Microorganisms and biotic interactions. In: *Environmental microbiology: fundamentals and applications*. Springer, pp 395–444.

Mohammad AW, Abdelazeem S, Munir A, Khalid E, Adel RAU (2020b) Extent of climate change in Saudi Arabia and its impacts on agriculture: a case study from Qassim region. In: Fahad S, Hasanuzzaman M, Alam M, Ullah H, Saeed M, Khan AK, Adnan M (eds) *Environment, climate, plant and vegetation growth*. Springer Publ Ltd, Springer Nature Switzerland AG. Part of Springer Nature, pp 635–658. doi:10.1007/978-3-030-49732-3

Mohammad AW, Ahmad M, Usman ARA, Akanji M, Muhammad IR (2020a) Advances in pyrolytic technologies with improved carbon capture and storage to combat climate change. In: Fahad S, Hasanuzzaman M, Alam M, Ullah H, Saeed M, Khan AK, Adnan M (eds) *Environment, climate, plant and vegetation growth*. Springer Publ Ltd, Springer Nature Switzerland AG. Part of Springer Nature, pp 535–576. doi:10.1007/978-3-030-49732-3

Mubeen M, Ashfaq A, Hafiz MH, Muhammad A, Hafiz UF, Mazhar S, Muhammad SD, Asad A, Amjed A, Fahad S, Wajid N (2020) Evaluating the climate change impact on water use efficiency of cotton-wheat in semi-arid conditions using DSSAT model. *J Water Climate Change*. doi:10.2166/wcc.2019.179/622 035/jwc2019179.pdf

Muhammad TQ, Amna F, Amna B, Barira Z, Xitong Z, Ling-Ling C (2020) Effectiveness of conventional crop improvement strategies vs. omics. In: Fahad S, Hasanuzzaman M, Alam M, Ullah H, Saeed M, Khan AK, Adnan M (eds) *Environment, climate, plant and vegetation growth*. Springer Publ Ltd, Springer Nature Switzerland AG. Part of Springer Nature, pp 253–284. doi:10.1007/978-3-030-49732-3

Muhammad Z, Abdul MK, Abdul MS, Kenneth BM, Muhammad S, Shahen S, Ibadullah J, Fahad S (2019) Performance of *Aeluropus lagopoides* (mangrove grass) ecotypes, a potential turfgrass, under high saline conditions. *Environ Sci Pollut Res*. doi:10.1007/s11356-019-04838-3

Munné-Bosch S, Alegre L (2004) Die and let live: leaf senescence contributes to plant survival under drought stress. *J Funct Plant Biol* 31:203–216.

Ngwene B, Boukail S, Söllner L, Franken P, Andrade-Linares D (2016) Phosphate utilization by the fungal root endophyte *Piriformospora indica*. *J Plant Soil* 405:231–241.

Noor M, Naveed R, Ajmal J, Fahad S, Muhammad A, Fazli W, Saud S, Hassan S (2020) Climate change and coastal plant lives. In: Fahad S, Hasanuzzaman M, Alam M, Ullah H, Saeed M, Khan AK, Adnan M (eds) *Environment, climate, plant and vegetation growth*. Springer Publ Ltd, Springer Nature Switzerland AG. Part of Springer Nature, pp 93–108. doi:10.1007/978-3-030-49732-3

Nowicka B, Ciura J, Szymańska R, Kruk J (2018) Improving photosynthesis, plant productivity and abiotic stress tolerance–current trends and future perspectives. *J Plant Physiol* 231:415–433.

O'Banion BS, O'Neal L, Alexandre G, Lebeis SL (2020) Bridging the gap between single-strain and community-level plant-microbe chemical interactions. *Mol Plant Microbe Interact* 33:124–134.

Odelade KA, Babalola OO (2019) Bacteria, fungi and archaea domains in rhizospheric soil and their effects in enhancing agricultural productivity. *Int J Environ Sci* 16:3873.

Ortíz-Castro R, Contreras-Cornejo HA, Macías-Rodríguez L, López-Bucio J (2009) The role of microbial signals in plant growth and development. *Plant Signal Behav* 4:701–712.

Pandey P, Irulappan V, Bagavathiannan MV, Senthil-Kumar M (2017) Impact of combined abiotic and biotic stresses on plant growth and avenues for crop improvement by exploiting physio-morphological traits. *Front Plant Sci* 8:537.

Pascale A, Proietti S, Pantelides IS, Stringlis IA (2020) Modulation of the root microbiome by plant molecules: the basis for targeted disease suppression and plant growth promotion. *Front Plant Sci* 10:1741.

Paul D, Lade H (2014) Plant-growth-promoting rhizobacteria to improve crop growth in saline soils: a review. *Agron Sustain Dev* 34:737–752.

Pinto RS, Lopes MS, Collins NC, Reynolds MP (2016) Modelling and genetic dissection of staygreen under heat stress. *Theor Appl Genet* 129:2055–2074.

Qamar Z, Zubair A, Muhammad Y, Muhammad ZI, Abdul K, Fahad S, Safder B, Ramzani PMA, Muhammad

N (2017) Zinc biofortification in rice: leveraging agriculture to moderate hidden hunger in developing countries. *Arch Agron Soil Sci* 64:147–161. doi:10.1080/03650340.2017.1338343

Qian D, Xiang Y (2019) Actin cytoskeleton as actor in upstream and downstream of calcium signaling in plant cells. *Int J Mol Sci* 20:1403.

Rashid M, Qaiser H, Khalid SK, Mohammad AW, Zhang A, Muhammad A, Shahzada SI, Rukhsanda A, Ghulam AS, Shahzada MM, Sarosh A, Muhammad FQ (2020) Prospects of biochar in alkaline soils to mitigate climate change. In: Fahad S, Hasanuzzaman M, Alam M, Ullah H, Saeed M, Khan AK, Adnan M (eds) *Environment, climate, plant and vegetation growth*. Springer Publ Ltd, Springer Nature Switzerland AG. Part of Springer Nature, pp 133–150. doi:10.1007/978-3-030-49732-3

Rehman M, Fahad S, Saleem MH, Hafeez M, Muhammad HR, Liu F, Deng G (2020) Red light optimized physiological traits and enhanced the growth of ramie (*Boehmeria nivea* L.). *Photosynthetica* 58 (4): 922–931.

Rodriguez PA, Rothballer M, Chowdhury SP, Nussbaumer T, Gutjahr C, Falter-Braun P (2019) Systems biology of plant-microbiome interactions. *Mol Plant* 12:804–821.

Rolly NK, Imran QM, Lee IJ, Yun B-W (2020) Salinity stress-mediated suppression of expression of salt overly sensitive signaling pathway genes suggests negative regulation by *AtbZIP62*. Transcription factor in *Arabidopsis thaliana. Int J Mol Sci* 21:1726.

Rosier A, Medeiros FH, Bais HP (2018) Defining plant growth promoting rhizobacteria molecular and biochemical networks in beneficial plant-microbe interactions. *J Plant Soil* 428:35–55.

Sabagh AEL, Hossain A, Barutçular C, Aamir Iqbal M, Sohidul Islam M, Fahad S, Sytar O, Çig F, Swaroop Meena R, Erman M (2020) Consequences of salinity stress on the quality of crops and its mitigation strategies for sustainable crop production: an outlook of arid and semi-arid regions. In: Fahad S, Hasanuzzaman M, Alam M, Ullah H, Saeed M, Khan AK, Adnan M (eds) *Environment, climate, plant and vegetation growth*. Springer Publ Ltd, Springer Nature Switzerland AG. Part of Springer Nature, pp 503–534. doi:10.1007/978-3-030-49732-3

Sadam M, Muhammad TQ, Ghulam M, Muhammad SK, Faiz AJ (2020) Role of biotechnology in climate resilient agriculture. In: Fahad S, Hasanuzzaman M, Alam M, Ullah H, Saeed M, Khan AK, Adnan M (eds) *Environment, climate, plant and vegetation growth*. Springer Publ Ltd, Springer Nature Switzerland AG. Part of Springer Nature, pp 339–366. doi:10.1007/978-3-030-49732-3

Saima S, Li G, Wu G (2018) Effects of drought stress on hybrids of *Vigna radiata* at germination stage. *Acta Biol Hung* 69:481–492.

Sajid H, Jie H, Jing H, Shakeel A, Satyabrata N, Sumera A, Awais S, Chunquan Z, Lianfeng Z, Xiaochuang C, Qianyu J, Junhua Z (2020) Rice production under climate change: adaptations and mitigating strategies. In: Fahad S, Hasanuzzaman M, Alam M, Ullah H, Saeed M, Khan AK, Adnan M (eds) *Environment, climate, plant and vegetation growth*. Springer Publ Ltd, Springer Nature Switzerland AG. Part of Springer Nature, pp 659–686. doi:10.1007/978-3-030-49732-3

Sajjad H, Muhammad M, Ashfaq A, Waseem A, Hafiz MH, Mazhar A, Nasir M, Asad A, Hafiz UF, Syeda RS, Fahad S, Depeng W, Wajid N (2019) Using GIS tools to detect the land use/land cover changes during forty years in Lodhran district of Pakistan. *Environ Sci Pollut Res*. doi:10.1007/s11356-019-06072-3

Saleem MH, Fahad S, Adnan M, Mohsin A, Muhammad SR, Muhammad K, Qurban A, Inas AH, Parashuram B, Mubassir A, Reem MH (2020a) Foliar application of gibberellic acid endorsed phytoextraction of copper and alleviates oxidative stress in jute (*Corchorus capsularis* L.) plant grown in highly copper-contaminated soil of China. *Environ Sci Pollution Res*. doi:10.1007/s11356-020-09764-3

Saleem MH, Fahad S, Shahid UK, Mairaj D, Abid U, Ayman ELS, Akbar H, Analía L, Lijun L (2020b) Copper-induced oxidative stress, initiation of antioxidants and phytoremediation potential of flax (*Linum usitatissimum* L.) seedlings grown under the mixing of two different soils of China. *Environ Sci Poll Res*. doi:10.1007/s11356-019-07264-7

Saleem MH, Rehman M, Fahad S, Tung SA, Iqbal N, Hassan A, Ayub A, Wahid MA, Shaukat S, Liu L, Deng G (2020c) Leaf gas exchange, oxidative stress, and physiological attributes of rapeseed (*Brassica napus* L.) grown under different light-emitting diodes. *Photosynthetica* 58(3):836–845.

Saleh J, Maftoun M (2008) Interactive effect of NaCl levels and zinc sources and levels on the growth and mineral composition of rice. *J Agric Sci Tech* 10:325–336.

Saman S, Amna B, Bani A, Muhammad TQ, Rana MA, Muhammad SK (2020) QTL mapping for abiotic stresses in cereals. In: Fahad S, Hasanuzzaman M, Alam M, Ullah H, Saeed M, Khan AK, Adnan M (eds) *Environment, climate, plant and vegetation growth*. Springer Publ Ltd, Springer Nature Switzerland AG. Part of Springer Nature, pp 229–252. doi:10.1007/978-3-030-49732-3

Samaniego-Gámez BY, Garruña R, Tun-Suárez JM, Kantun-Can J, Reyes-Ramírez A, Cervantes-Díaz L (2016) *Bacillus* spp. inoculation improves photosystem II efficiency and enhances photosynthesis in pepper plants. *Chil J Agric Res* 76:409–416.

Sander EL, Wootton JT, Allesina S (2015) What can interaction webs tell us about species roles? *J PLoS Comput Biol* 11:e1004330.

Santhanam R, Groten K, Meldau DG, Baldwin IT (2014) Analysis of plant-bacteria interactions in their native habitat: bacterial communities associated with wild tobacco are independent of endogenous jasmonic acid levels and developmental stages. *PLoS One* 9:e94710.

Saud S, Chen Y, Fahad S, Hussain S, Na L, Xin L, Alhussien SA (2016) Silicate application increases the photosynthesis and its associated metabolic activities in Kentucky bluegrass under drought stress and post-drought recovery. *Environ Sci Pollut Res* 23(17):17647–17655. doi:10.1007/s11356-016-6957-x

Saud S, Chen Y, Long B, Fahad S, Sadiq A (2013) The different impact on the growth of cool season turf grass under the various conditions on salinity and drought stress. *Int J Agric Sci Res* 3:77–84.

Saud S, Fahad S, Cui G, Chen Y, Anwar S (2020) Determining nitrogen isotopes discrimination under drought stress on enzymatic activities, nitrogen isotope abundance and water contents of Kentucky bluegrass. *Sci Rep* 10:6415. doi:10.1038/s41598-020-63548-w

Saud S, Fahad S, Yajun C, Ihsan MZ, Hammad HM, Nasim W, Amanullah Jr, Arif M, Alharby H (2017) Effects of nitrogen supply on water stress and recovery mechanisms in Kentucky bluegrass plants. *Front Plant Sci* 8:983. doi:10.3389/fpls.2017.00983

Saud S, Li X, Chen Y, Zhang L, Fahad S, Hussain S, Sadiq A, Chen Y (2014) Silicon application increases drought tolerance of Kentucky bluegrass by improving plant water relations and morph physiological functions. *Sci World J* 2014:1–10. doi:10.1155/2014/368694

Schirawski J, Perlin MH (2018) *Plant–microbe interaction 2017—the good, the bad and the diverse*. Multidisciplinary Digital Publishing Institute.

Schulz-Bohm K, Martín-Sánchez L, Garbeva P (2017) Microbial volatiles: small molecules with an important role in intra-and inter-kingdom interactions. *Front Microbiol* 8:2484.

Senol C (2020) The effects of climate change on human behaviors. In: Fahad S, Hasanuzzaman M, Alam M, Ullah H, Saeed M, Khan AK, Adnan M (eds) *Environment, climate, plant and vegetation growth*. Springer Publ Ltd, Springer Nature Switzerland AG. Part of Springer Nature, pp 577–590. doi:10.1007/978-3-030-49732-3

Shafi MI, Adnan M, Fahad S, Fazli W, Ahsan K, Zhen Y, Subhan D, Zafar-ul-Hye M, Martin B, Rahul D (2020) Application of single superphosphate with humic acid improves the growth, yield and phosphorus uptake of wheat (*Triticum aestivum* L.) in calcareous soil. *Agronomy* 10:1224. doi:10.3390/agronomy10091224

Shah F, Lixiao N, Kehui C, Tariq S, Wei W, Chang C, Liyang Z, Farhan A, Fahad S, Huang J (2013) Rice grain yield and component responses to near 2°C of warming. *Field Crop Res* 157:98–110.

Shou H, Bordallo P, Wang K (2004) Expression of the *Nicotiana* protein kinase (NPK1) enhanced drought tolerance in transgenic maize. *J Exp Bot* 55:1013–1019.

Sklensky DE, Davies PJ (2011) Resource partitioning to male and female flowers of *Spinacia oleracea* L. in relation to whole-plant monocarpic senescence. *J Exp Bot* 62:4323–4336.

Suarez C et al (2015) Plant growth-promoting effects of *Hartmannibacter diazotrophicus* on summer barley (*Hordeum vulgare* L.) under salt stress. *J Appl Soil Ecol* 95:23–30.

Subhan D, Zafar-ul-Hye M, Fahad S, Saud S, Martin B, Tereza H, Rahul D (2020) Drought stress alleviation by ACC deaminase producing *Achromobacter xylosoxidans* and *Enterobacter cloacae*, with and without timber waste biochar in maize. *Sustainability* 12(6286). doi:10.3390/su12156286

Tariq M, Ahmad S, Fahad S, Abbas G, Hussain S, Fatima Z, Nasim W, Mubeen M, Rehman MH, Khan MA, Adnan M (2018) The impact of climate warming and crop management on phenology of sunflower-based cropping systems in Punjab, Pakistan. *Agric Forest Met* 15(256):270–282.

Thomas H, Ougham H (2014) The stay-green trait. *J Exp Bot* 65:3889–3900.

Torabi M, Drahansky M, Paridah M, Moradbak A, Mohamed A, Owolabi FJI (2016) We are IntechOpen, the world's leading publisher of Open Access books Built by scientists, for scientists TOP 1%

Tyagi S, Mulla SI, Lee KJ, Chae JC, Shukla P (2018) VOCs-mediated hormonal signaling and crosstalk with plant growth promoting microbes. *Crit Rev Biotechnol* 38:1277–1296.

Ulrich DE, Sevanto S, Ryan M, Albright MB, Johansen RB, Dunbar JM (2019) Plant-microbe interactions before drought influence plant physiological responses to subsequent severe drought. *Int J Sci Rep* 9:1–10

Unsar N, Muhammad R, Syed HMB, Asad S, Mirza AQ, Naeem I, Muhammad HR, Fahad S, Shafqat S (2020) Insect pests of cotton crop and management under climate change scenarios. In: Fahad S, Hasanuzzaman M, Alam M, Ullah H, Saeed M, Khan AK, Adnan M (eds) *Environment, climate, plant and vegetation growth*. Springer Publ Ltd, Springer Nature Switzerland AG. Part of Springer Nature, pp 367–396. doi:10.1007/978-3-030-49732-3

Vaishnav A, Shukla AK, Sharma A, Kumar R, Choudhary DK (2019) Endophytic bacteria in plant salt stress tolerance: current and future prospects. *J Plant Growth Regul* 38:650–668.

Wahid F, Fahad S, Subhan D, Adnan M, Zhen Y, Saud S, Manzer HS, Martin B, Tereza H, Rahul D (2020) Sustainable management with mycorrhizae and phosphate solubilizing bacteria for enhanced phosphorus uptake in calcareous soils. *Agriculture* 10(334). doi:10.3390/agriculture10080334

Wajid N, Ashfaq A, Asad A, Muhammad T, Muhammad A, Muhammad S, Khawar J, Ghulam MS, Syeda RS, Hafiz MH, Muhammad IAR, Muhammad ZH, Muhammad HR, Veysel T, Fahad S, Suad S, Aziz K, Shahzad A (2017) Radiation efficiency and nitrogen fertilizer impacts on sunflower crop in contrasting environments of Punjab. *Pakistan Environ Sci Pollut Res* 25:1822–1836. doi:10.1007/s11356-017-0592-z

Waters BM, Uauy C, Dubcovsky J, Grusak MA (2009) Wheat (*Triticum aestivum*) NAM proteins regulate the translocation of iron, zinc, and nitrogen compounds from vegetative tissues to grain. *J Exp Bot* 60:4263–4274.

Wu CH, Bernard SM, Andersen GL, Chen W (2009) Developing microbe–plant interactions for applications in plant-growth promotion and disease control, production of useful compounds, remediation and carbon sequestration *Microb Biotechnol* 2:428–440.

Wu C, Tang S, Li G, Wang S, Fahad S, Ding Y (2019) Roles of phytohormone changes in the grain yield of rice plants exposed to heat: a review. *PeerJ* 7:e7792. doi:10.7717/peerj.7792

Wu C, Kehui C, She T, Ganghua L, Shaohua W, Fahad S, Lixiao N, Jianliang H, Shaobing P, Yanfeng D (2020) Intensified pollination and fertilization ameliorate heat injury in rice (*Oryza sativa* L.) during the flowering stage. *Field Crops Res* 252:107795.

Xiao BZ, Chen X, Xiang CB, Tang N, Zhang QF, Xiong LZ (2009) Evaluation of seven function-known candidate genes for their effects on improving drought resistance of transgenic rice under field conditions. *J Mol Plant* 2:73–83.

Xue D, Zhang X, Lu X, Chen G, Chen ZH (2017) Molecular and evolutionary mechanisms of cuticular wax for plant drought tolerance. *Front Plant Sci* 8:621.

Yang Z, Zhang Z, Zhang T, Fahad S, Cui K, Nie L, Peng S, Huang J (2017) The effect of season-long temperature increases on rice cultivars grown in the central and southern regions of China. *Front Plant Sci* 8:1908. doi:10.3389/fpls.2017.01908

Yi HS, Ahn YR, Song GC, Ghim SY, Lee S, Lee G, Ryu CM (2016) Impact of a bacterial volatile 2, 3-butanediol on Bacillus subtilis rhizosphere robustness. *Front Microbiol* 7:993.

Zafar-ul-Hye M, Tahzeeb-ul-Hassan M, Abid M, Fahad S, Brtnicky M, Dokulilova T, Datta R, Danish S (2020b) Potential role of compost mixed biochar with rhizobacteria in mitigating lead toxicity in spinach. *Sci Rep* 10:12159. doi:10.1038/s41598-020-69183-9

Zafar-ul-Hye M, Muhammad N, Subhan D, Fahad S, Rahul D, Mazhar A, Ashfaq AR, Martin B, Jiří H, Zahid HT, Muhammad N (2020a) Alleviation of cadmium adverse effects by improving nutrients uptake in bitter gourd through cadmium tolerant rhizobacteria. *Environments* 7(54). doi:10.3390/environments7080054

Zahida Z, Hafiz FB, Zulfiqar AS, Ghulam MS, Fahad S, Muhammad RA, Hafiz MH, Wajid N, Muhammad S (2017) Effect of water management and silicon on germination, growth, phosphorus and arsenic uptake in rice. *Ecotoxicol Environ Saf* 144:11–18.

Zeilinger S et al (2016) Friends or foes? Emerging insights from fungal interactions with plants. *FEMS Microbiol Rev* 40:182–207.

Zengler K (2009) Central role of the cell in microbial ecology. *Microbiol Mol Bio Rev* 73:712–729.

Zhang J et al (2018) Effect of drought on agronomic traits of rice and wheat: a meta-analysis. *Int J Environ Res* 15:839.

Index